ELEMENTS

OF

NATURAL PHILOSOPHY.

BY

W. H. C. BARTLETT, LL. D.,

PROFESSOR OF NATURAL AND EXPERIMENTAL PHILOSOPHY IN THE UNITED STATES
MILITARY ACADEMY AT WEST POINT.

SECTION I.

MECHANICS.

NEW YORK:

PUBLISHED BY A. S. BARNES AND COMPANY,

NO. 51 JOHN-STREET.

CINCINNATI: H. W. DERBY AND COMPANY.

1850

STEREOTYPED BY
BILLIN & BROTHERS,
No. 10 North William Street, New York.

F. C. GUTIERREZ, Printer,
No. 51 John Street, cor. of Dutch

PREFACE.

The present volume is the first of three in which its author desires to offer, to academies and colleges, a course of Natural Philosophy, including Astronomy. It embraces the subject of Mechanics—the ground-work of the whole. It is intended to be complete within itself, and to have no necessary dependence, for the full comprehension of its contents, upon those which are to follow. In its preparation, constant reference was made to the admirable labors of M. Poncellet, and much valuable assistance was derived from the work of M. Peschel.

Large type, marginal notes, tables of reference, and numerous diagrams, often repeated, have swollen the volume beyond the limits originally intended; but whatever of inconvenience may thence arise, will, it is hoped, be more than compensated by the facilities which these sources of increased size cannot fail to bring to the aid both of the teacher and student.

CONTENTS.

INTRODUCTION.

PART FIRST.

MECHANICS OF SOLIDS.

I.

II.

III.

IV.

V.

PART SECOND.

MECHANICS OF FLUIDS.

ELEMENTS

OF

NATURAL PHILOSOPHY.

INTRODUCTION.

THE term nature is employed to signify the assemblage of all the bodies of the universe; it includes whatever exists and is the subject of change. Of the existence of these bodies we are rendered conscious by the impressions they make on our senses. Their condition is subject to a variety of changes, whence we infer that external causes are in operation to produce them; and to investigate nature with reference to these changes and their causes, is the object of *Physical Science.* Nature. Bodies. Physical science.

All bodies may be distributed into three classes, viz.: *unorganized* or *inanimate*, *organized* or *animated*, and the *heavenly bodies* or *primary organizations.* Classification of bodies.

The *unorganized* or *inanimate* bodies, as minerals, water, air, form the lowest class, and are, so to speak, the substratum for the others. These bodies are acted on solely by causes external to themselves; they have no definite or periodical duration; nothing that can properly be termed life. Inanimate bodies, no definite period, no life.

The *organized* or *animated* bodies, are more or less perfect individuals, possessing *organs* adapted to the performance of certain appropriate functions. In consequence of an innate principle peculiar to them, know as *vitality*, bodies of this class are constantly appropriating to themselves unorganized matter, changing its properties, and Animated bodies, organs, vitality.

deriving, by means of this process, an increase of bulk. They also possess the faculty of reproduction. They retain only for a limited time the vital principle, and, when life is extinct, they sink into the class of inanimate bodies. The animal and vegetable kingdoms include all the species of this class on our earth.

Reproduction, and limited duration.

Animal and vegetable kingdoms.

Celestial bodies;

The *celestial bodies*, as the fixed stars, the sun, the comets, planets and their secondaries, are the gigantic individuals of the universe, endowed with an organization on the grandest scale. Their constituent parts may be compared to the organs possessed by bodies of the second class; those of our earth are its continents, its ocean, its atmosphere, which are constantly exerting a vigorous action on each other, and bringing about changes the most important.

organs—continents, ocean, atmosphere.

Earth existed long before plants and animals.

The earth supports and nourishes both the vegetable and animal world, and the researches of Geology have demonstrated, that there was once a time when neither plants nor animals existed on its surface, and that prior to the creation of either of these orders, great changes must have taken place in its constitution. As the earth existed thus anterior to the organized beings upon it, we may infer that the other heavenly bodies, in like manner, were called into being before any of the organized bodies which probably exist upon them. Reasoning, then, by analogy from our earth, we may venture to regard the heavenly bodies as the primary organized forms, on whose surface both animals and vegetables find a place and support.

Heavenly bodies—the support of animals and vegetables.

Natural philosophy, external changes.

Natural Philosophy, or *Physics,* treats of the general properties of *unorganized* bodies, of the influences which act upon them, the laws they obey, and of the *external* changes which these bodies undergo without affecting their *internal constitution.*

Chemistry;

Chemistry, on the contrary, treats of the *individual* properties of bodies, by which, as regards their constitution, they may be distinguished one from another; it also investigates the transformations which take place in the interior of a body—transformations by which the substance of the body is altered and remodelled; and lastly, it detects and classifies the laws by which chemical changes are regulated. Internal changes.

Natural History, is that branch of physical science which treats of organized bodies; it comprises three divisions, the one *mechanical*—the anatomy and dissection of plants and animals; the second, *chemical*—animal and vegetable chemistry; and the third, *explanatory*—physiology. Natural History—anatomy, chemistry, physiology.

Astronomy teaches the knowledge of the celestial bodies. It is divided into *Spherical* and *Physical* astronomy. The former treats of the appearances, magnitudes, distances, arrangements, and motions of the heavenly bodies; the latter, of their constitution and physical condition, their mutual influences and actions on each other, and generally, seeks to explain the causes of the celestial phenomena. Astronomy, spherical and physical.

Again, one most important use of natural science, is the application of its laws either to technical purposes—*mechanics, technical chemistry, pharmacy, &c.;* to the phenomena of the heavenly bodies—*physical astronomy;* or to the various objects which present themselves to our notice at or near the surface of the earth—*physical geography, meteorology*—and we may add *geology* also, a science which has for its object to unfold the history of our planet from its formation to the present time. Application of physical laws.

Natural philosophy is a science of *observation* and *experiment*, for by these two modes we deduce the varied information we have acquired about bodies; by the former we notice any changes that transpire in the condition or relations of any body as they spontaneously arise Natural philosophy, a science of observation and experiment.

without interference on our part; whereas, in the performance of an experiment, we purposely alter the natural arrangement of things to bring about some particular condition that we desire. To accomplish this, we make use of appliances called *philosophical* or *chemical apparatus*, the proper use and application of which, it is the office of *Experimental Physics* to teach.

Apparatus; experimental physics.

If we notice that in winter water becomes converted into ice, we are said to make an observation: if, by means of freezing mixtures or evaporation, we cause water to freeze, we are then said to perform an experiment.

Observation, experiment.

These experiments are next subjected to calculation, by which are deduced what are sometimes called *the laws of nature*, or *the rules that like causes will invariably produce like results*. To express these laws with the greatest possible brevity mathematical symbols are used. When it is not practicable to represent them with mathematical precision, we must be contented with inferences and assumptions based on analogies, or with probable explanations or *hypotheses*.

re.

Hypotheses and probability of their truth.

A hypothesis gains in probability the more nearly it accords with the ordinary course of nature, the more numerous the experiments on which it is founded, and the more simple the explanation it offers of the phenomena for which it is intended to account.

PHYSICS OF PONDERABLE BODIES.

Physical properties; the senses.

§ 1.—The *physical properties* of bodies are those external signs by which their existence is made evident to our minds; the senses constitute the medium through which this knowledge is communicated.

All the senses not equally employed.

All our senses, however, are not equally made use of for this purpose; we are generally guided in our decisions by the evidence of sight and touch. Still sight alone is frequently incompetent, as there are bodies which cannot be perceived by that sense, as, for example, all colorless gases; again, some of the objects of sight are not substantial, as, the shadow, the image in a mirror, spectra formed by the refraction of the rays of light, &c. Touch, on the contrary, decides indubitably as to the existence of any body.

Touch.

The properties of bodies may be divided into *primary* or *principal*, and *secondary* or *accessory*. The former, are such as we find common to all bodies, and without which we cannot conceive of their existing; the latter, are not absolutely necessary to our conception of a body's existence, but become known to us by investigation and experience.

Primary and secondary properties of bodies.

PRIMARY PROPERTIES.

§ 2.—The primary properties of all bodies are *extension* and *impenetrability*.

Extension is that property in consequence of which every body occupies a certain limited space. It is the condition of the mathematical idea of a body; by it, the *volume* or size of the occupied space, as well as its boundary, or *figure*, is determined. The extension of bodies is expressed by three dimensions, length, breadth, and thickness. The computations from these data, follow geometrical rules.

Extension; length, breadth, and thickness.

Impenetrability is evinced in the fact, that one body cannot enter into the space occupied by another, without previously thrusting the latter from its place.

Impenetrability.

A body then, is whatever occupies space, and possesses extension and impenetrability. One might be led to imagine that the property of impenetrability belonged only to solids, since we see them penetrating both air and water; but on closer observation it will be apparent that this property is common to all bodies of whatever nature.

Body defined.

Air and water impenetrable.

If a hollow cylinder into which a piston fits accurately, be filled with water, the piston cannot be thrust into the water, thus showing it to be impenetrable. Invert a glass tumbler in any liquid, the air, unable to escape, will prevent the liquid from occupying its place, thus proving the impenetrability of air. The diving-bell affords a familiar illustration of this property.

Experiment.

The difficulty of pouring liquor into a vessel having only one small hole, arises from the impenetrability of the air, as the liquid can run into the vessel only as the air makes its escape. The following experiment will illustrate this fact:

Experiment.

In one mouth of a two-necked bottle insert a funnel *a*, and in the other a siphon *b*, the longer leg of which is immersed in a glass of water. Now let water be poured into the funnel *a*, and it will be seen that in proportion as this water descends into the vessel *F*, the air makes its escape through the tube *b*, as is proved by the ascent of the bubbles in the water in the tumbler.

Fig. 1.

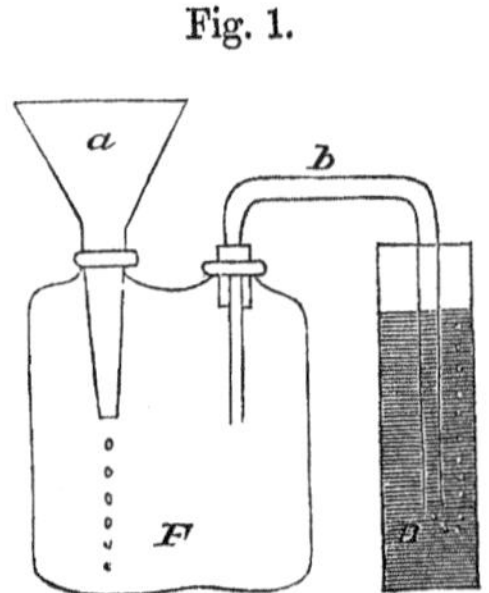

SECONDARY PROPERTIES.

Secondary properties.

The secondary properties of bodies are *compressibility, expansibility, porosity, divisibility,* and *elasticity.*

Compressibility, expansibility.

§ 3.—*Compressibility* is that property of bodies by virtue of which they may be made to occupy a smaller space; and *expansibility* is that in consequence of which they may be made to fill a larger, without in either case altering the quantity of matter they contain.

Both changes are produced in all bodies, as we shall presently see, by change of temperature; many bodies may also be reduced in bulk by pressure, percussion, &c. **Change of temperature, pressure, percussion.**

§ 4.—Since all bodies admit of compression and expansion, it follows of necessity, that there must be interstices between their minutest particles; and that property of a body by which its constituent elements do not completely fill the space within its exterior boundary, but leaves holes or pores between them, is called *porosity*. **Porosity.** The pores of one body are often filled with some other body, and the pores of this with a third, as in the case of a sponge containing water, and the water in its turn, containing air, and so on till we come to the most subtle of substances, *ether*, which is supposed to pervade all bodies and all space. **Pores filled with other bodies.** **Ether pervades all bodies and all space.**

In many cases the pores are visible to the naked eye; in others they are only seen by the aid of the microscope, and when so minute as to elude the power of this instrument, their existence may be inferred from experiment. Sponge, cork, wood, bread, &c., are bodies whose pores are noticed by the naked eye. The human skin appears full of them, when viewed with the magnifying glass; the porosity of water is shown by the ascent of air bubbles when the temperature is raised. **Visible and invisible pores.**

§ 5.—The *divisibility* of bodies is that property in consequence of which, by various mechanical means, such as beating, pounding, grinding, &c., we can reduce them to particles homogeneous to each other, and to the entire mass; and these again to smaller, and so on. **Divisibility.**

By the aid of mathematical processes, the mind may be led to admit the infinite divisibility of bodies, though their practical division, by mechanical means, is subject to limitation. Many examples, however, prove that it may be carried to an incredible extent. We are furnished with numerous instances among natural objects, **Infinite divisibility; practical limitation.** **Smallness of some natural objects.**

whose existence can only be detected by means of the most acute senses, assisted by the most powerful artificial aids; the size of such objects can only be calculated approximately.

Mechanical subdivisions in the arts

Mechanical subdivisions for purposes connected with the arts are exemplified in the grinding of corn, the pulverizing of sulphur, charcoal, and saltpetre, for the manufacture of gunpowder; and Homœopathy affords a remarkable instance of the extended application of this property of bodies.

Divisibility of gold.

Some metals, particularly gold and silver, are susceptible of a very great divisibility. In the common gold lace, the silver thread of which it is composed is covered with gold so attenuated, that the quantity contained in a foot of the thread weighs less than $\frac{1}{6000}$ of a grain. An inch of such thread will therefore contain $\frac{1}{72000}$ of a grain of gold; and if the inch be divided into 100 equal parts, each of which would be distinctly visible to the eye, the quantity of the precious metal in each of such pieces would be $\frac{1}{7200000}$ of a grain. One of these particles examined through a miscroscope of 500 times magnifying power will appear 500 times as long, and the gold covering it will be visible, having been divided into 3,600,000,000 parts, each of which exhibits all the characteristics of this metal, its color, density, &c.

Divisibility of dyes.

Dyes are likewise susceptible of an incredible divisibility. With 1 grain of blue carmine, 10 lbs. of water may be tinged blue. These 10 lbs. of water contain about 617,000 drops. Supposing now, that 100 particles of carmine are required in each drop to produce a uniform tint, it follows that this one grain of carmine has been subdivided 62 millions of times.

In the spider's thread, thread of the silkworm.

According to Biot, the thread by which a spider lets herself down is composed of more than 5000 single threads. The single threads of the silkworm are also of an extreme fineness.

In blood.

Our blood which appears like a uniform red mass, consists of small red globules swimming in a transparent fluid called serum. The diameter of one of these globules does not exceed the 4000th part of an inch: whence it follows that one drop of blood, such as would hang from the point of a needle, contains at least one million of these globules.

In the Infusoria.

But more surprising than all, is the microcosm of organized nature in the Infusoria, for more exact acquaintance with which we are indebted to the unwearied researches of Ehrenberg. Of these creatures, which for the most part we can see only by the aid of the microscope, there exist many species so small that millions piled on each other would not equal a single grain of sand, and thousands might swim at once through the eye of the finest needle. The coats-of-mail and shells of these animalcules exist in such prodigious quantities on our earth that, according to Ehrenberg's investigations, pretty extensive strata of rocks, as, for instance, the smooth slate near Bilin, in Bohemia, consist almost entirely of them. By microscopic measurements 1 cubic line of this slate contains about 23 millions, and 1 cubic inch about 41,000 millions of these animals. As a cubic inch of this slate weighs 220 grains, 187 millions of these shells must go to a grain, each of which would consequently weigh about the $\frac{1}{187}$ millionth part of a grain. Conceive further that each of these animalcules, as microscopic investigations have proved, has his limbs, entrails, &c., the possibility vanishes of our forming the most remote conception of the dimensions of these organic forms.

Ehrenberg's investigations.

Microscopic measurement,

weight.

Divisibility detected by smelling.

In cases where our finest instruments are unable to render us the least aid in estimating the minuteness of bodies, or the degree of subdivision attained; in other words, when bodies evade the perception of our sight and touch, our olfactory nerves frequently detect the presence of matter in the atmosphere, of which no chemical analysis could afford us the slightest intimation.

Instance of musk.

Thus, for instance, a single grain of musk diffuses in a

2

large and airy room a powerful scent that frequently lasts for years; and papers laid near musk will make a voyage to the East Indies and back without losing the smell. Imagine now, how many particles of musk must radiate from such a body every second, in order to render the scent perceptible in all directions, and you will be astonished at their number and minuteness.

Oil of lavender.

In like manner a single drop of oil of lavender evaporated in a spoon over a spirit-lamp, fills a large room with its fragrance for a length of time.

§ 6.—*Elasticity* is the name given to that property of bodies, by virtue of which they resume of themselves their figure and dimensions when these have been changed or altered by any extraneous cause. Different bodies possess this property in very different degrees, and retain it with very unequal tenacity. The measure of a body's elasticity, is the ratio obtained by dividing the capacity of restitution inherent in the body, by the capacity of the cause producing the change, both being supposed measurable. Thus, if R denote the capacity of restitution, F that of the extraneous cause, and e the elasticity, then will

Elasticity, its measure.

$$e = \frac{R}{F}.$$

When F and R are equal, the body is said to be perfectly elastic; when R is zero, the body is said to be non-elastic. These limits embrace all bodies in nature, there being none known to us which reach either extreme.

Examples of elastic bodies.

The following are a few out of a large number of highly elastic bodies; viz., glass, tempered steel, ivory, whalebone, &c.

Experiment with ivory.

Let an ivory ball fall on a marble slab smeared with some coloring matter. The point struck by the ball shows a round speck which will have imprinted itself on the surface of the ivory without its spherical form being at all impaired

Fluids under peculiar circumstances exhibit considerable elasticity; this is particularly the case with melted metals, more evidently sometimes than in their solid state. The following experiment illustrates this fact with regard to antimony and bismuth.

Elasticity of some melted metals.

Place a little antimony and bismuth on a piece of charcoal, so that the mass when melted shall be about the size of a peppercorn; raise it by means of a blowpipe to a white heat, and then turn the ball on a sheet of paper so folded as to have a raised edge all round. As soon as the liquid metal falls, it divides itself into many minute globules, which hop about upon the paper and continue visible for some time, as they cool but slowly; the points at which they strike the paper, and their course upon it, will be marked by black dots and lines.

Melted bismuth and antimony.

The recoil of cannon-balls is owing to the elasticity of the iron and that of the bodies struck by them.

Recoil of cannon-balls.

FORCE.

§ 7.—Whatever tends to change the actual state of a body, in respect to rest or motion, is called a force. If a body, for instance, be at rest, the influence which changes or tends to change this state to that of motion is called *force*. Again, if a body be already in motion, any cause which urges it to move faster or slower, is called *force*.

Forces.

Of the actual nature of forces we are ignorant; we know of their existence only by the effects they produce, and with these we become acquainted solely through the medium of the senses. Hence, while their operations are going on, they appear to us always in connection with some body which, in some way or other, affects our senses.

Ignorant of their nature; existence known by their effects on bodies.

Universal forces, attractions, and repulsions.

§ 8.—We shall find, though not always upon superficial inspection, that the approaching and receding of bodies or of their component parts, when this takes place apparently of their own accord, are but the results produced by the various forces that come under our notice. In other words, that the universally operating forces are those of *attraction* and of *repulsion.*

Atomical action; attraction of gravitation.

§ 9.—Experience proves that these universal forces are at work in two essentially different modes. They are operating either in the interior of a body, amidst the elements which compose it, or they extend their influence through a wide range, and act upon bodies in the aggregate; the former distinguished as *Atomical* and *Molecular action,* the latter as the *Attraction of gravitation.*

Force of cohesion and of dissolution.

§ 10.—Molecular forces and the force of gravitation, often co-exist, and qualify each other's action, giving rise to those attractions and repulsions of bodies exhibited at their surfaces when brought into sensible contact. This resultant action is called the force of *cohesion* or of *dissolution*, according as it tends to unite different bodies, or the elements of the same body, more closely, or to separate them more widely.

Inertia,

§ 11.—*Inertia* is that principle by which a body resists all change of its condition, in respect to *rest* or *motion.* If a body be at rest, it will, in the act of yielding its condition of rest, while under the action of any force, oppose a resistance; so also, if a body be in motion, and be urged to move faster or slower, it will, during the act of changing, oppose an equal resistance for every equal amount of change. We derive our knowledge of this principle solely from experience; it is found to be common to all bodies; it is in its nature conservative, though passive in character, being only exerted to preserve the rest or particular motion which a body has, by resisting

Known by experience; passive in character.

all variation in these particulars. Whenever any force acts upon a body, the inertia of the latter reacts, and this action and reaction are, as we shall see in the proper place, equal and directly opposed to each other. Action equal to reaction.

§ 12.—Molecular action chiefly determines the forms of bodies. All bodies are regarded as collections or aggregates of minute elements, called *atoms*, and are formed by the attractive and repulsive forces acting upon them at immeasurably small distances. Forms of bodies determined by molecular action.

Several hypotheses have been proposed to explain the constitution of a body, and the mode of its formation. The most remarkable of these was by Boscovich, about the middle of the last century. Its great fertility in the explanations it affords of the properties of what is called tangible matter, and its harmony with the laws of motion, entitle it to a much larger space than can be found for it in a work like this. Enough may be stated, however, to enable the attentive reader to seize its leading features, and to appreciate its competency to explain the phenomena of nature. Constitution of bodies; Boscovich.

1. All matter consists of indivisible and inextended *atoms*. First postulate.

2. These atoms are endowed with attractive and repulsive forces, varying both in intensity and direction by a change of distance, so that at one distance two atoms attract each other, and at another distance they repel. Second postulate.

3. This law of variation is the same in all atoms. It is, therefore, mutual; for the distance of atom a from atom b, being the same with that of b from a, if a attract b, b must attract a with *precisely* the same force. Third postulate.

4. At all considerable or *sensible* distances, these mutual forces are attractive and sensibly proportional to the square of the distance inversely. It is the attraction called *gravitation*. Fourth postulate.

5. In the small and insensible distances in which sensible contact is observed, and which do not exceed the Fifth postulate.

1000th or 1500th part of an inch, there are many alternations of attraction and repulsion, according as the distance of the atoms is changed. Consequently, there are many situations within this narrow limit, in which two atoms neither attract nor repel.

Sixth postulate.

6. The force which is exerted between two atoms when their distance is diminished without end, and is just vanishing, is an insuperable repulsion, so that no force whatever can press two atoms into mathematical contact.

Such, according to Boscovich, is the constitution of a material atom and the *whole* of its constitution, and the immediate efficient cause of *all* its properties.

Molecule, particle, body.

Two or more atoms may be so situated, in respect to position and distance, as to constitute a *molecule.* Two or more molecules may constitute a *particle.* The particles constitute a *body.*

Add inertia.

Now, if to these centres, or loci of the qualities of what is termed matter, we attribute the property called inertia, we have all the conditions requisite to explain, or arrange in the order of antecedent and consequent, the various operations of the physical world.

Exponential curve.

Boscovich represents his law of atomical action by what may be called an exponential curve. Let the dis-

Fig. 2.

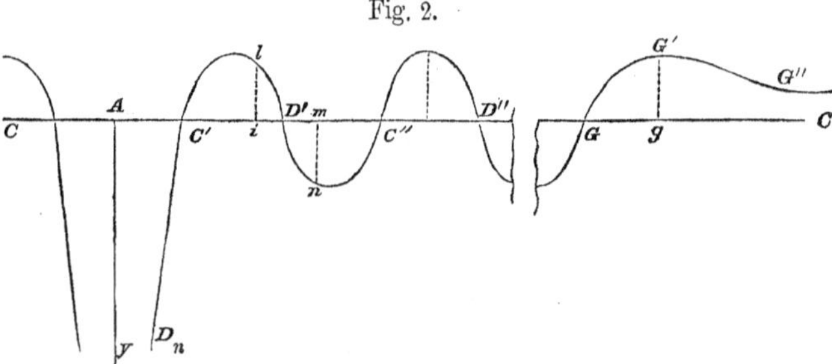

tance of two atoms be estimated on the line *C A C*, *A* being the situation of one of them while the other is placed anywhere on this line. When placed at *i*, for example, we may suppose that it is attracted by *A*, with a certain intensity. We can represent this intensity by the length of the line *i l*, perpendicular to *A C*, and can express the direction of the force, namely, from *i* to *A*, because it is attractive, by placing *i l* above the axis *A C*. Should the atom be at *m*, and be repelled by *A*, we can express the intensity of repulsion by *m n*, and its direction from *m* towards *G* by placing *m n* below the axis.

Attractive ordinates above.

Repulsive ordinates below.

This may be supposed for every point on the axis, and a curve drawn through the extremities of all the perpendicular ordinates. This will be the exponential curve or scale of force.

As there are supposed a great many alternations of attractions and repulsions, the curve must consist of many branches lying on opposite sides of the axis, and must therefore cross it at *C′*, *D′*, *C″*, *D″*, &c., and at *G*. All these are supposed to be contained within a very small fraction of an inch.

Curve on opposite sides of axis.

Beyond this distance, which terminates at *G*, the force is always attractive, and is called the force of *gravitation*, the maximum intensity of which occurs at *g*, and is expressed by the length of the ordinate *G′g*. Further on, the ordinates are sensibly proportional to the square of their distances from *A*, inversely. The branch *G′ G″* has the line *A C*, therefore, for its asymptote.

Force of gravitation.

Within the limit *A C′* there is repulsion, which becomes infinite, when the distance from *A* is zero; whence the branch $C' D_n$ has the perpendicular axis, *A y*, for its asymptote.

An atom being placed at *G*, and then disturbed so as to move it in the direction towards *A*, will be repelled, the ordinate of the curve being below the axis; if disturbed so as to move it from *A*, it will be attracted, the corresponding ordinates being above the axis. The point *G* is therefore a position in which the atom is neither attracted nor repelled, and to which it will tend to return when slightly removed in either direction, and is called the *limit of gravitation.*

Position of indifference

Limit of gravitation.

Limits of cohesion.

If the atom be at C', or C'', &c., and be moved ever so little towards A, it will be repelled, and when the disturbing cause is removed, will fly back; if moved from A, it

Fig. 3.

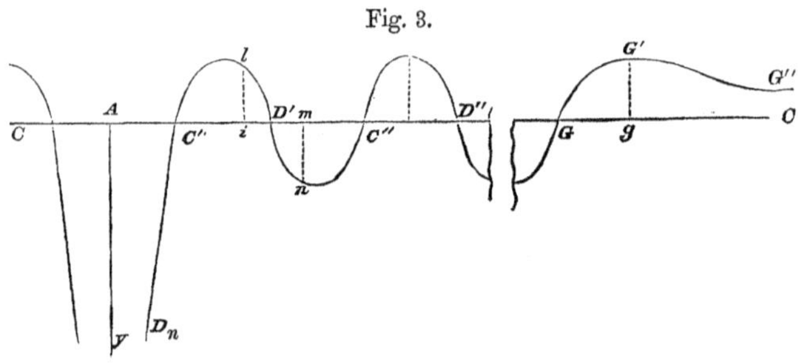

will be attracted and return. Hence C', C'', are positions similar to G, and are called *limits of cohesion*, C' being termed the *last limit of cohesion.* An atom situated at any one of these points will, with that at A, constitute a *permanent molecule* of the simplest kind.

Permanent molecule.

On the contrary, if an *atom* be placed at D', or D'', &c., and be then slightly disturbed in the direction either from or towards A, the action of the atom at A will cause it to recede still further from its first position, till it reaches a limit of cohesion. The points D', D'', &c., are also positions of indifference, in which the atom will be neither attracted nor repelled by that at A, but they differ from G, C', C'', &c., in this, that an atom being ever so little removed from one of them has no disposition to return to it again; these points are called *limits of dissolution.* An atom situated in one of them cannot, therefore, constitute, with that at A, a permanent molecule, but the slightest disturbance will destroy it.

Positions of indifference.

Limits of dissolution.

Molecules of different orders; particles.

It is easy to infer, from what has been said, how three, four, &c., atoms may combine to form molecules of different orders of complexity, and how these again may be arranged so as by their action upon each other to form particles. Our limits will not permit us to dwell upon these points, but we cannot dismiss the subject without suggesting a consequence which the reader will find of interest when he comes to the subjects of *light* and *heat*. We allude to those characteristics of the sun by which he is the main source of these principles to the inhabitants of the earth.

Inference—light and heat of sun.

It results from the laws of gravitation, that every atom in a spherical solid body is attracted towards the centre by a force directly proportional to its distance from that point. The pressure towards the centre will, therefore, increase as the magnitude of the sphere increases, and may ultimately become so great as to force the atoms near enough to each other to bring them within the last limits of cohesion, in which case, the mass, composed of atoms thus urged into close proximity, becomes perfectly elastic. The magnitude of this elastic mass will be greater in proportion as the whole sphere is greater. Every body falling upon the sphere will, on reaching its position at the surface, send the motion with which it arrived towards the centre to agitate the atoms of the elastic mass. These being once disturbed will, under the forces thus called into play, vibrate indefinitely about their positions of rest by virtue of their inertia.

Attraction of spherical masses.

Production of elasticity.

Effect of a falling body.

It is only necessary therefore to suppose, that the heavenly bodies have been formed by the gravitation of the particles of a vast *nebula* towards its centre, and to adopt the hypothesis which modern discoveries have revived and forced upon us, viz., that heat and light are but the effects of vibratory motion, to account for the *incandescent and self-luminous* character of the sun. The same principle furnishes an explanation of the internal heat of our earth which, together with all the heavenly bodies, would doubtless appear self-luminous were the acuteness of our sense of sight increased beyond its present limit in the same proportion that the sun exceeds the largest of these bodies. The sun far transcends all the other bodies of our system in regard to heat and light, simply because of his vastly greater size.

Nebular hypothesis.

Light and heat, effects of motion.

Incandescence and luminosity of the sun.

Those of the sun greater because of his greater size.

§ 13.—The molecular forces are the effective causes which hold together the particles of bodies. Through them, the molecules approach to a certain distance where they gain a position of rest with respect to each other. The power with which the particles adhere in these relative positions, is called, as we have seen, *cohesion.* This force is measured by the resistance it offers to mechanical separation of the parts of bodies from each other.

Effects of molecular action.

Measure of cohesion.

Three states of aggregation.

On the degree of this force, the three states or aggregate forms called solid, liquid, and gaseous depend. These different states of matter result from certain definite relations under which the molecular attraction and repulsion establish their equilibrium; there are three cases, viz. two extremes and one mean. The first extreme is that in which attraction predominates among the atoms; this produces the *solid* state. In the other repulsion prevails, and the *gaseous* form is the consequence. The mean obtains when neither of these forces is in excess, and then matter presents itself under the *liquid* form.

Solid, gas, liquid.

Let A represent the attraction and R the repulsion, then the three aggregate forms may be expressed by the following formulæ:

Formulæ.

$$A > R \quad \text{solid,}$$
$$A < R \quad \text{gas,}$$
$$A = R \quad \text{liquid.}$$

These three forms or conditions of matter may, for the most part, be readily distinguished by certain external peculiarities; there are, however, especially between solids and liquids, so many imperceptible degrees of approximation, that it is sometimes difficult to decide where the one form ends and the other begins. It is further an ascertained fact that many bodies, (perhaps all,) as for instance water, are capable of assuming all three forms of aggregation.

External peculiarities of bodies; subject to change.

Thus, supposing that the relative intensity of the molecular forces determines these three forms of matter, it follows from what has been said above, that this term may vary in the same body. Change of molecular action in same body.

The peculiar properties belonging to each of these states will be explained when solid, liquid, and aëriform bodies come severally under our notice.

§ 14.—The molecular forces may so act upon the atoms of dissimilar bodies as to cause a new combination or union of their atoms. This may also produce a separation between the combined atoms or molecules in such manner as to entirely change the individual properties of the bodies. Such efforts of the molecular forces are called *chemical action;* and the disposition to exert these efforts, on account of the peculiar state of aggregations of the ultimate atoms of different bodies, *chemical affinity.* Action of molecular forces between dissimilar bodies. Chemical action. Chemical affinity.

§ 15.—Beyond the last limit of gravitation, atoms attract each other: hence all the atoms of one body attract those of another, thus giving rise to attractions between bodies of sensible magnitudes through sensible distances. The intensities of these attractions are directly proportional to the number of attracting atoms, and inversely as the squares of their distances apart. Attraction of bodies of sensible magnitude. Intensity of this attraction.

The term *universal gravitation* is applied to this force when it is intended to express the action of the heavenly bodies on each other; and that of *terrestrial gravitation* or simply *gravity*, where we wish to express the action of the earth upon the bodies forming with itself one whole. The force is always of the same kind however, and varies in intensity only by reason of a difference in the number of atoms and their distances. Its effect is always to generate motion when the bodies are free to move. Universal gravitation. Terrestrial gravity. Effects of this force.

Gravity, then, is a property common to all terrestrial bodies, since they constantly exhibit a tendency to approach the earth and its centre. In consequence of this Gravity common to all bodies. Its consequences.

tendency, all bodies, unless supported, fall to the surface of the earth, and if prevented by any other bodies from doing so, they exert a pressure on these latter.

This is one of the most important properties of terrestrial bodies, and the cause of many phenomena, of which a fuller explanation will be given presently.

SECTION I.

MECHANICS.

§ 16.—That branch of Natural Philosophy which treats of the action of forces on bodies, is called *Mechanics*. Mechanics is usually divided into *Statics*, which treats of the mutual destruction of forces when applied to solid bodies; *Hydrostatics*, when applied to fluids; *Dynamics*, which treats of the motions of solid bodies; and *Hydrodynamics* which investigates the motions of fluids. *Statics* and *Dynamics* will be treated together, under the general head, *Mechanics of Solids*, as will also *Hydrostatics* and *Hydrodynamics*, under the head, *Mechanics of Fluids*.

Mechanics, Statics, Hydrostatics, Dynamics, Hydrodynamics. Mechanics of solids, and of fluids.

PART FIRST.

MECHANICS OF SOLIDS.

I.

SPACE, TIME, MOTION, AND FORCE.

§ 17.—*Space* is indefinite extension, without limit, and contains all bodies. Space.

§ 18.—*Time* is any limited portion of duration. We may conceive of a time which is longer or shorter than a given time. Time has, therefore, magnitude, as well as lines, areas, &c. Time; has magnitude.

To *measure a given time*, it is only necessary to obtain equal times which succeed each other without intermission, to call one of these equal times unity, and to express, by a number, how often this unit is contained in the given time. When we give to this number the particular name of the unit, as *hour*, *minute*, *second*, &c., we have a complete expression for time. Time measured. Units of time.

The *Instruments* usually employed in measuring time are *clocks, chronometers,* and *common watches*, which are too well known to need a description in a work like this. Time instruments.

The smallest division of time indicated by these time-pieces is the *second*, of which there are 60 in a minute,

Performance of chronometers.

3600 in an *hour*, and 86400 in a day; and chronometers, which are nothing more than a species of watch, have been brought to such perfection as not to vary in their rate a half a second in 365 days, or 31536000 seconds.

Thus the number of hours, minutes, or seconds, between any two events or instants, may be estimated with as much precision and ease as the number of yards, feet, or inches between the extremities of any given distance.

Time represented by lines.

Time may be represented by lines, by laying off upon a given right line AB, the equal distances from 0 to 1, 1 to 2, 2 to 3, &c., each one of these equal distances representing the unit of time.

Fig. 3.

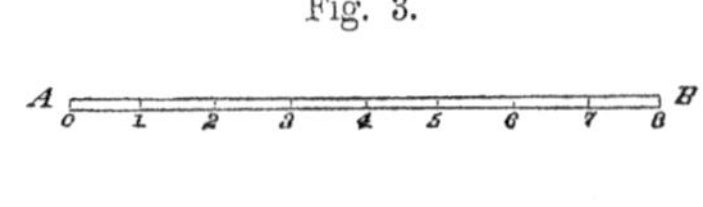

Rest; absolute and relative.

§ 19.—A body is in a state of *absolute rest* when it continues in the same place or position in space. There is perhaps no body absolutely at rest; our earth being, without cessation, in motion about the sun, nothing connected with it can be at rest. In what follows, rest must, therefore, be considered but as a *relative* term. A body is said to be at rest, when it preserves the same position in respect to other bodies which we may regard as fixed.

Example of relative rest.

A body, for example, which continues in the same place in a boat, is said to be at rest in relation to the boat, although the boat itself may be in motion in relation to the banks of a river on whose surface it is floating.

Motion, like rest, is relative.

§ 20.—A body is in *motion* when it occupies successively different positions in space. Motion, like rest, is but relative. A body is in motion when it changes its place in reference to those which we may regard as fixed.

It is continuous.

Motion is essentially *continuous;* that is, a body cannot pass from one position to another without passing through a series of intermediate positions; the motion of a point describes, therefore, a continuous line.

When we speak of the path described by a body, we are to understand that of a certain point connected with the body. Thus, the path of a ball, is that of its centre, &c. Path of a body.

§ 21.—The motion of a body is *curvilinear* or *rectilinear*, according as the path described is a *curve* or *right line*. When the motion is curvilinear, we may consider it as taking place upon a polygon, of which the sides are very small and sensibly coincide with the curve. The prolongation of any one of these sides will be a tangent to the curve, and will indicate the *direction* of the body's motion while upon this side. Curvilinear and rectilinear motion. Direction of a body's motion.

Conceive the time employed by a body to pass from one position to another, to be divided into a number of very small and equal parts. If the portions of the path successively described in these equal times be equal, the motion is said to be *uniform*. If otherwise, the motion is said to be *varied*. It is *accelerated* when these elementary paths are greater and greater; *retarded*, when less and less in the order of time. Uniform motion. Varied motion; accelerated and retarded.

§ 22.—*Velocity* is the *rate* of a body's motion. The rapidity or slowness of motion is indicated by the greater or less length of the path described by the body, during each of the small and equal portions of time into which the whole time is divided. This length is taken as the *measure* of the velocity when the small portion of time is made to denote the unit of time. Velocity; its measure.

The velocity is *constant* in uniform motion: it is *variable* in accelerated and retarded motion. Constant and variable.

§ 23.—In *uniform motion*, the small spaces described in equal consecutive portions of time being equal, it is obvious that the space described in any given time will Uniform motion.

contain as many equal parts of space as there are equal parts of time. Consequently, in uniform motion, *equal spaces will be described in equal times*, whatever be the rate of motion, and the spaces will be proportional to the times employed in describing them.

Relation of space to the time.

Denote by S the length of space described during the time T; s the length of the space described in the small portion of time t, then, from what precedes, we have

$$S : T :: s : t$$

$$\frac{S}{T} = \frac{s}{t} \quad . \quad . \quad . \quad . \quad . \quad (1).$$

a constant ratio.

Velocity measured by the space described in any unit of time.

§ 24.—Since in uniform motion, the spaces are proportional to the times employed in describing them, the velocity may be measured by the space described in any time whatever, for example in a *second*, *minute*, an *hour*, &c. Thus we say the velocity is 2 feet a second, or 120 feet a minute, or 7200 feet an hour, or $\frac{2}{10}$ of a foot in $\frac{1}{10}$ of a second, &c; all of which amounts to the same thing, since the ratio of the space to the time is not changed.

Rule for finding velocity.

When a body describes uniformly a certain space in a given number of units of time, as the second, for example, which is usually taken as the unit, the velocity is found by dividing the whole space by the whole time, for if we make $t =$ one second in equation (1), s becomes the velocity, § 22, and denoting this by V we have

$$V = \frac{S}{T} \quad . \quad . \quad . \quad . \quad . \quad (2).$$

Example.

Example: The space described in 1 minute and 5 seconds or 65^s being 260 feet, the space described in 1^s, or the velocity, is given thus:

$$V = \frac{S}{T} = \frac{260^f}{65^s} = 4^f.$$

Reciprocally, if the velocity be multiplied by the number of units of time, the space will result.

Periodical motion.

§ 25.—It frequently happens in practice that the velocity is not constant, although the spaces described at the end of certain equal intervals are equal. Such for instance is the case in all *periodical* movements of which the different changes are executed in the same interval of time, although the velocity is continually varying within this interval. The motion of a carriage and that of a pedestrian, are examples of this; the spaces described in certain intervals, are often the same, while the motion is sometimes accelerated and sometimes retarded.

Instance—carriage and pedestrian.

Relation of space and time, represented geometrically.

§ 26.—Conceive a table consisting of two vertical columns, in one of which are arranged *the numbers* expressive of the intervals of time elapsed since any given instant, and in the other, on the same horizontal lines, the numbers which designate the spaces described by any body in these intervals. Draw an indefinite right line OB; assume any linear dimension, as an inch, to represent the unit of time, and let the same length represent the unit of space; with a scale of equal parts, lay off a distance $O\,t_4$ representing an interval of time given by the table; upon a perpendicular to OB at the point t_4, lay off a distance $t_4\,e_4$ representing the distance passed over by the body in the time $O\,t_4$. Do the same for the other times and corresponding spaces of the table, and we obtain the points e_1, e_2, e_3, &c.,

In any kind of motion.

Fig. 4.

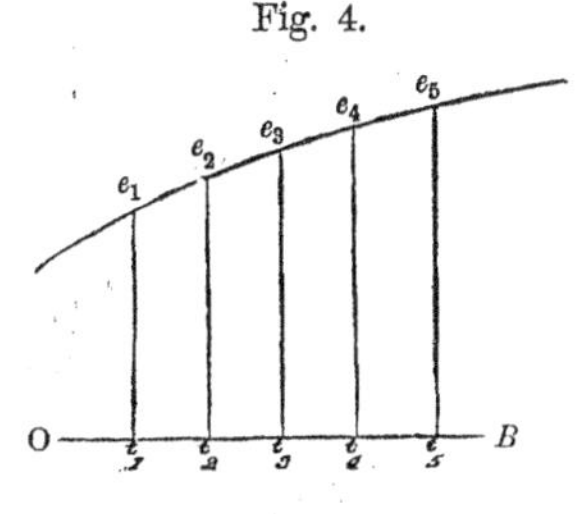

which, being united two and two by right lines, will give a polygon. This polygon will not differ sensibly from a curve when the intervals of time are small and differ very little from each other. $O\,t_1$, $O\,t_2$, $O\,t_3$, &c., are the abscisses, and $t_1\,e_1$, $t_2\,e_2$, $t_3\,e_3$, &c., the ordinates of this curve, of which the origin is O. It is obvious that by means of the curve we may obtain, as by the table, the space described during any given interval; so that this curve gives the relation which connects the spaces with the times, whatever be the nature of the motion.

In uniform motion.

In *uniform motion* the spaces increase in the direct ratio of the times, and the ordinates $t_1\,e_1$, $t_2\,e_2$, $t_3\,e_3$, &c., are therefore proportional to the abscisses $O\,t_1$, $O\,t_2$, $O\,t_3$, &c.; hence the curve becomes a right line. Let the axis OB, of *times*, be divided into any number of equal and very small parts; through the points of division draw the ordinates or *spaces*, and through the extremities of the ordinates draw the lines $e_1\,b_2$, $e_2\,b_3$, $e_3\,b_4$, &c., parallel to the axis of times, we shall thus form a series of small right-angled triangles $O\,t_1\,e_1$, $e_1\,b_2\,e_2$, &c., similar to the triangle $O\,t_4\,e_4$, and because $e_3\,b_4 = t_3\,t_4$, we have

Fig. 5.

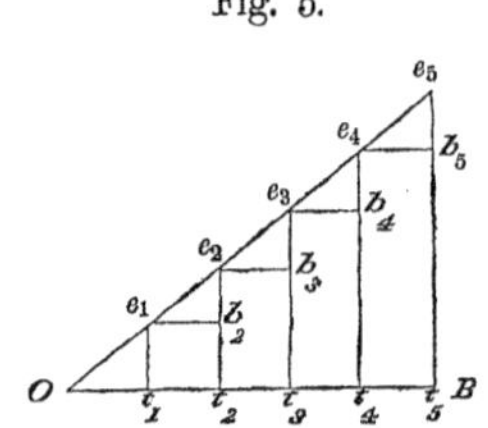

$$t_4\,e_4 \;:\; O\,t_4 \;::\; b_4\,e_4 \;:\; t_3\,t_4,$$

whence

$$\frac{t_4\,e_4}{O\,t_4} = \frac{b_4\,e_4}{t_3\,t_4};$$

Relation of spaces to the times.

but $b_4\,e_4$ is the space s, described in the small time $t_3\,t_4 = t$, and $t_4\,e_4$ the space S described in the time $O\,t_4 = T$, and the above may be written

$$\frac{S}{T} = \frac{s}{t},$$

and making $t = 1$, s becomes the measure of the velocity V, and we have

$$V = \frac{S}{T},$$

Velocity equal to the ratio of the space to the time.

the same as before, equation (1).

Or, $O t_4$ may be taken as the unit of time, in which case, $t_4 e_4$ becomes the velocity V, and we have

$$V = \frac{s}{t}.$$

Same for any space and time.

Varied motion.

In *varied motion*, the spaces not being proportional to the times, the line $O e_1$, $e_1 e_2$, $e_2 e_3$, &c., is not straight, and the small spaces $e_2 b_2$, $e_3 b_3$, &c., described in the elementary times $t_1 t_2$, $t_2 t_3$, &c., are not equal. The velocity must, therefore, vary at every instant. For the case represented by the figure, the motion is accelerated, because the spaces $e_2 b_2$, $e_3 b_3$, &c., described in the equal elementary times, continually increase. Now let it be supposed that at the point e_3 the motion ceases to be accelerated, and that it becomes uniform with the velocity which the body had at this instant. The law of the motion afterward will be represented by the right line $e_3 m$, the prolongation of $e_3 e_4$, and since, at the instant we are considering, the body describes a space equal to $e_4 b_4$ in the elementary time $e_3 b_4 = t_3 t_4$, it will, in virtue of

Fig. 6.

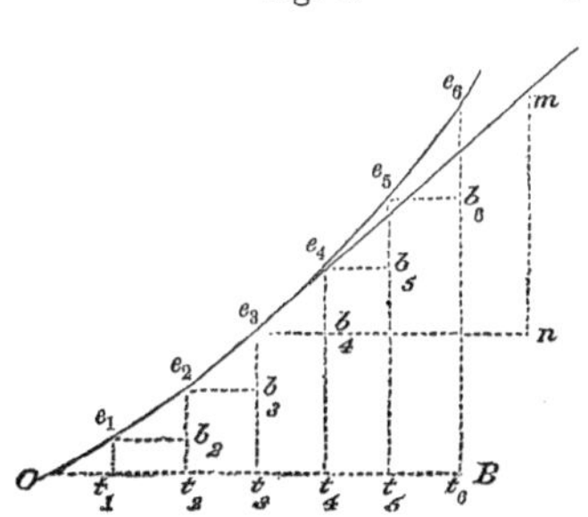

Accelerated motion, represented geometrically.

Motion ceases to be accelerated;

becomes uniform.

its uniform motion, describe in a unit of time a space equal to mn, obtained by laying off from the point e_3, on $e_3 b_4$ produced, a distance $e_3 n$ equal to the unit of time. But the space described in a unit of time, at a constant rate, is the measure of the velocity corresponding to the point e_3, or at the end of the time $O t_3$. From the figure we obtain

Fig. 6.

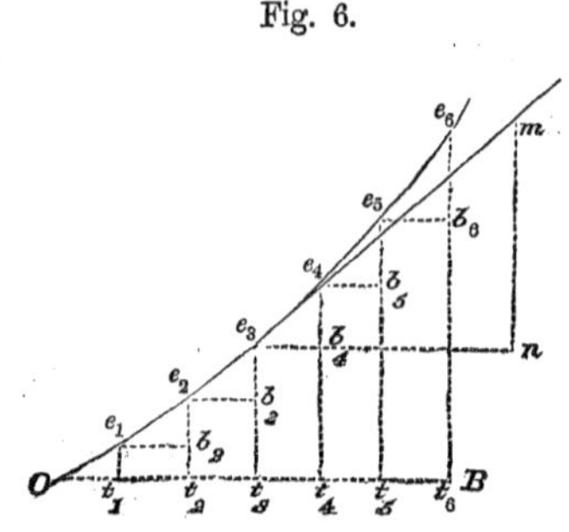

Measure of the velocity at any instant;

$$e_4 b_4 : e_3 b_4 :: mn : e_3 n;$$

or making

$$e_4 b_4 = s, \quad e_3 b_4 = t, \quad mn = V, \quad e_3 n = 1,$$

we have

$$s : t :: V : 1;$$

equal to the ratio of the element of the space, to the element of the time.

whence

$$V = \frac{s}{t}$$

as before.

If we suppose the element of time $t_3 t_4$ sufficiently small, the line $e_3 e_4$ will coincide with the curve to which $e_3 m$ will become a tangent at the point e_3. This tangent being constructed geometrically, will give, in the manner above indicated, the velocity corresponding to the point of the curve to which it is drawn, or the velocity at the end of the time $O t_3$.

Tangent line; will give the velocity.

Periodical motion, such as has been defined in § 25, will be represented by a waved line *EEE*, &c, whose undulations are regularly disposed about the right line e_1, e_2, e_3, &c., which represents the law of *uniform motion.*

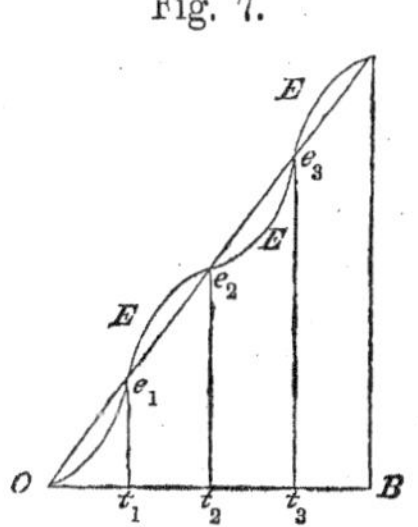

Fig. 7.

Geometrical representation of periodical motion.

It may be important to remark that the curves which have just been described, and which connect the lengths of the spaces and the times, in any kind of motion, must not be confounded with the actual path described by the body. In this last, the tangent simply gives the direction of the motion; and to obtain the velocity, the elementary portion of the curve, or of the tangent line, must be divided by the time during which this element is described.

Distinction between the line giving the law of the motion, and the path described by the body.

§ 27.—Matter in its unorganized state, is *inanimate* or *inert.* It cannot give itself motion, nor can it change of itself the motion which it may have received. A body at rest will forever remain so unless disturbed by something extraneous to itself; or if it be in motion in any direction, as from *a* to *b*, it will continue, after arriving at *b*, to move towards *c* in the prolongation of *ab;* for having arrived at *b*, there is no reason why it should deviate to one side more than another. Moreover, if the body have a certain velocity at *b*, it will retain this velocity unaltered, since no reason can be assigned why it should be increased rather than diminished in the absence of all extraneous causes.

Fig. 8.

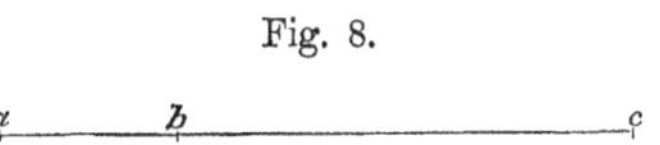

Inanimate bodies cannot change their state of rest or of motion.

If a billiard-ball, thrown upon the table, seem to diminish its rate of motion till it stops, it is because its

Apparent exception explained.

motion is resisted by the cloth and the atmosphere. If a body thrown vertically downward seem to increase its velocity, it is because its weight is incessantly urging it onward. If the direction of the motion of a stone, thrown into the air, seem continually to change, it is because the weight of the stone urges it incessantly towards the surface of the earth. Experience proves that in proportion as the obstacles to a body's motion are removed, will the motion itself remain unchanged.

Consequences of inertia.

It results, from what has been said, that when a body is put in motion and abandoned to itself, its *inertia* will cause it to move in a straight line and preserve its rate of motion unchanged. If, from any extraneous cause the body is made to describe a curve AB, and this cause be removed at the point B, the inertia will cause the body to move along the tangent BC, and to preserve the velocity which it had at B.

Fig. 9.

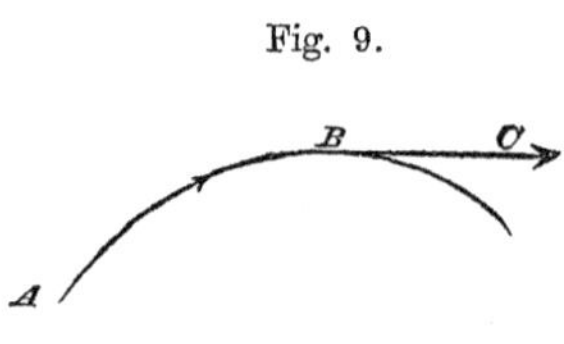

Forces; weight and heat.

§ 28.—A *force* has been defined to be that which changes or tends to change the state of a body in respect to rest or motion. *Weight* and *Heat* are forces. A body laid upon a table, or suspended from a fixed point by means of a thread, would move under the action of its weight, if the resistance of the table, or that of the fixed point did not continually destroy the effort of the weight.

Illustration.

A body exposed to any source of heat, expands, its particles recede from each other, and thus the state of the body is changed.

Forces produce various effects.

§ 29.—Forces produce various effects according to circumstances. They sometimes leave a body at rest, by destroying one another, through its intervention; sometimes they change its form or break it; sometimes they impress upon it motion, they accelerate or retard that which it has, or change its direction; sometimes these effects are produced gradually, sometimes abruptly, but however produced they require some *definite time,* and are effected by *continuous degrees.* If a body is sometimes seen to change suddenly its state, either in respect to the direction or the rate of its motion, it is because the force is so great as to produce its effect in a time so short as to make its duration imperceptible to our senses, yet some definite portion of time is necessary for the change. A ball fired from a gun, will break through a pane of glass, a piece of board, or a sheet of paper when freely suspended, with a rapidity so great that the parts torn away have not time to propagate their motion to the rest. A cannon freely suspended at the end of a vertical cord will throw its ball to the same point as though it were on its carriage, which proves that the piece does not move sensibly till the ball leaves its mouth, though afterward it recoils to a considerable distance. In these several cases the effects are obvious, while the times in which they are accomplished are so short as to elude the senses: and yet these times have had some definite duration, since the changes, corresponding to these effects, have passed in succession through their different degrees from the beginning to the ending.

These effects require definite portions of time.

A ball fired from a cannon.

Fig. 10.

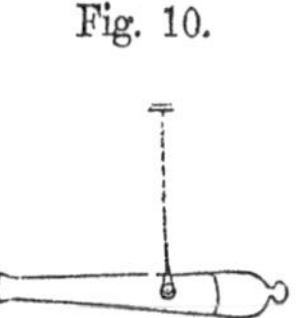

Effects obvious, while the times are not.

Forces which give motion to bodies are called *motive forces;* they are *accelerating* when they accelerate the motion at each instant, and retarding when they retard it.

Motive forces; accelerating and retarding.

§ 30.—We may form from our own experience a clear idea of the mode in which forces act; when we push or pull a body, be it free or fixed, we experience a sensation denominated *pressure, traction,* or in general, *effort.* This effort is analogous to that which we exert in raising a

Idea of the action of forces obtained from experience.

Forces are real pressures; unit of force. Equal forces.

weight, and thus forces are to us real pressures. Pressure may be strong or it may be feeble; it therefore has *magnitude*, and may be expressed in numbers by assuming a certain pressure as *unity*, which may easily be done if we can find pressures that are equal to each other.

Two forces are equal when, substituted, one for the other, in the same circumstances, they produce the same effect, or when, being directly opposed, they destroy each other.

Forces measured by weights. Double, triple, &c., force. Unit of force a pound weight.

Conceive a body W, suspended from the extremity of a thread; the thread will assume a vertical direction, and an effort will be necessary to support it; if two forces, applied successively to the thread and in the same manner, maintain the body at rest, these forces are equal to each other and to the weight of the body. A double, triple, &c., force, will support two, three, &c., bodies, similar to the first, suspended one above another on the same thread; taking one of these forces, that, for instance, which supports $\frac{1}{62,5}$th of a cubic foot of distilled water at the temperature of 60° Fahrenheit, and of which the weight is called a pound, for unity, any force will be expressed by a number which indicates how many pounds it will support.

Fig. 11.

Fig. 12.

Forces compared by the balance.

§ 31.—Weights are measured and compared by means of an instrument called a *balance*, and of which we shall speak hereafter. By the definition given above of equal forces, it will be easy to find the weights of bodies whatever be the merits or defects of such an instrument. We have but to require that these bodies substituted for a certain number of standard units of weight, shall produce, under the same circumstances, the same effect upon the balance. Under this point of view, many devices may be employed to measure the weights of bodies and consequently the magnitudes of forces.

Use of spring balance to measure forces.

Springs, among others, in supposing they preserve unimpaired for a long time their elasticity, may be, and indeed are, used in practice, for this purpose. Of such is the spring balance, a sketch of which is given in the figure. In using this instrument, it is necessary to determine previously the accuracy of its divisions by means of standard weights, and to change the values of its graduations if the elasticity of the spring shall be found to have undergone a change since its construction.

Fig. 13.

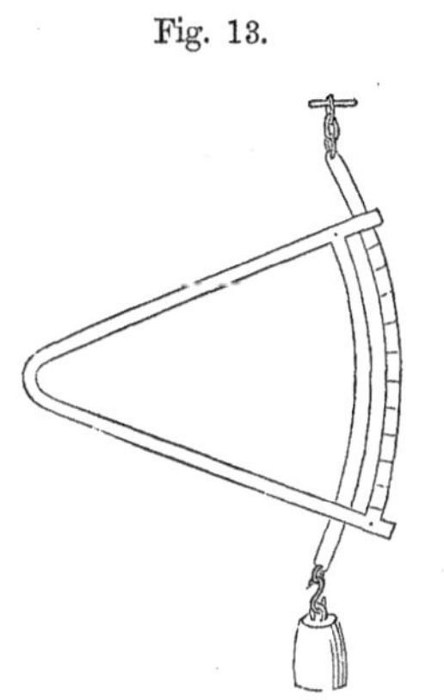

Verification of the elasticity.

Variation in force of gravity, small within moderate limits.

§ 32.—It is known from observation that the action of the force of gravity diminishes as the bodies upon which it is exerted are elevated above the surface of the earth. The same body, therefore, which will cause by its weight a spring to bend through a certain angle at the surface of the sea, will cause it to bend through a less angle when weighed at the top of a mountain, and thus the absolute weight of the body, or magnitude of the force which sustains it, is diminished. But this diminution for the height of three miles does not exceed $\frac{1}{750}$ of the total weight. Experience also shows that the weight of a body diminishes as it approaches the equator, but for an extent of territory equal to that of the state of New York this variation is scarcely appreciable.

Acts in parallel directions within the limits of ordinary bodies.

The directions of two plumb-lines being normal to the surface of the earth, cannot be perfectly parallel, since they converge to a point near its centre and which is therefore distant about 4000 miles from the place of observation. These lines when separated by a distance of 600 yards on the surface of the earth, will form with each other an angle not to exceed 6″, which is inappreciable to common instruments. It hence follows, that, within ordinary limits, *the force of gravity may be regarded as constant, and acting in parallel directions.*

Force of gravity constant, and acts in parallel directions.

II.

ACTION OF FORCES, EQUILIBRIUM, WORK.

Action of exterior forces on bodies;

§ 33.—When a force acts against a point in the surface of a body, it exerts a pressure which crowds together the neighboring particles; the body yields, is compressed and its surface indented; the crowded particles make an effort, by their molecular forces, to regain their primitive places, and thus transmit this crowding action even to the remotest particles of the body. If these latter particles are fixed or prevented by obstacles from moving, the result will be a compression and change of figure throughout the body. If, on the contrary, these extreme particles are free they will advance, and motion will be communicated by degrees to all the parts of the body. This internal motion, the result of a series of compressions, proves that a certain time is necessary for a force to produce its entire effect, and the absurdity of supposing that a finite velocity may be generated instantaneously. The same kind of action will take place when the force is employed to destroy the velocity which a body has already acquired; it will first destroy the velocity of the molecules at and nearest to the point of action, and then, by degrees, that of those which are more remote in the order of distance.

when some of the particles are fixed.

When none of the particles are fixed.

Definite velocity cannot be generated instantaneously.

Reaction equal and contrary to action.

§ 34.—As the molecular springs cannot be compressed without reacting in a contrary direction, and with the same effort, the *agent* which presses a body will experience an equal pressure. This is usually expressed by saying that *reaction* is equal and contrary to *action*. In pressing the finger against a body, in pulling it with a thread, or pushing it with a bar, we are pressed, drawn, or pushed in a contrary direction, and with the same effort. Two

Fig. 14.

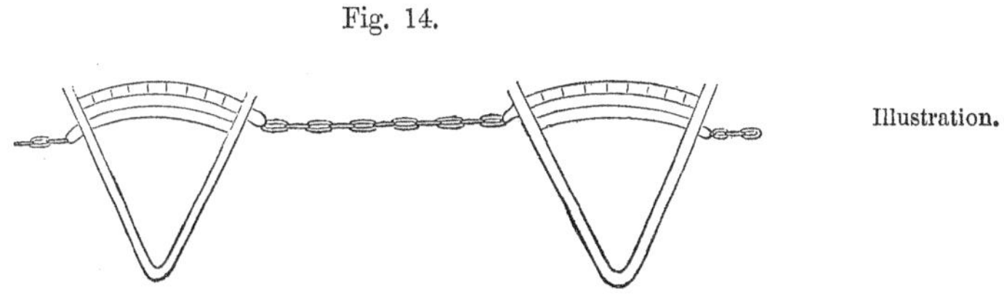

Illustration.

weighing springs attached to the extremities of a thread or bar, will indicate the same degree of tension, and in contrary directions when made to act upon each other through the intervention of the thread or bar.

Point of application, taken at any point in line of direction.

§ 35.—In every case, the action of a force is transmitted through a body to the ultimate point of resistance, by a series of equal and contrary actions and reactions which destroy each other, and which the molecular springs of all bodies exert at every point of the right line, limited by their boundaries, along which the force acts. It is in virtue of this property of bodies, that the action of a force may be supposed to be exerted at *any point in its line of direction.*

Bodies used to transmit the action of forces.

§ 36.—Bodies being more or less extensible and compressible, a thread or bar, interposed between the power and resistance, will be stretched or compressed to a certain degree, depending upon the energy with which these forces act; but as long as the power and resistance remain the same, the thread or bar, having attained its new length, will cease to change. On this account, bodies, regarded as rigid and inextensible; which are usually employed to transmit the action of forces from one point to another, may be regarded as perfectly inextensible or rigid, especially as such bodies are chosen and applied so as not to yield under this action.

Inertia measured by means of forces; § 37.—We have just seen that when a force acts upon a body to give it motion or to destroy that which it has, the body will react or oppose a resistance equal to the force. This resistance measures the inertia of the matter of the body. It is obvious that for the same body, this resistance increases with the degree of velocity imparted or destroyed; we shall presently find that it is proportional to this velocity, and that it also increases in the action of inertia on a thread; direct ratio of the quantity of matter in the body. If a body, free to move, be drawn by a thread, the thread will stretch and even break if the action be too violent, and this will the more probably happen in proportion as the body is more massive. conduct of a spring when under the action of inertia; If a body be suspended by means of a vertical cord, and a weighing spring be interposed in the line of traction, the graduated scale of the spring will indicate the weight of the body when the latter is at rest; but if we suddenly elevate the upper end of the thread, the spring will immediately bend more in consequence of the resistance opposed by the inertia of resistance to all changes of motion; the body. The motion once acquired by the body and become uniform, the spring will resume and preserve the degree of flexure or tension which it had when the body was at rest. If, now, the body being in motion, the velocity of the upper end of the thread be diminished, the

Fig. 15.

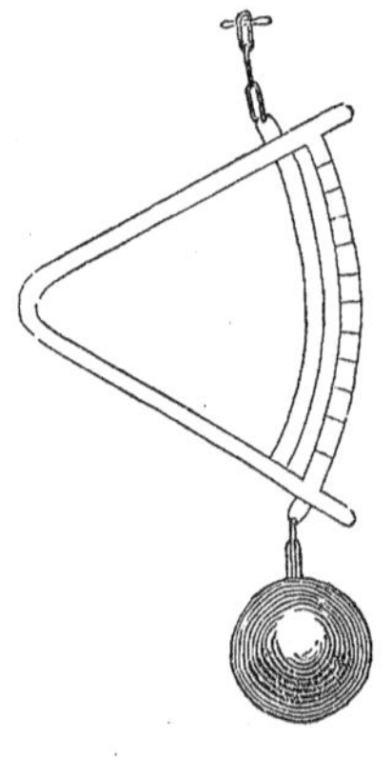

spring will unbend and the scale will indicate a pressure less than the weight of the body. The oscillations of the spring may therefore serve to measure the variations in the motions of a body, and the energy of its force of inertia, which acts against or with a power exerted in the direction of the motion, according as the velocity is increased or diminished.

Oscillations of a spring may indicate the changes in motion.

§ 38.—The *effect* of every force depends, 1st, upon its *point of application;* that is, the point to which it is directly applied: 2d, upon the *position of the line* along which it acts or the straight line which its point of application would describe if perfectly free: 3d, upon the *direction* in which it tends to solicit its point of application along this line, whether backward or forward: 4th, upon its *absolute intensity*, measurable in pounds or any other unit of weight.

Effect of a force; point of application, line of direction, and intensity.

§ 39.—Let A be the point of application of a force which acts upon the line AB; from A, lay off upon the direction in which the force acts, a distance AP, containing as many linear units, say inches, as there are pounds in the intensity of the force; the force will be fully represented. Commonly the direction of the action is indicated by an arrow, and the intensity of the force by some letter as P, for the sake of brevity. Thus, we say a force P or AP, a force Q or AQ, as we say a force of 5 pounds, a force of 8 pounds. In this way the investigations in mechanics are reduced to those of geometrical figures.

Graphical representation of a force;

by length of line or by symbol.

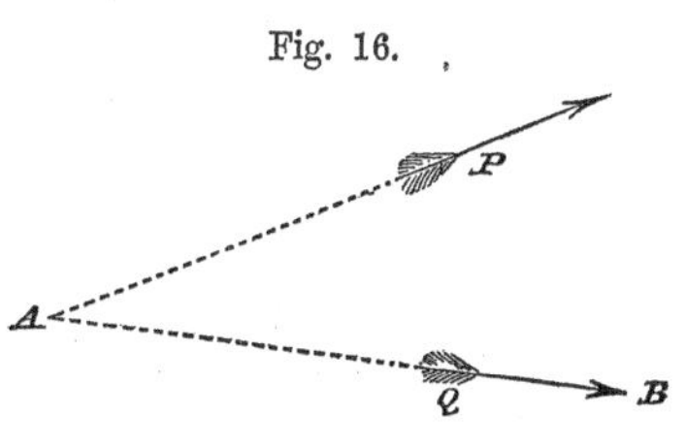

Fig. 16.

Equilibrium of forces; statical and dynamical. Illustration—two men.

§ 40.—When the forces applied to any body balance, or mutually destroy each other, so as to leave the body in the same state as before their application, these forces are said to be in *equilibrio*. The equilibrium may be *statical* or *dynamical*. In the first case, the forces finding the body at rest, will leave it so; in the second case, the forces being applied to the body in motion, will in no respect alter the motion. Two men pulling with equal strength at the opposite ends of a cord, will be a case of statical equilibrium if the men be at rest, and a case of dynamical equilibrium if they be in motion.

No case of absolute rest. Earth's motion. Repose not necessary to equilibrium.

In reality there is no case of absolute statical equilibrium, since the earth's motion involves that of every body connected with it, in the same way that a boat moving over the surface of the water carries every thing on board along with it. The idea of repose is not necessary to that of an equilibrium, which only requires the mutual destruction of all the forces which act at the same instant upon a body.

Forces in equilibrio; not in equilibrio when the motion changes. Effect of inertia on equilibrium of forces. Illustration—horse and carriage.

§ 41.—When a body, subjected to the action of several extraneous forces, preserves its motion perfectly uniform, notwithstanding these forces, these latter will, from the definition above, be in equilibrio. If the velocity however augment or diminish, the extraneous forces will not be in equilibrio; but if we take into account the force of inertia of the different particles of the body, and introduce among the extraneous forces one equal to it and capable of preventing the modification of the motion, there will again be an equilibrium among all the extraneous forces. A horse which draws a carriage along a road, destroys at each instant all resistances which are opposed to his action; if the motion is perfectly uniform, these resistances arise only from the ground, the different frictions, &c. If the velocity increases at each instant in consequence of an increased effort of the horse, the inertia of the carriage will come into action and add to the other resistances above named, and the effort of the horse during this increase of velocity, will be in equilibrio with all these forces; if, on the contrary, the velocity diminish, the inertia of the carriage, which tends to preserve its motion uniform, will add its action to that of the horse to overcome all the resistances, or to maintain the equilibrium.

Inertia always ready to establish an equilibrium among forces.

Thus inertia stands always ready to maintain an equilibrium among forces of whatever nature; and hence the distinction between the equilibrium of bodies and of forces. Forces are ever in equilibrio, while bodies are not necessarily so. If, for example, a material point be acted upon by a force, it will move in the direction of this force, while the force itself is maintained in equilibrio by the inertia developed during the yielding of the point. *Action and reaction are equal and contrary.*

Reaction equal and contrary to action.

§ 42.—When an equilibrium exists among several forces, as O, P, Q, &c., one of them, as O, may be considered as preventing the effect of all the others. If, then, we conceive a force R, equal and directly opposed to O, at the same point of application C, this force will destroy of itself the force O, and will therefore produce the same effect upon the body as the forces P, Q, &c., taken together. This force R is called the *resultant* of the forces, P, Q, &c., and these latter the *components* of the force R.

Fig. 17.

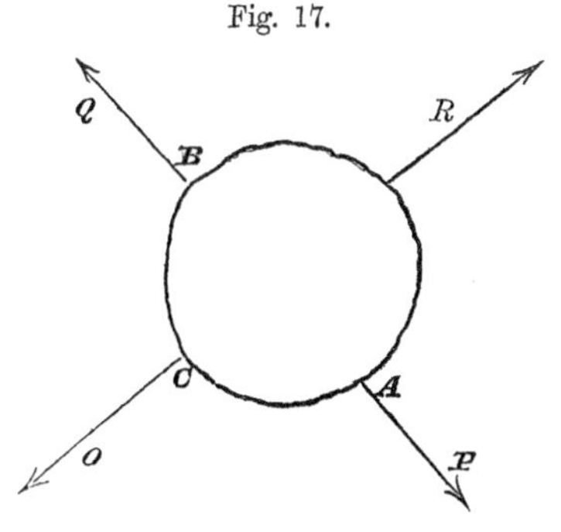

Resultant of forces, components of a force.

Reciprocally, if to the resultant R of several forces P, Q, &c., an equal force O, be immediately opposed, there will be an equilibrium between this force and the several

Resultant and components defined.

forces P, Q, &c.: hence, the *resultant* is a single force which will produce the same effect as two or more forces; the *components* are two or more forces which will produce the same effect as a single force.

Resultant of several forces acting along the same line.

§ 43.—When several forces act along the same straight line and in the same direction, their joint effect will obviously be the same as that of a single force equal to their sum, which single force will be their resultant. If some of the forces act in one direction, and others in an opposite direction, the resultant will be a single force equal to the excess of the sum of those which act in one direction over the sum of those which act in the contrary direction; and it will act in the direction of those forces which give the greater sum, for when two unequal forces are directly opposed, the smaller will destroy in the larger a portion equal to itself. Three men pulling in the same direction a cord, with efforts 10, 17, and 25 pounds, and two others pulling in the opposite direction with efforts 12 and 19 pounds, the effect to move the cord will be the same as though it were solicited by a single force $52 - 31 = 21$ pounds, acting in the direction of the first men.

Mechanical work of forces.

§ 44.—The most simple case of equilibrium, is that in which two equal and opposing forces destroy each other, and it is this to which the employment of force in the mechanic arts is always reduced. To *work*, is to destroy or overcome, in the service of the arts, resistances, such as the force of adhesion of the molecules of bodies, the strength of springs, the weight of bodies, their inertia, &c., &c. To polish a body by friction, to divide it into parts, to elevate weights, to draw a carriage along a road, to bend a spring, to throw stones, balls, &c., &c., is to work, to continually overcome resistances incessantly recurring.

Resistance overcome and reproduced.

Mechanical work not only supposes a resistance overcome, but a resistance reproduced along the path described by the point at which the resistance is exerted, and in the direction of this path. To take away from a body a portion of its matter with a tool, for example, we must not only overcome the resistance opposed by the matter removed, but also cause the point of action of the tool to advance in the direction of the line along which the resistance incessantly recurs. The further the tool advances, the greater will be the length of the removed portion; on the other hand, the broader and thicker this portion, the greater the resistance, and consequently, the greater the effort to overcome it. *The work performed, therefore, at each instant, increases with the intensity of the effort and the length of the path described by its point of application in the direction of the effort.*

Work increases with the effort and path described by the point of application.

§ 45.—Let us suppose a constant resistance and, therefore, a constant effort which is equal and directly opposed to it, that is, they are the same at each instant; it is obvious, from what precedes, that the work produced will be proportioned to the length of the path described by the point of application of the effort—double, if the path is double, triple, if the path is triple, &c.; so that, if we take for unity the work which consists in overcoming a resistance over a length of 1 foot, the total work will be measured by the number of feet passed over. But if for another work, the constant resistance is double, triple, &c. of what it was in the first case, for an equal length of path, the work will be double, triple, &c. of what it was before. If, for example, the resistance were 1 pound in the first case, and 2, 3, 4, &c. pounds in the second, the work for each foot of path would be 2, 3, 4, &c. times that of 1 pound. In assuming, then, the work which consists in overcoming a resistance of 1 pound, through a distance of 1 foot, for the unit of work, we shall have for the measure of the work, of which the object is to overcome a constant resistance, *the number of pounds which measures this resistance repeated as many times as there are feet in the*

Measure of the work when the resistance is constant.

Rule.

path described by the point of application of the resistance.

Illustration. For example, suppose a motive force employed to draw a body on a horizontal plane; the work will be, to overcome the resistance of the constant friction exerted between the body and plane. Let this friction be 37.5 pounds, and the path described 64 feet, the total work will be

$$37.5 \times 64 = 2400 \text{ pounds,}$$

or equal to 2400 pounds over 1 foot, or 1 pound over a distance of 2400 feet.

In general, then, denoting by Q, the quantity of work performed; by P the constant resistance, or its equal, the effort necessary to overcome it; and by S, the space described by the point of action, we shall have

Equation of the quantity of work.

$$Q = P. S. \quad . \quad . \quad . \quad . \quad . \quad (3).$$

Geometrical representation of the quantity of work. To represent this geometrically, assume any linear unit, as the inch, to represent 1 pound, and the same to represent the unit of linear length; lay off from O on the indefinite right line $O B$, the distance $O e_1$, equal to the length of path described by the point of action, and at e_1, the perpendicular $e_1 r_1$, containing as many inches as the constant effort contains pounds; then will the number of square inches in the rectangle $O e_1 r_1 r$, express the quantity of work.

Fig. 18.

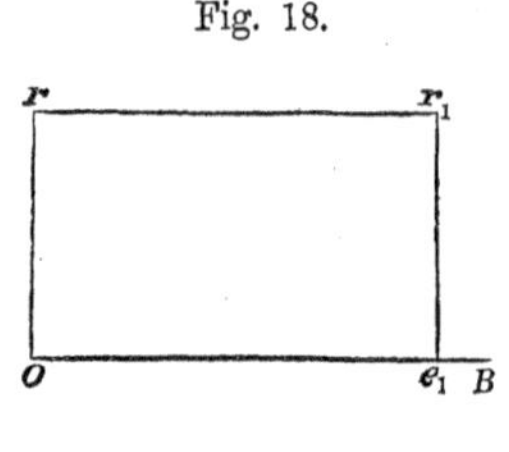

Work when the resistance is variable. § 46.—If the resistance, or the equal effort which destroys it, instead of being the same at each instant, varies incessantly, as is most frequently the case, the quantity of work will not be given by the simple rule above; but, as the effort, however variable, may, during the description of a very small portion of the path, be regarded as constant, the corresponding portion of work will still be measured by this constant effort into this small portion of the path.

Elementary quantity of work.

The total work, being composed of all its elements, will be measured by the sum of all these elementary products.

Draw the curve r, r_1, r_2, r_3, &c., of which the abscisses Oe, Oe_1, Oe_2, Oe_3, &c., shall represent the spaces described by the point of action of the resistance up to certain given successive instants of time, and of which the ordinates er, $e_1 r_1$, $e_2 r_2$, $e_3 r_3$, &c., shall represent the corresponding resistances. Let ee_1, $e_1 e_2$, $e_2 e_3$, &c., be the equal and very small spaces described in successive portions of time. The elementary portions of work during these intervals of time, having for their measures the products of the small spaces by the corresponding resistances, regarded as constant for each one, that is, by the products

Fig. 19.

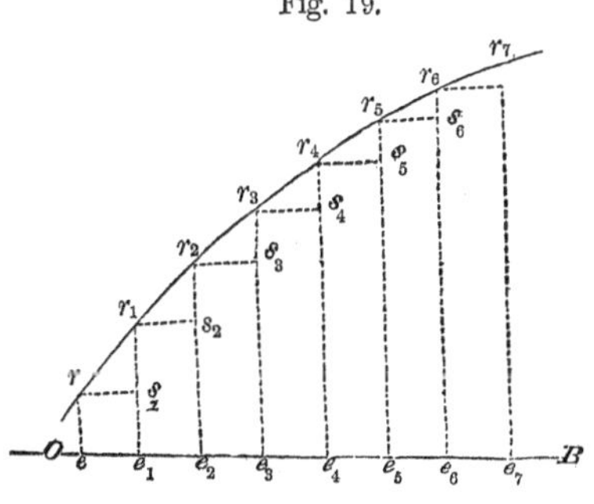

Represented by geometry.

$$ee_1 \times er, \quad e_1 e_2 \times e_1 r_1, \quad e_2 e_3 \times e_2 r_2,$$

these elementary portions of work are represented respectively by the elementary areas

$$e\,r\,s_1\,e_1, \quad e_1\,r_1\,s_2\,e_2, \quad e_2\,r_2\,s_3\,e_3, \quad \text{\&c.},$$

and the total work will be represented by the sum of all these rectangles. But if we multiply suitably the points of division e_1, e_2, e_3, &c., by diminishing the distances $e\,e_1$, $e_1\,e_2$, $e_2\,e_3$, &c., it is obvious that the sum of the rectangles will not sensibly differ from the area included by the curve $r\,r_1\;r_2 \ldots r_7$, the whole path $e\,e_7$ described by the point of action, and the two ordinates $e\,r$ and $e_7\,r_7$ drawn through its extremities.

Represented by an area.

Hence we see, that when we know from experience, the law which connects the variable resistance with the length of path described by its point of action, to compute the amount of work performed, is but to construct by points, or otherwise, the curve of this law, and to calculate the area included by the curve, the total length of path described and the extreme ordinates. When the unit of length employed to construct the ordinates is the same as that by which the length of path is measured, it is plain that the unit of area will represent the work performed by a *unit of effort*, as a pound, *through a unit of length*, say a foot.

Rule for finding the area.

To find this area, divide the path described into any *even* number of parts, and erect the ordinates at the points of division, and at the extremities; number the ordinates in the order of the natural numbers; *add together the extreme ordinates, increase this sum by four times that of the even ordinates and twice that of the uneven ordinates, and multiply by one third of the distance between any two consecutive ordinates.*

Fig. 20.

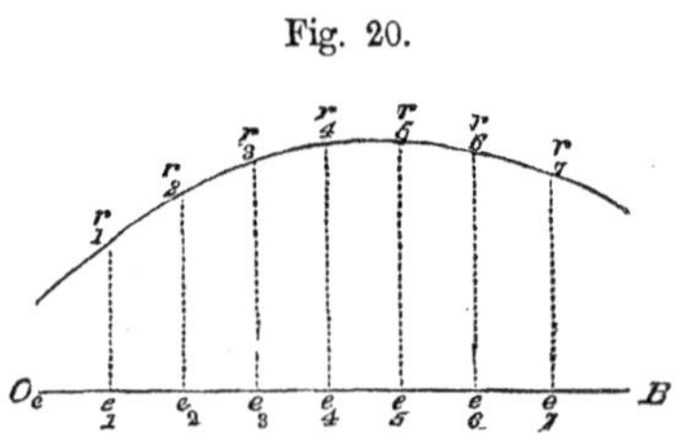

Demonstration: To compute the area comprised by a curve, any two of its ordinates and the axis of abscisses, by plane geometry, it is usual to divide it into elementary areas, by drawing ordinates, as in the last figure, and to regard each of the elementary figures, $e_1\,e_2\,r_2\,r_1$, $e_2\,e_3\,r_3\,r_2$, &c., as trapezoids; and it is obvious that the error of this supposition will be less, in proportion as the number of trapezoids between given limits is greater. Take the first two trapezoids of the preceding figure, and divide the distance $e_1\,e_3$ into three equal parts, and at the points of division, erect the ordinates $m\,n$, $m_1\,n_1$; the area computed from the three trapezoids $e_1\,m\,n\,r_1$, $m\,m_1\,n_1\,n$, $m_1\,e_3\,r_3\,n_1$, will be more accurate than if computed from the two $e_1\,e_2\,r_2\,r_1$, $e_2\,e_3\,r_3\,r_2$. Demonstration of the rule.

Fig. 21.

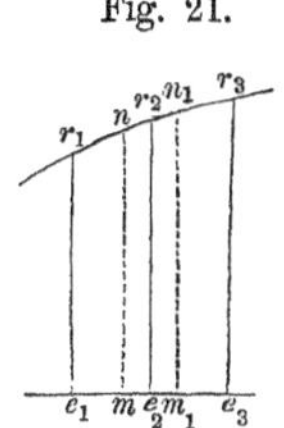

The area by the three trapezoids is

$$e_1\,m \times \frac{e_1\,r_1 + m\,n}{2} + m\,m_1\frac{m\,n + m_1 n_1}{2} + m_1\,e_3\frac{m_1\,n_1 + e_3\,r_3}{2}.$$

But by construction,

$$e_1\,m = m\,m_1 = m_1\,e_3 = \tfrac{1}{3}\,e_1\,e_3 = \tfrac{2}{3}\,e_1\,e_2,$$

and the above may be written,

$$\tfrac{1}{3}\,e_1\,e_2\,(e_1\,r_1 + 2\,m\,n + 2\,m_1\,n_1 + e_3\,r_3),$$

but in the trapezoid $m\,m_1\,n_1\,n$,

$$2\,m\,n + 2\,m_1\,n_1 = 4\,e_2\,r_2, \text{ very nearly;}$$

whence the area becomes

$$\tfrac{1}{3}\,e_1\,e_2\,(e_1\,r_1 + 4\,e_2\,r_2 + e_3\,r_3);$$

the area of the next two trapezoids in order, of the preceding figure, will be

$$\tfrac{1}{3}\,e_1\,e_2\,(e_3\,r_3 + 4\,e_4\,r_4 + e_5\,r_5);$$

and similar expressions for each succeeding pair of trapezoids. Taking the sum of these, and we have the whole area bounded by the curve, its extreme ordinates, and the axis of abscisses; or

Algebraic expression of the rule.

$$Q = \frac{1}{3} e_1 e_2 [e_1 r_1 + 4 e_2 r_2 + 2 e_3 r_3 + 4 e_4 r_4 + 2 e_5 r_5 + 4 e_6 r_6 + e_7 r_7];$$

whence the rule.

Mean resistance; equal to the entire work divided by the entire path.

§ 47.—When the value of the mechanical work of a variable resistance for any distance passed over by the point of action, is found by the method just explained, if this value be divided by the distance, the quotient will be a *mean resistance*, or the constant effort which, exerted through the entire path, will produce the same quantity of work; for we have seen that for a constant resistance, the quantity of work is measured by the product arising from multiplying this resistance into the path described by its point of action.

Examples of mechanical work; that of a force bending a spring,

§ 48.—When a motive force is employed to bend a spring, it will develop, at each instant, an effort which is greater in proportion as its point of action describes, in the direction of the effort, a greater path; an effort which we have seen may be measured for each position of the spring or point of action. The curve which gives the law of these efforts may be constructed by the method just given, and the area determined by the rule in § 46 will give the total mechanical work performed by the force.

of the draft of a horse, of the effort of a man.

We have already taken as an example the work produced by a constant force in drawing a body over a horizontal plane, and above we have taken the work which arises from the action of a variable force in bending a spring; the reasoning applied to these is applicable to all kinds of work employed in the arts. Does a horse pull upon the shaft of a mortar mill; a man draw water from a well;

an artificer saw, plane, file, polish; a turner fashion his materials in the lathe; the quantity of work performed is measured by the product of the effort, which is always equal and contrary to the resistance opposed by the matter to the tool, into the path described by the point of action, if the resistance is constant, or by the sum of the partial products which measure the elementary portions of work, if the resistance is variable. Of the manipulations of an artificer, obtained by the same rule.

§ 49.—In seeking to appreciate different kinds of work, we must be careful not to confound that which is really expended by the motive force, with that which is actually effective in accomplishing an object. It is to this last that are to be applied the foregoing considerations and measurements. We shall presently examine the mode of action of motive forces, the circumstances which modify the result of this action, and the waste which may attend it. Distinction to be observed in work.

§ 50.—To show the complication incident to certain kinds of mechanical work, take, for example, the work of a filer: it is necessary 1st, to press upon the file to make it take hold; 2d, to support continually its weight; 3d, to push it along the surface of the body; 4th, to move it with a certain velocity back and forth, and therefore to overcome the inertia of the file as well as that of the matter removed. The quantity of work is the result of these different circumstances; but this complication may be made to disappear by separating from the result of the work, every thing not indispensable to it, in considering only what takes place where the metal is removed by the file: there, we only perceive a resistance which is opposed to an equal and contrary effort in the direction of the path described by the points of action of the file, and of which the quantity of work is measured in the manner already described. The work of the operator may be reduced to this, by supposing the file placed upon a level surface, loaded with a given weight, and the Complication inherent in certain kinds of work. The work reduced and measured as before.

operator or motive power only employed in drawing it uniformly in the direction of its length.

What must be understood by mechanical work of a force;

§ 51.—In general, then, we must henceforth understand by mechanical work, that which results from the simple action of a force upon a resistance which is immediately opposed to it, and which is continually destroyed in causing the point of action to describe a path on the line of direction of this resistance. The force must be considered as a simple agent, producing an effort or pressure measurable in pounds, and acting in a single direction, as described in § 38; and we must be careful not to confound, as is frequently done, the terms *work* and *force*, with those by which we vaguely designate all the effects, more or less complicated, arising from the action of animate or inanimate agents upon resistances: thus we should not speak of the force of a horse, of a man, of a machine, without indicating the point of action of this force, its intensity, and its direction; we should not speak of the *mechanical work* of a force, without specifying the same things of the resistance which it overcomes at each instant, in each particular case of its application.

work of the resistance.

§ 52.—The most simple work, that which conveys at once an idea of its measure, is the elevation of a weight through a vertical height, if we omit the consideration of inertia. The work in this case obviously increases as the weight and vertical height increase, and is measured by the product of the two, agreeably to what is said in § 45 and § 46; here the unit of work, is the unit of weight raised through a unit of height.

Invariable standard by which to estimate the quantity of work;

The utility of this measure is its great simplicity, and the ease it affords of estimating the pressure or effort in pounds, and the path described by the point of action in feet. We might, to be sure, take any other standard unit, as, for instance, the quantity of work necessary to grind 1, 2, or 3 pounds of corn, which is the old standard of

utility of this standard.

millers and the proprietors of mills. But a given weight of corn will present different degrees of resistance, according to its quality and the kind of tool or machine employed to grind it; so that not only is it impossible for people generally to understand what the millers mean by their standard, but for the millers to understand each other. It is hence indispensable to have some standard which does not admit of variation, and of being interpreted differently by different people; of such a nature is the standard which results from the consideration of the effort, and the path described by its point of action in the direction of the effort.

Standard of millers; objections to it.

It will remain to be found how many pounds of corn this unit of work is capable of grinding, how many square yards of boards it will saw, &c.: all this must come from careful observation and experiment. It is, above all, essential that there shall be nothing arbitrary in the mode of estimating the quantity of mechanical work.

Means of comparing different standards.

§ 53.—Different authors have given different names to mechanical work, which should be carefully distinguished from the object accomplished, this latter being but its effect.

SMEATON calls it *mechanical power;* CARNOT, *moment of activity;* MONGE and HACHETTE, *dynamic effect;* COULOMB, NAVIER, and others, *quantity of action;* and this last expression is now generally adopted. It will hereafter be employed, and will always signify the *quantity of work—mechanical work.*

Different names given to mechanical work;

Sometimes the mechanical work has been called *quantity of motion*, and sometimes *living force*, both of which are but simple *effects* of mechanical work upon a body free to move. We shall explain, in the proper place, the meaning to be attached to these terms.

sometimes called quantity of motion and living force.

All work is judged of by the quantity of each particular species of result, or useful effect, which it produces; but we have seen that this quantiy of result is propor-

Work judged of by the useful effect.

tional to the quantity of mechanical work necessary to its production, and hence mechanical work or *quantity of action* is what pays in forces.

To express the continued work in numbers;

§ 54.—When a motive force acts with a constant effort, and its point of action moves uniformly during any considerable portion of time, it will be sufficient to express the work done in a unit of time, as a day, an hour, a minute, or second. This will avoid the use of multiplicity of figures in comparing the effects of different forces with each other, while it will enable us easily to obtain the value of the whole work, by simply multiplying the work in the unit of time, by the number of units of time during which the force has acted or been working. The duration of the work must, therefore, be noted. Thus, we say the mechanical work of a particular horse is 120 pounds raised through a vertical height of 3 feet in one second, or 120 pounds raised through 180 feet in one minute, this work being continued during 8 entire hours each day.

work in unit of time; note the duration of the effort.

The path described in a second is usually taken;

Ordinarily, we take for the length of path, that which is described in one second, this latter being taken as the unit of time. But this distance, according to the definition of uniform motion, is the measure of the velocity of the point of action, which we have supposed constant; by this coincidence, the mechanical work happens to be measured by the product of a constant effort into the velocity of its point of action: which has misled many persons in causing them, as we shall see further on, to confound the *quantity of work or of action* with the *quantity of motion*, although their measures are in fact very different.

the consequences of this.

All units arbitrary;

In the same way that the unit of time is arbitrary, so also are the units of effort or weight and distance, and consequently the unit of work, which is always equal to the unit of effort or weight, exerted through the unit of distance. We shall take for the unit of effort 1 pound, and for the unit of distance 1 foot, so that the unit of

unit of effort, one pound; unit of distance, one foot;

work will be, as before, *the effort one pound exerted through a distance of one foot.* unit of work, the product of these;

Suppose, for example, that the effort 75 pounds is exerted through the distance 4 feet, then will

$$4 \times 75 = 300 \text{ units of work,}$$

of which each one is equivalent to an effort of one pound exerted through a distance of one foot. This is ordinarily expressed thus, and has no reference to time.

$$300^{\text{lbs. } f.};$$

and is read, 300 pounds raised through 1 foot. And this has no reference to the time in which the work is performed.

§ 55.—Mechanicians long felt the necessity of some well defined unit by which to express the work performed, or capable of being performed, by a motive force, in a given time, and several were proposed; but these ill according among themselves, there seemed as little likelihood of a general agreement in this respect as in regard to the unit of velocity, which depends upon the units assumed for time and space. Different units of work proposed, when time is considered.

After the introduction of the steam-engine, the *horse-power* was proposed, and is now generally adopted as the measuring unit. By *horse-power* is meant, the quantity of work, measurable in pounds and feet, which a horse is capable of performing in a given time; but this would obviously be indefinite, since horses differ in strength and endurance, were it not that some fixed value has been agreed upon, according to the principle explained in § 51, as the standard of horse-power. This value is the mean of the results of a great many trials with different horses, and is set down at 550 pounds raised through a vertical height of 1 foot in 1 second, or 33,000 pounds raised through 1 foot in 1 minute, or 1,980,000 pounds raised Horse-power adopted; 550 lbs. through 1 foot in 1 second;

through 1 foot in 1 hour; all of which amount to the same thing.

Example.

When, then, we are told that a machine or engine is of 30-horse power, or has a power equal to 30, for instance, we are to understand that it will do work which is equivalent to raising $550 \times 30 = 16,500$ pounds through one foot in 1 second, or $33,000 \times 30 = 990,000$ pounds through one foot in 1 minute, &c.

Error of considering the greatest effort alone;

§ 56.—We can now appreciate the error we should commit, if, in estimating the productive power of a motive force or machine, we confine ourselves to the greatest absolute effort it is capable of exerting, without regard to the space described by its point of action; if, for example, in estimating the productive effort of a man, we only consider the greatest burden he is capable of supporting at rest under the action of its weight; or, of a horse, we consider alone the greatest effort, as indicated by a spring balance, he can exert while pulling against a fixed obstacle.

this effort may be replaced by a fixed obstacle;

We can conclude nothing from these in respect to the quantity of action; we must also have the path described in a continuous manner. Simply to support a weight or exert an effort, is not to work usefully; and this is rendered clear from the consideration that we may in all such cases replace the *moter* by an inert body, as a prop, a post, &c.; the action and reaction being equal and contrary, unaccompanied by any motion, there is no balance of work either in favor of the effort or resistance.

error of considering the path alone.

It would be equally impossible to infer any work or quantity of action from the path described by the point of action, without taking into account the effort exerted at each instant. It is obvious that a man or horse, running at full speed, without exerting any effort except that which he is capable of impressing upon himself, is producing no useful effect; he overcomes no resistance external to himself, which it can be an object to destroy.

In a word, the productive effect of every motive force is measured, at each instant, by the product of the effort into the path described in the direction of the effort; so that, if either the effort or path be zero, the quantity of action will also be zero.

Productive abilities measured by the product of the effort into the path.

§ 57.—It must be remarked, however, that, since all bodies are more or less extensible and compressible, a motive force cannot act against what are usually called fixed obstacles, without producing a certain quantity of action or mechanical work, such as we have defined it: for the point to which the force is applied will yield to a greater or less extent, and the body will be flattened or elongated; the molecular springs will oppose a resistance; there will be a small path described in the direction of the force. At first the efforts of the equal and contrary resistances are nothing; afterward they augment by degrees till the effort of the power attains its maximum, and the body its greatest change of shape; after this the action is reduced to maintaining the body or obstacle at its state of tension and repose, without producing henceforth any mechanical action.

Always some work, even on fixed obstacles;

§ 58.—Construct, in the manner before described, the curve $O r_1 r_2 \ldots r_6$, of which the abscisses $O e_1$, $e_1 e_2$, &c., represent the spaces described by the point of action in each successive instant of time in the direction of the force, and the ordinates, the corresponding pressures or resistances opposed by the body in a contrary direction. The quantity of work destroyed while the point of action is describing any one of the small paths, as $e_2 e_3$, is the

Fig. 22.

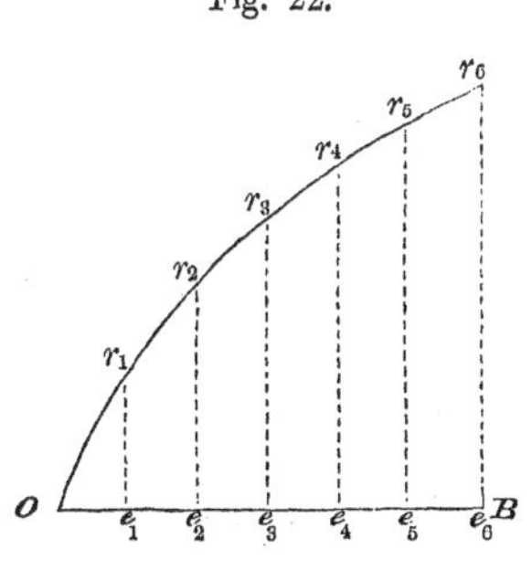

its value represented geometrically.

area of the trapezoid $e_2 e_3 r_3 r_2$, and the total quantity of action destroyed by the molecular action of the body during its entire change of figure, is the area comprised by the curve, its greatest ordinate $e_6 r_6$, which denotes the maximum resistance, and the axis of abscisses. If, then, it should happen that the body or obstacle is either compressed or extended by any appreciable quantity as $O e_6$, which is the path described by the point of action, and the greatest resistance $e_6 r_6$ should be considerable, this quantity of work must be taken into account in certain circumstances which will be explained.

This work may in most cases be neglected;

§ 59.—But in general the bodies employed to receive and transmit the action of forces, are selected with special reference to their capacity to resist all change of figure; so that when well chosen and judiciously disposed in combinations, the work referred to in the preceding article, becomes so small a fraction of that developed by the force when it produces motion, or when the space described by the point of action is considerable in comparison with that which measures the linear change of figure, that it may, and indeed is in practice, neglected. It is under this point of view only that the work developed by a force, applied to a fixed obstacle can be said to be nothing.

especially when action and the motion are at right angles to each other.

This work may also be neglected when the force which develops it, acts in a direction perpendicular to the path which the body is, by its connection with others, compelled to describe. The force in this case will only compress or stretch the body uselessly, without adding to or subtracting from the work in the direction of the motion. A man who pushes against the side of a carriage in a direction perpendicular to the path along which it is moving, neither aids nor hinders the horses: and although he actually develops a quantity of work by the compression of the carriage, it must be totally neglected in making an estimate of the useful effect.

§ 60.—These considerations are important, as they prove, in general, that forces may work without producing any useful effect. If the different pieces, for example, which compose a machine, and which serve to transmit motion and work, in acting upon each other, become compressed or stretched, it is obvious that, even though the point of action moves in the direction of the force, this latter must first expend a certain quantity of work in changing the figure of the pieces before the motion can become regular or uniform throughout. And it may happen that this first work of the power will be totally lost, if the pieces, on ceasing to be compressed or stretched, retain their altered shape: that is to say, if they be not *elastic*, or, more generally, if the molecular springs do not contribute to augment the work when the effort of the force is relaxed, as they did to diminish it when the action began.

Motive forces may work without useful effect; the pieces transmitting the work not being perfectly elastic.

§ 61.—We also see that if the action of the force or moter, or the resistance occasioned by the work, undergo frequent alterations, in becoming sometimes feeble and sometimes stronger; in a word, if the pieces are often compressed and distended, the loss of work thence arising may bear a considerable ratio to the total work of the power, which could not take place if the action of the latter were constantly the same from the beginning to the end of the work.

Loss greater in proportion as the force changes.

§ 62.—The shock of bodies develops considerable pressure, and produces sensible changes of figure; the quantity of action destroyed or generated will, therefore, always be appreciable. On this account it becomes indispensable, in the application of mechanics, to pay the strictest attention to the influence of concussions which may occur during the performance of mechanical work.

Still greater in the case of shocks.

§ 63.—And hence we perceive the advantage arising

Advantages of

stiff and elastic materials.

from the use of very stiff and very elastic materials in the construction of those pieces which are employed to receive and transmit the action of forces, and to regulate the motions they produce.

Elastic bodies restore, in expanding, the work absorbed in being compressed.

§ 64.—To obtain a clear idea how the molecular springs of a body may develop or restore a certain quantity of mechanical work, we have but to consider what takes place at the instant when a body begins to resume, progressively, its primitive figure after it has been changed, and to recall what was said of the measure of the quantity of work of a force, employed to bend a spring, to compress or distend a body. Indeed, we have only to estimate, in pounds, the different pressures corresponding to each state of the body, from that of greatest compression or distention to that of restitution, or to some intermediate state which the body will retain of itself. If the body resume, at last, precisely the form which it had before the change; if, also, the pressures which correspond to the same degree of tension—to the same shape and size of the body, are the same, if, in a word, the body be perfectly elastic, the quantity of work produced during the process of restitution against a resistance opposed to it, will be equal to that required to compress or distend it, since the curve, which gives the law of the pressures and spaces, will be the same in the two cases.

Loss of work when the bodies are not perfectly elastic.

If, on the contrary, the body be not perfectly elastic, it will not return to its former figure; the pressures will be less during the process of restitution, there will be a loss of space described by the point of action, and, consequently, less work performed than in the first change of figure, there will be a certain quantity of action lost.

Examples of elastic bodies;

There are scarcely any perfectly elastic bodies except the gases and vapors, and these must be confined in a close vessel or reservoir and acted upon by a piston. Such contrivances, together with springs made of the most elastic solids, serve to store up mechanical work for future use; forces are employed to compress or bend them, in which state they are retained by mechanical contrivances till the work thus expended is required for other purposes; the restraint is then removed and the work transferred to some other body, which, in its turn, communicates it to something else, and so on to the ultimate object to be attained. The balistas, catapultas, and bows of the ancients, throwing arrows, stones, and other missiles are examples of this; the air-gun, in which the motive power is but a reservoir of compressed air, is well known; and every body is familiar with the steam-engine, in which, by the application of heat, water is expanded into vapor whose molecular spring or elasticity is capable of performing any amount of work, by the simple alternations of heating and cooling. No one is ignorant of the terrible effects of steam and gunpowder, when overheated, and yet, when properly managed, these agents admit of being pent up in inert bodies or vessels, and made to do the work not only of the lower animals, such as horses, oxen, &c., but almost of intelligent beings. It is by means of this principle of elasticity, that clocks and watches, are kept in motion for days and entire months.

Their use.

Examples—balistas, bows, air-gun;

steam and gunpowder.

§ 65.—Weight also affords the means of storing up mechanical work, and of rendering it available when wanted. When a motive force has elevated a body through a certain height, in expending upon it a quantity of work, measured by the product of its weight into the height, this body, employed afterward to overcome a resistance either directly or by means of a machine, may restore, in its descent, precisely the same quantity of work which had been before expended upon it. It is in this way that motion is communicated to clocks, spits, &c., &c.

Weight as a means of storing mechanical work.

By the action of heat, water assumes at the surface of the ocean the form of vapor, ascends to elevated regions

Elevation of water by heat.

in the atmosphere, whence it is precipitated in the form of rain, is collected into natural reservoirs, and becomes, by its weight, a source of motion to mills, machinery, &c.

Work employed to break, &c., not reproduced.

This reproduction does not obtain, however, when the work is employed to divide, to break, to polish, to rub, to destroy, in a word, the natural affinity of bodies. The quantity of work thus expended is, in a mechanical point of view, totally annihilated; it cannot be restored by the body after it has undergone this change of state.

Portability of springs, animals, and combustibles.

Springs, like animals, and combustibles which give heat, have this peculiarity, viz.: they are very portable, and may be even used as a motive power for vehicles. Thus carriages have been put in motion by springs attached, as boats are put in motion by animals on board, and by the vapor of heated water. But springs are never perfect, and being subjected to the action of foreign resistances, never restore the whole of the mechanical work which they have received. Finally, animals, and heat even, the primitive source of all the mechanical work employed in the arts, require a certain expense in nourishment and fuel which, according to the beautiful theory of Leibig, are the same in principle. This nourishment and fuel become, therefore, the representatives of a certain amount of mechanical work, so that it is really impossible to create a motive force, without having previously incurred an equivalent expenditure.

Nourishment and fuel representatives of mechanical work.

Inertia a source of reproduction of mechanical work.

§ 66.—Thus far we have only examined the work of forces when employed to overcome the weight of bodies, the resistance inherent to their state of aggregation or force of affinity, their elasticity, &c. It remains to speak of the resistance which all bodies oppose to a change of their state in respect to motion or rest, by reason of their inertia, of which no estimate has been made in what has gone before, and from which it is impossible to separate the other species of resistance in all questions affecting quantity of work. It has already been remarked that the artificer must overcome the inertia of the matter of which his tool is made; the draft-horse, that of the carriage, and of the load it bears, &c. But independently of this, it is very important to be able to estimate the quantity of work which a body will absorb in acquiring a certain degree of velocity, for this is often the only useful object in view, as in the case of throwing projectiles by the elastic force of gases or solids, which gives rise to the art of *balistics*, employed in war. Besides, it very often happens that instead of applying a force directly to the object in view, we cause it to act upon a free body, and subsequently, by the aid of its inertia, concentrate the quantity of action absorbed by it to do the work at a blow, as in the example of the *pile-ram, common hammer*, &c.; the inertia of bodies is thus made, like weight, elasticity, &c., to restore the quantity of work which has been expended in subduing it; and we now proceed to the consideration of the action of forces employed to overcome inertia and to produce motion.

Examples—pile-ram and common hammer.

III.

VARIED MOTION.

Varied motion; constant force.

§ 67.—We will begin with the most simple case of varied motion, viz: that in which a body is pressed by a constant force, that is to say, one which does not change the intensity of its action, and which is equal and contrary to the resistance opposed by the inertia in the line of direction of the motion.

It is clear that, the pressure being the same at each instant, the small increments or decrements of velocity will, for the same body, also be the same; and thus the velocity will increase or decrease with the time; in other words, the velocity will be proportional to the time

uniformly varied, accelerated, and retarded.

elapsed since the commencement of motion. This is called *uniformly varied motion* in general; which becomes *uniformly accelerated* or *uniformly retarded*, according as the force increases or diminishes the velocity of the body.

Uniformly accelerated;

§ 68.—First, take the case of uniformly accelerated motion, and recall to mind that the *velocity acquired* at any instant is, § 25, measured by the space described by the body in the unit of time succeeding this instant, if, the force having ceased its action, the body continue to move uniformly in virtue of its inertia; this velocity we have seen how to calculate by means of the law which connects the time with the spaces.

graphical representation of this motion.

Let O be the point of starting. Draw the line $O v_1 v_2 \ldots v_6$, of which the abscisses $O t_1$, $O t_2$, &c., represent the time elapsed from the origin or beginning of the motion, and of which the ordinates $t_1 v_1$, $t_2 v_2, \ldots t_6 v_6$, represent the velocities acquired at the end of the times $O t_1$, $O t_2, \ldots O t_6$.

Fig. 23.

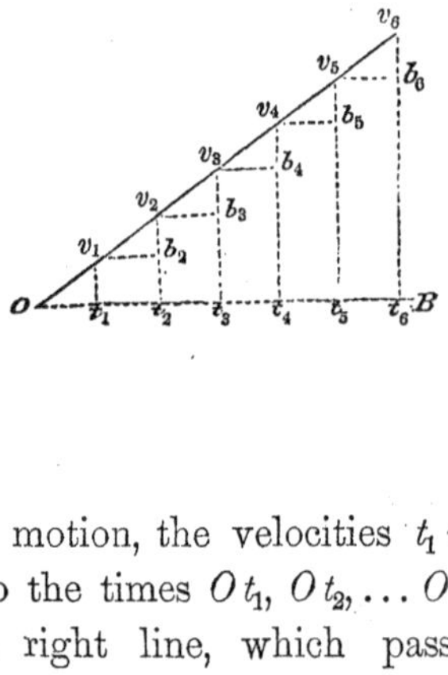

Since in uniformly varied motion, the velocities $t_1 v_1$, $t_2 v_2, \ldots t_6 v_6$ are proportional to the times $O t_1$, $O t_2, \ldots O t_6$, the line $O v_1 v_2 v_3 \ldots v_6$, is a right line, which passes through the point O from which the body takes its departure; for at this point, the velocity and time are zero together, at the instant of starting. The distances $O t_1$, $t_1 t_2$, $t_2 t_3$, &c., being equal, if through the points v_1, v_2, $v_3, \ldots v_6$, lines be drawn parallel to the axis $O B$ of times, there will be formed a series of right-angled triangles, $O t_1 v_1$, $v_1 b_2 v_2, \ldots v_5 b_6 v_6$, all equal to each other. The sides $t_1 v_1$, $v_2 b_2$, $v_3 b_3, \ldots v_6 b_6$, will represent the successive increments of velocity, which are equal and constant, by the definition of uniformly varied motion, since the corresponding intervals of time $O\,t_1$, $v_1\,b_2$, $v_2\,b_3, \ldots . v_5\,b_6$, are equal.

The successive intervals of time $O\,t_1$, $t_1\,t_2$, $t_2\,t_3$, &c., being supposed very small, we may regard the body as moving uniformly during any one of them as $t_3\,t_4$ or its equal $v_3\,b_4$, and with the velocity $t_3\,v_3$ acquired at its commencement. But by virtue of uniform motion, the path described by the body contains as many linear units as the rectangle of the time into the velocity contains superficial units, and, in this sense, the distance passed over by the body in the time $t_3\,t_4$, will have for its measure the product of this elementary portion of time by the velocity $t_3\,v_3$, or the area of the rectangle $t_3\,t_4\,b_4\,v_3$: for another interval $t_4\,t_5$, the path described will have for the measure of its length, the area $t_4\,t_5\,b_5\,v_4$, and so on; so that the total length of path described by the body during the time $O\,t_6$, will be the sum of all the partial rectangles $t_1\,t_2\,b_2\,v_1$, $t_2\,t_3\,b_3\,v_2, \ldots . t_5\,t_6\,b_6\,v_5$; which sum will not differ sensibly from the area of the triangle $O\,t_6\,v_6$, when the points of division t_1, $t_2, \ldots t_5$, are greatly multiplied.

Path represented by half the rectangle of velocity into time.

From this fact, viz.: that the length of the path described by a body in uniformly varied motion, is represented by the area of a triangle whose base is the time during which the motion takes place, and altitude the velocity acquired at the end of this time, we easily deduce several important consequences, called the laws of uniformly varied motion.

Since the area of the triangle $O\,t_6\,v_6$, has for its measure, the half of its base into its altitude, and as the base into the altitude, or the entire rectangle, represents the length of path described in the time $O\,t_6$, with a constant velocity $t_6\,v_6$, acquired at the end of this time, it follows,

Laws of uniformly varied motion.

1st. *In uniformly accelerated motion, the path described at the end of any time, is half that which the body would*

First law.

describe in the same time, if it were to move uniformly with the velocity acquired during this time.

Since the paths described during any two times, as $O\,t_3$, $O\,t_5$, are represented by the triangles $O\,t_3\,v_3$, $O\,t_5\,v_5$, respectively, and since these triangles are similar and their areas are to each other as the squares of their homologous sides, it also follows,

Second law.

2d. *In uniformly accelerated motion, the paths described at the end of any two times, are to each other as the squares of these times.*

Third law.

3d. *That these paths are to each other, as the squares of the velocities acquired at the end of the corresponding times.*

Formulas to compute the circumstances of this motion.

When in uniformly accelerated motion, the velocity $t_5\,v_5$, acquired at the end of a given time $O\,t_5$, say one second, taken as the unit of time, is given, the law of the motion or the right line $O\,v_6$, which represents it, is completely determined, and we may compute the velocity and space which correspond to any other time.

Denote by e_1 and v_1, the length of path and velocity which correspond to the first second, and by S and V, the path and velocity corresponding to any other time, as T; we have by the first law,

Space in unit of time;

$$e_1 = \tfrac{1}{2} v_1 \times 1^s = \tfrac{1}{2} v_1 \quad . \quad . \quad . \quad (4),$$

relation of space, time, and velocity;

$$S = \tfrac{1}{2}\,V\,T \quad . \quad . \quad . \quad . \quad . \quad . \quad . \quad . \quad (5);$$

and by the second law,

$$e_1 \;:\; S \;::\; 1^s \times 1^s \;:\; T \times T \;::\; 1^s \;:\; T^2;$$

whence,

space in any time.

$$S = e_1 \times T^2 \quad . \quad . \quad . \quad . \quad . \quad . \quad (6);$$

and replacing e_1 by its value, Eq. (4),

$$S = \tfrac{1}{2} v_1 T^2. \quad . \quad . \quad . \quad . \quad . \quad (7).$$

From the third law, Space in any time;

$$e_1, \quad \text{or} \quad \tfrac{1}{2} v_1 : S :: v_1^2 : V^2;$$

whence

$$V^2 = 2 v_1 S \quad . \quad . \quad . \quad . \quad . \quad (8).$$

Velocity due to any space.

By the definition of uniformly varied motion, we have,

$$v_1 : V :: 1^s : T;$$

whence

$$V = v_1 T. \quad . \quad . \quad . \quad . \quad . \quad (9).$$

Velocity due to any time.

In what precedes, we have supposed the body to start from rest, so that the right line, which gives the law of the motion, passes through the point of departure O. But if the body have already a velocity $O v_0$, acquired previously, this right line will pass through v_0, the extremity of the ordinate which represents the velocity of the body at the instant from which the time is reckoned. The velocity $O v_0$, is called the *initial velocity*.

The body has already an acquired velocity; initial velocity.

Fig. 24.

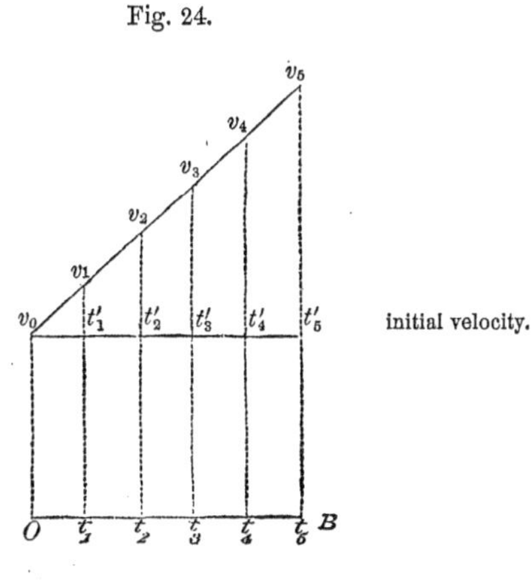

By drawing $v_0 t'_5$, parallel to OB, we see that the velocity $t_3 v_3$, which corresponds to the time $O t_3$, is composed of two parts, viz. $t_3 t'_3$, and $t'_3 v_3$; the first is equal to the initial velocity $O v_0$, and the second to the

Formulas to compute the circumstances of the motion;

velocity which the body would acquire in the time $v_0 t'_3$, equal to $O t_3$, under the action of the constant force, had it moved from the point v_0 with no initial velocity, as in the preceding case; for the line $v_0 v_5$ gives, in reference to the line $v_0 t'_5$, the law of acceleration. Knowing, then, the velocity which the force is capable of impressing upon the body in a unit of time when moved from a state of rest, it is easy to construct the line $v_0 v_5$, in relation to $v_0 t'_5$ or its parallel $O t_5$, and to deduce from it all the circumstances of the motion.

Let it be required, for example, to find the length of path described by the body in the time $O t_4$. This path will contain as many linear units as the trapezoidal area $O t_4 v_4 v_0$ contains superficial units. We perceive at once, that this length will be composed of two parts, viz.: that described uniformly in virtue of the initial velocity $O v_0$, and represented by the rectangle $O t_4 t'_4 v_0$, and that described in virtue of the constant force and represented by the triangle $v_0 t'_4 v_4$. But, denoting by a the initial velocity, and by T the time, we have for the measure of the rectangle

$$a T;$$

and for the measure of the triangle, Eq. (7),

$$\tfrac{1}{2} v_1 T^2;$$

and if we denote by S the total length of path actually described by the body, we have

value of the space;

$$S = a T + \tfrac{1}{2} v_1 T^2 \quad . \quad . \quad . \quad . \quad (10):$$

and because the actual velocity at the end of any time, is the initial velocity increased by that due to the action of the constant force during this time, we have, Eq. (9),

value of the velocity.

$$V = a + v_1 T \quad . \quad . \quad . \quad . \quad (11).$$

§ 69.—If we now suppose the constant force, instead of increasing the initial velocity of the body, to diminish it, the motion becomes *uniformly retarded*, and the line $v_0 v_4$ gives the law of the motion.

Uniformly retarded motion;

By drawing $v_0 t'_5$ parallel to $O t_5$, we see that the velocity $v_3 t_3$, which corresponds to the time $O t_3$, is nothing else than the initial velocity $O v_0$ diminished by the velocity $t'_3 v_3$, which the body would acquire under the action of the constant force at the end of the time $O t_3$ had it moved from rest. The length of path described is now represented by the trapezoidal area $O t_3 v_3 v_0$; and is equal to that which would be *uniformly described in the same time, with the initial velocity $O v_0$, diminished by that which would be described in the same time, if moved from rest under the action of the constant force, by a motion uniformly accelerated; that is to say, the length of path is represented by the rectangle $O t_3 t'_3 v_0$ diminished by the triangle $v_0 v_3 t'_3$.*

Fig. 25.

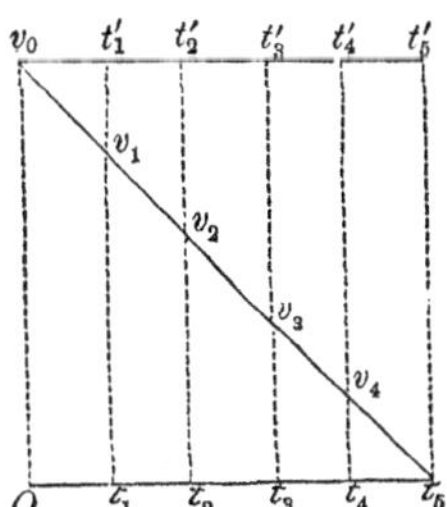

graphical representation.

The equations (10) and (11), which appertain to uniformly accelerated motion, become, therefore, applicable to uniformly retarded motion, by simply changing the sign of the velocity generated by the constant force, and that of the area of the triangle, which represents the path due to the action of this force; hence,

Formulas to compute the circumstances of this motion.

$$S = a T - \tfrac{1}{2} v_1 T^2 \quad . \quad . \quad . \quad . \quad (12),$$

Value of space;

$$V = a - v_1 T \quad . \quad . \quad . \quad . \quad . \quad (13).$$

of velocity.

Let us suppose that, among other things, we desire the time required for the force to destroy all the initial velocity; we have only to make $V = 0$, and equation (13) becomes

$$a - v_1 T = 0,$$

whence

Time required to destroy all a body's velocity.

$$T = \frac{a}{v_1} \quad . \quad . \quad . \quad . \quad . \quad . \quad (14);$$

from which we conclude that the time required for a constant force to destroy all the velocity a body may have, is equal to the quotient arising from dividing the value of this velocity, by the velocity which the force can generate in the body in one unit of time.

To find the length of path described by the body during the extinction of its velocity, substitute the value of the time above found in equation (12), and we have

The path described during the destruction of its velocity;

$$S = \frac{a^2}{2 v_1} \quad . \quad . \quad . \quad . \quad . \quad (15);$$

that is to say, the space through which a body will move during the entire destruction of its velocity by the action of a constant force, is equal to the square of the velocity destroyed, divided by twice the velocity which this force can generate in the body during a unit of time.

It is important to remark, that if the force continue to act after having destroyed all the velocity, the body will return along the path already described, and pass in succession and in reverse order, as to time, through its previous positions, at each of which it will have the same velocity it had there before; for while the body is losing its velocity, it may be regarded as beginning its motion at any point of its path with its remaining velocity or that yet to be destroyed, which, in such case, is denoted by a, and when all its velocity is destroyed, it returns from a state of rest or begins to move backward with no initial velocity; so that equations (4) to (9) become applicable to

after the velocity is destroyed, the body will return;

this latter motion, while equations (14) and (15) are to the former. But from equation (8) we have

$$V = \sqrt{2\,v_1\,S},$$

and substituting for S its value given by equation (15) we get

$$V = \sqrt{\frac{2\,v_1\,a^2}{2\,v_1}} = a;$$

and have at its previous positions the same velocity as before.

that is to say, the velocity V, which the body has acquired in moving backward through a space S, is equal to the velocity a, with which it began to describe the same space in its forward motion.

Motion of falling bodies;

§ 70.—One of the most important examples of uniformly accelerated motion, is that presented by the vertical fall of heavy bodies; but, before discussing it, we will make known some of the circumstances which accompany and modify this motion at the surface of the earth.

causes which modify this motion;

We have already seen, § 32, that the force of gravity may be considered as constant within ordinary limits. But at the surface of our globe, all bodies are plunged into the atmosphere, and this atmosphere is itself a material body, which, by its inertia and impenetrability, opposes with greater or less energy all kinds of motion of bodies; this opposition is named *atmospheric resistance*. Experiment shows us that this resistance increases as the velocity of the body and the extent of its surface increase; thus, in striking the air with a light flat board, the resistance which we experience is greater in proportion as the motion is more rapid, while it is scarcely sensible when the motion is very slow; and again, the resistance will be less if, instead of striking the air with the broad surface, we present to it the edge of the board.

influence of velocity and extent of surface;

It is plain, therefore, that the presence of the air must modify the laws of the vertical fall of bodies subjected to the action of their weight. In permitting bodies to fall through the air, from the same height, it is observed that those which weigh most under the same volume, or those which present the least surface in the direction of the motion, arrive soonest at the bottom; thus, a ball of lead will fall sooner than a ball of equal volume of common wood, and a ball of common wood sooner than one of cork, &c. But if made to fall in vacuo, or in a long hollow cylinder from which the air has been removed, experiment shows that all bodies fall equally fast, and therefore will reach the bottom at the same instant if they start together. This is called the *guinea and feather experiment*, from the fact that a guinea and feather will fall under the action of their respective weights in vacuo, with the same velocity and, therefore, will reach the bottom in the same time.

influence of air on the fall of bodies;

bodies which weigh most and have least surface, fall most rapidly;

in vacuo all bodies fall equally fast;

Fig. 26.

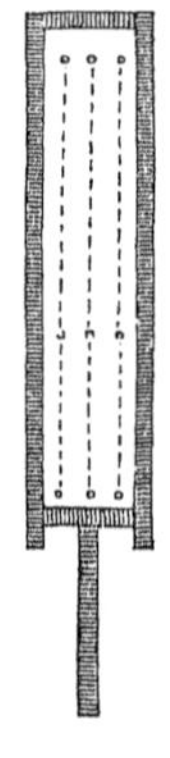

From this it follows, that *the force of gravity acts indiscriminately upon every particle of matter, and impresses upon each, at every instant, the same degree of velocity in vacuo,* a fact which it is important to remember.

gravity acts on the interior and exterior particles of a body alike;

We may easily assure ourselves that the force of gravity acts on the interior as well as on the exterior particles of all bodies, by observing that the same body weighs just as much by the weighing spring whether placed in the open air, or in a close chamber; which proves that the force of gravity acts through this chamber envelope without undergoing any change.

distinction between the weight of a body and the force of gravity.

The weight of a body, is the resultant of all the actions of the force of gravity upon its elementary particles; we must be careful, therefore, not to confound the *weight* with the *force of gravity* itself, which is, in fact, only the elementary force impressed upon each particle.

§ 71.—Finally, it is important to remember that the denser bodies, such as gold, lead, iron, &c., are those which, under equal volumes, or equal surfaces, will fall most rapidly in the air, because the resistance of the latter is weaker when considered *in reference* to the weight; and this resistance may become relatively so small that we may neglect it, particularly when the fall of the body is not very rapid. **Gold, lead, &c. fall most rapidly in the air;**

Galileo, an Italian philosopher, was the first to investigate, experimentally, the laws which govern the motion of bodies falling under the action of their own weight, in vacuo; and he found the motion to be *uniformly accelerated.* The force of gravity is, therefore, within the limits of experiment, a *constant accelerating* force, acting with an equal intensity at each instant whatever be the velocity acquired. Atwood, an English philosopher, in resuming the experiments of Galileo, with greatly improved means, obtained the same results. **the motion of falling bodies uniformly accelerated.**

§ 72.—Hence, when a body falls from rest through a certain height, in vacuo, **Laws of the motion of falling bodies;**

1st. The velocities acquired are proportional to the times elapsed since the beginning of the motion. **first law;**

2d. The total spaces passed over, or the heights of the fall, are proportional to the squares of the times elapsed. **second law;**

3d. These heights are proportional to the squares of the velocities acquired at the end of each. **third law;**

4th. The velocity acquired at the end of the first unit of time, is measured by double the height of fall passed over during this time. **fourth law.**

Although the force of gravity, may, without sensible error, be regarded as constant at the same locality, it yet varies, as we have seen, from place to place, in going southward or northward, and cannot, therefore, generate as much velocity in one latitude as another. From careful experiments, made with a pendulum at different places, it is found that the length of path described by a **Force of gravity varies with the latitude;**

body in the first second of its fall from rest in vacuo, will be given by the following formula, viz:

space a body will describe under its action in first second;

$$e_1 = \overset{feet.}{16.0904} - 0.04105 \cos. 2 \psi \quad . \quad . \quad (16),$$

in which e_1 is the space, and ψ the latitude of the place.

In works on mechanics, the *velocity* which the force of gravity can generate in a second of time at the surface of the earth, is usually denoted by g; and as this velocity is equal to twice e_1, Eq. (4), as given by the above equation, we have,

velocity it can generate in one second;

$$g = \overset{feet.}{32.1808} - 0.0821 \cos. 2 \psi \quad . \quad . \quad (17);$$

hence all the circumstances of the motion of falling bodies at any place, will be given by equations (4) to (15) after substituting therein g for v_1.

Let H represent the height, in feet, through which the body has fallen in a given time denoted by T, and V the velocity acquired at the bottom of this height; then, from equations (5), (7), (8), and (9), we have

formulas which relate to the fall of bodies in vacuo;

$$H = \tfrac{1}{2} V T \quad . \quad . \quad . \quad . \quad . \quad (18),$$

$$H = \tfrac{1}{2} g T^2 \quad . \quad . \quad . \quad . \quad . \quad (19),$$

$$V^2 = 2 g H \quad . \quad . \quad . \quad . \quad . \quad (20),$$

$$V = g T \quad . \quad . \quad . \quad . \quad . \quad . \quad (21),$$

in which, for all ordinary cases we may take

$$g = 32.1808 \text{ feet} \quad . \quad . \quad . \quad (22).$$

application to examples;

Suppose we are required to find the velocity acquired and the path described at the end of 7 seconds; from equation (21), we have

$$V = 32.1808 \times 7 = 225.2656 \text{ feet},$$

from equation (19),

$$H = \frac{32.1808}{2} \times (7)^2 = 788.4296 \text{ feet};$$

that is to say, at the end of 7 seconds, the body will have a velocity which would carry it over a distance of 225.2656 feet during the 8th second, were its velocity at the end of the seventh second to become constant, and the space described during the seven seconds of fall, will be 788.4296 feet. results;

It must be remembered that, in the atmosphere, the body will not fall with the same velocity, on account of the resistance of this medium; but from what has already been remarked, this resistance will not have much influence if the falling body be very dense, as iron, lead, &c.; or if the surface of the body be small; or if the height of fall be not great, say sixty or seventy feet. We might, therefore, measure approximately, the height of towers, depth of wells, &c., &c., by noting the time, as indicated by a watch beating tenths or fifths of seconds, required by a body to fall through the height.

influence of the atmosphere; in the case of metals, if the surface of the body and height be small; application to find the height of towers and depth of wells.

If we have given the height through which a body has fallen, it is easy to find the velocity acquired; for from equation (20), we have

$$V = \sqrt{2gH}.$$

Suppose a body to fall through a height of 80 feet, then will

$$V = \sqrt{2 \times 32.1808 \times 80} = 71.75 \text{ feet}.$$

This proposition is of frequent occurrence in practical mechanics.

The quantity V is called, *the velocity due to a given height* H; and H, *the height due to a given velocity* V.

Velocity due to a given height; height due to a given velocity.

A body thrown vertically upward;

§ 73.—When a body, as the ball from a gun, for example, is thrown vertically upward, its weight acts at each instant with the same intensity to diminish by equal degrees its primitive velocity; the motion will be *uniformly* retarded; the velocity will be totally destroyed when the body attains a certain height, from which it will descend, in taking successively the different degrees of velocity which it had at the same places in its ascent, all of which is obvious from what was said in § 69. Thus, at the distance of 1, 5, 7, &c. feet from the place of starting, the body will have exactly the same velocity in ascending and descending; it will only have the direction of its motion changed. When it returns to its point of departure, its velocity will be the same as it was at starting.

Denote by H, the greatest height the body will attain; and V, the primitive or initial velocity; then will, equations (20) and (21),

greatest height to which it will ascend;

$$H = \frac{V^2}{2g} \quad . \quad . \quad . \quad . \quad . \quad (23),$$

time required to reach its greatest height.

$$T = \frac{V}{g} \quad . \quad . \quad . \quad . \quad . \quad (24).$$

That is to say, the greatest height to which a body will ascend, when thrown vertically upward, is equal to the square of its initial velocity, divided by twice the force of gravity; and the time of ascent will be equal to the initial velocity, divided by the force of gravity

Example;

Let the body, for example, leave the earth with a velocity of 150 feet a second, then will

$$H = \frac{(150)^2}{2 \times 32{,}1808} = 350.28 \text{ feet},$$

$$T = \frac{150}{32{,}1808} = 4.658 \text{ seconds}.$$

This is on the supposition that the air opposes no resistance. The body will not ascend so high in the air; and, moreover, will fall with less velocity than in vacuo. effect of atmospheric resistance.

§ 74.—We may now appreciate the quantity of work or of action which the weight of a body will expend, in impressing upon itself a certain velocity, or in *overcoming its inertia*. Denote by W, the weight of the body, expressed in pounds, or, in other words, the *absolute effort* which gravity exerts upon the body, and which is equal and contrary to that necessary to support it in a given position; this will measure the constant effort exerted upon the body during its descent through the height H. The quantity of work consumed during this fall will, § 45, be denoted by

Quantity of work of the weight, required to impress upon a body a given velocity;

quantity of work consumed during its fall;

$$W \times H,$$

and this quantity of work will have generated in the body the velocity V, computed by the equation

$$V^2 = 2gH;$$

from which we have

$$H = \frac{V^2}{2g};$$

and multiplying both members by W,

$$WH = \tfrac{1}{2}\frac{W}{g} \times V^2 \quad . \quad . \quad (25).$$

§ 75.—Thus, the quantity of work developed by the weight of a body to impress a certain degree of velocity upon itself, is equal to half the product obtained by multiplying the square of this velocity, by the weight of the body, divided by the velocity g, which the force of gravity is capable of impressing upon all bodies during

work required to impress a given velocity;

the first second of their fall. This product,

$$\frac{W}{g} \cdot V^2,$$

living force; equal to double the quantity of action necessary to produce it;

is what mechanicians call the *living force* of the body whose weight is W. We see, therefore, that *the quantity of action expended by the weight of a body, is half the living force impressed*; or that *the living force impressed, is double the quantity of action expended by the weight.*

half the living force lost or gained, equal to the work that overcomes the inertia.

It is to be remarked, that when a body is thrown vertically upward with a certain velocity, the quantity of action of the weight, which is always measured by the product of the weight into the height to which this body has risen, is employed, on the contrary, to destroy this velocity, so that in the two cases of ascent and descent, the half of the living force lost or gained, measures the quantity of action or of work necessary to overcome the inertia of the body, whether the object of this action be to impress upon the body a certain velocity, or to destroy that which it already has.

This principle is, as we shall soon see, general, whatever be the motive force employed to communicate motion to a body, and whatever be the direction of the motion. But it is necessary first to remark upon certain terms employed in mechanics.

meaning of living force;

§ 76.—As the expression of *living force*, employed to designate the product,

$$\frac{W}{g} \cdot V^2,$$

may lead to error, it is proper to remark here, that it must not be regarded as the name of any force, any more than the name given to the product

$$W \cdot H,$$

not a force, but the result of a force's action;

or the *quantity of action*, designates a force; it is simply the result of the activity of a motive force, expressible in pounds, which has been employed to overcome the inertia of a body, to impress upon it a certain motion—a certain velocity. Under this point of view, the *living force* is but a *dynamic effect* of a force, or rather double this effect, since

a dynamic effect.

$$\frac{W}{g} \cdot V^2 = 2\,W \cdot H.$$

A body in motion, or a certain dynamic effect, may indeed become, in its turn, a source of work; as, for example, a body thrown vertically upward is elevated in virtue of its velocity to a certain height, as though it were taken there by the incessant action of an animated moter. But this is, in all respects, analogous to what takes place when a force has developed a certain quantity of work to bend or compress a spring; the inertia of the matter has been brought into play in the same manner that the molecular springs have in this latter case. This inertia, § 66, when it has been thus conquered, becomes capable of restoring the quantity of work expended upon it, as well as a compressed spring; in a word, inertia, like a spring, serves to store up a quantity of action, to transform it into living force, so that living force is a true disposable quantity of action. The same may be said of a body elevated to a certain height; this body solicited by its weight is the source of a quantity of action, of which we may subsequently dispose to produce a certain amount of mechanical work. But as we cannot say that this body, elevated to a certain height, is a *force*, that a compressed spring is a *force*, neither can we say that a body in motion, or that

A body in motion may be a cause of work;

but cannot be a force any more than an elevated body, or bent spring,

$$\frac{W}{g} \cdot V^2$$

is a *force*. It is the same of *men*, *animals* in general, of *caloric*, of *water-courses*, of *wind*, &c., &c.; these are but agents of work, or moters—*not simple forces*.

or animals, caloric, the wind, &c.

Object of mechanics as applied to the arts.

It is the object of mechanics, in its application to the arts of life, to study the different transformations or metamorphoses which the work of moters undergoes by means of machines and implements, to compare different quantities of work with each other, and to estimate their value in money, or in work of this or that kind.

In short, when we speak of *living force*, communicated to, or acquired by a body, it is only necessary to remember, that *it relates to a real motion of the body, and is equal to the product of the square of its velocity into its weight, divided by the force of gravity.*

The mass of a body;

§ 77.—Since the force of gravity acts indiscriminately upon all the particles of a body, and impresses upon them at each instant, the same degree of velocity at the same place, the weight of a body, which is the result of these partial actions, may give us an idea of the relative *quantity of matter* it contains, or of its *mass*, for it is plain that the mass must be proportional to the weight; often, indeed, the weight is taken for the mass. But as the intensity of the force of gravity varies from one locality to another, and as the quantity of matter in the same body or the mass remains absolutely the same, it is obvious that this latter would be but ill defined by its weight. Experience shows that the velocity impressed by the force of gravity, in one second of time, is directly proportional to the intensity of this force, and that therefore the ratio

force of gravity proportional to the velocity it may impress in one second.

$$\frac{W}{g},$$

must remain the same for all places, since the weight is also directly proportional to the force of gravity. Thus if W and W', be the weights of the same body at different places, and g and g' the intensities of the force of gravity at those places, respectively, then will

$$W : W' :: g : g';$$

whence

$$\frac{W}{g} = \frac{W'}{g'}.$$

This invarible ratio $\frac{W}{g}$, is taken, in mechanics, as the measure of the *mass* of a body. Ordinarily the mass is represented by M, whence

Measure of the mass of a body;

$$M = \frac{W}{g},$$

or

$$W = Mg \quad . \quad . \quad . \quad . \quad . \quad (26),$$

measure of the weight.

in which W expresses the effort or pressure exerted by the weight of the body, and g the velocity which this weight can impress upon the body in a second of time.

§ 78.—By substituting the value of the weight, as given by equation (26), in the expression for the living force, we find

$$\frac{W}{g} V^2 = MV^2;$$

Living force in terms of the mass and velocity;

that is to say, the living force of a body in motion, is equal to *the product of its mass into the square of its velocity.*

Finally, mechanicians have agreed to call the product of the mass of a body, as above defined, into its velocity, or MV, *the quantity of motion of the body;* and this it must be remarked is very different from the quantity of action or of work. To understand what is meant by this new expression, denote the quantity of motion by Q, then will

quantity of motion;

$$Q = \frac{W}{g} V = MV \quad . \quad . \quad . \quad (27);$$

or, which is the same thing,

its meaning;

$$Q : W :: V : g.$$

But W, is the weight of the body, and g, the velocity which this weight can generate in this body, in one second of time; hence Q must designate either a weight or an equivalent effort, which can generate in the body, the velocity V, in one second.

it is a pressure, like weight;

We see also that the living force,

$$M V^2, \text{ or } M V V = Q V,$$

living force equal to the quantity of motion into the velocity.

is the product of this effort, by the velocity V, or by the path described uniformly by the body in a unit of time in virtue of its acquired velocity.

These observations show the distinction between the *quantity of motion* of any body and its *living force*, and the identity between this latter and double the *quantity of action*.

Use of the denominations mass and quantity of motion.

§ 79.—It is principally to abridge and simplify the computations and reasonings, that the denominations mass and quantity of motion, are employed in mechanics; and they might easily be dispensed with. But as authors generally have used them, it becomes important to understand their precise significations.

A force is proportional to the velocity it can generate in a given time, only when constant.

When the force is variable, it is proportional to the small degree of velocity imparted at a given instant.

§ 80.—We have just seen that the force of gravity will impress upon a body, during one second of time, velocities which are constantly proportional to its intensity, or to the absolute weight of the body in each locality. But this property arises only from the fact, that the weight remains constant during the fall, so that the total velocity at the end of the fall, is proportional to the equal degrees of velocity impressed at each instant. When the motive force, instead of being constant, varies at each instant, it is obvious that its intensity can no longer be measured by the velocity which it impresses upon the same body during a unit of time, and that its measure must depend upon the *small degree* of velocity which it communicates at a given *instant*.

By observing what takes place at the surface of the earth, and in our planetary system, it is found that the *motive forces or pressures are, in fact, proportional to the small degrees of velocity which they impress upon the same body in equal indefinitely small portions of time.* This fact serves as the basis of all dynamic investigations, and must be regarded as a general law of nature.

Forces proportional to the small degrees of velocity they can impress in a very small portion of time.

§ 81.—Accordingly, let F be the measure, in pounds, of a force of pressure; let v be the small degree of velocity which it can impress upon a body at any instant or epoch, during an indefinitely small interval of time, denoted by t; also, let W be the pressure exerted by the weight of a body at any given place, and v' the small degree of velocity which this weight can impress upon the body during the same short interval t. We shall have, from the principles already established, since F may be regarded as constant within the limited time t,

Measure of the motive force or of inertia by the velocity impressed in a small time.

$$F \ : \ W \ :: \ v \ : \ v';$$

Consequences of this law;

whence

$$F = \frac{W}{v'} \cdot v.$$

But from the first law of falling bodies

$$v' \ : \ g \ :: \ t \ : \ 1^{\text{sec.}};$$

whence

$$v' = gt;$$

therefore

measure for the intensity of any motive force;

$$F = \frac{W}{g} \times \frac{v}{t} = M . \frac{v}{t} \quad . \quad . \quad (28).$$

That is, the *intensity of any motive force, is measured by the product of the mass into the velocity it can generate while acting with a constant intensity, divided by the duration of the action.* Thus, when we know the small velocity v, impressed in the short interval of time t, by the force F, we may compute the value of this force, which is equal and contrary to the resistance opposed to motion by the inertia of the body. This resistance has been called by some the *force of inertia*, and by others *dynamic force*. The relation given by Eq. (28), shows us *that the force of inertia, which is equal and contrary to F, is directly proportional to the mass, and to the velocity v which this mass receives during the elementary time t.*

inertia exerted, proportional to the product of mass into the velocity imparted;

Let F' be the measure of a second force, which acts upon the mass M', impressing upon it in the same time t, the small velocity v', then will

$$F' = M' . \frac{v'}{t},$$

which, with Eq. (28), gives

relation of any two motive forces.

$$F : F' :: Mv : M'v'.$$

That is to say, *any two motive forces are to each other, as the quantities of motion they can impress in the same elementary portion of time.*

§ 82.—From Eq. (28), we find

Velocity impressed in any short time;

$$v = \frac{F . t}{M};$$

from which we perceive that the degree of velocity which a motive force impresses upon a body, during a short elementary portion of time, increases with the intensity of the force, and inversely as the mass, or weight.

proportional to the intensity of the force divided by the mass.

§ 83.—If now we suppose, at any instant, the force suddenly to cease to vary, and to continue to act upon the body with the intensity which it possessed at that instant, the velocity will increase or diminish, proportionally to the time, § 67, and the intensity of the force may be measured by the definite quantity of motion which it can impress upon the body during the first succeeding second.

Measure of inertia and of the equal and contrary motive force;

Designate by V_1 the velocity generated in the body during the first second succeeding the instant in which the force becomes constant, then will

$$V_1 \; : \; v \; :: \; 1^{\text{sec.}} \; : \; t;$$

whence

$$V_1 = \frac{v}{t},$$

which, in Eq. (28), gives

$$F = V_1 M \quad . \quad . \quad . \quad . \quad . \quad (29);$$

and, in general, *the motive force, equal and contrary to the force of inertia, is measured, at each instant, by the quantity of motion it can impress during one second, if, instead of varying, it retain unaltered the intensity it had at that instant.*

equal to the quantity of motion the latter can impress in a unit of time, when constant.

When the mass becomes the unit of mass, Eq. (29) becomes

$$F = V_1 . \quad . \quad . \quad . \quad . \quad . \quad . \quad (30);$$

the force in this case is called the *accelerating force*, or, more properly, the *acceleration* or *retardation due to the force*,

Accelerating force;

measured by the velocity impressed on a unit of mass in unit of time;

and *is always measured by the velocity it is capable of impressing on a unit of mass in a unit of time, acting with a constant intensity.*

And from Eq. (29), which gives,

$$V_1 = \frac{F}{M},$$

is equal to the motive force divided by the mass.

it appears that the acceleration or retardation due to the force, is, in every case, nothing more than that portion of the entire *motive force* which results from dividing the latter by the number of units in the mass acted on.

Geometrical illustration;

§ 84.—Trace, according to the method described for uniformly varied motion, § 68, the curve $v_0\,v_1\,v_2\,v_3$, &c., which represents the law of the times and velocities; let $t_3\,v_3$ and $t_4\,v_4$ represent the velocities which correspond to the end of the times $O\,t_3$ and $O\,t_4$, or at the beginning and end of the very small portion of time

Fig. 27.

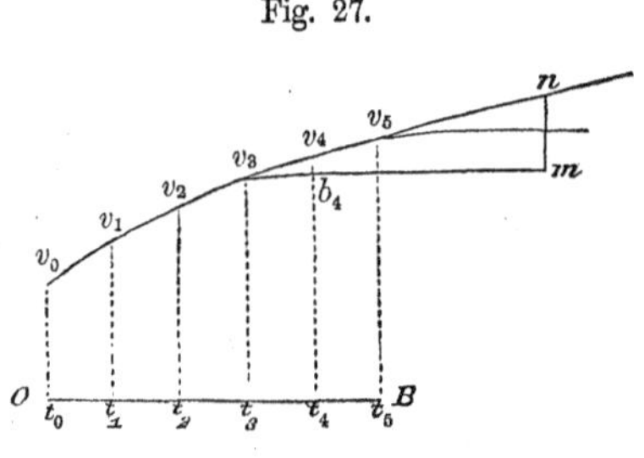

$$t_3\,t_4 = t.$$

Draw through v_3 the line $v_3\,b_4$, parallel to the axis $O\,B$ of times, and produce it till $v_3\,m = 1$ second; this line will meet the ordinate $t_4\,v_4$, and $b_4\,v_4$ will be the small portion of velocity $= v$, impressed by the force, during the small portion of time t. Now if, at the instant corresponding to the end of the time $O\,t_3$, the force become constant, it will subsequently impress upon the body equal increments of velocity during the equal intervals of time t, and the curve $v_3 v_4 v_5$ will become the straight line $v_3 n$, tangent to the curve at the point v_3. Drawing through m a line parallel to $t_4 v_4$, the portion $m n$ will represent the velocity V_1 impressed in one second, and the two similar triangles, $v_3 b_4 v_4$ and $v_3 m n$, will give

$$v_3 b_4 \ : \ b_4 v_4 \ :: \ v_3 m \ : \ m n;$$

or

$$t \ : \ v \ :: \ 1^{\text{sec.}} \ : \ V_1;$$

whence

$$V_1 = \frac{v}{t};$$

the value of the velocity at any instant;

as before found.

Thus, when we know the law which connects the velocity with the time, or the curve which represents this law, we may, at any instant, by drawing a tangent to the curve, determine the velocity V_1, and consequently compute the value of the intensity of the force from the equation,

found by the tangent line;

$$F = M V_1 = \frac{W}{g} \cdot V_1;$$

the measure of the motive force.

or, which is the same thing, the value of the equal and contrary resistance, opposed by the inertia of the body, at each instant during the action of the force.

§ 85.—Reciprocally, if we know the value of the intensity of the force F at each instant, we deduce from it the corresponding value of

Value of the accelerating force, equal to motive force divided by mass.

$$V_1 = \frac{F}{M};$$

Inclination of tangent to the curve.

or of the inclination of the tangent $v_3 n$, or that of the element of the curve of velocities to the axis $O B$ of times. The tangent of this inclination is given by

$$\frac{m n}{v_3 m} = V_1$$

Curve constructed by means of this tangent.

and if the initial velocity $O v_0$ be given, nothing is easier than to construct the curve, of which the ordinates shall be the successive velocities acquired under the action of the force; since, by means of the inclinations of the tangents or elements of the curve corresponding to each absciss of time, those elements may be drawn one after the other, thus forming a polygon, which will differ less and less from the curve, in proportion as the number of values of the force between given limits is greater.

Work necessary to impress a given velocity;

§ 86.—By the aid of what precedes, we may readily compute the quantity of work which must be expended against a body, whose weight is W, by a force F, equal and contrary to the force of inertia, to impress upon it a certain velocity V, or, more generally, to augment or diminish its velocity by a given quantity.

Fig. 28.

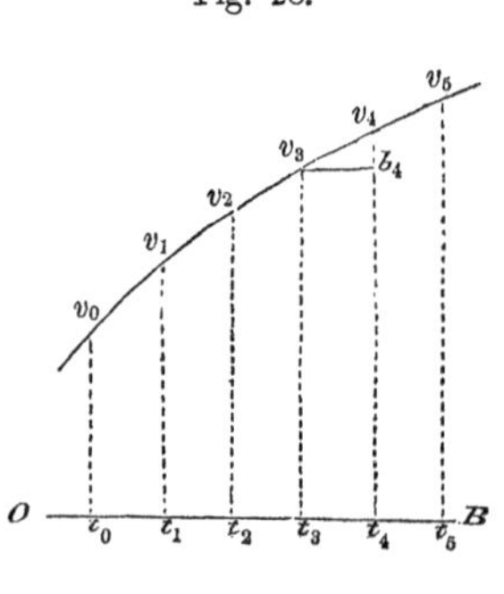

The quantity of work expended during any small interval of time t, has, for its measure, the product of the intensity of the force F, into the elementary portion of the path described by the body during this time. This small path is given by the area of the small rectangle $v_3\, t_3\, t_4\, b_4$, whose base is the element $t_3 t_4 = t$, and whose altitude is $t_3 v_3 = V$, § 67 and § 68;

that is to say, by the product Vt. Hence the elementary quantity of work is

$$F\ V\ t,$$

for each instant of time, or for each small increment $b_4 v_4$ of velocity V. But from Eq. (28) we have

$$F = M\frac{v}{t};$$

replacing F by this value, in the preceding expression, we have, for the elementary quantity of work,

$$M\ V\ v;$$

elementary quantity of work;

and it is the sum of all these partial quantities of work which composes the total quantity of work; this sum may be found thus:

geometrical method of finding the whole work.

From the point O, as an origin, lay off the distances $O\,w_1$, $w_1 w_2$, $w_2 w_3$, &c., to represent the different increments of velocity during the different successive elementary portions of time t, which have elapsed since the beginning of motion — increments that will not be equal in the case of a variable force; then will $O\,w_1$, $O\,w_2$, $O\,w_3$, &c., represent the velocities of the body at the corresponding instants: lay off these same lengths upon the ordinates $w_1 v_1$, $w_2 v_2$, $w_3 v_3$, &c., so that we shall have

Fig. 29.

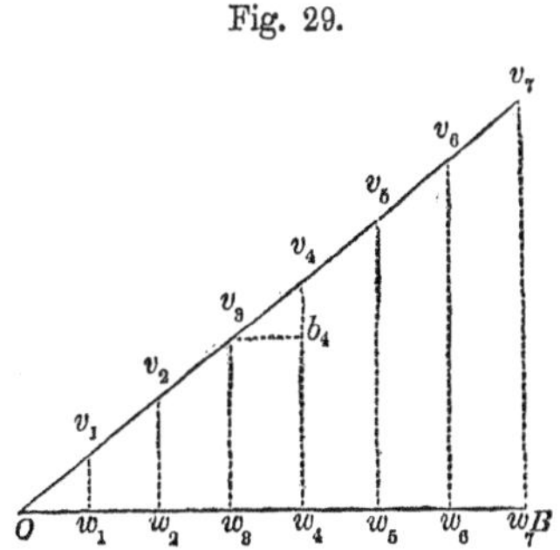

$$w_1 v_1 = O\,w_1, \quad w_2 v_2 = O\,w_2, \quad w_3 v_3 = O\,w_3, \text{ \&c.};$$

the series of points v_1, v_2, v_3, &c., will lie on a right line, inclined to the axis $O\,B$, in an angle of 45°. Consider now the velocity $v_3\,w_3 = V$, for instance, of which the increment $w_3\,w_4$ or $v_3\,b_4 = v_4\,b_4$, is called v. The product $V v$, will here be represented by the small rectangle $v_3\,w_3\,w_4\,b_4$, or by the trapezoid $v_3\,w_3\,w_4\,v_4$, to which it becomes sensibly equal when the increment of velocity or that of the time is very small. The sum sought, of all the partial products $V v$, has for its measure the sum of all the corresponding elementary trapezoids, or the area comprised within the right line $O\,v_7$, the axis $O\,w_7$, and the ordinate $w_7\,v_7$, which latter represent the velocity acquired from the beginning to the end of the time for which we wish to estimate the work done by the force.

The area of a triangle represents the sum of all the products $V v$.

Fig. 29.

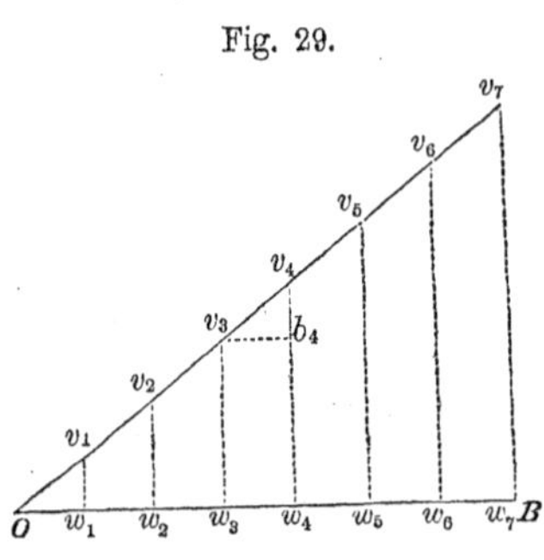

Work consumed when the body's motion is accelerated;

§ 87.—For example, if the body sets out from rest, and we desire to find the sum of the products of $V v$, corresponding to the acquired velocity $w_4\,v_4 = V'$, this sum being represented by the area of the triangle $O\,w_4\,v_4$, we shall have

$$\tfrac{1}{2}\,O\,w_4 \times w_4\,v_4 = \tfrac{1}{2}(w_4\,v_4)^2 = \tfrac{1}{2}\,V'^2;$$

hence the quantity of work corresponding to the velocity V', and consumed by the inertia of the body whose mass is M, will be measured by $\frac{1}{2}\,M\,V'^2$, or by half the living force communicated from the beginning of the motion, § 76. This principle obtains, therefore, for any kind of motion, or for a motive force different from the force of gravity.

equal to half the living force communicated;

For another velocity, $w_7\,v_7 = V''$, the consumption of work will be in like manner measured by $\frac{1}{2}\,M\,V''^2$, and consequently for the interval between the positions in which the body had the velocities $V'$and V'', the quantity of work consumed will be measured by the difference, or

$$\tfrac{1}{2} M V''^2 - \tfrac{1}{2} M V'^2,$$

corresponding to the trapezoid $w_4 w_7 v_7 v_4$. But $M V'^2$ and $M V''^2$ are the *living forces* at the beginning and end of the interval of time during which we are considering the work of the motive force; the expression above is, therefore, one half the *increment* of living force, or half the living force communicated in this interval; so that the principle applies to any two instants of the body's motion, and thus *the quantity of work expended has, in every case, for its measure, half of the living force communicated in the interval between these two instants.*

Work consumed in any interval, equal to half the difference of living force at beginning and end.

§ 88.—Finally, it must be remarked, that the preceding supposes the velocity of the body to increase incessantly; if it were otherwise, the force would be opposed to the motion, and would be a *retarding force.* But the reasoning remaining the same, would be applicable to this case, and we should find that the quantity of work or action developed by the resistance F, (equal and contrary to the force of inertia now become a power,) during the time necessary, to reduce the velocity from V' to V'', would have for its measure,

Work developed when the motion is retarded;

$$\tfrac{1}{2}(M V'^2 - M V''^2),$$

or half the living force *destroyed* or *lost.*

equal to half the difference of living force at the beginning and end of interval.

Thus, the *diminution* of the living force of a body between any two given instants, supposes that a quantity of work or of action equal to the half of this diminution, has been developed by the inertia of this body against obstacles or resistances, as its *augmentation* supposes, on the part of a power, a consumption of work equal to the half of this augmentation.

Inertia serves to transform work into living force, and living force into action;

§ 89.—We now clearly perceive how the inertia of a body, serves to transform work into living force, and living force into work; or, to use the expressions employed, § 76, on the occasion of the vertical motion of heavy bodies, we see that inertia will store up the work of moters by converting it into living force, and give this work out again when the living force comes to be destroyed against resistances.

examples in the mechanic arts;

The mechanic arts offer a multitude of instances in which these successive transformations take place, in operating by means of machinery, implements, &c., &c. The water contained in the reservoirs of grist-mills, for example, represents a certain quantity of disposable action, or work, which is changed into living force when the sluice gates are opened; in its turn, this living force acquired by the water, in virtue of its weight and descent from the reservoir, is changed into a certain quantity of work; this is communicated to the wheels of the mill, and these latter transmit it to the millstones which pulverize the corn. The air confined in the reservoir of an air-gun, represents the value of the mechanical work expended by a certain moter in compressing it; on opening the valve, the air acts upon the ball, impels it forward, and converts a certain quantity of work into living force. If this ball be thrown against a spring, or an elastic body, the latter will be compressed in opposing a greater or less resistance to the inertia of the former, and will finally have destroyed all its motion at the instant the quantity of work, developed by the spring, becomes equal to half the living force of the ball; the spring being retained by any means in its compressed state, the living force will be stored up as a quantity of disposable work, so that when the restraint is removed from the

example of the grist-mill;

the air-gun;

the action of the ball against a spring.

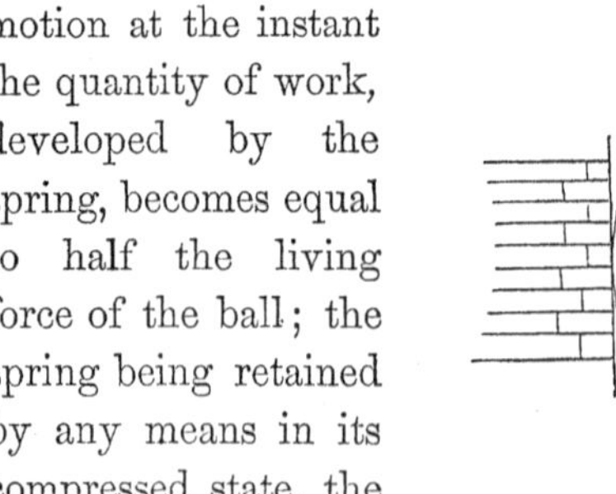

Fig. 30.

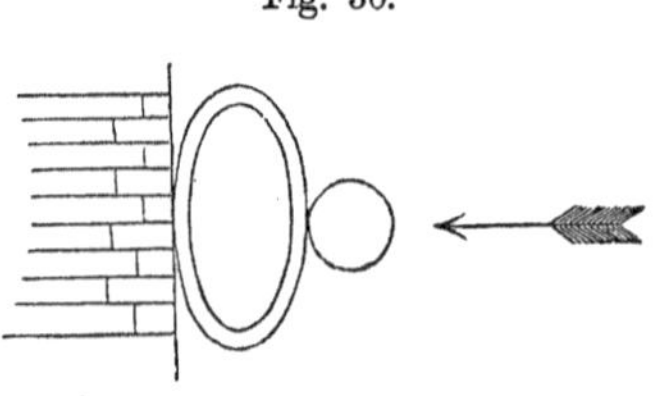

spring, the ball will be thrown back with a velocity such, that the living force will be double the quantity of action or of work, restored by the spring in unbending or expanding.

§ 90.—If, then, the spring be perfectly elastic, the velocity communicated to the ball, will be precisely equal to that impressed upon it by the air-gun in a contrary direction. Thus, in the example before us, the quantity of work has been alternately changed into *living force*, and *living force* into *quantity of work*, without any thing having been lost or gained. But if the spring be not perfectly elastic, a portion of the living force impressed upon the ball will be employed in destroying the molecular force of the spring, that is to say, in producing a permanent change in the arrangement of its particles.

Perfectly elastic bodies restore all the living force lost during an impact.

§ 91.—Hence, in the collision of bodies, not perfectly elastic, there will always be a loss of quantity of work, and this, from what has already been said, must be equal to half the living force destroyed. Few, if any, solid bodies are perfectly elastic, and as the vast majority are, to a considerable degree, deficient in this quality, the quantity of work uselessly consumed by the molecular forces will, in general, bear an appreciable ratio to that developed by inertia during the compression; and it therefore follows, that if this last force, or the velocity which occasions the collision, be considerable, there will take place, in a very short time, a great loss in the quantity of action; and this is why it is important, as before remarked, to avoid all shocks in the motion of machinery.

Living force is always lost in the collision of bodies not perfectly elastic.

§ 92.—We also see, from what precedes, that it is as impossible for the force of a spring to develop, in unbending, a living force greater than that consumed in bending it, as for the force of gravity, § 65, to give to a body while falling, a living force exceeding that destroyed in it, through the same height, while rising; indeed, the whole of the velocity will not, in general,

The work restored can never exceed that consumed in creating the moter.

be restored, and as the corresponding living force lost in the shock, has really been employed to overcome a certain resistance, and therefore to produce a certain quantity of work, it is true, as before stated, that inertia does actually perform an amount of work equivalent to that which has been employed in putting it into action; only it happens, that, in certain cases, a portion of this work is diverted from the object we desire to accomplish, and is not, on that account, regarded as forming a part of the useful effect, as was explained in § 50 with regard to the ordinary force of pressure.

What takes place in periodical motion;

§ 93.—We have shown, by examples, how the quantity of work or of action may be transformed alternately into living force, and living force into quantity of action, by means of springs and machines which store up and give them out successively. These transformations take place, in general, whenever the motion of a body solicited by a motive force varies, by insensible degrees, so as sometimes to be accelerated and sometimes retarded. This occurs, for example, in the periodical motion spoken of in § 25, and, in general, in all cases of forward and backward movement, usually called alternating, and in which the velocity becomes nothing from time to time. The motion of the pendulum and that of the plumb-bob are evident examples of this last kind. When the velocity of a body augments, it is a sign that some portion of the moter's work acts in the direction of the body's motion, and increases its living force by a quantity double this portion of work; the other portion being absorbed by resistances; if, on the contrary, the velocity of the body diminish, notwithstanding the power may be exerted in the direction of the motion, a certain portion of the living force acquired will be expended against the resistances, and will augment the work of the moter by a quantity equal to half the living force thus expended, and so on, according to the number of alternations.

when the velocity is increased, inertia opposes the force;

when the velocity diminishes, inertia aids the force.

§ 94.—From which we see, that when the velocity or living force of a body oscillates between certain limits, it is a proof that inertia has alternately absorbed and given out portions of the moter's work. The work absorbed by inertia will be the same for all equal velocities, and for the interval between the instants of equal velocities there will be nothing lost or gained, and the power must be considered as having been entirely employed to overcome resistances other than inertia. But, if in any interval of time, the velocity, after having undergone alternations, does not attain to what it was before, the half of the difference of the living forces which correspond to the beginning and end of this interval, measures the quantity of work which has really been consumed or given out by the inertia of the body. Consequently, if the body were to set out from rest, the quantity of work consumed by its inertia up to any instant, would be measured by half the living force possessed by the body at this instant; if the velocity had increased incessantly, the inertia of the body would have opposed the motive force without intermission; if the velocity had, during any part of the time, diminished, the inertia would have aided the force.

Within the intervals between instants of equal velocities, the moter is not employed to overcome inertia;

work absorbed or given out by inertia, equal to half the living force acquired or destroyed.

§ 95.—All of which may be made manifest by means of the second figure employed in § 86, in observing that when the velocity of the body diminishes, after having augmented during a certain time, so will the abscisses and ordinates of the right line Ov_7, which represent this velocity; the extreme ordinates $w_7 v_7$, after receding from the point O, while the velocity is increasing, will, on the contrary, approach this point while

Fig. 31.

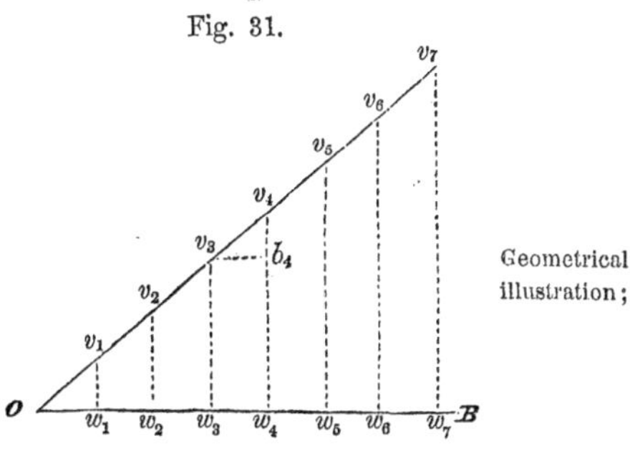

Geometrical illustration;

the velocity is diminishing, to keep the triangular area $O w_7 v_7$, constantly proportional to the quantity of work absorbed by the inertia, or to its equal, one half the living force. A carriage travelling at a variable rate, sometimes faster, sometimes slower, offers an example of this: at first, the horses exert a certain quantity of action to move the carriage with a trot; then, when the velocity is diminished, by an increase of resistance, or by feebler action on the part of the horses, the inertia of the carriage develops against the resistances to its motion, a portion of the work it had at first absorbed, equal to half the diminution of its living force: and this alternation will continue till the carriage is brought to rest, at which instant, the work restored by the inertia will be exactly equal to the quantity of work consumed, so that nothing will be lost. In what is here said, it is understood, however, that no diminution of velocity results from opposition or holding back of the horses, for in that case, the moter would be converted into resistance.

example of a carriage drawn by horses.

The same reflections apply to weight as well as to inertia.

§ 96.—The same reflections are applicable to the weight of a carriage in ascending and descending a hill. The quantity of work employed in overcoming the weight while ascending will be restored during the descent, provided the latter be not so steep as to cause the horses to hold back, by which a quantity of work would be consumed uselessly. And this consideration shows us one of the many advantages which results from giving gentle slopes to roads.

When a force is employed to raise a weight, inertia retains nothing of the moter's work;

§ 97.—When a moter is employed to raise a burden through a vertical height, it takes the body from a state of rest, and hence a quantity of work must be expended to overcome its inertia. Arrived at the desired height, the effort of the moter is relaxed to restore the body to a state of rest, and during this diminished action, a portion of the living force acquired is employed to destroy in part the effect of the body's weight, and the inertia will finally retain nothing of what it had absorbed. The same thing may be said of the operation of an artificer in filing, sawing, &c., since at the end of each oscillation of the tool, the velocity becomes nothing through insensible variations. This could not be the case if the motion were suddenly to change, or if concussions should take place between bodies not perfectly elastic; a portion of the living force would, in that case, be destroyed, or, which is the same thing, diverted from its intended purpose in producing a permanent change in the arrangement of the particles of the colliding bodies.

the same is true of the inertia of an artificer's tool.

§ 98.—Finally, in order to give a fuller idea of the part performed by inertia in the various operations of the mechanic arts, and to demonstrate how it may serve to explain an almost infinite variety of effects, we shall add a few special examples to those already mentioned.

Examples of the part performed by inertia;

To take from a plane-stock its chisel, the carpenter strikes the plane a blow on the back; a velocity is thus suddenly impressed upon the stock which the chisel and its wedge only partake of in part, because of their inertia and imperfect connection with the body of the plane, and are, therefore, left behind.

the chisel of a plane;

A *bung* is taken from a cask by striking, on either side of it, the stave in which it is inserted; the resistance which the inertia of the bung opposes to the sudden motion communicated to the stave, causes the separation.

the bung of a cask;

We often see a handle adjusted to a tool, as an axe or hammer, by striking it on the end in the direction of its length; the inertia of the handle and that of the tool tend to resist the sudden motion impressed by the blow, but the former yielding more than the latter, by reason of the slight connection, the handle becomes inserted.

handles of tools;

As an illustration of the agency of inertia, in transforming quantity of action into living force, take the common *sling*, from which a stone may be thrown with much

the common sling;

greater velocity than from the naked hand. Here, living force is accumulated in the stone, by whirling it through many accelerated turns about the hand before it is discharged. The common *top* turns and runs along the ground, in virtue of the living force acquired during an accelerated unwinding of the string from the coils of which it is thrown.

The common whirling top.

§ 99.—We would recommend to the reader, to consider attentively these examples, as well as all others of like nature which his observation and memory may furnish. They will aid his conceptions of the manner in which the inertia of bodies, like their weight and molecular spring, sometimes acts as a mere passive resistance, and sometimes as a real motive force, according to the circumstances.

Inertia sometimes a passive resistance; sometimes a real motive force.

It is, however, proper to remark, that the last example is mainly concerned with the inertia of a body having a motion of rotation, while, thus far, we have only spoken of the living force of a body possessing a motion of translation, in which all the particles have the same velocity; but we shall soon see, that the principles which connect the living force with the quantity of action, are universal and applicable to all kinds of motion.

IV.

OF FORCES, WHOSE DIRECTIONS MEET IN A POINT.

Forces whose directions meet in a point;

§ 100.—Thus far we have only considered the effect of a single force, directly opposed to an equal force, viz.: to *molecular spring* or elasticity, to *weight*, or to *inertia*. It often happens that several forces are applied to a body, in different directions, to overcome certain resistances through its intervention. When a body is thus subjected to the action of several forces, (*powers*, or *resistances*,) we say these forces are in equilibrio, when one of them destroys or prevents the effect which the others would produce, if the first did not exist. The body itself is in equilibrio, if the different forces applied to it, leave it at rest. This last kind of equilibrium can never be absolute, because all bodies connected with the earth partake of its continual motion through space, and there is, in fact, no rest for them. A body may, however, have relative rest, as when it retains the same place in reference to surrounding objects, such as mountains, houses, &c., which we are in the habit of regarding as fixed. Thus, the idea of equilibrium is not alone related to rest, and by no means excludes motion. From this results the distinction of *statical* and *dynamical* equilibrium; the former relating to the repose of the body, and the latter to the mutual destruction of the forces which solicit it. Thus, a body may be in motion while the forces acting upon it are in equilibrio, or it may be at rest under the same circumstances.

forces in equilibrio when one prevents the effects of all the others; no absolute equilibrium of bodies; statical and dynamical equilibrium.

§ 101.—It has already been stated, § 43, that when several forces act along the same right line and in the same direction, their effect will be equivalent to that of a single force equal to their sum, and which will therefore be their resultant. If these forces act in opposite directions, and along the same straight line, their resultant will be equal to the excess of the sum of those which act in one direction, over the sum of those which act in the opposite direction, and it will act in the direction of the greater of these sums. This is the case in which several forces are exerted in the direction of the same cord. The tension of the cord will be the same throughout, and it is not possible to draw its two ends with different efforts. The tension of a cord is *the effort by which any two of its consecutive portions are urged to separate from each other*,

Resultant of several forces; when acting along the same line, in same or in different directions; tension of a cord;

the effect of unequal forces acting upon a cord.

and this being the same throughout, the excess of the sum of the forces which act in one direction over that of those which act in the opposite direction, will be wholly employed in overcoming the cord's inertia and giving it motion.

§ 102.—When a body, or material point, moves from A to B, so as to describe the rectilineal path $A B$, each of the positions A and B may be projected upon the right lines $O M$ and $O N$, situated in the same plane with the line $A B$, by drawing parallels to these lines considered as axes, the place A giving the two co-ordinates $A A'$ and $A A''$, and the position B the two co-ordinates $B B'$ and $B B''$. The positions A' and A'', on the axes, are simultaneous with the position A; and those of B', B'', with the position B. The paths $A' B'$ and $A'' B''$, on the directions $O M$ and $O N$, are, therefore, described by the projections in the same time as the path $A B$ by the moving point. The first are called *component* or *relative* paths in such and such directions. Prolong the co-ordinates of the points A and B, till the parallelogram $A E B F$ is formed, and this principle will appear, viz.: *the rectilineal path described by a point, may always be resolved into two relative or component paths, in any two directions, and these component paths will be the sides of a parallelogram, constructed upon the path described by the point as a diagonal, and parallel to the assumed directions.* Reciprocally, when we have the relative paths in any two directions, the true path, called the *resultant,* will be that

Parallelogram of paths;

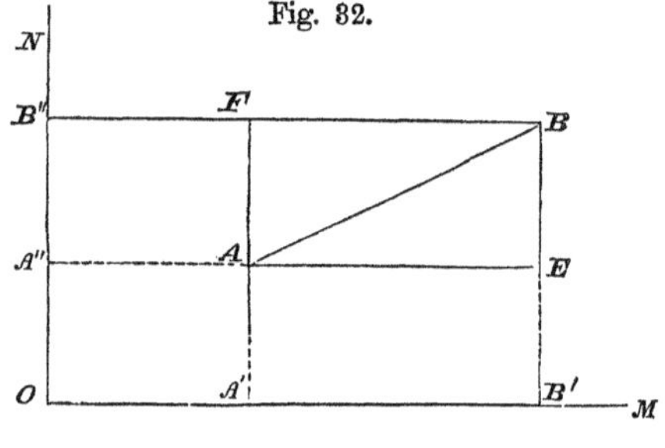

Fig. 82.

relative or component paths;

resolution of any path into component paths;

diagonal of the parallelogram constructed upon the relative paths which passes through their point of meeting. composition of the relative paths.

§ 103.—It has been shown, that the velocity of a body in motion, is represented by the length of path described uniformly in any very small portion of time, assumed as the unit of time, and that it is only in the case of uniform motion, that the interval of time during which the velocity is estimated, may be taken as great as we please. The path $A\,B$, in the last figure, being described by the body in the same time that its relative paths $A'\,B'$ and $A''\,B''$ are described by its projections on the directions $O\,M$ and $O\,N$, the first may be regarded as the point's true velocity, and the two last as its relative velocities. Hence *the true velocity of a body, is the diagonal of a parallelogram constructed upon its two relative velocities, estimated in any given directions whatever.*

Parallelogram of velocities; true and relative velocities; true velocity found from relative velocities, and the reverse.

§ 104.—If the motion be curvilinear, the rectilineal diagonal $A\,B$ can no longer represent, in general, the path described. Nor, if the motion be varied, can its length measure the velocity, when the time of description is considerable. In such cases, conceive a given interval of time divided into a great number of small and equal portions, and determine the relative paths described during each, by the projections of the moving point on the axes. Each pair of these relative paths will determine a parallelogram, of which the diagonal will be the corresponding elementary path described by the point itself. Any one of these diagonals,

Relative paths in curvilinear and varied motion; geometrical representation;

Fig. 33.

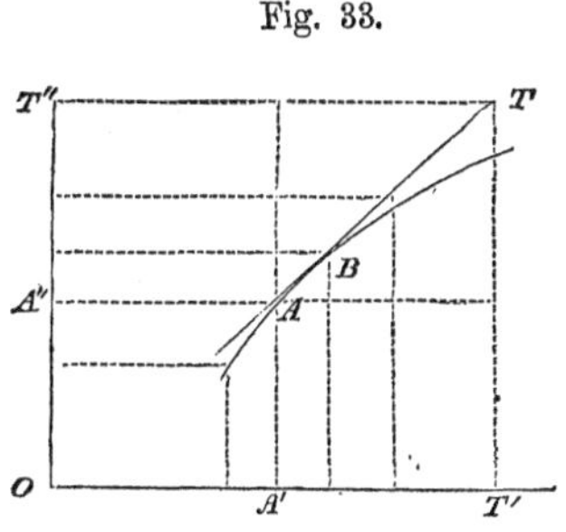

construction of the direction of the body's motion.

as $A B$, will sensibly coincide with an element of the curve, and its prolongation $A T$ will be tangent to the curvilinear path. This tangent will determine the direction of the body's motion at the instant, and may be drawn by laying off from the projections A' and A'' of the body's place, the distances $A' T'$ and $A'' T''$, equal respectively to double, triple, quadruple, or any number of times the body's relative velocities at the time, and drawing $T' T$ and $T'' T$, respectively, parallel to the directions $O T''$ and $O T'$.

Roberval's method of constructing the tangent;

§ 105.—When the law of a body's motion in two directions is known, it is always possible by the preceding method to draw a tangent to the path described. Take, for example, the ellipse: this curve is generated by fixing at two points F and F', called the foci, the ends of a thread $F A F'$, equal in length to a given line $M M'$, called the transverse axis, and moving the point of a pencil A to all positions in which it will keep the thread stretched. Since, in the motion of the describing point, the sum of the lengths $F A$ and $A F'$ is always the same, the portion $F A$ will increase just as much as the portion $A F'$ will diminish, and therefore the point A tends to describe equal relative paths, or will have equal relative velocities, in the two directions $A B$ and $A F'$. Hence, taking upon $F A$ produced, and upon $A F'$, the equal portions $A B$ and $A B'$, and completing the parallelogram $A B C B'$, the diagonal $A C$, passing through the position of the point, will be a tangent line to the path described. This method, which is due to Roberval, is very useful in

Fig. 34.

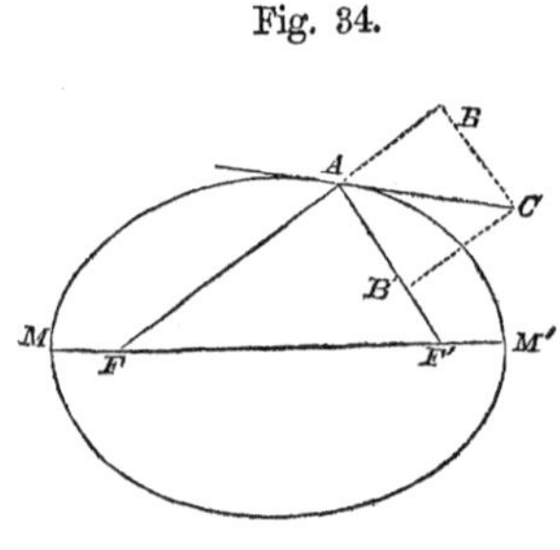

results from the law which determines the path.

all cases where we know the law by which the curve is described.

§ 106.—We have seen that any single motion may be resolved into two others, and the reverse. This arises from the simple fact, that a body may, in reality, be animated by two or more simultaneous velocities. To illustrate, let it be supposed that while a boat is crossing a river, a man walks from one side of the boat to the other, and that, starting from the point A, for example, he arrives at B at the moment the boat reaches a position such that the point A shall be at A', and the point B at B'. It is plain, that the man, though only conscious of having walked across the boat from A to B, will, in fact, have been carried from A to B' in reference to the surface of the river. He will have moved, at the same time, with the velocity which he impressed upon himself, and that impressed upon him by the boat. This being understood, it is easy to see that the result would be the same, if the boat first move from A to A', and afterward the man walk across it from A' to B'; or if the boat were stationary, while the man is crossing it from A to B, and then were to move from B to B'. But this is not all; the earth turns about its axis, while the boat floats along the surface of the water, and the man walks across the deck of the boat; add now the motion of the earth about the sun through space, and we shall find the man animated by four simultaneous velocities, of which it is easy to see that we shall find the resultant, in compounding, by the rule given in § 103, first, any two, then the resul-

Illustration of the coexistence of simultaneous velocities;

Fig. 35.

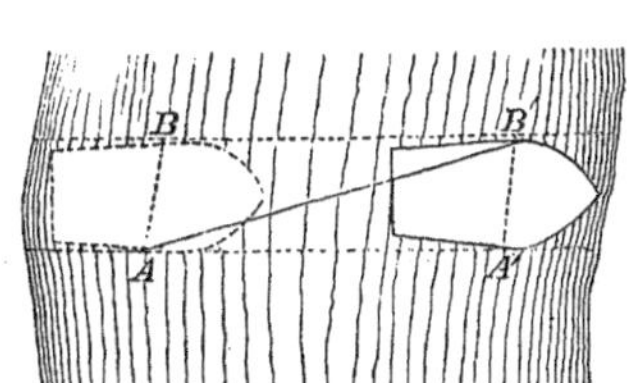

example of four simultaneous motions;

tant of these two with the third, and the resultant of the three with the fourth. In fact, when a body has several simultaneous motions, the effect is the same as if the body had received, one after the other, all the motions which it possesses at the same time. Hence, this rule, viz.: *The resultant of several simultaneous velocities is found by constructing a polygon, of which the sides are equal and parallel to the component velocities, and by joining, with a right line, the point of departure with the extremity of the last side. This right line will represent the resultant required.*

resultant of several simultaneous velocities;

rule;

Fig. 36.

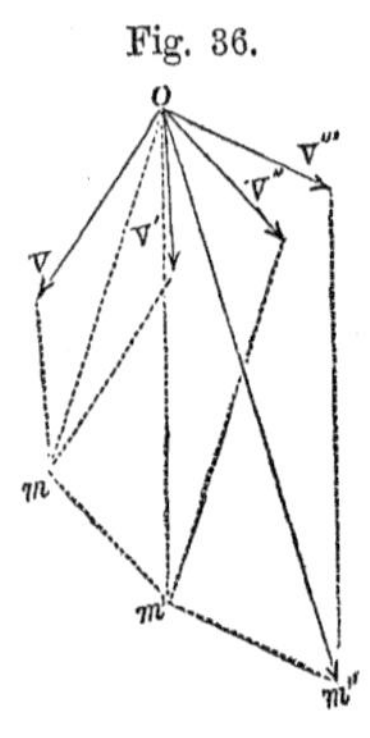

illustration.

Thus, let the point O have the simultaneous velocities $O\,V$, $O\,V'$, $O\,V''$, $O\,V'''$; from the extremity V of $O\,V$, draw Vm parallel, and equal to $O\,V'$; from m draw $m\,m'$ parallel, and equal to $O\,V''$; from m' draw $m'\,m''$ parallel, and equal to $O\,V'''$, and join O with m''; the line $O\,m''$ will be the resultant velocity.

Independence of the action of simultaneous forces;

§ 107.—The action of a force upon a body, whether at rest or in motion, is always the same, and impresses upon it the same degree of velocity. Let a body fall, for example, under the action of its own weight, gravity will impress upon it the same velocity in a given portion of time, whether it set out from rest or is projected downward by the action of some other force. For example, when a bombshell is thrown into the air, it describes a curve, under the joint action of the living force with which it leaves the mortar, and the incessant action of its

Fig. 37.

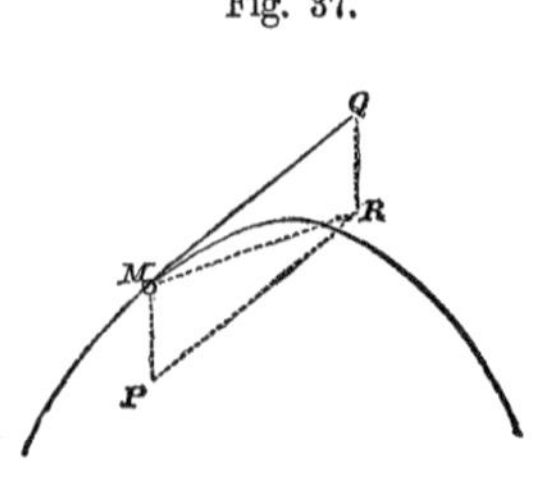

weight, and its velocity at any instant is the resultant MR, of the velocity MQ, which it has received at the instant immediately preceding that we are considering, and the small velocity MP, which its weight can impress upon it during the very short interval of time between these two instants. Thus, when two forces are applied to the same body, they impress upon it, at each instant, and simultaneously, the same degree of velocity which each would impress if acting alone. This degree of velocity, we have said, § 81, is, from the general law of nature, proportional to the intensities of the forces.

two forces impress simultaneously, the same velocity as if acting separately.

§ 108.—Accordingly, let a material point A be acted upon by the two forces P and Q, represented in intensity and direction by the lines AB and AC respectively. These forces will impress simultaneously, and in their respective directions, the same degrees of velocity Am and An, as though each acted separately. The resultant velocity will, § 107, be represented by the diagonal Ar of the parallelogram $Amrn$. Conceive a force X, to act upon the point along this diagonal, but in the opposite direction, or from r to A, and with such intensity as to destroy this velocity; no motion can take place, so that the force X, destroying the effect of the forces P and Q, will maintain these forces in equilibrio. Take, upon the diagonal, the distance $AD = X$, and conceive it to represent a force that acts upon the point A, from A towards D; it will produce the same effect as the forces P and Q, and will, therefore, be their resultant. Now, the forces P and Q, and their

Parallelogram of forces;

Fig. 38.

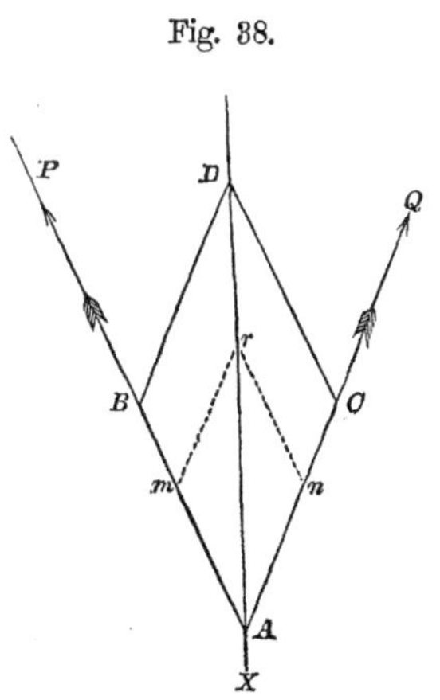

resultant $A\,D$, equal in intensity to X, are proportional to the velocities $A\,m$, $A\,n$, and $A\,r$, which they can simultaneously produce, and, therefore, $A\,D$ must be the diagonal of the parallelogram constructed upon the lines $A\,B$ and $A\,C$ as sides. Whence results this important principle, known under the denomination of *parallelogram of forces*, viz.: *the resultant of any two forces applied to the same point, is represented, in magnitude and direction, by the diagonal of a parallelogram, constructed upon the lines which represent, in intensity and direction, the two forces.* It must not be forgotten that a force is, in geometrical investigations of mechanics, always represented by a portion of its line of direction, containing as many linear units as there are pounds in the intensity of the force. It is plain, therefore, that forces may be combined by the same rules as velocities; and this is confirmed by experiment. If, for example, we attach to a cord $A\,C\,B$, fixed at its two ends, a weight $R =$ fifteen pounds, it is easy, by a balance-spring, to measure the efforts exerted in the directions $C\,A$ and $C\,B$. Laying off upon the vertical through C, and from the point C, a distance $C\,D$ equal to 15 inches, and completing the parallelogram by drawing $D\,a$ and $D\,b$ parallel respectively to $C\,B$ and $C\,A$, we shall find the number of inches in $C\,a$ and $C\,b$ to be the same as the number of pounds indicated by the balances A and B.

the resultant of any two oblique forces applied to a point;

represented by the diagonal of a parallelogram;

forces combined by the same rules as velocities;

experimental illustration of the parallelogram of forces.

Fig. 39.

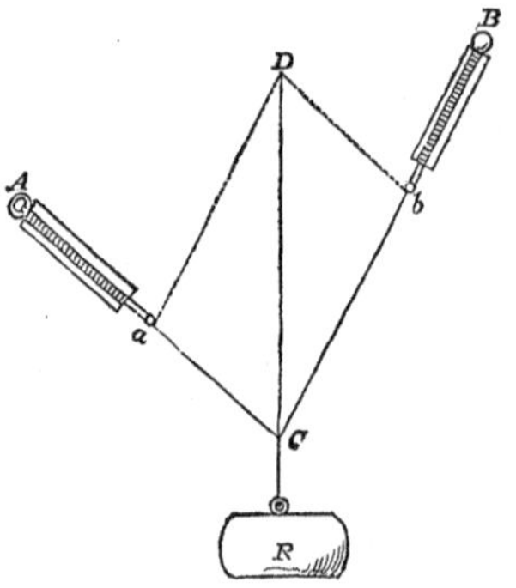

§ 109.—By the same principle that two forces, applied

to the same point, may, without change of effect, be replaced by a single one, may a single force be replaced by two others, acting in given directions. Let a given force, applied to the point O, be represented in direction and intensity by the line $O\,r$: its components, in any two assumed directions, as $O\,A$ and $O\,B$, are thus found. Through the point r, the extremity of $O\,r$, draw $r\,m$ and $r\,n$ parallel, respectively, to $O\,B$ and $O\,A$; the portions $O\,m$ and $O\,n$ will represent the components required.

Resolution of a force into two others;

Fig. 40.

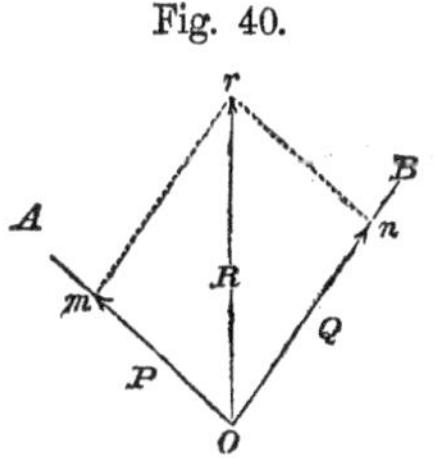

Make $O\,m = P$; $O\,n = Q$; $O\,r = R$; the angle $A\,O\,B = \varphi = r\,n\,B = 180° - r\,n\,O$. Then, in the triangle $O\,r\,n$, because $O\,m = r\,n = P$, we shall have

trigonometrical relation of resultant to its two components;

$$R^2 = P^2 + Q^2 + 2\,P\,Q\,\cos\varphi,$$

or

$$R = \sqrt{P^2 + Q^2 + 2\,P\,Q\,\cos\varphi} \quad . \quad (31);$$

value of resultant;

and because the angle $O\,r\,n$ is equal to the angle $r\,O\,m$, and $\sin r\,n\,O = \sin A\,O\,B$, we also have, from the same triangle,

$$R : Q :: \sin\varphi : \sin r\,O\,m,$$

$$R : P :: \sin\varphi : \sin r\,O\,n;$$

whence,

$$\left.\begin{aligned} \sin r\,O\,m &= \frac{Q\sin\varphi}{R} \\ \sin r\,O\,n &= \frac{P\sin\varphi}{R} \end{aligned}\right\} \quad . \; . \; . \quad (32).$$

its inclination to its components.

§ 110.—We have heretofore supposed the resistance

Quantity of work when the resistance is not immediately opposed to the force;

immediately opposed to the force destined to overcome it. Let us now consider the case in which the resistance is exerted in any line of direction other than that of the force, and in which the point of application of the force can only move along the line of direction of the resistance.

Let, for example, $A\ R$ represent a force applied to the point A, which can only move in the direction $A\ B$. Decompose this force, which denote by R, into two components P and Q—the first perpendicular to $A\ B$, and the other in the direction of that line, and, consequently, immediately opposed to the resistance that may be overcome. Since the point A cannot yield in a direction perpendicular to $A\ B$, the component P can only tend to press it, without producing any work. The component Q, is immediately opposed to the resistance, and, if $A\ a$ be the small path described by the point of application A, the product $Q \times A\ a$, will measure the elementary quantity of work necessary to overcome the elementary quantity of resistance over the same path; such will be the measure of the effective quantity of work of the force R.

Fig. 41.

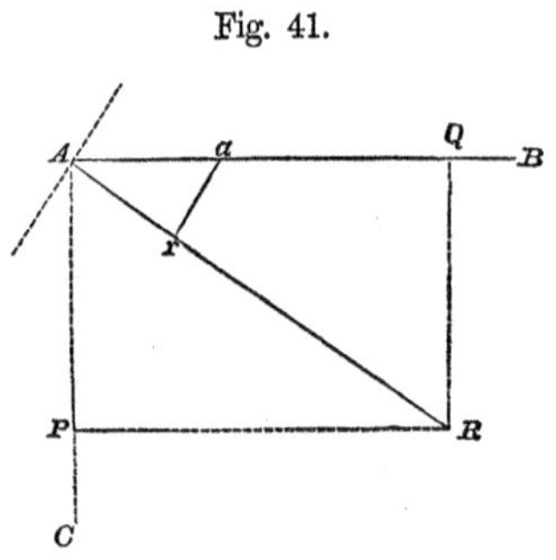

Draw from the point a, $a\ r$ perpendicular to $A\ R$; $A\ r$ will obviously be the length of path described by A in the direction of the force R, and we shall have, from the triangles $A\ a\ r$ and $A\ Q\ R$, which are similar. having a common angle A, and each a right angle,

equal to the product of the force into the path, estimated in direction of force.

$$A a : A r :: R : Q;$$

whence,

$$A a \times Q = A r \times R \quad . \quad . \quad . \quad (33);$$

which shows that *the quantity of work of a force, not immediately opposed to a resistance, is equal to the product of the force into the length of path described by its point of application, estimated in the direction of the force.*

Quantity of work of the weight of a body, moving on a curve;

§ 111.—When a heavy body is compelled to move upon the curve $A\,B\,C$, the elementary quantity of work expended by its weight W, in causing it to describe the elementary path $B\,C$, is, from what has just been shown, equal to the product $W \times b'\,c'$, estimated upon the vertical line $A\,D'$. It is also the measure of the quantity of work expended in the direction of the curve. Adding together all the elementary quantities of work by which the body is made to describe the whole curve, it is plain that the sum, or the whole quantity of work expended by the weight, must be equal to the weight multiplied into the sum of the elementary paths $b'\,c'$, which make up the whole height $A\,D' = H$; or to $W \times H$. This is also the measure of the quantity of work performed by the component of the weight, which acts in the direction of the motion, along the curve. But, from § 88, the double of this last quantity is equal to the living force of the body; that is to say, to the product

Fig. 42.

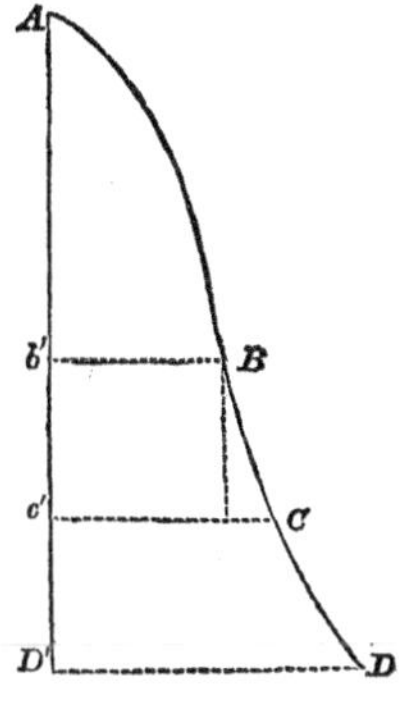

the same as that of the component of the weight in direction of curve.

$$\frac{W}{g} \times V^2;$$

in which V denotes the velocity of the body in the direction of the curve, at the instant the work terminates; whence

$$2\,WH = \frac{W}{g}\,.\,V^2,$$

or

$$V^2 = 2\,g\,H;$$

The velocity depends upon the height, and not on the path described.

that is to say, the velocity acquired by a body in moving down a curve, under the action of its own weight, is the same as though the body had fallen vertically through the same height. And we see, from this investigation, that the quantity of work which a moter must expend, in elevating a weight along any inclined surface, is always measured by the product of the weight of the body, into the vertical height to which it is raised.

Elementary quantity of work of two forces applied to a point;

§ 112.—It has just been shown, § 110, that the elementary quantity of work of a force, of which the point of application is moved in a direction different from that of the force, is measured either by the product of this force into the length of the path described, estimated in the direction of the force, or by the product of the real path into that one of the two rectangular components of the force, which acts in the direction of the motion; and it must here be remarked, that this component is nothing more nor less than the projection of the force on the direction of the motion. Accordingly, let us consider two forces, P and Q, applied to the point A, R their resultant, and $a\,A$ the small path described by the point of application. Let fall from the points Q, P, and R, the perpendiculars $Q\,Q'$, $R\,R'$, and $P\,P'$, upon $A\,a$ produced; the projection of the force P will be $A\,P'$, that of Q, $A\,Q'$, and that of the resultant R, $A\,R'$.

when the projections of components fall on opposite sides of point of application;

Fig. 43.

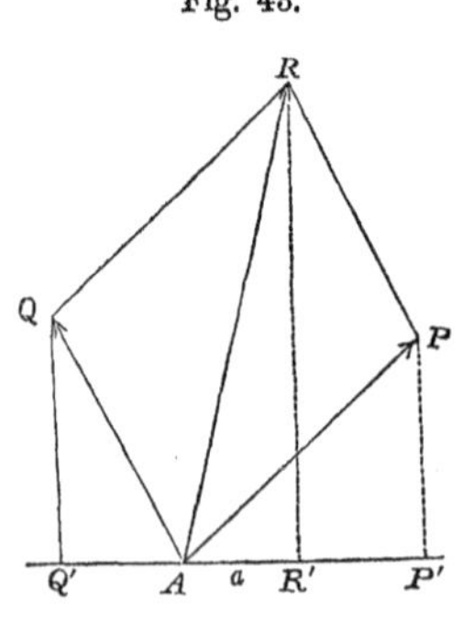

Now,

$$AR' = AP' - R'P',$$

but AQ and RP, being equal and parallel, their projections AQ' and $R'P'$ upon the same line, are equal, and hence

$$AR' = AP' - AQ,'$$

and multiplying both members by the path Aa, we have

work of resultant equal to difference of work of components;

$$AR' \times Aa = AP' \times Aa - AQ' \times Aa;$$

the first member is the elementary quantity of work of the resultant R, the first term of the second member is the elementary quantity of work of the component P, and the last term, the elementary quantity of work of the component Q. And it must be remarked that the component AP' acts in the direction of the motion, while the component AQ' acts in the opposite direction; so that the effective quantity of work of these components, which is the same as that of the components P and Q, § 110, is equal to the difference of the quantities of work taken separately.

when the projections fall on same side;

Had the motion taken place so as to cause the projections of the points Q and P to fall on the same side of the point A, a little consideration will show that the last equation would become

Fig. 44.

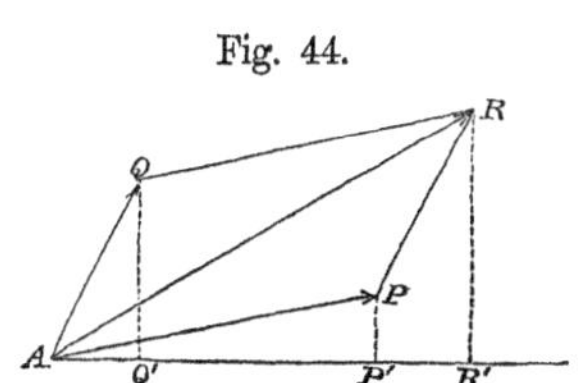

$$AR' \times Aa = AP' \times Aa + AQ' \times Aa,$$

the work of resultant equal to sum of that of components.

and that the effective quantity of action of the components AP' and AQ', would be the sum of the quantities taken separately, and the equation may be written, generally,

$$A R' \times A a = A P' \times A a \pm A Q' \times A a \,.\,. (34).$$

The work of resultant equal to the algebraic sum of the work of its components.

Hence, *the elementary quantity of work of the resultant of two forces, applied to a point, is equal to the algebraic sum of the quantities of work of the two components.*

When the projection of a force falls on the same side of the point of application as the path described, and we give the corresponding elementary quantity of work the positive sign, then when it falls on the opposite side, the work must have the negative sign.

Motion about a fixed point.

§ 113.—The small space $A\,a$, may be described in different ways. If we suppose, for example, that the point of application A is on an axle $A\,O$, which turns horizontally about some point O, taken arbitrarily in the plane of its motion, as in the case of a bark or mortar mill, the path $A\,a$ becomes the small arc of a circle, which we may regard as a small right line perpendicular to $A\,O$. From the point a, let fall the perpendiculars $a\,b$, $a\,d$, and $a\,c$, upon the directions of the forces P, Q, and their resultant R; then will the elementary quantities of work due to these forces be respectively $P \times A\,b$, $Q \times A\,d$, and $R \times A\,c$; and from § 112.

Fig. 45.

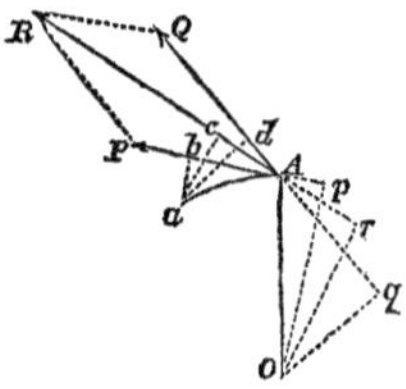

$$R \times A c = P \times A b \pm Q \times A d.$$

From the point O, about which the motion takes place, let fall the perpendiculars $O\,p$, $O\,q$, and $O\,r$, upon the directions of the forces P, Q, and R, respectively; the triangles $A\,O\,p$ and $A\,a\,b$ are similar, since each has a right angle, and the angle $A\,O\,p$, of the first, is equal to the angle $a A b$ of the second, the sides $A O$ and $O p$ being, respectively, perpendicular to the sides $A a$ and $A b$; hence,

$$A b : A a :: O p : A O;$$

whence,

$$A b = O p \times \frac{A a}{A O};$$

and, in like manner, from the similar triangles $A d a$ and $O A q$, we have

$$A d = O q \cdot \frac{A a}{A O};$$

and from the similar triangles $A c a$ and $A O r$,

$$A c = O r \times \frac{A a}{A O};$$

these values, substituted in the above equation, give, after omitting the common factors, and making $O r = r$, $O q = q$, and $O p = p$,

$$R r = P \times p \pm Q \times q \quad . \quad . \quad . \quad (35).$$

The *effective quantity of work* which a force is capable of performing, while its point of application is constrained to describe an elementary path $A a$, about a fixed centre O, is called the *moment of the force;* the fixed point O is called the *centre of moments;* and the perpendiculars p, q, and r, the *lever arms* of the forces P, Q, and R, respectively. **Moment of a force; the centre of moments; lever arms;**

The elementary quantities of work performed by the forces P, Q, and R, during the description of the path $A a$, are measured by the products $P p$, $Q q$, and $R r$, multiplied each by the constant ratio $\frac{A a}{A O}$; and if this

the relative measure of a moment;

constant ratio be omitted, these products may be taken as the relative measures of the elementary quantities of work. Hence, the relative measure of a moment, is *the product of the intensity of the force into its lever arm;* and from Eq. (35) we see that *the moment of the resultant of two forces, applied to a point, is equal to the algebraic sum of the moments of the components.*

the moment of the resultant of two forces.

§ 114.—In what precedes, the two forces, P and Q, have been supposed to be applied to the same point; if they be applied to different points C and B, it is evident that we may suppose two rigid bars, CA and BA, to be firmly attached to the body, and to coincide in direction with the given forces. These bars, if the forces act in the same plane, will meet at the point A, and the latter thus becoming invariably connected with the body, may be taken as the common point of application, without changing the effect of the forces. The resultant AR will be obtained by means of the diagonal of the parallelogram $APRQ$, and the point D, where it meets the surface, may be taken as its point of application. If, now, the body be constrained to move around any point, as O, the common point of application A, will describe the small arc of a circle, which may be regarded as a small right line, to be projected on the directions of the forces, as in the last article; and the same reasoning will show us, that in this case also, the moment of the resultant is equal to the algebraic sum of the moments of the components.

When the forces are not applied to the same point;

Fig. 46.

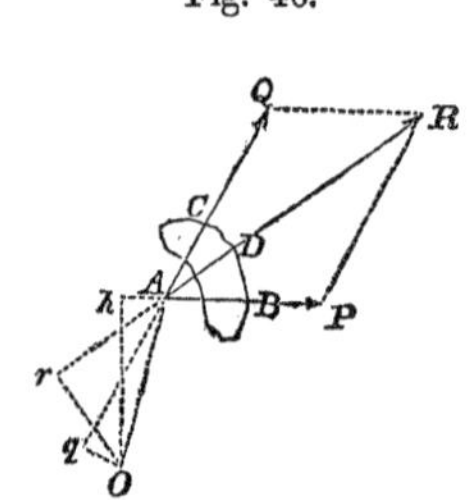

the moment of the resultant is still equal to the algebraic sum of the moments of the components.

§ 115.—The relations which have just been established between the quantities of work, and between the moments of forces and of their resultant, will always obtain wherever the point O be taken, since its selection was entirely arbitrary; but these relations were obtained by considering the motion of the point, common to the directions of the forces, this point being assumed as their common point of application. To show that they are equally true in regard to the motion of the true points of application B, C, and D, see the last figure, we have only to remark that the measure of the moment depends alone upon the intensity of the force, and the length of the perpendicular drawn from the centre of moments to its line of direction, and is wholly independent of the position of the point of application. The moment of the force P, for example, will be the same whether it be supposed applied at A, or at the point B, where its direction meets the surface of the body. The theorem of moments will be true, therefore, when the forces P and Q are not applied to the same point.

These relations equally true, wherever the centre of moments be taken;

or wherever the points of application.

§ 116.—If it be shown that the quantity of work of a force is the same, whatever point be taken on its line of direction as the point of application, it is obvious that the theorem of the quantity of work, estimated by the motion of the common point of union of two forces and their resultant, will be equally true of all cases in which the quantities of work of these forces are computed in reference to the motion of their respective points of application. Three cases may arise, according as the body has a motion of rotation, of translation, or of both combined.

Extension of the theorem of the quantity of work;

work estimated by any point of application on line of direction;

First case. The body and the direction A P, of the force P, being supposed to have a motion of rotation about the point O, any two points, as A and B of the

First—in motion of rotation;

Fig. 47.

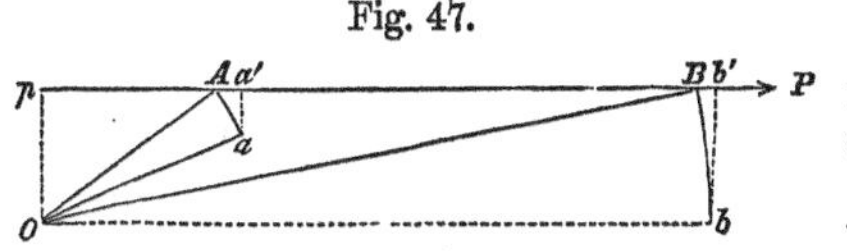

line $A P$, will describe arcs which are proportional to their distance, $O A$ and $O B$, from O; and we shall have

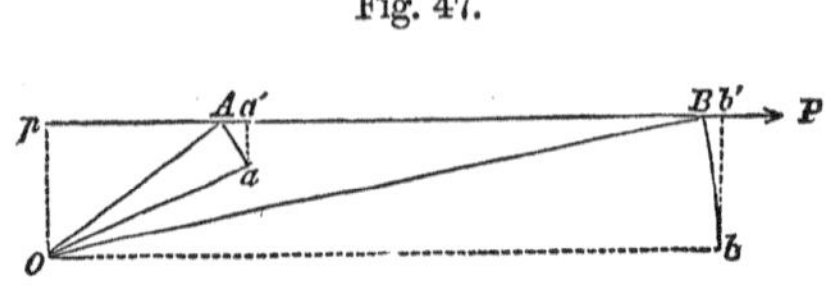

Fig. 47.

$$\frac{A a}{A O} = \frac{B b}{O B};$$

but the quantity of work of the force P, estimated by the motion of its point of application supposed at A, will have, § 113, for its measure,

$$P \times Op \times \frac{A a}{O A};$$

or estimated by the motion of its point of application, supposed at B, will be measured by

$$P \times Op \times \frac{B b}{O B}.$$

Hence, the quantities of work are equal, being measured by the product of the intensity P, the length of the perpendicular $O p$, and the equal factors $\frac{A a}{O A}$, and $\frac{B b}{O B}$.

Second—in motion of translation;

Second case. If the body only have a motion of translation, any two points of application, as A and B, will describe the equal and parallel paths $A a$ and $B b$, which will be projected upon the direction $A P$, in the equal paths $A a'$ and $B b'$; and the quantities of work in the two cases being $P \times A a'$ and $P \times B b'$, are equal to each other.

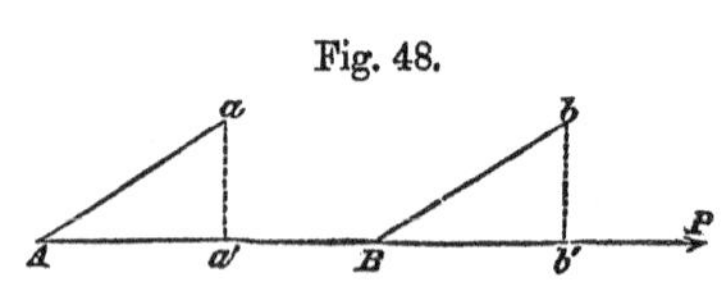

Fig. 48.

Third case. Suppose the line of direction $A P$ of the

force P, to take the position $A_1 B_1$, in virtue of the combined motion of rotation and translation, and the points A and B to be transferred to the positions a and b. This motion of the points A and B may be regarded as resolved into a motion of rotation around the point O, the centre of a circle, tangent to the two positions of the line of direction, supposed indefinitely near each other, and of translation along the second position of this line. By the first, the points A and B are carried in the arcs of circles to A_1 and B_1, and by the second, from these latter positions to a and b, thus making $A\,a$ and $B\,b$ the actual paths described. Projecting these latter paths on the primitive direction of the force by the perpendiculars $a\,a'$ and $b\,b'$, we shall have for the quantities of work, considered in reference to the motion of the points A and B, $P \times A\,a'$ and $P \times B\,b'$, respectively.

Third—when the motion is of translation and of rotation combined;

Fig. 49.

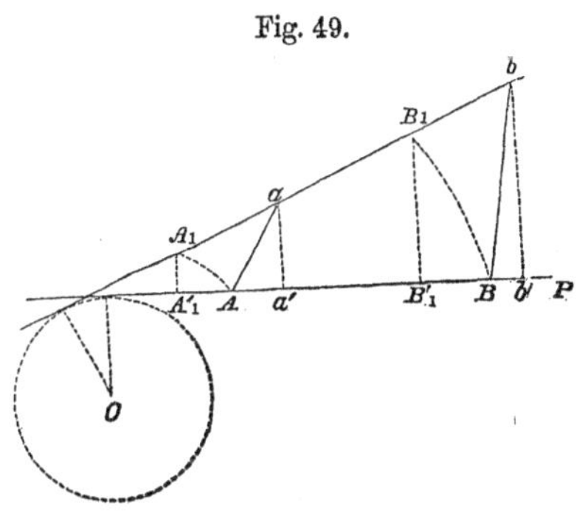

But by projecting the points A_1 and B_1 on the primitive direction, by the perpendiculars $A_1 A_1'$ and $B_1 B_1'$, we have

$$A\,a' = A_1'\,a' - A_1'A\,,$$

$$B\,b' = B_1'\,b' - B_1'\,B;$$

multiplying each equation by P,

$$P \times A\,a' = P \times A_1'\,a' - P \times A_1'\,A,$$

$$P \times B\,b' = P \times B_1'\,b' - P \times B_1'\,B.$$

Now $P \times A_1'\,a'$, and $P \times B_1'\,b'$, are the quantities of work, on the supposition of a simple motion of translation alone, in the direction $A\,B$, and these have been shown, in the *second case*, to be equal; whence,

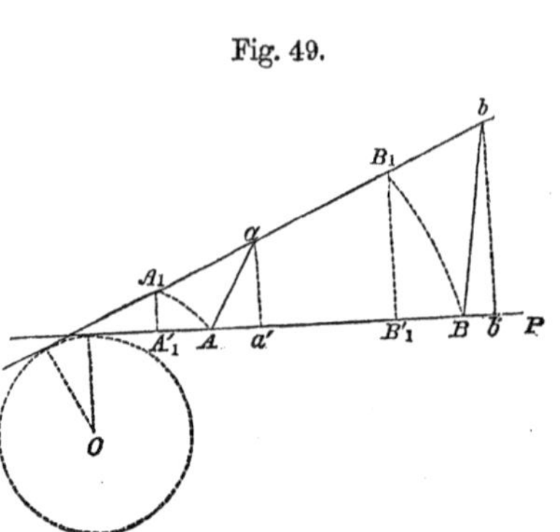

Fig. 49.

$$A_1{}'\,a' = B_1{}'\,b'.$$

no matter where the points of application be taken on the lines of direction; The products

$$P \times A_1{}'\,A,$$

and

$$P \times B_1{}'\,B,$$

measure the quantities of work due to the motion of A and B, on the supposition of a simple motion of rotation about O, which have been shown to be equal, in the *first case;* whence,

$$A_1{}'\,A = B_1{}'\,B;$$

and consequently,

$$P \times A\,a' = P \times B\,b'.$$

the work of the resultant, is equal to the algebraic sum of the quantities of work of the components. Thus, the relation given in § 112, between the quantity of work of the resultant of two forces, and the total quantities of work of the components, subsists in all cases, whatever be the points of application, and whatever be the nature of the motion.

§ 117.—Resuming Eq. (35),

$$Rr = Pp \pm Qq,$$

When the resultant is zero, or when its line of direction passes through a fixed point, in which r, p, and q, denote the lengths of the *lever arm* of the resultant R and of the two components P and Q, we see that the moment $R\,r$, of the resultant, can only reduce to zero when the moments of the components P and Q are equal and have contrary signs. But the product $R\,r$, becomes nothing, either when $R = 0$, or $r = 0$. In the first case, the resultant is nothing, and there will be an equilibrium independently of all other considerations. In the second case, the perpendicular r, which measures the distance of the line of direction of the resultant from the centre of moments, being nothing, indicates that the resultant passes through the fixed point. Again, the equality of the moments of the components, necessarily implies an equality in the quantity of work performed by each, and these quantities, having different signs, destroy each other; hence, there will be an equilibrium about a fixed point, when the resultant of the forces which act upon the body, passes through this fixed point.

there will be an equilibrium.

V.

OF FORCES WHOSE DIRECTIONS ARE PARALLEL.

§ 118.—It has been shown of two forces whose directions intersect: 1st, that the line of direction of the resultant, will intersect those of the components in the same point; 2d, that the moment of the resultant is equal to the sum or difference of the moments of the components, according as the components tend to turn the body upon which they act, in the same or in opposite directions about the centre of moments. Now, these properties, being entirely independent of the position of the point of meeting O, and of its distance from the body or centre of moments, will not cease to be true when

Theorem of the quantity of work, and of the moments equally true, when the forces are parallel.

Fig. 50.

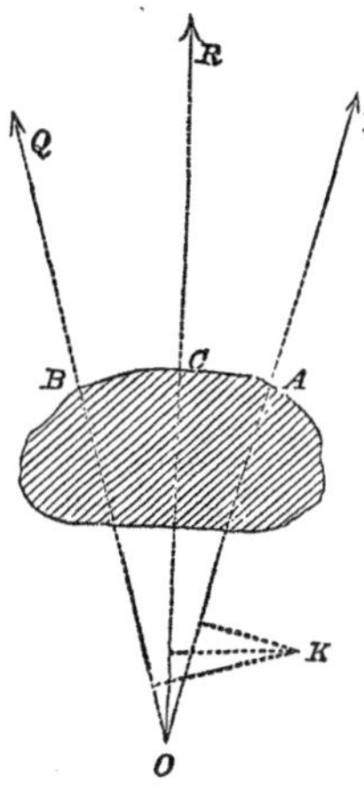

the point O is so far removed as to make the directions of the forces sensibly parallel: whence we must conclude, that the line of direction of the resultant of two parallel forces is in the plane of the forces, is parallel to the direction of the forces, and that the moment of the resultant, taken in reference to any point in the plane of the forces, is equal to the sum or difference of the moments of the components, according as they tend to turn the system in the same or opposite directions about the centre of moments.

Fig. 51.

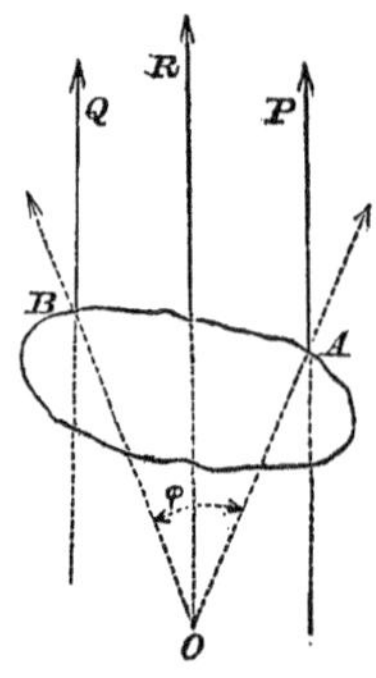

Resuming Eq. (31), and revolving the directions of the forces P and Q about their points of application A and B till they become parallel, and the forces act in the *same* direction, the angle φ will become zero, and we shall have

Value of resultant when the components act in same direction;

$$R = \sqrt{P^2 + Q^2 + 2PQ} = P + Q.$$

Fig. 52.

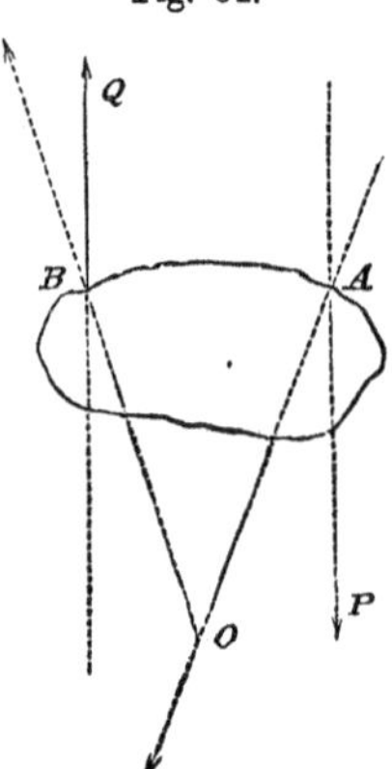

Again, revolving the directions as before, till they become parallel and the forces act in *opposite* directions, the angle φ will equal 180°, and Eq. (31) reduces to

$$R = \sqrt{P^2 + Q^2 - 2\,P\,Q} = P - Q;$$ value of resultant when components act in opposite directions;

whence we conclude, that *the intensity of the resultant* of two parallel components, *is equal to the sum or difference of the intensities of the components according as these latter act in the same or in opposite directions.* rule;

Now, resuming Eqs. (32), and changing the notation to suit the first figure in § 118, we have

$$\sin R\,O\,A = \frac{Q \sin \varphi}{R},$$

$$\sin R\,O\,B = \frac{P \sin \varphi}{R};$$

in which, if we make $\varphi = 0$, or $180°$, we obtain

$$\sin R\,O\,A = 0,$$

$$\sin R\,O\,B = 0;$$

that is to say, the angle which the direction of the resultant of two parallel forces makes with the directions of the components, is nothing; in other words, *the direction of the resultant of the parallel forces is parallel to that of the components, which is a confirmation of what we said above.* the direction of the resultant of two parallel components, is parallel to that of the components.

§ 119.—Passing thus to the limits of the case in which the directions of two forces P and Q, applied at the points A and B of any body, meet in a point; assume any point as K, in the plane of the forces, and let fall the perpendiculars $K\,a$, $K\,b$. Denote by R, the intensity of the resultant, supposed to act along the line $R\,c$, The theorem of moments true of parallel forces.

Fig. 53.

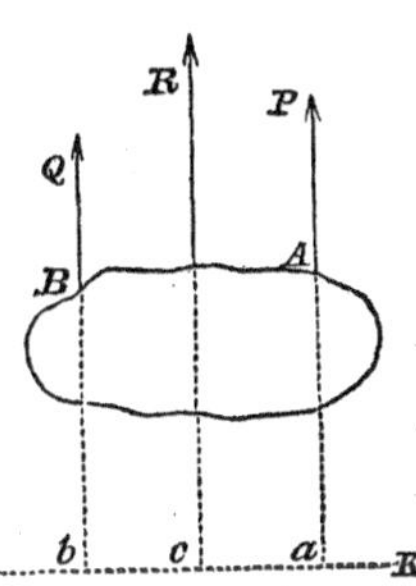

then, from the principle of moments, will

$$R \times Kc = P \times Ka \pm Q \times Kb;$$

the upper or lower sign being taken, according as the forces tend to turn the body in the same or opposite directions about the point K.

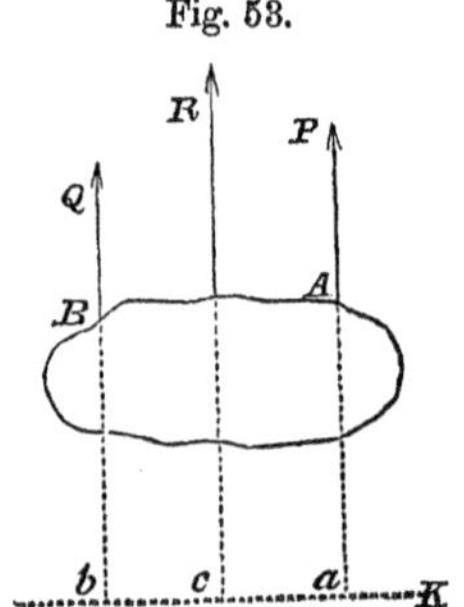

Fig. 53.

Relation of resultant to its two parallel components; Replacing R by its value $P \pm Q$, the above becomes

$$(P \pm Q) Kc = P \times Ka \pm Q \times Kb;$$

which, by an obvious reduction, becomes

$$P (Kc - Ka) = Q(\pm Kb \mp Kc);$$

but

$$Kc - Ka = ca; \quad \pm Kb \mp Kc = \pm bc;$$

whence

$$P \times ac = \pm Q \times bc,$$

or

$$P : Q :: bc : ac;$$

the distance of either component from resultant, proportional to the other component. that is to say, the line of direction of the resultant, divides the perpendicular distance between the lines of direction of the components, into parts which are reciprocally proportional to the forces.

Fig. 54.

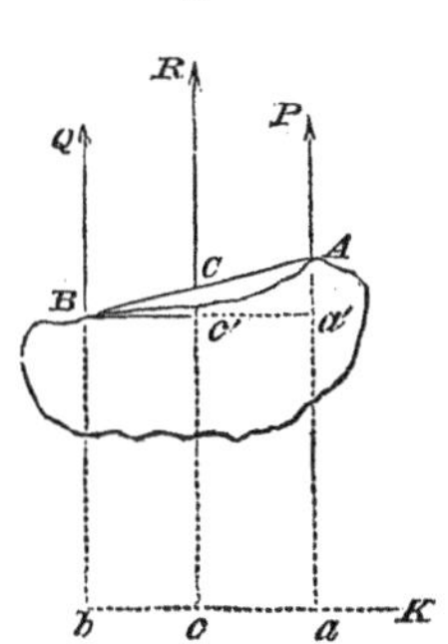

§ 120.—Let the parallel forces P and Q, be applied to the points A and B. Join A and B by a straight line, and draw Ba' parallel to ba, then will

$$Bc' = bc; \qquad c'a' = ca;$$

and because Cc' is parallel to Aa', the triangles $Bc'C$ and $Ba'A$, give the proportion,

$$Bc' : c'a' :: BC : CA;$$

whence

$$P : Q :: BC : AC;$$

Rule for position of resultant;

that is to say, *the line of direction of the resultant of any two parallel components, divides the line joining their points of application into parts which are reciprocally proportional to the intensities of the components.*

Fig. 55.

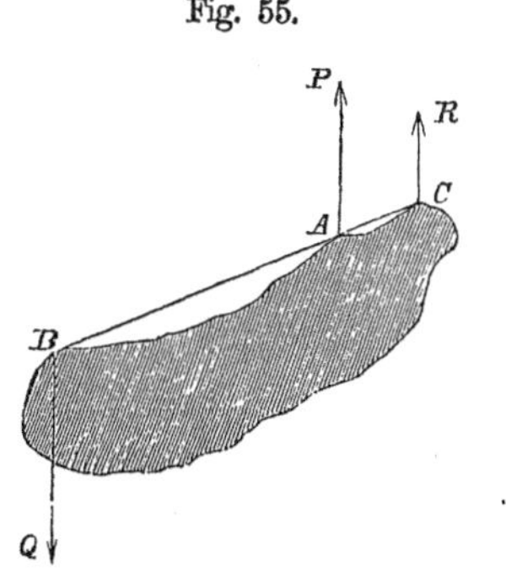

The above proportion gives by composition,

$$P \pm Q : P :: BC \pm AC : BC,$$

$$P \pm Q : Q :: BC \pm AC : AC;$$

or, replacing $P \pm Q$ by R, and $B\,C \pm A\,C$ by the whole line $B\,A$,

$$R : P :: AB : BC,$$

$$R : Q :: AB : AC;$$

relation of resultant to either component.

that is to say, *the resultant of two parallel components is to either component, as the length of the straight line joining the points of application of the components, is to the portion of this line between the point in which it is cut by the direction of the resultant, and the point of application of the other component.*

Moments of parallel forces in reference to an axis;

§ 121.—When two forces are parallel, their moments may not only be taken in reference to a point, but also in reference to a right line, supposed fixed. Thus, suppose the forces P, Q, and their resultant R, to act along the parallel lines $A\,P$, $B\,Q$, and $C\,R$, respectively. Assume any line, as $M\,L$, at pleasure; conceive a plane drawn through this line and perpendicular to the plane of the forces, and let $K\,L'$ be the intersection of these planes. From the point K, draw $K\,L''$ perpendicular to the direction of the forces; then, regarding K as the centre of moments, will

Fig. 56.

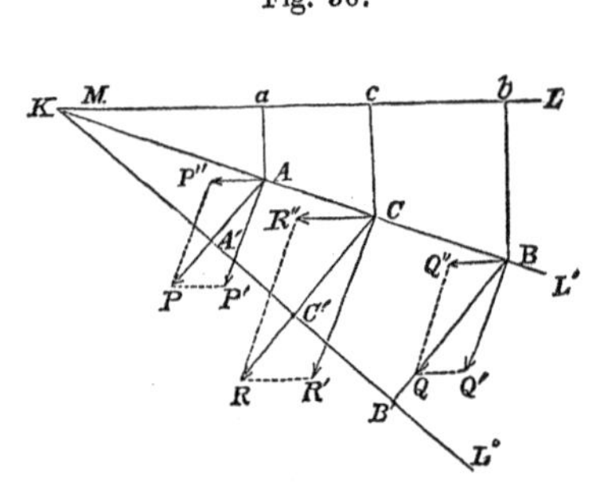

moments referred to a centre;

$$R \times KC' = P \times KA' + Q \times KB';$$

whence

$$R = P \times \frac{KA'}{KC'} + Q \times \frac{KB'}{KC'}.$$

But from the similar triangles, $KA'A$, $KB'B$, and $KC'C$, we have

$$\frac{KA'}{KC'} = \frac{KA}{KC},$$

$$\frac{KB'}{KC'} = \frac{KB}{KC};$$

which, substituted in the above equation, gives, on clearing fractions,

$$R \times KC = P \times KA + Q \times KB \ldots (36).$$

Dividing both members by $R \times KC$,

$$1 = \frac{P}{R} \times \frac{KA}{KC} + \frac{Q}{R} \times \frac{KB}{KC}.$$

From the points A, B, and C, draw the lines Aa, Bb, and Cc, perpendicular to the line KL. Also, resolve each of the forces P, Q, and R, supposed applied at A, B, C, respectively, into two components, one parallel, and the other perpendicular, to the line KL; and let AP'', BQ'', and CR'' be the former, and AP', BQ', and CR', the latter of these components.

forces replaced by their components;

In the similar triangles PAP', RCR', and QBQ', we have, denoting the components AP', CR', and BQ', by P', R', and Q', respectively,

$$\frac{P}{R} = \frac{P'}{R'},$$

$$\frac{Q}{R} = \frac{Q'}{R'};$$

and from the similar triangles KAa, KCc, and KBb,

$$\frac{KA}{KC} = \frac{Aa}{Cc},$$

$$\frac{KB}{KC} = \frac{Bb}{Cc};$$

which values, substituted in the foregoing equation, give, after clearing the fractions,

moments of components perpendicular to the axis;

$$R' \times Cc = P' \times Aa + Q' \times Bb \quad . \quad . \quad (37).$$

The effective quantity of work performed by each of the forces P, Q, and R, may be replaced by the algebraic sum of the quantities of work performed by its components; but the effective quantities of work of the components which are parallel to the line KL, will be zero, since the points of application are constrained to move in planes at right angles to this fixed line, and hence the terms in Eq. (37) will, for reasons explained in § 113, be the measures of the *relative* quantities of work of the forces P, Q, and R, being the products of the remaining components into the perpendicular distances of their respective lines of direction from points on the line KL.

Fig. 56.

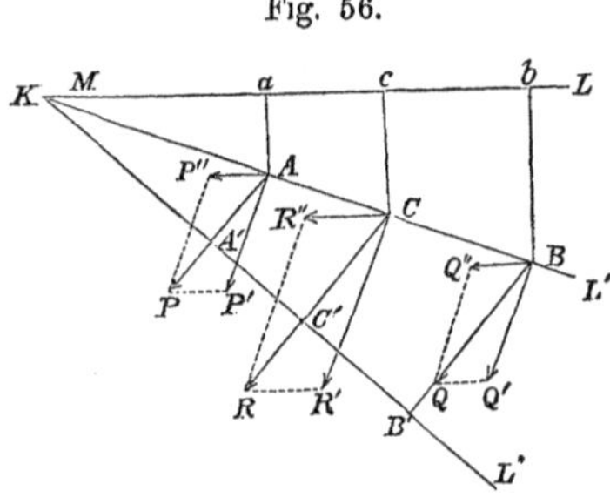

moments of the parallel components;

moment of a force in reference to a fixed axis, defined;

The moment of a force in reference to a line, is the effective quantity of work which the force is capable of performing while its point of application is constrained to describe an elementary path about this line, considered as fixed; and its relative measure is, the product of the component at right angles to the line, (the other being parallel to it,) into the shortest distance from the fixed line to that of the direction of the force.

The fixed line is called *the axis of moments.*

the axis of moments.

§ 122.—Dividing Eq. (36) by $K\,C$, we find

$$R = P \cdot \frac{K\,A}{K\,C} + Q \cdot \frac{K\,B}{K\,C},$$

and substituting the values of

$$\frac{K\,A}{K\,C} \text{ and } \frac{K\,B}{K\,C},$$

as given on the opposite page, we find, after clearing the fraction,

$$R \times Cc = P \times A\,a + Q \times Bb;$$

Relation of the forces to the distances of their points of application from a line, and plane.

from which we see, that the product of the resultant of two parallel forces into the perpendicular distance of its point of application from any given straight line, is equal to the sum of the products of the forces into the perpendicular distances of their respective points of application from the same line. It is easy to see that the same is equally true of any plane, since we have but to project the line joining the points of application of the forces upon the assumed plane, and take this projection as the axis of moments.

Resultant of any number of parallel forces;

§ 123.—Now let us suppose any number of parallel forces—for instance, five. Find the resultant of any two of them; compound this resultant with the third force, and the resultant of the first three with the fourth, and so on. The final resultant thus obtained, will be equal in intensity to the sum of the intensities of the forces which act in one direction, diminished by the sum of the intensities of those which act in the opposite direction. Its action will be in the direction of the greater sum. And the moment of the resultant will be equal to the algebraic sum of the moments of the components.

rule for finding;

Men pulling upon parallel ropes, horses drawing upon

examples of parallel forces.

their traces attached to whipple-trees, are examples of parallel forces.

The work performed by the resultant of parallel forces;

§ 124.—Suppose a body to be drawn in one direction by any number of parallel forces P, Q, R, &c., and in the opposite direction, by the parallel forces P', Q', R', &c. If the points of the body move in parallel lines, it is plain that the paths described by the points of application will be equal to each other, and thus the quantity of work of any force, will be given by the product of its intensity into the small path common to all the forces. The total work will be equal to the sum of the quantities of work performed by the forces P, Q, R, &c., diminished by the sum performed by the forces P', Q', R', &c.; that is to say, it will be equivalent to the product of the common path, multiplied into the algebraic sum of all the forces, or into the resultant. But this latter product is the quantity of work performed by the resultant. Hence, the quantity of work performed by the resultant of any number of parallel forces, is equal to the algebraic sum of the quantities of work performed by the components.

Fig. 57.

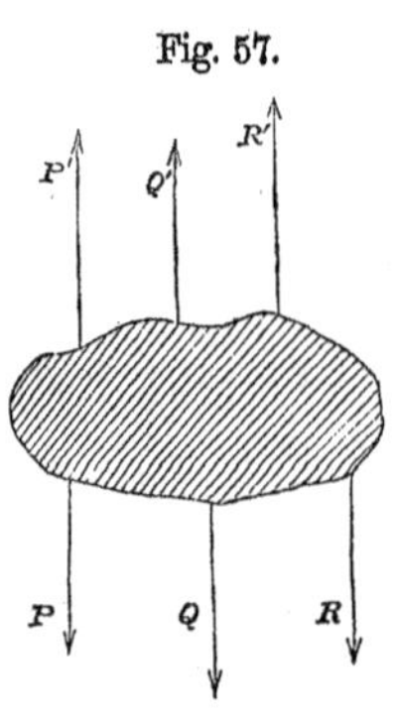

equal to the algebraic sum of the work of the components.

§ 125.—We have seen, § 122, that the product of the intensity of the resultant of several parallel forces into the perpendicular distance of its point of application from any plane, is equal to the sum of the products arising from multiplying the intensity of each force into the perpendicular distance of its point of application from the same plane. Denote this latter sum by K, the intensity of the resultant by R, and the perpendicular distance of its point of application from a given plane by r, then will

$$Rr = K,$$

Position of the resultant of parallel forces.

whence

$$r = \frac{K}{R};$$

and if the given plane be parallel to the direction of the forces, r will be the distance between it and a second plane containing the line of direction of the resultant. If we know the value of K, in reference to another plane, also parallel to the direction of the forces, the corresponding value of r, will give the position of a second plane, whose intersection with the first will give the line of direction of the resultant. Thus, the principle explained in § 122, may be employed to determine the line along which the resultant of several parallel forces acts.

Illustration of the principle of parallel forces by the steelyard.

§ 126.—To illustrate the principle of parallel forces, let us take the example of the common steelyard, an instrument employed to ascertain the weight of different substances. It consists of a bar MN, which turns freely about an axis C suspended from a fixed point; the substance Q to be weighed, is placed at one end A, while a constant weight P is placed at a suitable point B, towards the other end. In order that there may be an equilibrium, it is necessary that the resultant of the forces P and Q shall pass through the fixed point C; in other words,

Fig. 58.

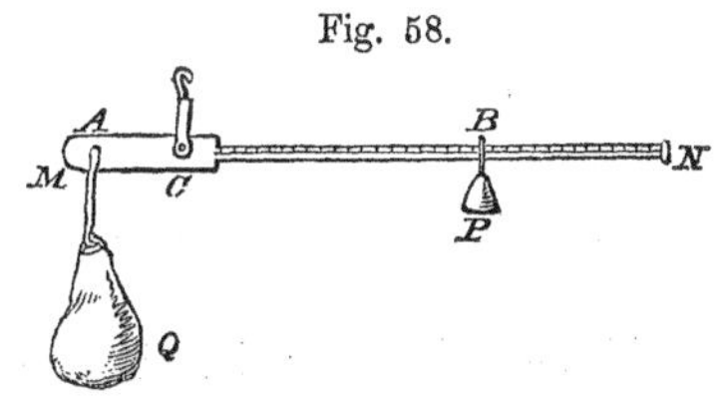

$$Q \times AC = P \times CB,$$

from which

$$BC = \frac{Q}{P} \times AC;$$

or, if P be taken equal to one pound, then will

$$B\,C = Q \times A\,C.$$

The scale of the steelyard constructed.

If Q be taken successively equal to 1, 2, 3, 4, &c. pounds, then will the corresponding values of $B\,C$, become $A\,C$, $2\,A\,C$, $3\,A\,C$, $4\,A\,C$, &c. Thus, if a scale of equal parts be constructed on the longer arm, having its zero at the point C, and the constant distance between the consecutive divisions equal to $A\,C$; the number of the division estimated from C, on which the weight P is placed to hold Q in equilibrio, will indicate the weight of the latter.

Fig. 58.

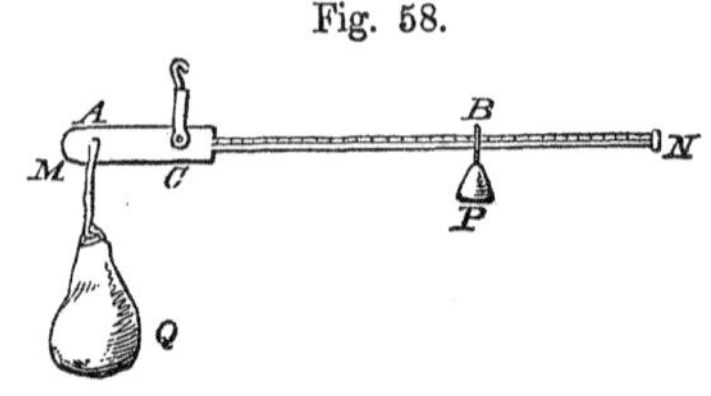

The construction of the steelyard depends, as we see, upon very simple principles; it gives rise, however, to considerations, which will be referred to when we come to treat of the lever.

VI.

CENTRE OF GRAVITY OF BODIES.

Point of application of resultant of parallel forces;

§ 127.—Whatever be the angle which two parallel forces, P and Q, make with the line $A\,B$, joining their points of application, the intensity of the resultant R, and the position of its point of application C, will always be the same, however the direction of the forces may revolve about their points

Fig. 59.

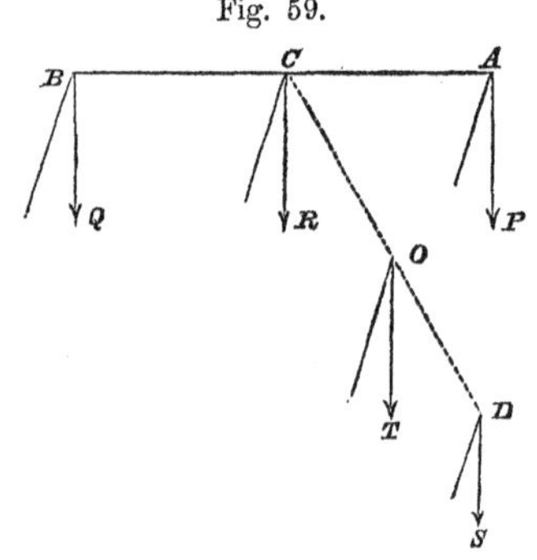

of application, provided the forces continue parallel to each other, and preserve unchanged the ratio of their intensities; for the intensity of the resultant is given by

$$R = P \pm Q,$$

and the point C, by

$$P \times AC = Q \times BC;$$

which are wholly independent of the angle which the common direction of the forces makes with the line AB. So, likewise, if there be three forces P, Q, and S, we may join the point of application D, of the third force S, with that of the resultant R, and show, in like manner, that the position of O, the point of application of the resultant T of R and S, (that is, of P, Q, and S,) is entirely independent of the inclination of the forces to the line CD. And as the same reasoning may be extended to any number of parallel forces, we conclude, that in every system of parallel forces, there is one point through which the resultant will always pass.

there is one point through which the resultant will always pass;

This point is called the *centre of parallel forces*.

the centre of parallel forces.

§ 128.—Every body is composed of an indefinite number of elementary heavy particles, which are the points of application of as many vertical or parallel forces. Their resultant is a force equal to their sum, and is called the *weight of the body*. The point of application of the weight is obtained by combining the parallel forces in the manner before explained; this point will be the centre of the system, and, because the forces are those which result from the action of gravity, it is called the CENTRE OF GRAVITY. The centre of gravity of any body may be defined, *the point through which the line of direction of the weight always passes.*

Weight of a body; centre of gravity.

§ 129.—The centre of gravity of a body being the centre of all the vertical forces which solicit its heavy particles, this point must remain invariable, while the forces, without ceasing to be parallel, revolve about the points of application. Instead of causing the forces to rotate, let the body revolve. In this motion, the forces will preserve their vertical direction, and the line of direction of the weight always passing through the centre of gravity, there will result two very simple methods of finding the position of this point as long as the figure of the body remains unchanged.

Two methods of finding centre of gravity;

first method—by suspension;

A body being suspended by means of a thread $A\ C$, from the point A, will take such a position, that the effort exerted along the thread to support it, will be in equilibrio with the weight, and thus, when the body comes to rest, the direction of the thread will pass through the centre of gravity G. If we change the point C, to which the thread is attached, to C', the body will assume a new position, and when it comes to rest again, we shall have a second line $C'\ G$, also passing through the centre of gravity, and whose intersection with the first, will determine the position of that point.

Fig. 60.

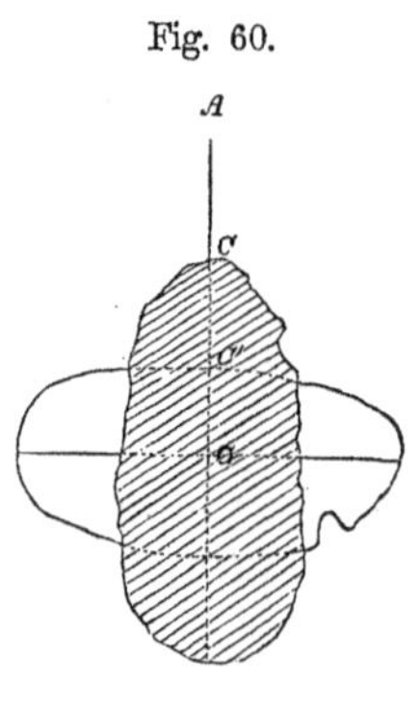

By the same reasoning it follows, that a body will be supported upon a point, whenever the vertical through the centre of gravity passes through this point; and all positions of the body which satisfy this condition, give as many lines intersecting at the centre of gravity. The upper and lower points, in which any two of these lines pierce the surface, being known, and connected by rectilineal openings, these openings will

second method—by poising;

Fig. 61.

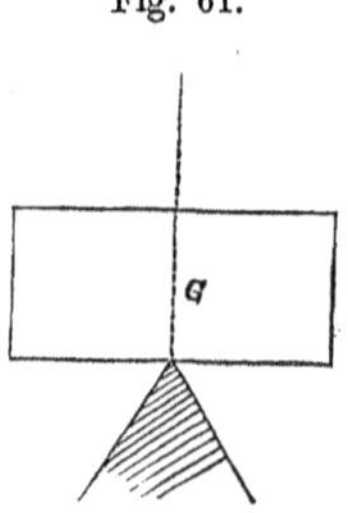

give, by their intersection, the centre of gravity of the body. To find these upper and lower points, suspend the body, by a thread or rope, and when it comes to rest, suspend a plummet on each side, and in such positions that the plane of their threads shall contain the suspension line of the body; then, with a pencil, trace upon the body the intersection of this plane with its surface. Next, suspend the body from some other point, and repeat the same operation; the intersections of the two traces will give two of the points required; and the same for others.

traces of the planes through the centre of gravity found.

§ 130.—This method becomes impracticable in the case of very heavy bodies, of those which are fixed, or of such as do not yet exist, and of which the construction is only in project. In general, when the form of a body is defined geometrically, or by a drawing, the centre of gravity is determined in this wise. Conceive the body to be divided into small portions by a series of planes; take the product of the weight of each portion into its distance from some assumed plane of reference, and take the sum of these products; this sum is, according to what we have seen of the principles of parallel forces, equal to the product of the entire weight of the body into the distance of its centre of gravity from the same plane. Hence, *the distance of the centre of gravity from any plane, is equal to the sum of the products obtained by multiplying the weight of each element of the body into its distance from this plane, divided by the whole weight of the body.*

Centre of gravity found by computation;

distance of centre of gravity from a plane;

Find the distance, given by this rule, from any three arbitrary planes, and the position of the centre of gravity becomes known. This method, which becomes long and tedious in many instances, may be abridged according to circumstances, particularly when the object is to find the centre of gravity of *homogeneous* bodies. A body is said to be *homogeneous*, when any two of its parts have the same weight under equal volumes.

from three assumed planes;

process may be abridged in the case of homogeneous bodies.

Centre of gravity of regular and homogeneous bodies;

§ 131.—Experience shows us that a bar A B, of wood, metal, or any other material, which is perfectly homogeneous, will remain in equilibrio in a horizontal position, if suspended by its middle point C; and hence the centre of gravity of this bar is situated at the middle of its length.

of a bar;

The bar is also found to remain in equilibrio when placed in a vertical position, if suspended by the central point of its end; and hence the centre of gravity is situated at the central point of its thickness.

of a bar with equal spheres at the ends;

If the bar support at its ends equal spheres, it will still remain in equilibrio when suspended by its middle point, if placed in a horizontal position.

Fig. 62.

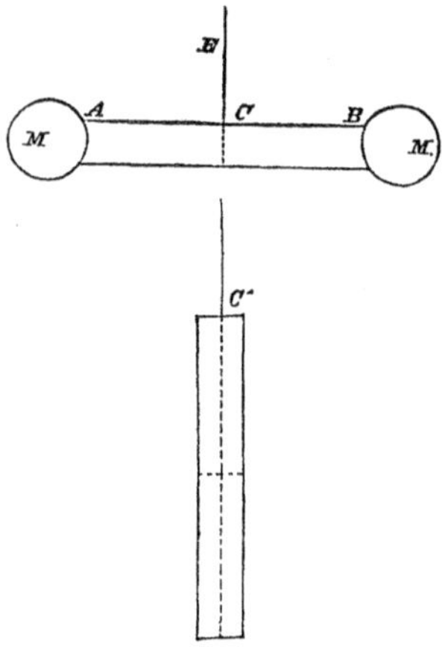

The centre of gravity of a sphere is at its centre of figure, for when suspended by any one of its points, the direction of the suspending thread always passes through that point.

centre of gravity of regular and homogeneous bodies, at the centre of figure; right prism; circle, &c.;

And it is a general principle, that the centres of gravity of all regular and homogeneous bodies are at their centres of figure. And, hence, a right prism or cylinder has its centre of gravity at the middle of its length, breadth, and thickness; a circle at its centre; and a right line at its middle point.

centre of gravity of a surface; of a line.

By the centre of gravity of a surface, is understood that of a body of extreme thinness, such as paper, tin-foil, gold-leaf, &c.; and by the centre of gravity of a line, is meant that of a body whose breadth and thickness are very small as compared with its length.

Body symmetrical in reference to a plane;

§ 132.—A body is said to be symmetrical in reference

to a plane, when the latter cuts into two equal parts every perpendicular which is drawn to it, and which is terminated by the opposite extremes of the body. This plane is called the *plane of symmetry*.

Plane of symmetry;

A body is symmetrical in reference to a line, when it has two planes of symmetry passing through the line. This line is called a *line of symmetry*.

symmetrical in reference to a line; line of symmetry;

A surface is symmetrical in reference to a line, when the latter cuts into two equal parts, all the perpendiculars to it which are terminated on opposite sides by the contour of the surface.

surface symmetrical in reference to a line;

In all cases, the centre of gravity of homogeneous symmetrical bodies, is situated in their planes, or lines of symmetry. Consider, for example, a curve having *A B* for its line of symmetry, and of which we have found the centres of gravity *G* and *G*, of the two halves *A M B* and *A M B*. These two halves being turned about the line of symmetry till one is applied to the other, their centres of gravity will coincide; that is to say, the centres of gravity *G* and *G*, were, before the motion, situated upon a right line *G G*, perpendicular to the line *A B*. Hence, if the curves be supposed concentrated at their respective centres of gravity, *G G* becomes a right line, terminated by two material points whose common centre of gravity is at the middle point *O*, on the line of symmetry. A similar reasoning may be applied to all bodies of symmetrical dimensions.

centre of gravity in planes and lines of symmetry;

Fig. 63.

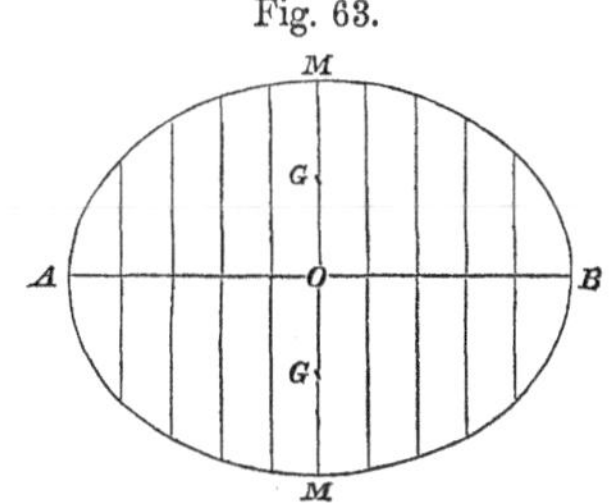

illustration in case of a symmetrical curve;

The centre of gravity of a surface which has two axes of symmetry, is at the intersection of these axes. The transverse and conjugate axes of the ellipse, for example, being axes of symmetry, cut each other at the

centre of gravity of a surface with two axes of symmetry;

case of the ellipse;

centre of gravity of the elliptical surface. For the same reason, the

rectangle;

centre of gravity of a rectangle is at the intersection of the right lines joining the middle points of its opposite sides.

Fig. 64.

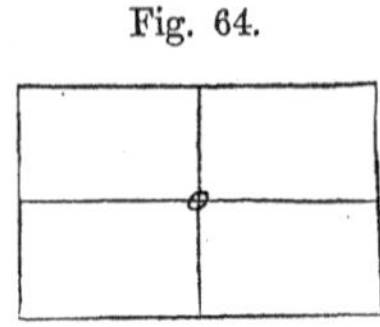

When a volume has a right line of symmetry, its centre of gravity is on this line. A

volume with one axis of symmetry;

right cylinder, with an elliptical base, has two planes of symmetry, determined by the longer and shorter axes of the ellipse, its centre of gravity is, therefore, on the line $G\ G$, joining the centres of gravity of the bases, and at its middle point O.

Fig. 65.

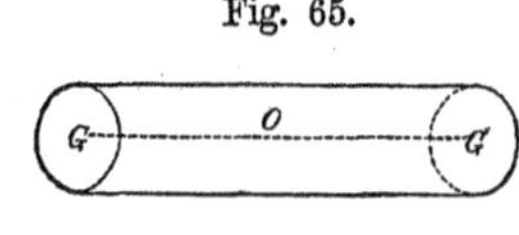

sphere many axes of symmetry.

Other bodies are divided symmetrically, in an infinity of ways. Such, for example, is the sphere of which all the planes of symmetry pass through the centre of figure; it is for this reason that this point is also its centre of gravity.

Centre of gravity of two homogeneous bodies, one within the other.

§ 133.—If the regular homogeneous body contain within its boundary another homogeneous body of different density, the centre of gravity of the whole mass is found, by first regarding it as of uniform density, and the same as that of the larger body; the centre of gravity O, obtained on this hypothesis, gives rise to a first approximation. We then conceive the weight w, of the body supposed homogeneous, to be concentrated at the centre of gravity O, and subtracting this weight w from the total weight W, we obtain a difference $W - w$, neglected in finding the point O. Let O' be the centre of gravity of the volume corresponding to this

Fig. 66.

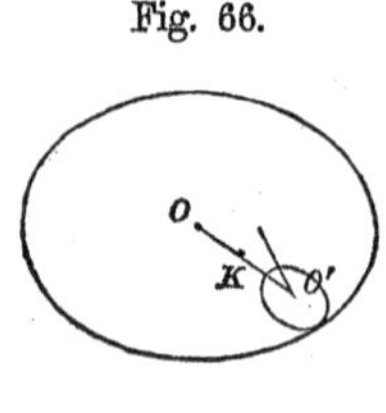

difference; join O with O' by a right line, and divide this line at the point K, so that

$$w \times OK = (W - w)\ KO';$$

the point K will be the common centre of gravity.

§ 134.—Whenever a body may be divided into parallel layers, and the centres of gravity of these are situated on a right line, the centre of gravity of the whole body is also upon this line. For compounding the weights of any two of these layers, supposed concentrated at their respective centres of gravity, and the resultant of these with the weight of a third, &c., it is easy to see, from the principle of parallel forces, that the point into which the whole weight must be concentrated will be on the line in question. **When the layers of a body have their centres of gravity on a right line.**

§ 135.—If, for example, the parallelogram $A\ B\ C\ D$, supposed to possess a small thickness, be divided by planes parallel to $C\ D$, into an indefinite number of strata or layers, the centre of gravity of each one will be at its middle point, and therefore on the line $F\ E$, joining the middle points of the opposite sides $C\ D$ and $A\ B$; the centre of gravity of the parallelogram will, § 134, also be on this line. In like manner, it may be shown to be on the line $I\ N$, joining the middle points of the opposite sides $C\ B$ and $D\ A$; it must, therefore, be at their intersection O. **Centre of gravity of a parallelogram;**

Fig. 67.

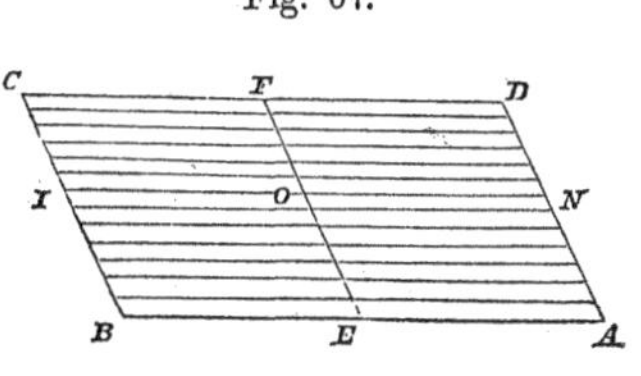

A similar reasoning will show that the centre of gravity of a parallelopipedon and cube, will be at the common intersection of three right lines joining the centres of gravity of their opposite faces. **of a parallelopipedon and cube.**

§ 136.—The triangle $A\,B\,C$, being divided into very thin layers, parallel to the side $A\,C$, it follows, from what has just been said, that the centre of gravity of each layer, and, therefore, of the whole triangle, will be situated upon the right line $B\,D$, drawn from the vertex B to the middle of the side $A\ C$. For the same reason, the centre of gravity of the triangle will also be on the line $A\,F$, drawn from the angle A to the middle of the opposite side $C\,B$; and hence it must be at the intersection G.

Centre of gravity of a triangle;

Fig. 68.

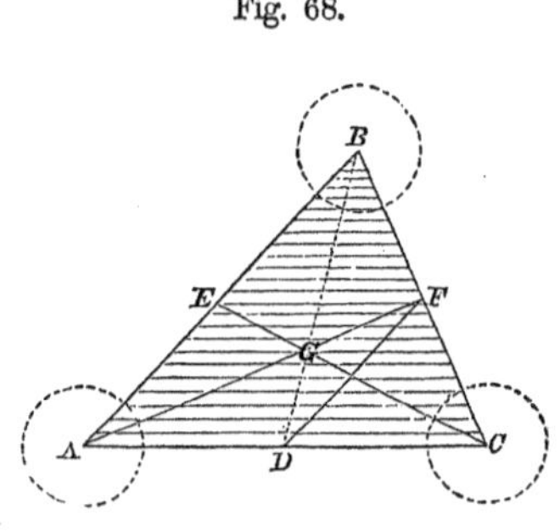

Join $F\ D$. Since the sides $A\ C$ and $B\ C$, are divided proportionally at the points D and F, the line $D\ F$ is parallel to $A\ B$; hence the triangles $A\ G\ B$ and $D\ G\ F$ are similar, and give the proportion

$$A\,G\ :\ G\,F\ ::\ A\,B\ :\ F\,D;$$

but, because the points F and D are at the middle of the lines $B\ C$ and $A\ C$, it follows that $F\ D$ is half of $A\ B$, and, therefore, from the above proportion, $F\ G$ is half $A\ G$; or $F\ G$ is one third of the whole line $A\ F$. Hence, the centre of gravity of a triangle, *is on a line drawn from one of the angles to the middle point of the opposite side*, and at a distance from this side equal to one third of the line.

where situated;

common centre of gravity of three equal balls.

This point is also the common centre of gravity of three equal balls, whose centres of gravity are situated at the angles of the triangle, for the centre of gravity of the balls A and C is at the middle point D, and this point being joined with B, the centre of gravity of the three balls will divide the line $B\ D$ at the point G, so that $B\ G$ shall be double $G\ D$.

§ 137.—To find the centre of gravity of any polygon, as $A\,B\,C\,D\,E\,F$, draw from any one of the angles, as A, the diagonals $A\,C$, $A\,D$, $A\,E$, &c., and thus divide the polygon into triangles. Find the centres of gravity g, g', g'', g''', &c. of each of these triangles by the rule above; join the points g and g' by the right line $g\,g'$, and denote the areas of the triangles $A\,B\,C$ and $A\,C\,D$ by a and a', respectively; then will the centre of gravity of the area $A\,B\,C\,D\,A$, be found by the proportion

Centre of gravity of a polygon.

Fig. 69.

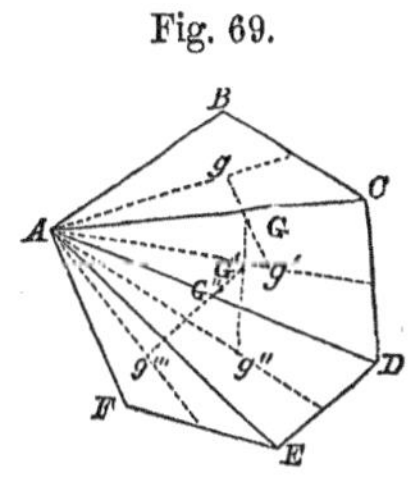

$$a + a' \;:\; a \;::\; g\,g' \;:\; g'\,G.$$

In like manner, joining G and g'' by a right line, and denoting the area of the triangle $A\,D\,E$ by a'', will the centre of gravity of the area $A\,B\,C\,D\,E\,A$ be found from the proportion,

$$a + a' + a'' \;:\; a'' \;::\; G\,g'' \;:\; G\,G';$$

and so on to the last triangle; the quantities $g'\,G$, $G\,G'$, &c., being the only unknown quantities become known from the proportions.

§ 138.—A series of planes parallel to the base $D\,B\,C$, of the triangular pyramid $A\,B\,C\,D$, will give rise to a series of strata or layers perfectly similar to the base, and all their centres of gravity will be situated upon a right line joining the centre of gravity of the base and the vertex, because they are all similarly situated to the base.

A pyramid divided into layers parallel to the base;

Fig. 70.

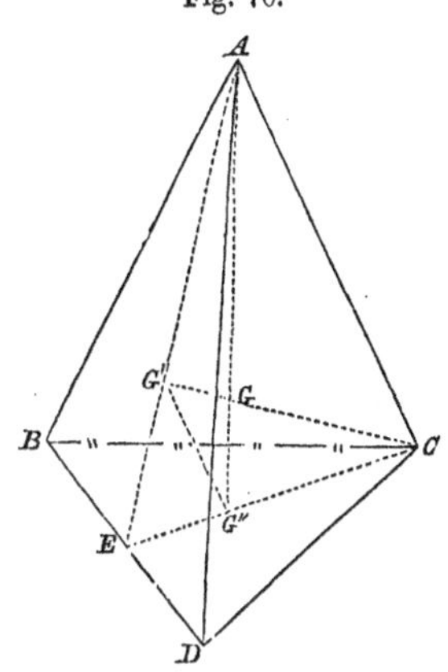

As either of the solid angles may be taken as a vertex and the opposite face as a base, and as the dividing planes may be passed parallel to each of the bases, it follows that the centre of gravity of the pyramid must be upon the four lines drawn from the solid angles to the centre of gravity of the opposite faces, and must, therefore, be at their common point of intersection.

its centre of gravity found;

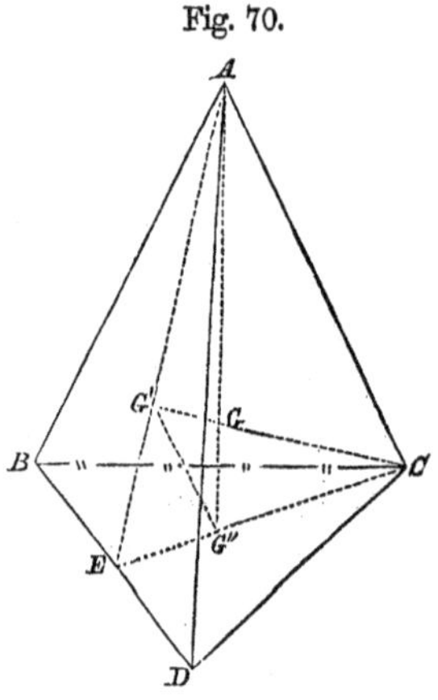

Fig. 70.

Let G' and G'' be the centres of gravity of the triangular faces $A\ B\ D$ and $B\ C\ D$; join these points with the opposite vertices by the right lines $A\ G''$ and $C\ G'$, their point of intersection G, will be the centre of gravity of the pyramid. Join G' and G''; then, because the lines $A\ E$ and $E\ C$ are divided proportionally at the points G' and G'', the line $G'\ G''$ is parallel to $A\ C$, the triangles $G\ G'\ G''$ and $G\ A\ C$ are similar, and give the proportion,

$$G'\,G'' \ : \ G\,G'' \ :: \ A\,C \ : \ A\,G;$$

where situated.

but $G'\ G''$ is one third of $A\ C$, and hence $G\ G''$ is one third of $A\ G$, or one fourth of $A\ G''$. *The centre of gravity of a triangular pyramid is, therefore, on a line joining one of the angles with the centre of gravity of the opposite face, and at a distance from this face, equal to one fourth of the line.*

The common centre of gravity of four equal balls.

The same result may be obtained for the common centre of gravity of four equal balls, whose centres of gravity are situated at the four vertices of the pyramid.

§ 139.—The foregoing reasoning is equally applicable to a pyramid, of which the base is any polygon. For the centre of gravity is on a line drawn from the vertex S to the centre of gravity of the base, because it contains the centres of gravity of all sections parallel to the base; and if we conceive the pyramid divided into triangular pyramids by planes through this line, and through the angles A, B, C, D, &c. of the base, the centres of gravity of these elementary pyramids, and therefore of the whole pyramid, will be situated in a plane parallel to the base, and at one fourth the distance from the base to the vertex; it must, therefore, be at the intersection of this line and plane. Hence, *to find the centre of gravity of any pyramid, join the vertex with the centre of gravity of the base, and lay off a distance from the base on this line equal to one fourth of its length.*

Centre of gravity of any pyramid; where situated.

Fig. 71.

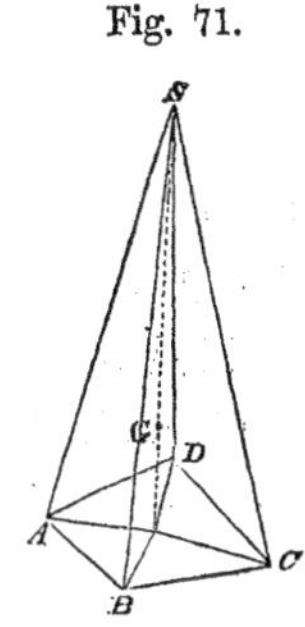

This rule is also applicable to a cone, which may be regarded as a pyramid of an indefinite number of sides.

Centre of gravity of a cone.

§ 140.—Since every polyhedron may be divided into triangular pyramids whose weights may be supposed to act at their respective centres of gravity, and since, from the principles of parallel forces, the sum of the products which result from multiplying the weight of each partial pyramid into the distance of its centre of gravity from any plane, is equal to the product of the entire weight of the polyhedron into the distance of its centre of gravity from the same plane, the distance of the centre of gravity from three planes may be found, and thus its position determined.

Of any polyhedron.

§ 141.—When a body is terminated by curved surfaces, by planes, or by curve lines, it may be divided into small elementary parts, similar to the figures which have been already considered—as right lines, triangles, parallelo-

Of a body of any form;

10

grams, pyramids, parallelopipedons, polyhedrons, &c.; the sum of the products which result from multiplying the weight of each into the distance of its centre of gravity from some assumed plane, or right line, must be found, and this sum divided by the entire weight of the body; the result will be the distance of the centre of gravity from the plane or line. Let it be required, for example, to determine the centre of gravity of any plane area $C a b F d c$; draw in its plane any right line $A B$, and divide the given area into a series of very thin layers, perpendicular to this right line. The layer $a c d b$, may be regarded as a small rectangle, and, supposing its density uniform, its centre of gravity is at its middle point O; denoting the density by D, and the force of gravity by g, one of the partial products will be

the partial products found;

the sum of these divided by the entire weight;

Fig. 72.

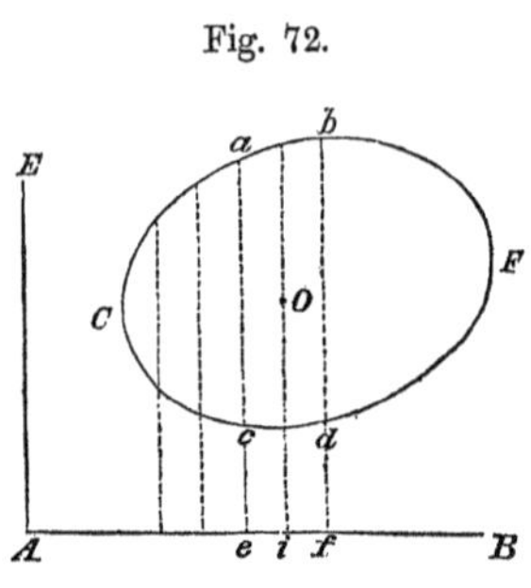

illustration;

$$D \times g \times \frac{a c + d b}{2} \times i O = D g \times \frac{a c + d b}{2} \times \frac{e a + e c}{2}.$$

The other partial products being found in the same way, and their sum divided by the product of $D g$ into the entire area $C c d F b a C$, determined by the method of § 46, will give the distance of the centre of gravity of this area from the line $A B$. Performing the same operation in reference to another line $A E$, the centre of gravity is completely determined, being the intersection of two right lines, parallel respectively to $A B$ and $A E$, and distant from them, equal to the results obtained by the above process.

when the force of gravity is constant and density uniform;

It is to be remarked, that when the force of gravity g is constant, and the density D is uniform throughout the body, these quantities strike out, and leave the distance of the centre of gravity from the line, or plane, equal to the sum of the products arising from multiplying the elementary *volumes* into the distances of their respective centres from the line or plane, divided by the entire volume. the partial products in terms of the volumes.

§ 142.—The consideration of the centre of gravity is very useful in computing certain volumes and surfaces, which are found with considerable difficulty by the ordinary process. The screw, the curbs of stair-ways, surfaces of revolution generated by the rotation of a plane curve $C D E$ about an axis $A B$, situated in its plane, are examples. Suppose, in the case of a volume, the generating area $C D E$ to be divided into small rectangles, of which the sides are parallel and perpendicular to the axis $A B$. Each rectangle will generate around the axis an elementary ring, and the sum of all these rings will give the volume of the solid of revolution. Let r denote the distance of the centre of gravity of one of these small rectangles from the axis; we know that the volume of the ring, of which the profile is the rectangle, is measured by the product of the area a of the rectangle, multiplied by the mean circumference of the ring, $2 \pi r$; for the annular base of such a ring being developed, will form a trapezoid, the half sum of whose parallel sides is equal to $2 \pi r$, and hence we shall have for the value of the ring the expression $2 \pi r a$. The volumes generated by the other rectangles, whose areas are a', a'', a''', &c., will be $2 \pi r' a'$, $2 \pi r'' a''$, $2 \pi r''' a'''$, &c. And denoting by V the total volume generated, we shall have

Use of the centre of gravity in computing volumes and surfaces;

Fig. 73.

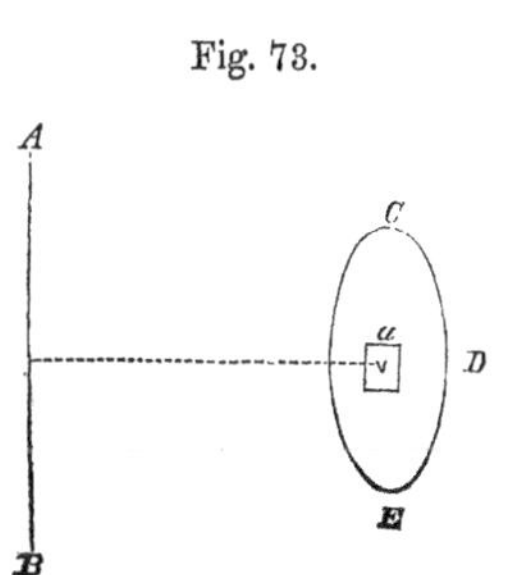

$$V = 2 \pi (a r + a' r' + a'' r'' + a''' r''' + \&c.);$$

but the quantity within the brackets, is the sum of the products which result from multiplying the elementary volumes of the generating area $C E D$, by the distances of their respective centres of gravity from the line $A B$, which we know to be equal to the product of the whole area $C E D$, into the distance of its centre of gravity from the same axis. Denoting the area $C E D$ by A, the distance of its centre of gravity from $A B$ by R, we, therefore, have

Fig. 73.

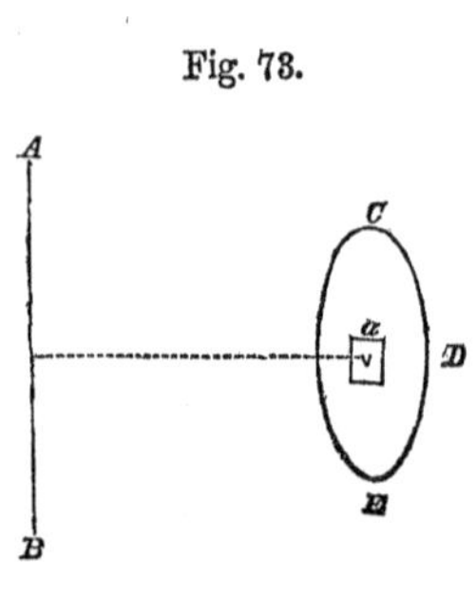

relation of volume to generatrix and path of centre of gravity;

$$V = 2 \pi R A \quad . \quad . \quad . \quad . \quad . \quad (38).$$

If, instead of an area, we had considered a plane curve $C E$, the quantities a, a', a'', &c., would represent the lengths of elementary portions of this curve, A would represent its entire length, R would be the distance of its centre of gravity G, from the line $A B$, and V would be the value of the surface generated by the entire curve about $A B$. Whence we derive this rule, viz.: *The volume generated by the motion of any plane, or surface generated by the motion of any line, is equal to the generatrix, multiplied by the path described by its centre of gravity;* the direction of the motion being perpendicular to the generatrix.

Fig. 74.

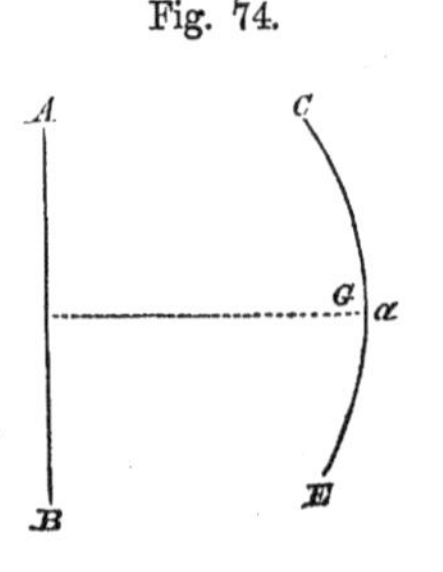

rule;

This rule supposes the body to possess a constant profile, of which the plane is perpendicular to the path of the centre of gravity.

Example 1*st*. Let it be required to find the volume generated by the rotation of the right-angled triangle $A\,B\,C$, about the side $A\,B$. The centre of gravity G, being found by the rule already explained, draw $G\,D$ perpendicular to $A\,B$. Then, in the triangles $E\,G\,D$ and $E\,B\,C$, we have

example—the volume of a cone;

Fig. 75.

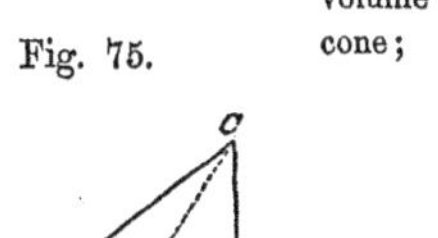

$$CB : GD :: CE : GE :: 3 : 1;$$

whence

$$GD = \tfrac{1}{3}\,CB;$$

and

$$2\,\pi\,G\,D = \tfrac{2}{3}\,\pi\,C\,B,$$

which is the length of the path described by the centre of gravity. The area of the triangle is

$$\tfrac{1}{2}A\,B \times CB;$$

whence the volume V becomes

$$V = \tfrac{1}{3}\,\pi\,CB^2 \times A\,B,$$

which is the usual measure of the volume of a cone.

Example 2*d*. Let it be required to find the surface generated by the rotation of the line $C\,D$, about $A\,B$. The centre of gravity of $C\,D$ is at its middle point G; and $G\,D'$, $C\,A$, and $D\,B$ being perpendicular to $A\,B$, we have

example—the surface of a conic frustum;

Fig. 76.

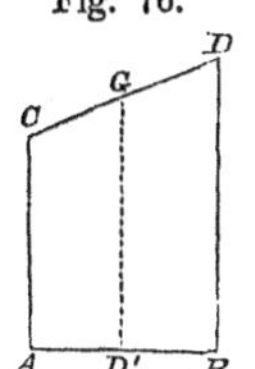

$$G\,D' = \tfrac{1}{2}(A\,C + BD);$$

and for the path described by G,

$$2\,\pi\,G\,D' = 2\,\pi\,\frac{(A\,C + BD)}{2};$$

and hence,

$$V = \frac{2\pi AC + 2\pi BD}{2} . CD;$$

which is the usual measure for the convex surface of a conic frustum.

example—the volume of a stairway curb;

Example 3d. Let it be required to find the volume of the curb of a stairway, of a helical form.

Fig. 77.

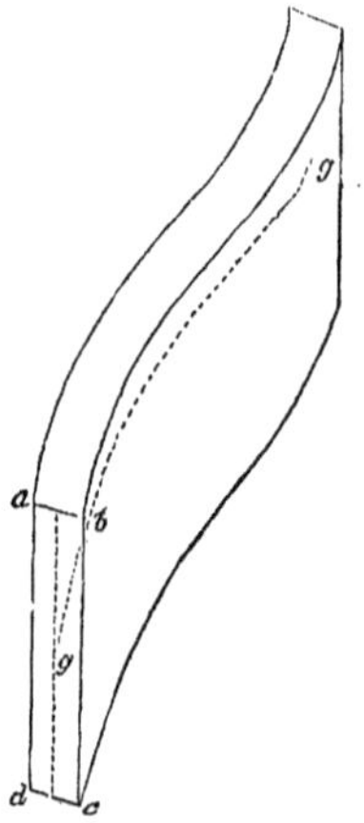

First, compute the area of a section $a\,b\,c\,d$, perpendicular to a mean helix $g\,g$, or that described by the centre of gravity; then multiply this section by the length of the mean helix.

excavation from ditches;

The excavation taken from a ditch, of which the profile is constant, may be estimated in the same way.

rule holds for any portion of an entire revolution.

In examples 1st and 2d, the centre of gravity is supposed to have described an entire circumference; but had it moved through only an eighth, tenth, or any other fractional portion of a circumference, the volume generated would still, as in example 3d, have been given by the area of the generatrix into the extent of the path described.

VII.

MOTION OF TRANSLATION OF A BODY OR SYSTEM OF BODIES.

Motion of translation;

§ 143.—A body, or system of bodies, is said to have a simple *motion of translation*, when all its elements describe, simultaneously, equal and parallel paths.

Denote by v the velocity which any motive force communicates to all parts of the system during any small interval of time t. The force of inertia f, of an element whose weight is p, will be given by the equation

$$f = \frac{p}{g} \cdot \frac{v}{t};$$

the measure of the inertia of an element;

and the force of inertia f', of an element whose weight is p', by

$$f' = \frac{p'}{g} \cdot \frac{v}{t};$$

and so of all the others, provided the degree of velocity impressed upon all the elements is the same during the time t. Moreover, as each force of inertia is exerted in the direction of the path along which the elements respectively move, and as these are supposed parallel, the forces of inertia are parallel, and give a resultant equal in intensity to their algebraic sum. Denoting the intensity of this resultant by F, we have

$$F = f + f' + f'' + \&c. = \frac{v}{t}\left(\frac{p + p' + p'' + p''' + \&c.}{g}\right);$$

and replacing the sum of the partial weights by the entire weight P, and $\frac{P}{g}$ by the entire mass M of the system, we shall finally have

$$F = M \cdot \frac{v}{t} \quad . \quad . \quad . \quad . \quad . \quad . \quad (39).$$

of that of the entire mass;

It remains to find the invariable point of application of F. This point is called the *centre of inertia*. The intensities of the forces f, f', f'', &c., are proportional to the weights p, p', p'', &c., to which they are respectively applied, and thus the point of application of F, will coincide with that of the resultant of the forces p, p', p'', &c.;

centre of inertia;

measure of inertia in words.

that is to say, with that of the entire weight P, which is the centre of gravity of the system. Hence, *the total force of inertia of a body, or system of bodies, having a simple motion of translation, is measured by the mass of the system, multiplied into the ratio which the small degree of velocity communicated bears to the time during which the velocity is impressed. And the total force of inertia has its point of application at the centre of gravity.*

The force of gravity being constant, the centre of gravity and of inertia coincide;

This coincidence of the centre of inertia with the centre of gravity, results from the assumption that the force of gravity is the same in its action upon the different parts of the system. Had it been otherwise, that is to say, had the force of gravity varied in intensity from one element to another, the centre of inertia, being always at the centre of mass, would be different from the centre of gravity.

these centres sensibly the same in bodies on the earth.

The intensity of the force of gravity being regarded as the same within the limits of a body on the earth's surface, the centre of inertia and of gravity may be regarded as coinciding, and hence these terms will be used indiscriminately.

Quantity of motion of a body;

§ 144.—Let V represent the velocity of a body having a motion of translation, supposed uniform at any instant; the quantity of motion of any one of its elements whose weight is p, is measured by

$$\frac{p}{g} V,$$

and of an element whose weight is p',

$$\frac{p'}{g} V,$$

and so for the other elements; and as these motions are parallel, their sum will give the quantity of motion of the entire body. Designating this quantity by Q, we shall have

$$Q = \frac{p + p' + p'' + \&c.}{g} V = MV \quad . \; . \; (40).$$ its measure.

Thus the total quantity of motion, in any body having a motion of translation, is measured by the mass of the body into its velocity.

§ 145.—When a certain degree of velocity v, is impressed upon *all* the elements of a body during a very short interval of time t, we have seen that the total force of inertia is given by, Eq. (39),

Motion of a body when the direction of the motive force passes through the centre of gravity;

$$F = M \times \frac{v}{t};$$

We have seen, also, that this force of inertia is exerted in the direction of the body's motion, and through the centre of gravity. If, therefore, we suppose that at the instant in which the body has acquired the velocity v, a force equal to F is applied in a direction contrary to the motion, and at the centre of gravity, it will destroy the motion. This being supposed, if we apply at the centre of gravity of the body, a motive force X, it will communicate to it a simple motion of translation. For this force X will be equal and directly opposed to the force of inertia F, which it develops. This latter force F will be resolved into as many partial forces of inertia f, f', f'', &c., as there are elementary portions of the body, and the intensities of these partial forces will be proportional to the respective weights of these elements. Denoting the masses of the elements by m, m', m'', &c., we shall have,

Fig. 78.

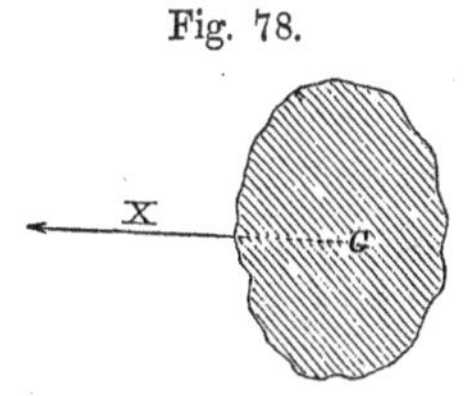

$$f = \frac{m}{M} F, \qquad f' = \frac{m'}{M} F, \qquad f'' = \frac{m''}{M} F, \text{ \&c.}$$

The degree of velocity which each of these forces impresses upon the part on which it acts, will, § 82, be measured by

$$\frac{f\,.\,t}{m}, \qquad \frac{f'\,.\,t}{m'}, \qquad \frac{f''\,.\,t}{m''}, \text{ \&c.};$$

or, replacing f, f', f'', &c., by their values as given above, simply by the expression

$$\frac{F\,.\,t}{M};$$

and as this measure is the same as that before deduced, Eq. (39), for the degree of velocity impressed on the centre of gravity by the force F, or its equal X, we see that, *to impress a simple motion of translation upon any body, it is necessary that the line of direction of the motive force, or the resultant of the motive forces, when there are several, must pass through the centre of gravity;* and, reciprocally, *if the line of direction of the force, or that of the resultant, in the case of several forces, pass through the centre of gravity, the body will have a simple motion of translation.*

will be that of simple translation.

§ 146.—If the force X, were applied along the right line $A\ B$, not passing through the centre of gravity G, it is easy to see that the motion cannot be one of simple translation. For, if this latter motion obtained, the partial forces of inertia would have a resultant of which the line of direction would, from what we have seen, pass through the centre of gravity G; and if this resultant were replaced by an equal force F, applied along the same line and directly opposed to the motion, the latter would be destroyed, and an equilibrium would result. But it is impossible that two forces X and F, applied to the extremities of a physical line or bar $A\ G$, can produce an

Motion when the force does not pass through the centre of gravity;

Fig. 79.

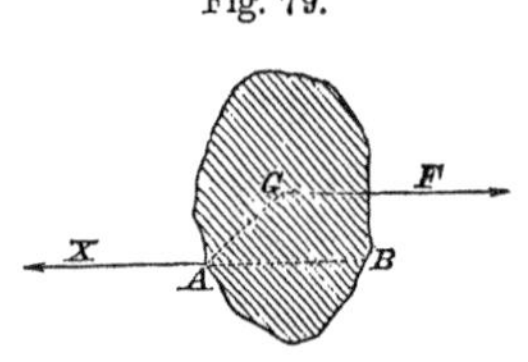

equilibrium, unless they act in the direction of the bar. Hence, when a body receives the action of a force, of which the direction does not pass through the centre of gravity, its motion will not be that of simple translation, but will be compounded of a motion of translation and of rotation; that is to say, some one of its elements will move, for the instant, in a right line, while the others will rotate about it as a centre.

will be that of translation and of rotation at the same time;

To find this element C, conceive a plane to be drawn through it, parallel to the direction of its motion, and perpendicular to the planes in which the other elements, for the instant, rotate, and let $A\,B$ be its trace upon that one of these planes which contains the point C, and its rectilineal path. Let m_1 be the projection of some one element m' upon this latter plane, and take $C\,C_1$ to represent the velocity v of translation, and $m_2 m_3$ the velocity of rotation acquired by the element m', in the small time t. Make $m_1 m_2$ equal and parallel to $C\,C_1$; then would $m_1\,m_2$ represent the velocity acquired by m', had the body moved with a simple motion of translation; but by virtue of the motion of rotation, the actual velocity acquired by m', in the direction of C's motion, is $m_1\,m_2$, diminished or increased by the projection of $m_2\,m_3$ upon the line $C\,C_1$ according to the direction of the rotation.

Fig. 80.

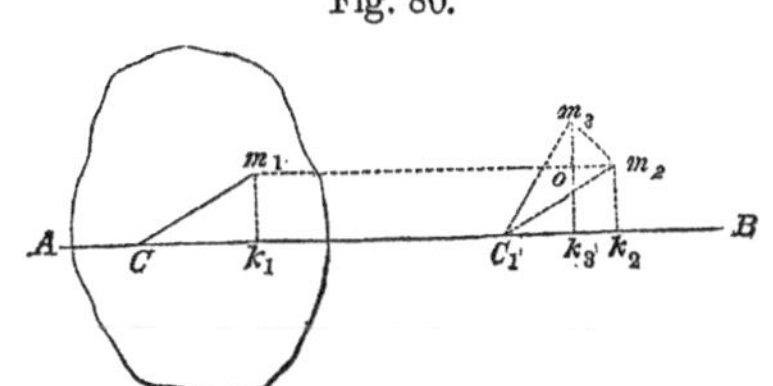

position of the element having a motion of translation.

Project the points m_1, m_2, and m_3, upon $A\,B$, by the perpendiculars $m_1\,k_1$, $m_2\,k_2$, $m_3\,k_3$; then will the actual velocity v', acquired by m', be $m_1\,m_2 - m_2\,o$, or

$$v' = v - m_2\,o;$$

but

$$m_2\,o = m_3\,m_2 \times \cos m_3\,m_2\,o = m_3\,m_2 \times \cos C_1\,m_2\,k_2;$$

denoting $C m_1 = C_1 m_2 = C_1 m_3$ by r, $m_1 k_1 = m_2 k_2$ by $y_{,}$, and the velocity of rotation acquired by a point at the unit's distance from C by V_1, then will

$$m_2 m_3 = V_1 r,$$

and

$$\cos C_1 m_2 k_2 = \frac{y_{,}}{r},$$

Relative velocity of two elements of a solid body in motion;

which substituted above, give

$$v' = v - V_1 y_{,} \quad . \quad . \quad . \quad . \quad (41).$$

Moreover, $m_3 o$ is the velocity of the element m' perpendicular to the direction of C's motion; and calling this velocity v'', and the distance $C_1 k_2$, $x_{,}$, we shall have

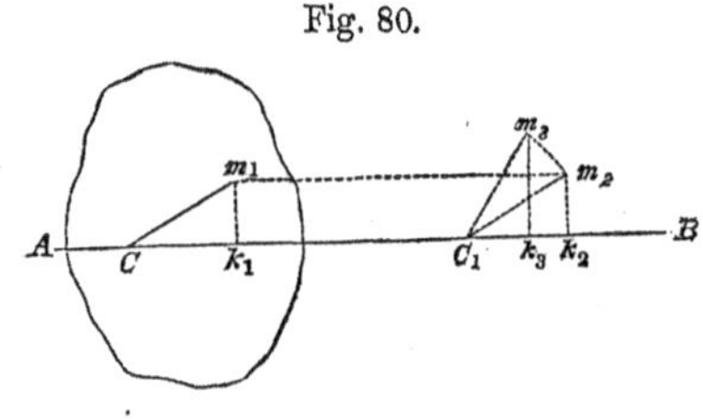

Fig. 80.

the same in a direction perpendicular to the former;

$$v'' = V_1 x_{,} \quad . \quad . \quad . \quad . \quad . \quad (42).$$

Denoting, as before, the weight of the element m' by p', and its force of inertia in the direction $C C_1$ by f', we have

$$f' = \frac{p'}{g} \cdot \frac{v'}{t} = \frac{p'}{gt}(v - V_1 y_{,}),$$

and similar expressions for the inertia of the other elements. Taking the sum of these, and representing the inertia of the entire mass by F, we have, from the principle of parallel forces,

$$F = \frac{1}{t} \cdot v \cdot \left(\frac{p'}{g} + \frac{p''}{g} + \frac{p'''}{g} + \&c.\right)$$

$$- \frac{1}{t} \cdot V_1 \cdot \left(\frac{p'}{g} \cdot y_{,} + \frac{p''}{g} \cdot y_{,,} + \frac{p'''}{g} y_{,,,} + \&c.\right);$$

or, denoting the entire mass of the body by M, and the masses of the several elements by m', m'', m''', &c., this reduces to

$$F = M \cdot \frac{v}{t} - \frac{V_1}{t} \cdot (m' y_{,} + m'' y_{,,} + m''' y_{,,,} + \&c.).$$

value for the force of inertia of a body;

Now the first term of the second member, which alone involves the motion of the point C, is wholly independent of the figure of the body and of the distribution of its elements.

It will, therefore, remain the same whatever changes take place in its figure and size, provided its quantity of matter remain the same. The place of C, as determined from any supposition consistent with this last condition, will, therefore, be its position generally.

This being understood, conceive the whole body to contract gradually in all directions till it is concentrated in a single point; this point must, from necessity, be the centre of gravity which alone remains undisturbed during contraction, as it will during an expansion, being the centre of mass. The point C, and the centre of gravity, not being disturbed by this change of volume, must coincide, and hence must always remain one and the same point.

But when the plane in reference to which the products $m' y_{,,}$ $m'' y_{,,}$, &c., are taken, passes through the centre of gravity, we have

$$m' y_{,} + m'' y_{,,} + m''' y_{,,,} + \&c. = 0;$$

and the above equation reduces to

$$F = M \cdot \frac{v}{t};$$

always equal to the mass into ratio of the increment of velocity to that of the time.

which is identical with Eq. (39).

The body will have a motion of translation; it will also rotate about the centre of gravity;

We conclude, therefore, 1st, *that when a body is acted upon by one or more forces, its centre of gravity will move as though the forces were applied directly to it, provided their directions remain unchanged;* 2d, *that when the line of direction of the force, or that of the resultant of several forces, does not pass through the centre of gravity, the body will, in addition, rotate about this centre.*

The law which regulates the motion of the centre of gravity results from the above equation, for if X represent the resultant of all the forces, and F the total force of inertia, we have from the equality of action and reaction, $X = F$, which value of F, substituted above gives, after reduction,

value of the velocity of translation.

$$v = \frac{X \cdot t}{M} \quad . \quad . \quad . \quad . \quad . \quad (42)';$$

in which v is the velocity impressed in the very short interval t, from which we may pass to the velocity acquired at the expiration of any time, and thence to the space described.

§ 147.—What has been before explained, applies also to the total *living force* possessed by a body having a simple motion of translation. For v being the common velocity of any one element, $\frac{p}{g} \times v^2$, will be the living force of that whose weight is p; $\frac{p'}{g} \times v^2$ the living force of that whose weight is p', &c.; so that the sum of all these living forces, or the total living force, denoted by L, will be $v^2 \times \frac{p + p' + p'' + \&c.}{g}$; and representing the entire mass of the system by M, as before,

Living force in a simple motion of translation.

$$L = Mv^2.$$

If the body have a motion of rotation as well as of

translation, then will the living force of m', in the direction of the motion of translation be, Eq. (41), If the body have also a motion of rotation;

$$m' v'^2 = m' (v - V_1 y_{,})^2 = m' v^2 - 2 v . V_1 m' y_{,} + V_1^2 m' y_{,}^2;$$

and in the direction perpendicular to the motion of translation, Eq. (42),

$$m' v''^2 = m' V_1^2 x_{,}^2;$$

and similar expressions for the elements whose masses are m'', m''', &c. Taking the sum of these, denoting the living force, as before, by L, and reducing by the equations

$$m' y_{,} + m'' y_{,,} + \&c. = 0,$$

$$y_{,}^2 + x_{,}^2 = r_{,}^2,$$

$$y_{,,}^2 + x_{,,}^2 = r_{,,}^2,$$

$$\&c. \quad \&c. = \&c.,$$

$$m' + m'' + m''' + \&c. = M;$$

we find

$$L = M v^2 + V_1^2 (m' r_{,}^2 + m'' r_{,,}^2 + \&c.);$$

or, making

$$m' r_{,}^2 + m'' r_{,,}^2 + \&c. = \Sigma . m r^2,$$

$$L = M v^2 + V_1^2 . \Sigma m r^2 \quad . \quad . \quad . \quad (43).$$

the living force is equal to that due to translation, increased by that due to rotation.

§ 148.—The considerations which have now been developed, show that in the motion of translation of a body or system of bodies, the computations may be greatly simplified, since we are permitted to disregard the shape of bodies, to suppose them concentrated about their centres of gravity, and to reason upon these points as upon the total masses.

General theorem—the quantity of work of weights;

§ 149.—We have seen that in all questions affecting the circumstances of simple motion of translation, we may regard the mass as concentrated about its centre of gravity. But when the different parts of a body receive motions which differ from each other, this concentration is generally inadmissible, since the partial forces of inertia not being parallel, their resultant will no longer be equal to their sum. If, however, we desire, in any case of the coexistence of various motions, to estimate the work performed by the weights of the parts of a body, during a given time, the action exerted by these latter forces being parallel, and their resultant or the total weight always passing through the centre of gravity, we may still reason upon the motion of this point as though the mass were concentrated at it, and disregard the motion of rotation of the other parts of the body about it. In this case, the quantity of work expended in every instance, will be obtained by taking the product of the weight into the path described by the centre of gravity, estimated in a vertical direction. If, for example, the centre of gravity of any body, as a bomb-shell, pass from the position G to G', describing the curve $G\,G'$, we obtain the work done by the weight during the interval of time occupied in passing from one of these positions to the other, by multiplying the weight of the shell into $G'\,R$, the projection of the path $G\,G'$ on the vertical through G'.

equal to the weight, into the projection of path of the centre of gravity.

Fig. 81.

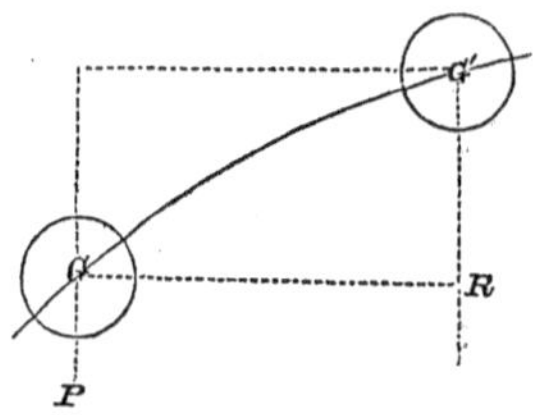

This theorem, in regard to the work performed by the weight, is by no means restricted to the motion of a single body, but extends to a collection of pieces, such as wheels, bars, levers, &c., connected with each other after the manner of ordinary machinery. If the quantity of work performed by each piece be computed, and the algebraic sum be taken, it will be found to be equal to the quantity of work performed by the weight of the whole system, acting at its centre of gravity, computed by the same rule. Applies to all kinds of machinery;

In general, let p, p', p'', &c., be the weights of the several pieces connected together; h, h', h'', &c., the vertical distances passed over by their respective centres of gravity, in passing from one position to another, by virtue of their connection; P, the sum of all the weights or the weight of the entire system; and z, the vertical space described by the common centre of gravity: then will

$$P z = p\,h + p'\,h' + p''\,h'' + \text{\&c.} \quad . \quad . \quad . \quad (44).$$

mathematical expression of the rule;

To demonstrate this, let m, m', m'', &c., be several bodies so connected as to be acted upon by each other's weights. Let P denote the weight of the entire system; p, p', p'', &c., the weights of the several bodies m, m', m'', &c.; Z, the distance of the common centre of gravity from a horizontal plane $A\,B$; and H, H', H'', &c., the distances of the centres of gravity of the bodies m, m', m'', &c., from the same plane. Then, from the principle of the centre of gravity, will

Fig. 82.

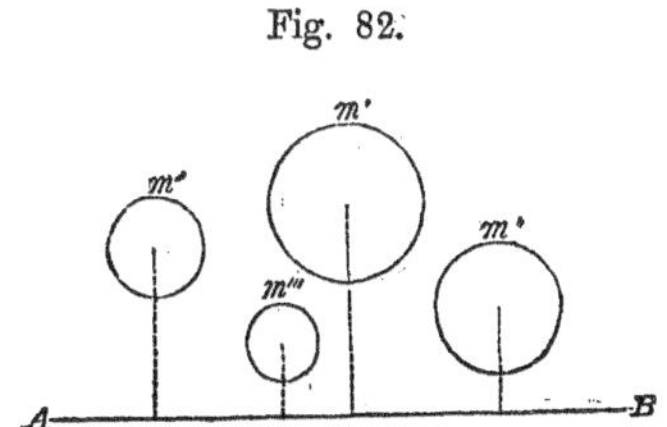

$$P\,Z = p\,H + p'\,H' + p''\,H'' + \text{\&c.};$$

demonstration of the rule;

and for a second position of the system,

$$P\,Z_{,} = p\,H_{,} + p'\,H_{,}' + p''\,H_{,}'' + \text{\&c.};$$

and subtracting the first from the second,

$$P(Z_{,}-Z)=p(H_{,}-H)+p'(H_{,}'-H')+p''(H_{,}''-H'')+\&c.$$

And supposing the horizontal plane of reference to be below both positions of the entire system, $Z_{,}-Z$ is the vertical distance z, through which the common centre of gravity has ascended or descended, according as $Z_{,}$ is greater or less than Z; $H_{,}-H$, $H_{,}'-H'$, $H_{,}''-H''$ &c., are the corresponding distances h, h', h'', through which the centres of gravity of the bodies m, m', m'', &c., have ascended or descended. Moreover, the products $P(Z_{,}-Z)$, $p\,(H_{,}-H)$, $p'\,(H_{,}'-H')$, &c., are the quantities of work due to the entire weight and to the partial weights. Whence this rule, viz.: *The total quantity of work due to the action of the entire weight of any system, is equal to the sum of the quantities of work of the weights which ascend, diminished by the sum of the quantities of work of the weights which descend.*

conclusion and rule.

VIII.

EQUILIBRIUM OF A SYSTEM OF HEAVY BODIES.

Equilibrium of heavy bodies;

§ 150.—If the system of heavy bodies be so connected, and in such condition that the common centre of gravity continue on the same horizontal line, while the bodies are made to take different positions, then will $Z_{,}-Z=z=0$, and Eq. (44) becomes,

$$ph+p'h'+p''h''+\&c.=0 \quad . \quad . \quad (45);$$

partial quantities of work destroy each other;

hence, the partial quantities of work of the several bodies destroy each other, and, therefore, there must be an equilibrium in the system, and the least extraneous effort will impart motion. Such is the condition of equilibrium of a system of bodies acted upon only by their own weights. This equilibrium presents itself under different states according to the positions of the system. If the position be such that in a slight derangement the common centre of gravity descend, it will tend to descend more and more, and a certain quantity of work will be requisite to restore it to its primitive position. Such an equilibrium is said to be *unstable*, because the system tends of itself, on slight derangement, to depart from it. On the contrary, if on slightly displacing the system, the common centre of gravity ascend, this displacement will require the expenditure of a certain quantity of work which the weight of the system tends to restore; the equilibrium is then said to be *stable*, because the system is urged by its own weight to return to its primitive state when abandoned or left to itself. Finally, if during a slight derangement, the centre of gravity neither ascend nor descend, the quantity of work expended by the system is always nothing, the system will have no tendency of itself to return to, or depart from its first position, and the equilibrium is said to be *indifferent*.

The slightest effort sufficient to give motion.

Unstable equilibrium;

stable equilibrium;

equilibrium of indifference.

§ 151.—Take a rod suspended at one end so as to turn freely about a horizontal axis A, and supporting at the other a body which is symmetrical in reference to a line drawn from the axis A to the common centre of gravity G. It is obvious that there will be an equilibrium when the rod is vertical. It is moreover *stable*; for in deflecting the system, the centre of gravity will ascend while

Fig. 83.

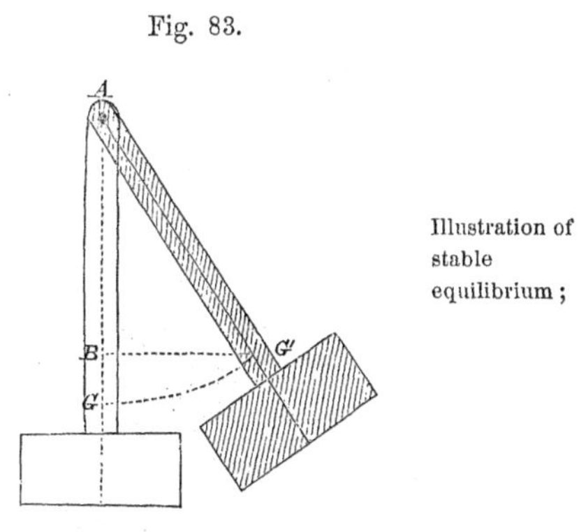

Illustration of stable equilibrium;

describing the arc $G G'$, about A as a centre, and a certain amount of work will be expended which the weight will restore as soon as the deflecting cause is removed. Indeed, the system will, when abandoned, perform a series of oscillations, whose amplitude about the vertical A G, will diminish continually till it comes to rest.

Illustration of unstable equilibrium;

Now suppose the system inverted; if the rod be perfectly vertical, the line of direction of the weight will pass through the point of support A, and there is no reason why the system should move one way rather than another. It will therefore be in equilibrio, but the equilibrium will be unstable; for, however slight the derangement, the centre of gravity G will descend along the circular path $G G'$, described about A as a centre, and a certain amount of work will be requisite to bring it back to its primitive position.

Fig. 84.

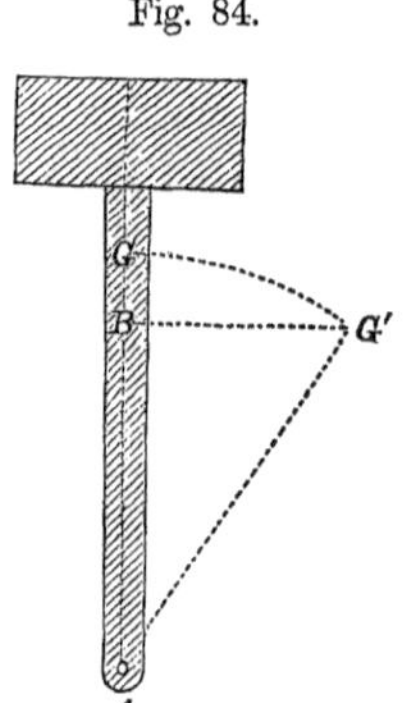

both kinds illustrated by means of the cone;

When a cone $A B C$, resting upon its base $B C$, is inclined to the position $A' B' C$, its centre of gravity G will ascend and describe an arc $G G'$, and if, in this inclined position, it be abandoned by the disturbing force, it will return. When the cone is placed upon its vertex, with its centre of gravity directly above that point, it will also be in equilibrio as it was when resting on its base, but the slightest motion will cause the centre of gravity to descend. The first position is one of *stable*, the second of *unstable* equilibrium.

Fig. 85.

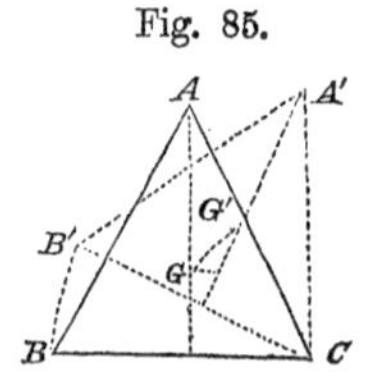

Fig. 86.

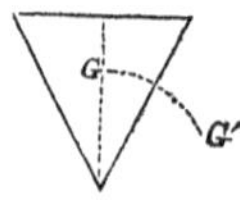

An elliptical cylinder placed upon a horizontal plane

is in stable or unstable equilibrium, according as the smaller or longer axis of its elliptical base, is perpendicular to the plane.

also by an ellipsoid of revolution;

Fig. 87.

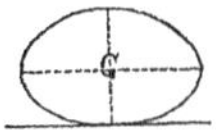

Fig. 88.

A spherical ball upon a horizontal plane, is an example of *equilibrium of indifference.* The centre of gravity remaining at the same level however the ball may be displaced, provided it preserve its contact with the plane, the quantity of work necessary to displace it will always be inappreciable, and the ball will, in consequence, have no tendency either to recede from or return to its primitive position. A perfectly circular cylinder on a horizontal plane is an example of the same kind.

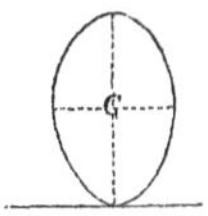

Fig. 89.

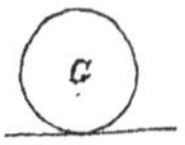

equilibrium of indifference exemplified by the sphere;

Some varieties of *draw-bridges* are but collections of pieces in a state of equilibrium of indifference. And to insure this state, it is only necessary that the common centre of gravity of the bridge and appendages, shall preserve the same level during the motion, in which case, the system will be in equilibrio in all possible positions.

by some varieties of draw-bridges;

Wagons and carriages should, in strictness, require no work to move them on a horizontal plane, except to overcome their inertia, and should, therefore, be so constructed as to preserve their centres of gravity always on the same level.

If, during the motion of a wheel, it is seen sometimes to quicken and sometimes to slacken its motion, it is because the centre of gravity G is out of the axis of motion A, and, therefore, alternately rises and falls during the rotation. A wheel whose centre of gravity is out of the axis of motion, passes

Fig. 90.

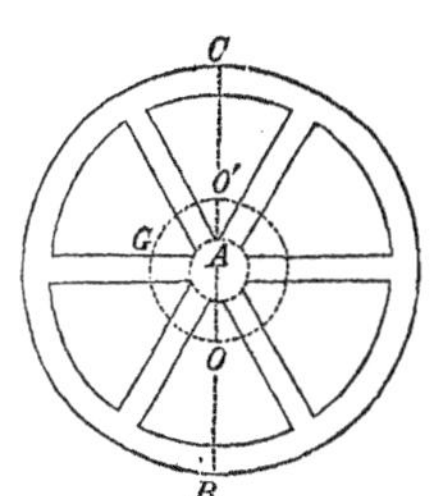

effect of throwing the centre of gravity out of the axis of a wheel.

in the course of a single revolution through the conditions of stable and unstable equilibrium, the former occurring when the centre of gravity G crosses the vertical line $B\ C$, through the axis A, at the lowest point O, and the latter when it crosses the same line at the highest point O', of its path.

The common balance; The common balance consists of a horizontal arm $A\ B$, mounted upon a knife-edge D, resting upon the surface of a circular opening made in the end of a vertical frame C, which is supported by a hook attached to a fixed point E. The ends of the balance carry basins of equal weights, one of which receives a substance to be weighed, and the other the standard weights previously determined. The balance may be *stable*, *unstable*, or *indifferent*, the position of its centre of gravity; according as it tends to return to a horizontal position when deflected from it, to overturn, or to retain any position in which it may be placed. Referring the entire system to any horizontal plane $A'\ B'$, and taking the sum of the products which result from multiplying the weight of each piece by the distance of its centre of gravity from this plane, and dividing this sum by the weight of the entire balance; the quotient will give the distance of the common centre of gravity of the moveable part of the apparatus from the plane $A'\ B'$. If this distance be less than $F\ D$, the distance of the knife-edge above the plane of reference, the balance will be stable; if greater, the balance will be unstable; and if equal to this distance, the

Fig. 91.

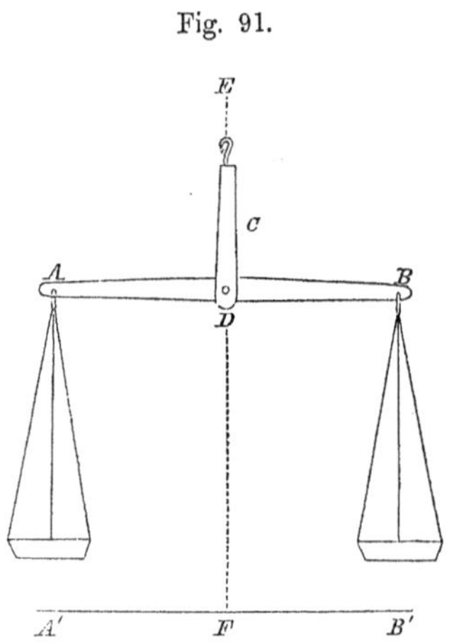

when stable; unstable; indifferent;

balance will be indifferent. All of which supposes the common centre of gravity to fall somewhere on the vertical line FD, passing through the knife-edge.

centre of gravity on the vertical through the point of support.

IX.

EQUILIBRIUM OF SEVERAL FORCES, VIRTUAL VELOCITIES, AND MOTION OF A SOLID BODY.

§ 152.—To find the conditions of equilibrium of several forces, P, Q, R, S, &c., applied to different points of a solid body, take in the interior of the body three points a, b, c, and regard these points as the vertices of an invariable triangle abc; resolve each force into three components whose directions shall pass through the given point of application and the vertices a, b, and c. In this way we shall be able to replace the given forces by three groups of components, the directions of each group having a common point at a, b, or c. Each of these groups, having a common point, may be replaced by a single resultant, and thus, the equilibrium of the given forces is reduced to that of three forces. Call the resultant of the group having the common point a, X; that of the group having the common point b, Y; and that of the group having the common point c, Z. These three forces being in equilibrio, the equilibrium will not be affected by supposing the three

Equilibrium of forces acting upon a free body;

Fig. 92.

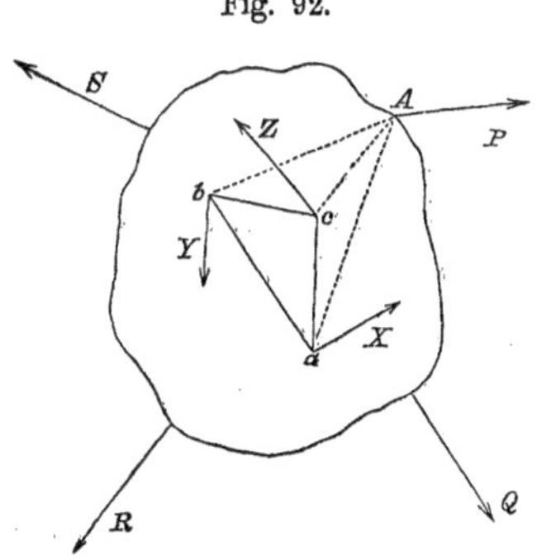

the given forces may be replaced by three groups of components;

and these by three single forces;

lines $a b$, $b c$, and $c a$, to become fixed in succession. The line $a b$ being fixed, the forces X and Y, whose directions intersect it, will be destroyed by its resistance, and if the third force Z, does not act in the plane $a b c$, it will cause the system to turn about $a b$; the same may be shown of the forces X and Y. The forces, X, Y, Z, must, therefore, act in the same plane; and in order that they may be in equilibrio, the resultant of either two of them must be equal and directly opposed to the third; that is to say, the resultant of the three must be zero. If the resultant be zero, the quantity of work is zero. The quantity of work of X, Y, or Z, is equal to the algebraic sum of the quantities of work of the group of which it is the resultant, and thus the sum of the quantities of work of X, Y, and Z, may be replaced by that of the quantities of work of the forces grouped about a, b, and c. But these last, taken three by three, give the quantities of work of the proposed forces P, Q, R, S, &c.; so that the sum of the quantities of work of the forces X, Y, and Z, is the same as the algebraic sum of the quantities of work of the forces P, Q, R, S, &c. Whence we conclude, *that several forces, acting upon the different points of a free body, will be in equilibrio, when the algebraic sum of the quantities of work of the forces is equal to zero.*

an equilibrium requires these three to act in the same plane;

the resultant zero;

the forces will be in equilibrio when the algebraic sum of the quantities of work is zero.

Fig. 92.

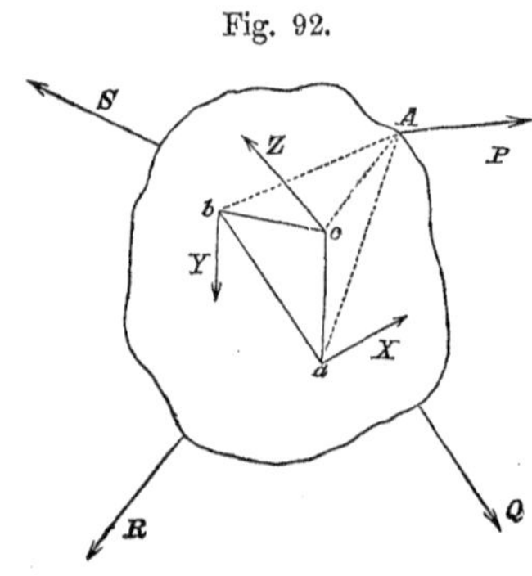

Now suppose the body to be slightly deranged from its state of rest, and let $A A'$ be the path described by the point of application A, of

Fig. 93.

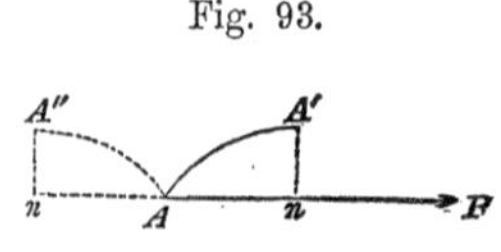

the force P, in an indefinitely short time t. Draw $A'n$ perpendicular to $A\ P$; $A\,n$ will be the space described by the point of application in the direction of the force, and the quantity of work performed by P during the derangement will be $P \times A\,n$, or Pp, denoting $A\,n$ by p.

The path $A\ A'$ is called the *virtual velocity* of the force P; $A\,n = p$, the *projection* of the virtual velocity; and the product Pp, the *virtual moment* of the force P. **Virtual velocity, projection of virtual velocity and virtual moment;**

Denoting by q, r, s, &c., the projections of the virtual velocities of the forces Q, R, S, &c., the quantities of work, or the virtual moments of these forces, will be, respectively, $Q\,q$, $R\,r$, $S\,s$, &c; and if the system be in equilibrio, we have, from the rule just demonstrated,

$$Pp + Q\,q + Rr + Ss + \&c. = 0 \quad . \quad . \quad (46).$$

This equation is but the mathematical expression of the *principle*, known under the name—*virtual velocities*, which consists in this, viz.: *when several forces are in equilibrio, the algebraic sum of their virtual moments is equal to zero.* **Principle of virtual velocities.**

§ 153.—Any mechanical device that receives the action of a force or power at one point, and transmits it to another, is called a *machine*. **A machine;**

Conceive a machine, composed of wheels whose axes are sustained by supports, and which communicate motion to each other, either by teeth, chains, or straps, on their circumferences. Suppose a force or power to be applied so as to turn the first wheel; this wheel will experience a resistance from the second; this resistance, in its turn, becomes, for the second **composed of wheels;**

Fig. 94.

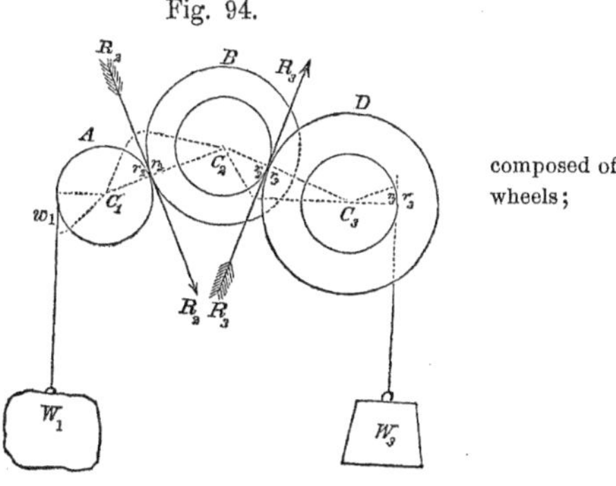

wheel, a power which causes it to rotate also; the second will experience a resistance from the third wheel, which resistance becomes a power to give it motion, and so on to the end. But each wheel experiences a reaction at the points of support which keep it in position, and it is this reaction that becomes the means of transmitting the power to the following wheel; for if these points were unsupported, the wheels would cease to act upon each other and the power first applied could not be transmitted.

the process by which the action of a moter is transmitted;

Fig. 94.

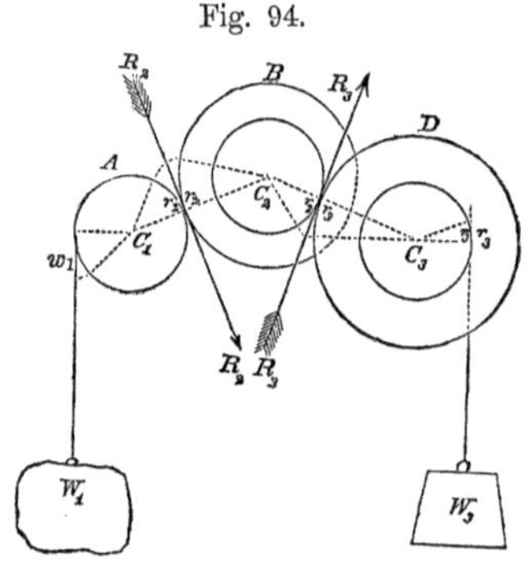

points of support replaced by active forces;

Now, replace the supports, by the efforts of reaction which they exert: each piece or wheel will become a free body subjected to the action of the preceding piece, the resistance of the following, and the force of reaction by which we have replaced its point of support; and if the piece be in equilibrio, the algebraic sum of the virtual moments of this action, resistance, and reaction, must be equal to zero.

Represent the power applied to give motion to the first wheel A by W_1, the resistance of the second wheel B by R_2, and the reaction at the point of support of the first wheel, by C_1; the projection of the virtual velocity of W_1 by w_1, that of R_2 by r_2, and that of C_1 by c_1; then will

sum of the virtual moments for first piece;

$$W_1 w_1 + C_1 c_1 + R_2 r_2 = 0;$$

denoting the resistance of the third wheel D by R_3, the reaction at the centre of the second wheel by C_2; and the projections of their virtual velocities by r_3 and c_2, respectively,

$$R_2 r_2 + C_2 c_2 + R_3 r_3 = 0;$$ same for second;

and thus we may continue throughout the entire combination till we finally arrive at the last wheel, to which is opposed, as a final resistance, the work to be done. Denoting this resistance by W_e, the resistance of the last wheel to the action of the preceding by R_e, the reaction of the support of the last wheel by C_e; and the projections of the corresponding virtual velocities by w_e, r_e, and c_e, respectively, we shall finally have,

$$R_e r_e + C_e c_e + W_e w_e = 0.$$ also for the last;

But from the nature of the connection, the points of support must not move; their virtual velocities, and therefore the projections, must be zero. Hence, $C_1 c_1 = 0$, $C_2 c_2 = 0$, . . . $C_e c_e = 0$, and the preceding equations become

$$\left.\begin{aligned} W_1 w_1 + R_2 r_2 &= 0, \\ R_2 r_2 + R_3 r_3 &= 0, \\ \ldots\ldots &= 0, \\ R_e r_e + W_e w_e &= 0, \end{aligned}\right\} \quad . \quad . \quad (47).$$ virtual moments of points of support, zero;

Subtracting the second from the first, and adding the third, subtracting from this result the fourth and adding the fifth, and so on to the last, we finally obtain

$$W_1 w_1 + W_e w_e = 0 \quad . \quad . \quad . \quad (48);$$ relation of motive force to the final resistance;

22

which shows us that the quantity of work of the final resistance is equal to the quantity of work of the power, or that no work is lost. In other words, the *quantity of work* of the forces which tend to turn the system in one direction is exactly equal to *the quantity of work* of those which tend to turn it in the opposite direction. no work lost.

An examination of Eqs. (47), will show that the same remark is applicable to each piece of the combination taken separately, and thus starting from the piece which first receives the action of the force, and proceeding to that which does the work, and which, on this account, is called the *tool*, we see that the quantity of work of the power is equal to that of the *tool*. In a word, where forces work upon bodies through the medium of machinery, we must distinguish the powers from resistances, and we shall always find the work of the first to be equal to that of the second.

Tool; work of power equal to that of tool;

If the bodies press against each other in a way to produce a change of figure and friction, new resistances arise which must be taken into account, and the work of these must be subtracted from that of the forces to obtain the work of the tool, and hence such resistances are, in general, a hinderance to the final work to be accomplished.

the work of friction and that employed to change figure;

If the equilibrium is to be maintained while the machine is at rest, then must the quantity of work be estimated by the aid of a *supposed* displacement, as in that case, the influence of inertia will be avoided.

work estimated by a supposed displacement, to avoid inertia;

If the equilibrium is to exist during a uniform motion of the machine, the quantity of work must be computed from the actual motion of the points of application, for then the inertia will again be excluded.

the same in uniform motion;

But if the equilibrium is to take place during an acceleration or retardation of the motion, the inertia of the pieces will no longer be zero, and must be comprehended among the powers and resistances. The conditions of the motion must, however, always be the same; that is to say, the work of the powers must be equal to that of the resistances, augmented by the work of inertia when the motion is accelerated, and diminished by the same work when the motion is retarded.

when the motion is variable, the work of inertia comes into the account.

§ 154.—Whenever the forces applied to a body accelerate or retard its motion, the inertia of the body is developed; and by virtue of the principle that action is equal and contrary to reaction, this inertia must be in equilibrio with the forces; that is, the quantity of work of inertia will be equal to the sum of the simultaneous quantities of work of the forces which urge the body in one direction, diminished by the quantity of work of those which urge it in the opposite direction. But we have seen, § 85, that when the body takes, at different instants of time, two velocities which differ from each other, the work of inertia is measured by half the difference of the living forces possessed by the body at these instants, or by half the living force gained or lost in the interval, according as the motion has been accelerated or retarded. Hence, *the total work of several forces acting upon a body, during any time, is always equal to half of the living force gained or lost by the body during the same time.*

Relation of the work of inertia to the work of all other forces;

the total work of several forces, equal to half the living force lost or gained;

Suppose, for example, a projectile whose weight is P, to leave the point A with an initial velocity V. If its weight did not act, the body would pursue its primitive rectilineal path $A\,T$. But by virtue of the weight, which would act alone in vacuo, the projectile is continually deflected from this path, and will, in consequence, describe a curve line $A\,B\,D$; and we know, § 112, that when a body describes any curve under the action of its weight alone, the work is equal to the weight of the body into the difference of level of its two positions. Thus, in the case before us, while the projectile

illustration—the case of a projectile;

Fig. 95.

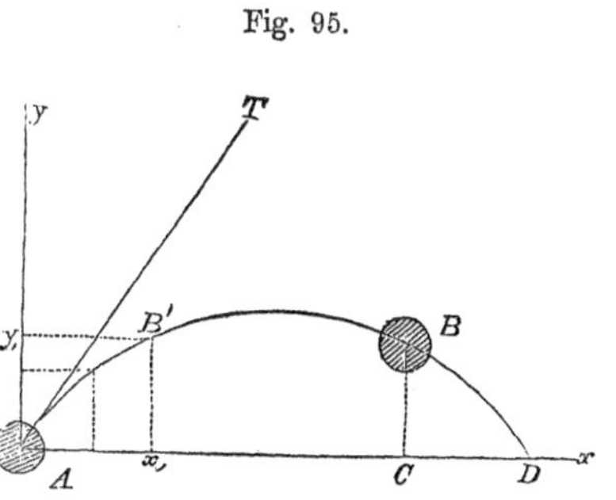

is passing from A to B, the work expended by its weight will be $P \times BC$, or $P \times H$, by making $BC = H$.

the living force at two points;

Denoting by V', the velocity of the projectile at B, its living force at this point will be

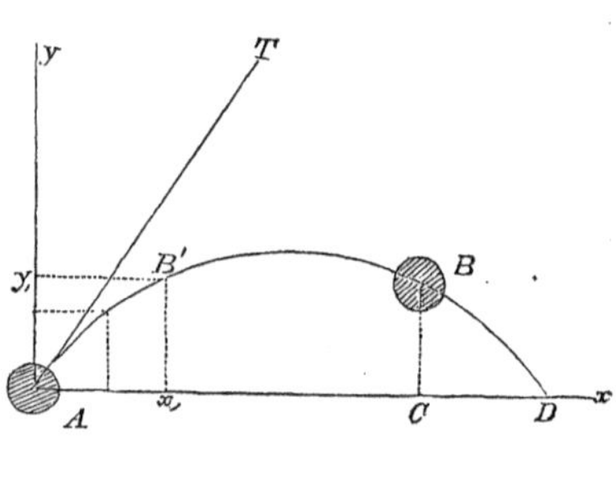

Fig. 95.

$$\frac{P}{g} \times V'^2,$$

and at A, it was

$$\frac{P}{g} V^2,$$

and its loss of living force, in passing from A to B,

loss or gain of living force;

$$\frac{P}{g}(V^2 - V'^2),$$

the half of which is the quantity of work of the extraneous forces, (in this case the body's weight,) in the same time, and hence

equal to double the work of the force;

$$\tfrac{1}{2} \frac{P}{g}(V^2 - V'^2) = PH,$$

or

$$V^2 - V'^2 = 2gH \quad . \quad . \quad . \quad (49).$$

relation of velocity to difference of level;

Thus the difference of the squares of the velocities in any two positions of the projectile, moving in vacuo, is equal to the difference of level of the two positions, multiplied by twice the force of gravity. When the projectile arrives at D, then will

$$H = 0; \quad \text{and} \quad V^2 - V'^2 = 0;$$

that is to say, the velocity will be equal to what it was at A. velocity same on the same level;

From Eq. (49), it is obvious that while the projectile is on the ascending branch of the curve, its velocity diminishes, and while on the descending branch its velocity, on the contrary, increases. velocity on ascending and descending branch of curve;

The description of the trajectory or curve $A\,B\,D$ in vacuo, is obtained by very simple considerations, founded upon the independence of the motions of the same body, and of the action of forces which solicit it in the directions of these motions, (§ 105 and 108.) The body may be regarded as animated by two motions, one horizontal in the direction $A\,x$, the other vertical in the direction $A\,y$. The initial velocities in the directions of these motions are the components of the initial velocity V, computed by the principle of the parallelogram of velocities. After the body leaves the point A, it will be subjected to the action of no motive force in the horizontal direction; the horizontal component of its velocity will be constant, and the spaces described in this direction in equal times will be equal. Denote the angle $x\,A\,T$ by α; the space described in the horizontal direction $A\,x$ by x, and the time required for its description by t, then will the curve in vacuo found;

Fig. 96.

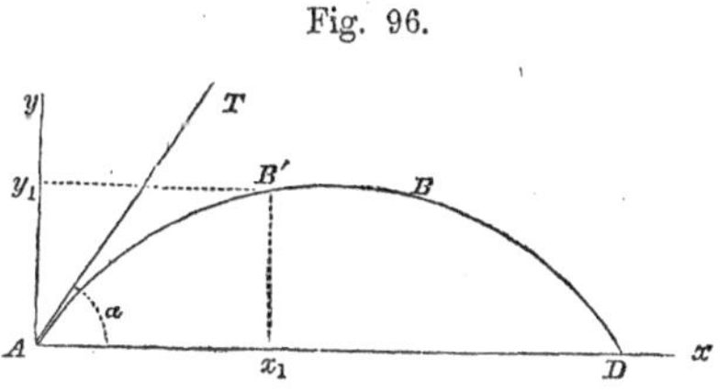

$$x = V \cos \alpha\, t \;.\;.\;.\;.\;.\; (50).$$

space described horizontally in the time t;

But in the vertical direction, the weight will, during equal times, diminish the component of the initial velocity, in that direction, by equal degrees; the motion will be uniformly varied, and the spaces described in the direction of the vertical $A\,y$, in the time t, will be given by Eq. (12), after substituting $V \sin \alpha$ for a, t for T, and

space in same time in vertical direction;

y for V_1. Hence, denoting the vertical space by y, we get

$$y = V \sin \alpha \, t - \tfrac{1}{2} g t^2 \quad . \quad . \quad . \quad (51).$$

the true position of the projectile at any instant;

The true positions of the projectile, which are but points of the curve $A\,B\,D$, are given by the intersections of a vertical and horizontal line drawn at distances from A, equal to the spaces $y = A\,y_1$, and $x = A\,x_1$, simultaneously described in these two directions. To find these distances, it will be sufficient to substitute a given value of t, in equations (50) and (51).

Eliminating t from these same equations, and reducing, we find

the equation of the curve described—a parabola;

$$y = \tan \alpha \, . \, x - \frac{g}{2\,V^2 \, . \cos^2 \alpha} x^2 \quad . \quad . \quad (52);$$

which is an equation of a *parabola*. Hence, the curve described by a body when thrown in a direction oblique to the horizon, and acted upon alone by its own weight, is a parabola.

The horizontal distance intercepted between the point of projection A, and the point D where the projectile attains the same level, is called the *range*. The angle $x\,A\,T = \alpha$, is called the angle of projection.

range; angle of projection;

To find the range, make $y = 0$, in Eq. (52), and find the corresponding value of x. Making $y = 0$, we have

$$0 = \tan \alpha \, x - \frac{g}{2\,V^2 \cos^2 \alpha} x^2,$$

whence

$$x = 0,$$

$$x = \frac{2\,V^2 \sin \alpha \, . \cos \alpha}{g} = A\,D = \text{range};$$

and representing by h, the height, due to the velocity V, we have

$$V^2 = 2\,g\,h \quad . \quad . \quad . \quad . \quad . \quad (53);$$

and denoting the range by R, and recollecting that

$$2 \sin \alpha \,.\, \cos \alpha = \sin 2\,\alpha,$$

we have, finally,

$$R = 2\,h \,.\, \sin 2\,\alpha \quad . \quad . \quad . \quad . \quad (54).$$

the value of the range;

This value for the range will be a maximum when $\alpha = 45°$, in other words, the greatest range corresponds to an angle of projection equal to 45°.

Fig. 97.

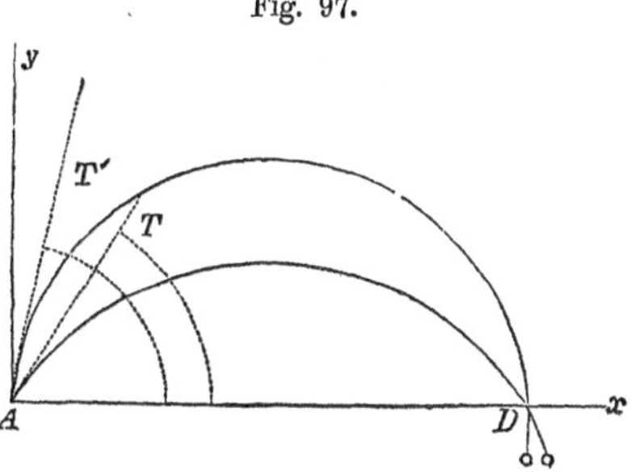

Since

$$\sin 2\,\alpha = \sin 2\,(90° - \alpha),$$

complementary angles give the same range;

it follows that the same range may be attained by two angles DAT and DAT', which are complements of each other.

If in Eq. (54), we make $\alpha = 45°$, then will

$$R = 2\,h,$$

greatest range given by an angle of projection equal to 45°;

whence

$$h = \tfrac{1}{2}\,R;$$

and this in Eq. (53), will give

$$V = \sqrt{R\,g} \quad . \quad . \quad . \quad . \quad . \quad (55).$$

value of initial velocity in terms of greatest range;

That is to say, if the range corresponding to an angle of 45° be measured, the initial velocity may be readily found, being equal to the square root of the product of this range into the force of gravity. Squaring the above equation, we obtain

$$V^2 = R\,g;$$

and denoting by W, the weight of the projectile, its living force on leaving the mouth of the piece from which it is thrown becomes

$$\frac{W}{g} \cdot V^2 = R\,W;$$

and the effective quantity of action impressed, denoted by Q,

effective quantity of action impressed upon the projectile;

$$Q = \tfrac{1}{2} R\,W \quad . \quad . \quad . \quad . \quad . \quad (56).$$

It is from this relation that are obtained the results of the *eprouvette*, a small mortar constructed to test the relative strength of different samples of gunpowder. For this purpose, a heavy solid ball is projected from it under an angle of 45°, with small but equal charges of different kinds of powder, and the relative strength is inferred from the effective quantity of action impressed.

eprouvette;

For example, suppose equal charges of two different samples of powder, give $R = 1050$ feet, and $R = 1086$ feet; these values substituted successively in Eq. (56) give

examples of its use;

$$Q = W \,.\, 525$$

$$Q = W \,.\, 543;$$

so that, the weights of the projectiles being the same, the strengths of the two samples of powder will be to each other as 525 to 543.

these results but approximations in air;

This supposes the motion to take place in vacuo. If the trajectory be described in the air, the resistance of this fluid will diminish the velocity of the projectile, the curve will cease to be a parabola, and the results above will be but approximations to the truth. But as the resistance to the motion of the same body in air varies as the square of the velocity, these approximations may be made as close as we please by using small charges and very dense projectiles.

Taking the general case, without limitation as regards the velocity of a body in air, the curve may still be described, provided we have a table giving, in pounds or any other unit of weight, the resistances corresponding to different velocities of different calibres.

Thus, knowing the initial velocity and its two components, find from this table, in pounds, the value of the initial resistance, and its horizontal and vertical components at the commencement of the motion. Of these components, one is the motive force in the horizontal, and the other, added to the weight of the projectile, the motive force in the vertical direction. With these forces, supposed constant during a very short time, compute by the laws of uniformly varied motion, the loss of velocity in these two directions during this short interval; subtract from the primitive components of the initial velocity, the loss in their respective directions; the remainders will be new component velocities, of which, find the resultant, and take from the tables the corresponding resistance. This new resistance treated in the same manner as that due to the initial velocity, will give a third resistance, and this a fourth, and so on indefinitely. We thus obtain a series of components, forces acting for a short time with constant intensity in the horizontal and vertical directions; with these compute, by the laws of uniformly varied motion, the corresponding spaces described in their respective directions by the projectile. The total spaces simultaneously described, obtained by adding together the spaces corresponding to the same number of consecutive intervals from the beginning of the motion, will give the distances, $A\,x_1$ and $A\,y_1$, which determine the points of the curve. The actual space described by the trajectory will be the development of this curve.

these approximations made close by giving small velocities;

general case in which the projectile is thrown into the air;

tables of atmospheric resistance;

successive steps by which to obtain the place of the projectile at any instant and the curve.

X.

MOTION AND EQUILIBRIUM OF A BODY ABOUT AN AXIS.

The work of forces which turn a body about a fixed axis

§ 155.—The principle demonstrated in § 113, of the work of forces acting upon a body, may be extended to any case whatever. Let us now apply it to that of a body which is free to turn about a fixed axis with which it is invariably connected.

Conceive a force R, acting upon the point A of a body free to turn about a fixed axis LM; resolve this force into two others, the one Q, parallel to LM, the other P in a plane perpendicular to this line, and passing through the point of application A. Doing the same with regard to all the other forces acting upon the body, the system will be reduced to two groups of forces, of which one will be parallel to the axis, and the other in planes at right angles to it. The algebraic sum of the quantities of work of the components is equal to that of the resultants. But the work of the first group, is equal to the product of their resultant, multiplied by the path described by the body in the direction of this resultant, that is to say, in the direction of the axis; but as the body is invariably connected with the axis, it cannot move in that direction, and the path described by the point of application of the resultant of the parallel group is nothing, and therefore the quantity of work is nothing. Thus, the total quantity of work of the given forces is

Fig. 98.

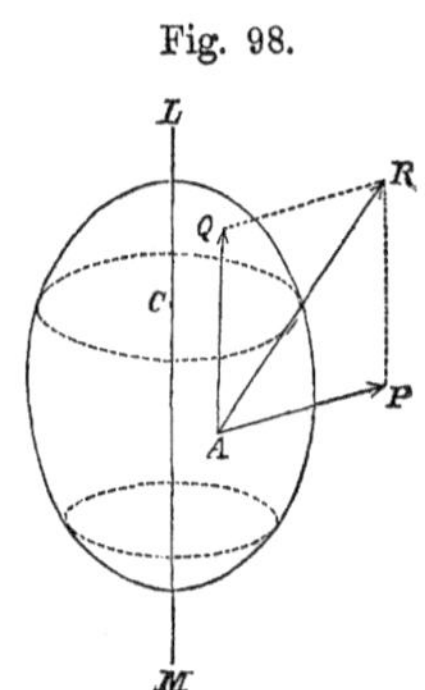

is reduced to that of their components in planes perpendicular to the axis.

reduced to that of their components, in planes perpendicular to the axis, and passing through the points of application.

§ 156.—The quantity of work of forces applied to a body which can only have a motion of rotation is always, as we have just seen, reduced to that of their components in planes perpendicular to the axis, or, which is the same thing, to that of the projections of the forces on these planes. It remains to determine this work.

The work of the components perpendicular to the axis;

Let P be one of these components, A its point of application upon the body, C the point of the axis in which it is cut by the perpendicular plane containing the component P. Let fall upon PA, the perpendicular CD, and recall what has been demonstrated in § 116, viz.: that the quantity of work of a force is always the same wherever its point of application be taken upon its line of direction. The quantity of work of P, estimated by the path described by the point D, is the same as that estimated by the path of A. But the point D describes, in the short interval of time t, an arc S, of which CD is the radius, and, hence, the quantity of work of P will be $P \,.\, S$.

Fig. 99.

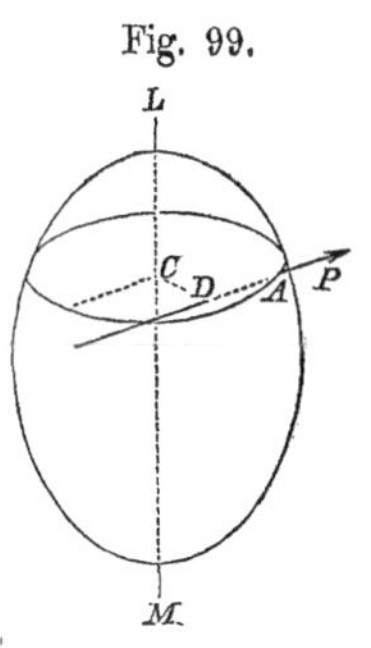

As all the points of the body are invariably connected with the axis and with each other, they will describe simultaneously equal angles, and consequently arcs proportional to their distances from the axis; hence if S_1 denote the length of arc described at the unit's distance, and r the distance of the point D from the axis, then will

$$S = r\,S_1,$$

and the quantity of work of P becomes

the quantity of work of a single component;

$$P\,r\,S_1,$$

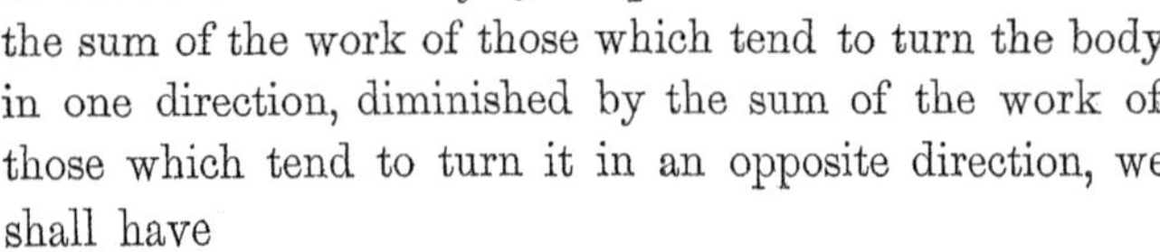

Fig. 99.

and for forces of which P', P'', &c., are the projections, at distances from the axis equal to r', r'', &c., respectively, we have the quantities of work measured by,

the same for other components;

$$P'\,r'\,S_1, \quad P''\,r''\,S_1, \text{ \&c. \&c.}$$

Knowing that the total quantity of effective work of the given forces, which we will denote by Q, is equal to the sum of the work of those which tend to turn the body in one direction, diminished by the sum of the work of those which tend to turn it in an opposite direction, we shall have

the effective work of all the components;

$$Q = S_1 (P\,r + P'\,r' + P''\,r'' + \text{\&c.}) \quad . \quad . \quad (57).$$

But we recognize $P\,r$, as the moment of the component P in reference to the axis, and the same of $P'\,r'$, $P''\,r''$, &c.; *whence, the effective work of the component, and consequently of the force itself, is equal to the product arising from multiplying the arc described at the unit's distance from the axis, into the moment, in reference to the same line, of the projection of the force on the perpendicular plane;* and

conclusion;

Eq. (57) shows that the *effective quantity of work of several forces, applied to turn a body about an axis, is equal to the arc described at the unit's distance multiplied by the algebraic sum of the moments of the projections of the forces on planes perpendicular to the axis.*

signs of the moments.

The sign of the moments of those forces which tend to turn the body in one direction, must be different from the sign of those which tend to turn it in an opposite direction; in other words, if the sign of the first be positive, that of the latter must be negative.

§ 157.—If the given forces be in equilibrio about the axis, their total work will be zero, whether the body be at rest or in motion; a condition that can only be fulfilled by making, in Eq. (57),

$$Pr + P'r' + P''r'' + \&c. = 0 \quad . \quad . \quad (58);$$

that is to say, several forces will be in equilibrio about a fixed axis, *when the algebraic sum of the moments, in reference to this axis, of the projections of the forces on perpendicular planes, is zero.*

Forces in equilibrio in reference to a fixed axis.

§ 158.—When forces are applied to a body to turn it about an axis, the motion of its particles can only take place in planes perpendicular to the axis; if the forces be not in equilibrio, the motion will be either accelerated or retarded, and will give rise to forces of inertia which act in the direction of the motion, and of which the quantity of work will be equal to that developed in the same time by the motive forces. When all the points of the body have simultaneously the same velocity, the total quantity of work of inertia is equal to the product arising from multiplying half the mass into the difference of the squares of the common velocity at the beginning and end of the interval for which the estimate is made. But when the different points have different velocities during the same time, which is always the case in a motion of rotation, it is necessary to estimate at the beginning and the end of the interval, the living force of each element of the body, to take the sum of those at the beginning, and the sum of those at the end; the difference of these sums will be the total increment or decrement of living force during the interval. The half of this living force being the work of inertia, and this latter being equal to that developed by the motive forces, or rather by their projections on planes perpendicular to the axis, it is easy to perceive that in the motion of rotation of a body, the work of the perpendicular components of the forces is half of the increment of the living force of the body. The process of estimating the living force

Extension of the principle of living forces to a motion of rotation;

in rotation, the work of the perpendicular components, is half the increment of living force.

of a body having a motion of rotation will now be given.

Estimate of the living force of a body turning about a fixed axis;

§ 159.—Consider an element m of a body, situated at a distance r from an axis of rotation LM. Denote by V the velocity which it has at any instant, and by p its weight, m its mass $= \frac{p}{g}$. Then will its living force be $\frac{p}{g} . V^2$ or $m V^2$.

Fig. 100.

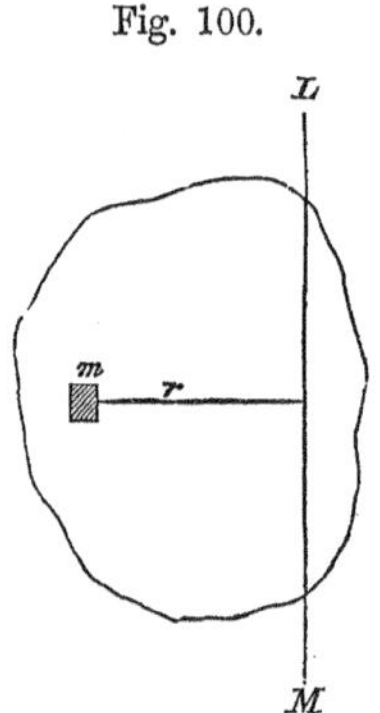

If S denote the space described by m during a very short interval of time t, and S_1 the space described in the same time by a point at the unit's distance from the axis, we shall have

$$S = r . S_1,$$

and dividing both members by t,

$$\frac{S}{t} = r \cdot \frac{S_1}{t} \quad . \quad . \quad . \quad . \quad (59);$$

but we have seen that, in any motion whatever, the velocity is equal to the space described, during a very short interval of time, divided by this interval, hence

the angular velocity;

$$\frac{S}{t} = V,$$

$$\frac{S_1}{t} = V_1;$$

in which V_1 is the velocity of the point at the unit's distance from the axis—in other words, the angular velocity; and Eq. (59) becomes

relation of angular to absolute velocity;

$$V = r . V_1,$$

and the living force of m becomes $m r^2 V_1^2$. The simultaneous living force of m', is $m' r'^2 V_1^2$, and so on of others; and the total living force of the entire body, denoted by L, becomes

$$L = V_1^2 (m r^2 + m' r'^2 + m'' r''^2 + \&c.) \quad . \; . \; (60).$$

value of the living force of a rotating body;

In which it is to be remarked, that if the living force changes, the factor V_1 will alone vary, while the factor $(m r^2 + m' r'^2 + m'' r''^2 + \&c.)$ will remain constant, and of course, appear in the estimate of the new living force. This quantity, which has been called *the moment of inertia*, let us designate by I, and we have

$$I = m r^2 + m' r'^2 + m'' r''^2 + \&c. \; . \; . \quad (60)'$$

$$L = V_1^2 I \; . \; . \; . \; . \; . \; . \quad (60)'';$$

whence we see, that *the living force of a body which turns about an axis, is equal to the product of the square of its angular velocity, multiplied by its moment of inertia.*

equal to the square of the angular velocity into the moment of inertia;

Let us suppose that at the end of a certain interval, the angular velocity becomes V_1', the living force L', will be

$$L' = V_1'^2 I;$$

and subtracting the preceding equation from this one, we get

$$L' - L = I . (V_1'^2 - V_1^2) \quad . \; . \; . \quad (61),$$

increment of living force during any interval;

for the increment of the living force during this interval, which is double the quantity of work produced by the motive forces, or their perpendicular components, during the same interval. Denote by F, the resultant of these components, and by E, the path described by its point of application, estimated in its own direction during the interval in question; then will

equal to twice the quantity of work of the motive forces in same time.

$$I.(V_1'^2 - V_1^2) = 2F.E \quad . \quad . \quad . \quad (62).$$

From this expression it is easy to deduce the nature of the quantity I. For if we suppose the change in the angular velocity to be unity,

$$V_1'^2 - V_1^2 = 1,$$

and

$$I = 2F.E;$$

What is meant by the moment of inertia? its measure?

whence we conclude, *that the moment of inertia of any body, is twice the quantity of work exerted by its inertia, during a change in its angular velocity equal to unity.* It is measured by the sum of the products which arise from multiplying each elementary mass into the square of its distance from the axis, Eq. (60)′.

Example illustrative of the preceding principle;

§ 160.—By the aid of what has just been explained, we may find the intensity of a motive force which causes a body to rotate about an axis, when we know the angular velocity at any two given instants of time, and the path described by the point of application in the interval between them. And reciprocally, if the force and the path described by the point of application be given, we may deduce the angular acceleration. Suppose a wheel, for example, mounted upon a horizontal *arbor* and turned around its axis by a weight P, suspended from a cord wound around the arbor; required

Fig. 101.

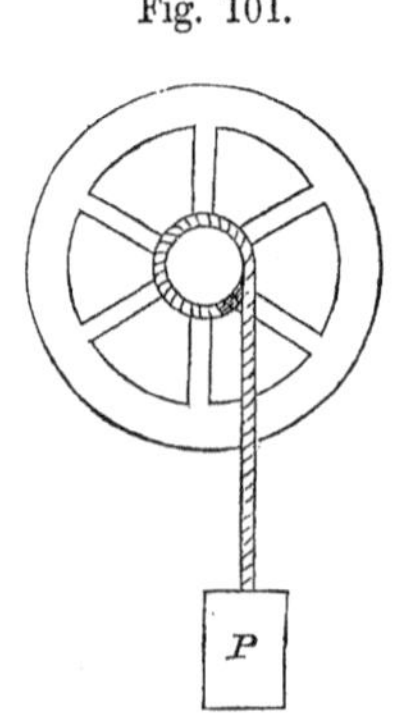

the angular velocity V_1 of the wheel when, moving from a state of repose, the weight shall have descended through a vertical height H. Let I denote the moment of inertia of the wheel, then will the living force acquired be $I\,V_1^2$, and we shall have, wheel turned by the action of a weight;

$$I \,.\, V_1^2 = 2\,P \,.\, H;$$

whence

$$V_1^2 = \frac{2\,P \,.\, H}{I};$$

and consequently

$$V_1 = \sqrt{\frac{2\,P \,.\, H}{I}}.$$

value of the angular velocity;

§ 161.—The *fly-wheel* is a large ring, usually of metal, of which the circumference is thrown to a considerable distance from the arbor upon which it is mounted by means of radial arms, and is used to collect the work of a moter when the effort of the latter is greater than that required to overcome a given resistance, to be given out again when the resistance becomes greater than the effort of the moter. It is a kind of store-house in which to husband work for future use. application to the fly-wheel;

Conceive one or more forces to act upon such a wheel during an interval separating two given instants at which the angular velocities are V_1 and V_1'. The increment of the living force of the fly-wheel will be equal to double the effective quantity of work of the moter, and we shall have, retaining the notation of § 159,

$$I\,(V_1'^2 - V_1^2) = 2\,FE;$$

increment of living force in any interval;

or

$$V_1'^2 - V_1^2 = \frac{2\,F \,.\, E}{I};$$

which gives the difference of the squares of the angular velocities. If the quantity of work developed by the moter remain the same during the interval, and, by changing the wheel, the moment of inertia increase, the fraction

the value of the difference of the squares of the velocities;

$$\frac{2F.E}{I},$$

and consequently the difference of the angular velocities at the beginning and end of the interval, will be less. And, as the moment of inertia is in the direct ratio of the mass into the square of its distance from the axis, it is plain that it is always possible so to construct a wheel as to make its motion approximate to uniformity, even though the motive force be very great.

the motion may be made to approach uniformity;

While the motion is accelerated, it is obvious that the work of the moter will exceed that of the resistance; the fly-wheel will acquire an increase of living force which it will retain till, on the contrary, the motion is retarded, when it will be again given out in aid of the moter, which now becomes less than the resistance.

use of fly-wheel;

There are certain machines whose tool cannot perform its work without the fly-wheel. This is strikingly exemplified in the instance of the common saw-mill, in which it is obvious that the work during the ascent and descent of the saw is very different; the work of the moter exceeds that of the tool or saw during one semi-oscillation, while the reverse takes place during the other; in the first case, the saw is merely elevated and the fly-wheel absorbs living force; in the second, this living force is given out to aid the moter in overcoming the resistance opposed to the saw, which, in its descent, sinks into the wood and is thus made to perform its work.

exemplified in the common saw-mill.

§ 162.—If the elementary mass m, receive in the short interval t, the velocity V, and we denote by f its inertia, we shall have, Eq. (28),

$$f = \frac{m\,V}{t};$$

Distance from the axis at which the resultant inertia of a rotating body is exerted;

and for any other masses m', m'', &c., whose acquired velocities in the same time are V', V'',

$$f' = \frac{m'\,V'}{t},$$

$$f'' = \frac{m''\,V''}{t}, \text{ \&c.}$$

If, moreover, the masses m, m', m'', &c., form parts of a body which has a motion of rotation, their velocities will be proportional to their respective distances from the axis. Denoting these distances by r, r', r'', &c., and by V_1 the small degree of velocity impressed upon the point at the unit's distance from the axis, we shall have

$$V = r\,V_1; \quad V' = r'\,V_1; \quad V'' = r''\,V_1;$$

which in the above equations give,

$$f = m\,r\,.\frac{V_1}{t}; \quad f' = m'\,r'\,.\frac{V_1}{t}; \quad f'' = m''\,r''\frac{V_1}{t}; \text{ \&c.}$$

value of the partial forces of inertia;

But, if this *increment* of angular velocity V_1, has been impressed upon the body by a force F, whose direction is perpendicular to the axis, and applied at a distance from it equal to R, this force is the measure of the inertia of the body, and will be in equilibrio with all the partial forces of inertia f, f', f'', &c. But these latter act in directions tangent to the circles described by the masses m, m', m'', &c., about the axis, and hence, § 157,

$$F\,R - (f\,r + f'\,r' + f''\,r'' + \text{\&c.}) = 0;$$

equilibrium of these with the motive force equal to their resultant;

or

$$F\,R - \frac{V_1}{t}\,(m\,r^2 + m'\,r'^2 + m''\,r''^2 + \text{\&c.}) = 0;$$

but the expression within the brackets is the moment of inertia I, and therefore

the moment of the inertia actually exerted;

$$F \,.\, R = \frac{V_1}{t} \,.\, I \quad . \quad . \quad . \quad . \quad (63);$$

whence we see, that the moment of the inertia exerted by a body while receiving a motion of rotation about an axis, is equal to its *moment of inertia* in reference to the same axis, multiplied into the quotient arising from dividing the small degree of angular velocity communicated, by the element of the time during which it is impressed. Notwithstanding the close analogy which exists between the *moment of the inertia* of a body, and what has been called the *moment of inertia*, they must not be confounded with each other. The former is converted into the latter by making $\frac{V_1}{t}$ equal to unity.

distinction between this and what is usually called the moment of inertia;

From Eq. (63) we find

value for the angular velocity;

$$V_1 = \frac{F \,.\, R \times t}{I} \quad . \quad . \quad . \quad . \quad (64),$$

from which, having given the motive force that impresses a motion of rotation upon a body about an axis perpendicular to its direction, we may find, at each instant of time, the angular velocity communicated, provided we can calculate the moment of inertia of the body in reference to the same axis. And from this, it is possible, by means of a curve which has for its abscisses the series of times t, and for its ordinates the velocities V_1 acquired, to determine all the circumstances of the motion of rotation.

how used.

§ 163.—The moment of inertia of a body with reference to any axis, we have seen, is measured by the sum of the products which arise from multiplying each elementary mass into the square of its distance from the axis.

Measures of the moments of inertia;

Of all the moments of inertia of the same body, those are easiest obtained which refer to axes through the centre of gravity. It is, therefore, important to be able to find the moment of inertia with reference to any axis, by means of that taken with reference to a parallel axis through the centre of gravity.

those in reference to axes through centre of gravity easiest obtained;

Fig. 102.

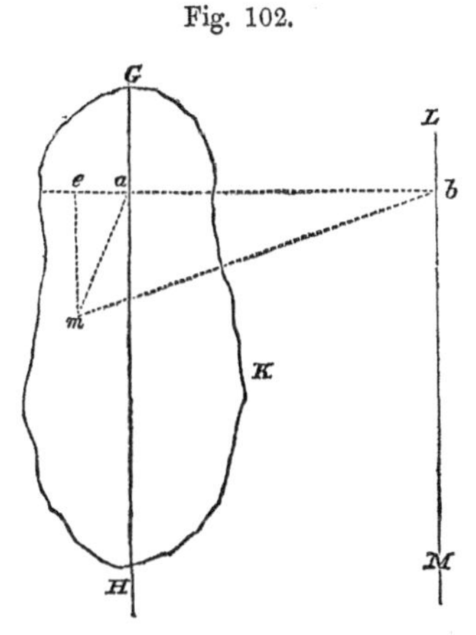

Let GH be this latter axis, LM any parallel axis, m an elementary mass of the body GKH, through which element conceive a plane to be passed perpendicular to the axes, and cutting them at the points a and b. Join m with a and b, and let fall from m, the perpendicular me upon ab. Designate mb by r, ma by $r_{\prime}$, ab by D, and ae by d; we shall have

that in reference to any axis, in terms of the moment in reference to a parallel axis through centre of gravity;

$$r^2 = r_{\prime}^2 + D^2 + 2Dd,$$

and multiplying by the mass m,

$$mr^2 = mr_{\prime}^2 + mD^2 + 2mDd;$$

and for the masses m', m'', m''', &c.,

$$m'r'^2 = m'r_{\prime}'^2 + m'D^2 + 2m'Dd',$$

$$m''r''^2 = m''r_{\prime}''^2 + m''D^2 + 2m''Dd'',$$

&c., &c., &c.

D, which is the distance between the two axes, remaining obviously the same in all.

Adding these equations together, and denoting the moment of inertia in reference to the axis GH by I_1, and that in reference to LM by I, we find

the sum of all the partial moments;

$$I = I_1 + D^2(m + m' + m'' + \&c.) + 2D(md + m'd' + m''d'' + \&c.);$$

but $m + m' + m'' + \&c.$ is the entire mass of the body, and $md + m'd' + m''d'' + \&c.$ is the sum of the products which result from multiplying each mass into its distance from a plane through the centre of gravity, which sum is equal to zero. Hence, designating the mass by M, we have

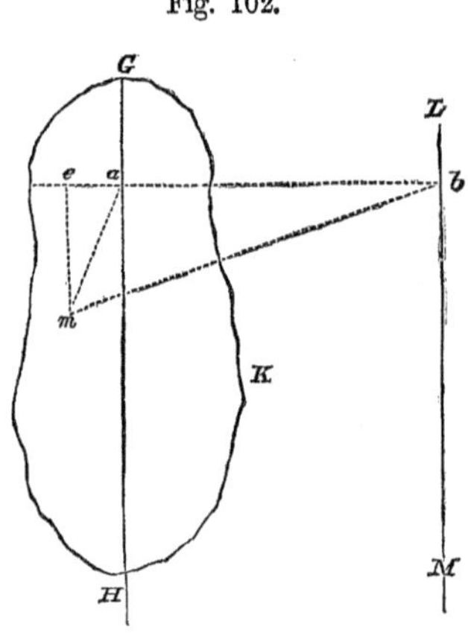

Fig. 102.

resulting value;

$$I = I_1 + MD^2 \quad . \quad . \quad . \quad . \quad (65);$$

conclusion.

whence we conclude that, *the moment of inertia of a body, taken with reference to any axis, is equal to the moment of inertia taken with reference to a parallel axis passing through the centre of gravity, increased by the product of the entire mass of the body into the square of the distance from the centre of gravity to the first axis.*

Value, when the linear dimensions of the bodies are small, in comparison with their distances from the axis;

It follows from this theorem, that if the distances of the particles of the body from its centre of gravity be small in comparison with the distance of this point from the axis of rotation, we may take, for the moment of inertia, simply the product of the mass into the square of the distance of the axis from the centre of gravity.

Finally, if Eq. (65) be multiplied by the square of

the angular velocity, V_1, with which the body turns about the axis $L M$, we shall have

$$V_1^2 . I = V_1^2 I_1 + M . D^2 V_1^2 \quad . \quad . \quad (66);$$

value of the living force;

but $V_1^2 I$ is the living force of the body; $V_1^2 I_1$ is the living force it would have, if it rotated about a parallel axis through the centre of gravity with the same angular velocity V_1; $M . D^2 V_1^2$ is the living force of the same body supposed concentrated at its centre of gravity. Whence, *the living force of a body which rotates about any axis, is equal to the living force of the same body concentrated at its centre of gravity, augmented by that which it would possess if it turned, with the same angular velocity, about a parallel axis through the same centre.*

expressed in words;

Finally, when the body is so small that $I_1 V_1^2$ may be neglected in comparison with $M . D^2 V_1^2$, we have simply

$$V_1^2 I = M . D^2 V_1^2 \quad . \quad . \quad . \quad (66)';$$

value when the linear dimensions of the body are very small as compared with its distance from the axis.

that is to say, the living force of the body is equal to the product of its mass into the square, $D^2 V_1^2$, of the velocity of its centre of gravity.

§ 164.—Thus far the moment of inertia of a body has been expressed in terms of its elementary masses. If the body be homogeneous and the specific gravity or weight of a unit of its volume be denoted by δ, its elementary volumes by a, a', a'', &c., and masses by m, m', m'', &c., we shall have

Moment of inertia of bodies in terms of their linear dimensions and density;

$$m = \frac{\delta\, a}{g}; \quad m' = \frac{\delta\, a'}{g}; \quad m'' = \frac{\delta\, a''}{g}, \text{ \&c.};$$

and these in the general expression I, of the moment of inertia, give

$$I = \frac{\delta}{g} (a\, r^2 + a' r'^2 + a'' r''^2 + \text{\&c.});$$

its general value;

rule; that is to say, to find the moment of inertia of any homogeneous body, find the value of $a r^2 + a' r'^2 + a'' r''^2 + \&c.$, and multiply it by the quotient arising from dividing the specific gravity, or weight of a unit of its volume, by the force of gravity.

moment of inertia of a straight bar, in reference to a perpendicular axis through its middle;

§ 165.—1st. The moment of inertia of a straight bar whose length is a and cross section b, in reference to an axis passing through its middle point A, and perpendicular to its length, is given by

Fig. 103.

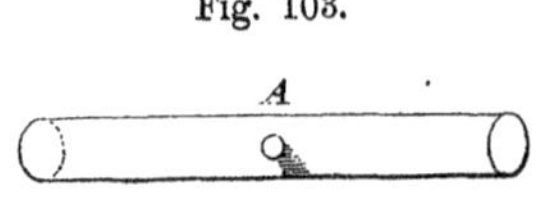

$$I_1 = b \cdot (\tfrac{1}{12} a^3), \frac{\delta}{g} \text{ very nearly.}$$

that of a right cylinder, in reference to its own axis;

2d. The moment of inertia of a right cylinder having a circular base, with respect to an axis through its centre of gravity, and coinciding with its axis of figure, is given by the equation

Fig. 104.

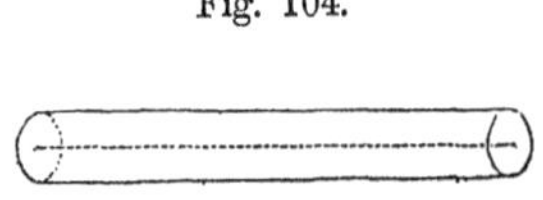

$$I_1 = \frac{\pi}{2} c r^4 \times \frac{\delta}{g},$$

in which r is the radius of the base, c the length of the cylinder, and π the ratio of the circumference to the diameter of a circle.

that of a circular ring, in reference to a perpendicular axis through its centre;

3d. The moment of inertia of a circular ring, whose section by a plane through its centre of figure is rectangular, taken with reference to an axis through its centre of gravity and perpendicular to its plane, gives

$$I_1 = 2 \pi r a b \left(r^2 + \frac{b^2}{4}\right) \times \frac{\delta}{g};$$

in which r is the mean radius, or that of a circle whose circumference is midway between the inner and outer surface of the ring, a the thickness parallel to the axis, and b the thickness in the direction of the radius.

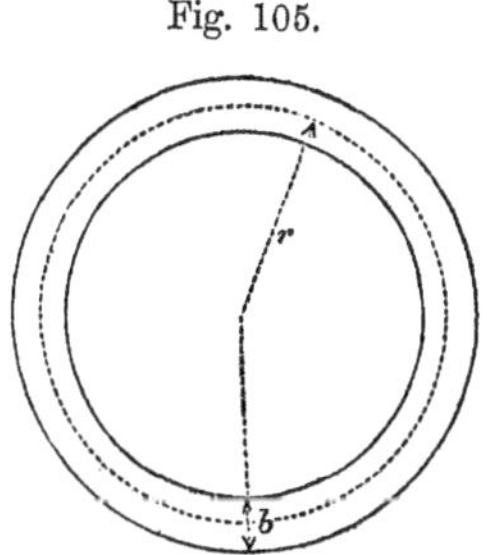

Fig. 105.

Fig. 106.

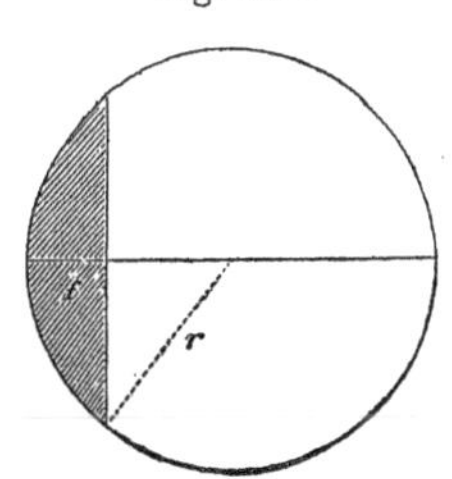

4th. That of a spherical segment taken in reference to a diameter passing through its centre of gravity, or middle, gives

that of a spherical segment, in reference to its versed sine;

$$I_1 = \pi f^3 \left(\tfrac{2}{3} r^2 - \tfrac{1}{2} f r + \tfrac{1}{10} f^2\right) \times \frac{\delta}{g},$$

in which f denotes the versed sine of the segment, and r the radius of the sphere; and for the entire sphere,

of a sphere, in reference to a diameter;

$$I_1 = \tfrac{8}{15} \pi r^5 \times \frac{\delta}{g}.$$

Fig. 107.

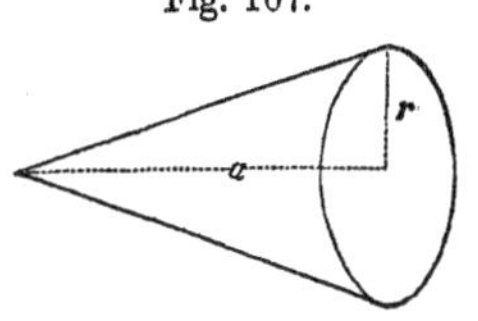

5th. That of a right cone having a circular base, taken with reference to the axis of figure gives,

that of a cone, in reference to its axis;

$$I_1 = \frac{\pi}{10} a r^4 \times \frac{\delta}{g};$$

Fig. 108.

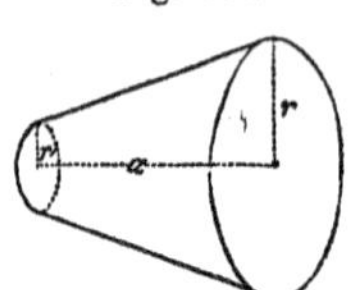

that of a truncated right cone;

and that of a truncated right cone, having circular bases,

$$I_1 = \frac{\pi}{10} a \frac{r^5 - r'^5}{r - r'} \times \frac{\delta}{g};$$

in which r and r' are the radii of the greater and smaller bases respectively, and a the altitude.

Fig. 109.

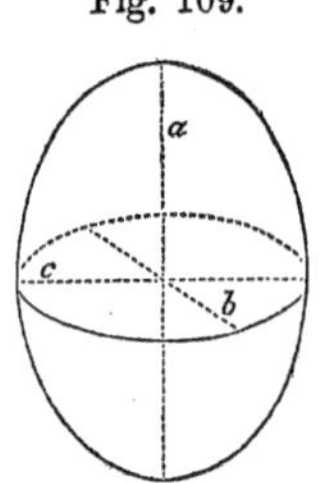

6th. The moment of inertia of an ellipsoid is given by

that of an ellipsoid;

$$I_1 = \tfrac{4}{15} \pi \, a \, b \, c \, (b^2 + c^2) \times \frac{\delta}{g};$$

in which a, b, and c denote the three axes, and the moment being taken with reference to the axis a.

Fig. 110.

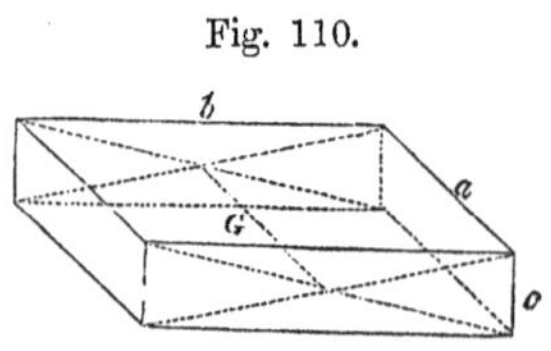

that of a rectangular parallelopipedon;

7th. That of a rectangular parallelopipedon, of which the three contiguous edges are a, b, and c, taken with reference to an axis passing through its centre of gravity G, and parallel to the edge a,

$$I_1 = \tfrac{1}{12} a \, b \, c \, (a^2 + b^2) \times \frac{\delta}{g}.$$

The same taken in reference to an axis through the middle of the face $a \, b$ and parallel to a,

the same for a different axis;

$$I = \tfrac{1}{12} a \, b \, c \, (b^2 + 4 c^2) \cdot \frac{\delta}{g}.$$

8th. That of a right prism having a trapezoidal base of which the greater and less parallel sides are respectively b and b' and distance between them c, the altitude of the prism being a, and the moment taken with reference to an axis through the middle of the side b, and parallel to the altitude a, — that of a right prism with trapezoidal base;

Fig. 111.

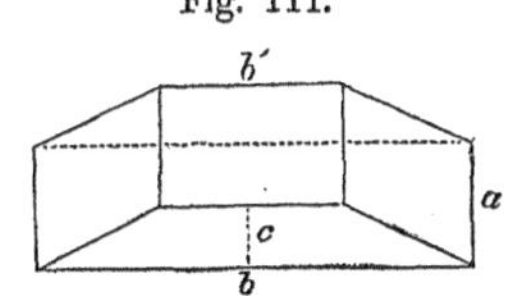

$$I = a\,c\left(\frac{b + b'}{2}\right) \cdot \left(\frac{b^2 + b'^2}{24} + \frac{c}{6} \cdot \frac{b + 3\,b'}{b + b'}\right) \times \frac{\delta}{g}.$$

9th. If the trapezoidal base of the above prism be replaced by a segment of a parabola, of which c is the length of the transverse axis, and b that of the chord perpendicular to it, and which terminates the parabola, the moment of inertia, with reference to an axis parallel to the altitude and passing through the middle of b, is given by — the same when the base becomes a segment of a parabola.

Fig. 112.

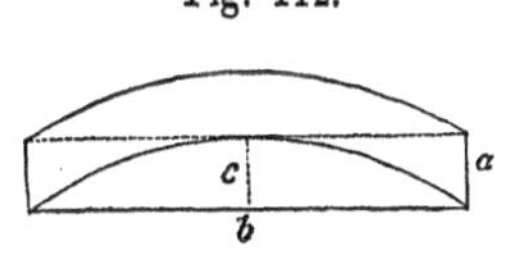

$$I = \tfrac{2}{3}\,a\,b\,c\left(\frac{3\,.\,5\,.\,b^2 + 16\,c^2}{70}\right) \times \frac{\delta}{g}.$$

§ 166.—We shall close this subject with an example for the sake of illustration, and we shall first take that of a trip-hammer, whose weight is P, mounted upon a handle in the shape of a rectangular parallelopipedon which turns freely about an axis O, at right angles to its length. Denote by R, the distance of the centre of gravity of the head B from the axis O. — Application to examples; that of a common trip-hammer;

Fig. 113.

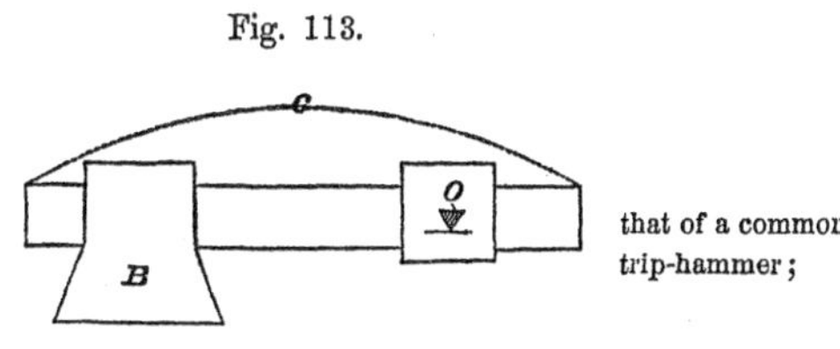

If the linear dimensions of the head be small compared with this distance, its moment of inertia will not differ much from

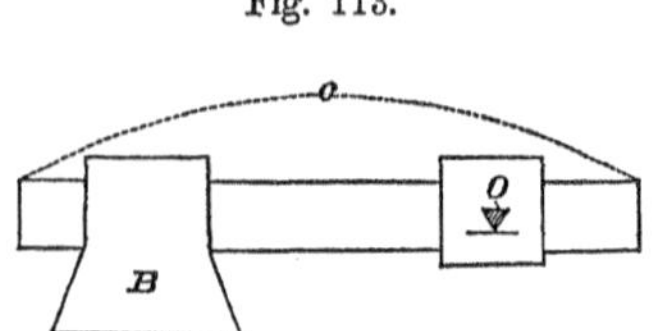

Fig. 113.

the moment of inertia of the head;

$$\frac{P}{g} \times R^2,$$

and that of its handle is given in reference to an axis through its centre of gravity by the 7th case, or

that of the handle with reference to its centre of gravity;

$$\tfrac{1}{12}\, a\, b\, c\, (a^2 + b^2) \cdot \frac{\delta}{g};$$

and denoting by K, the distance of the centre of gravity of the handle from the axis, its moment of inertia, with reference to the axis O, becomes, Eq. (65),

$$\frac{\delta}{g} \times \frac{a\, b\, c}{12}\,(b^2 + a^2) + \frac{\delta}{g}\, a\, b\, c\, K^2,$$

or

with reference to the axis;

$$\frac{P'}{g}\,\left(K^2 + \frac{b^2 + a^2}{12}\right);$$

since $abc\,\delta = P'$, the weight of the handle. The total moment of inertia is, therefore, given by

the moment of the entire hammer;

$$I = \frac{P}{g}\, R^2 + \frac{P'}{g}\,\left(K^2 + \frac{b^2 + a^2}{12}\right).$$

moment of inertia of the fly-wheel;

The process for finding the moment of inertia of the fly-wheel is much simplified by the fact that all its parts are nearly at the same distance from the axis. Thus, by calling R the mean radius of the wheel, we may take $\frac{P}{g} R^2$ for $mr^2 + m'r'^2 + m''r''^2 +$ &c.; and hence,

$$I_1 = \frac{P}{g} R^2;$$ its value;

and denoting the angular velocity of the wheel by V_1, its living force will be, § 159,

$$\frac{P}{g} \times R^2 \,.\, V_1^2.$$ living force of the fly-wheel;

To find the angular velocity of the wheel, count the number of its revolutions in a given time, multiply this number by 2π, and divide the product by the number of seconds in the given time; the quotient will be the angular velocity. Let V_1 equal 9 feet; the weight P of the wheel 2000 pounds, and the mean radius R, 6 feet; omitting the fraction in the value of g, the expression for the moment of inertia becomes

angular velocity found experimentally;

$$I_1 = \frac{2000}{32} \times 36 = 2250;$$

and for the living force,

example.

$$2250 \times 81 = 182{,}250;$$

the half of which, or 91,125 pounds, raised through one foot, is the quantity of work absorbed by the inertia of the wheel, to be given out when the moter ceases to act.

§ 167.—Resuming Eq. (60)′, we may make

$$m r^2 + m' r'^2 + m'' r''^2 + \&c. = M K^2,$$ Centre and radius of gyration;

in which M is the entire mass of the body, and

$$K = \pm \sqrt{\frac{mr^2 + m' r'^2 + m'' r''^2 + \&c.}{M}}.$$

But this is equivalent to *concentrating the entire mass into a single point whose distance from the axis is K, without* *changing the value of the moment of inertia.* This point is called the *centre of gyration*, and the distance K, is called the *radius of gyration*. As the moment of inertia varies with the position of the axis, there will be an infinite number of centres and radii of gyration, or as many of each as there are possible positions for the axis. When the axis passes through the centre of gravity, they are called the *principal centre and radius of gyration.*

definition;

principal centre and radius of gyration;

Denoting the principal radius of gyration by K', we may write MK'^2 for I_1, in Eq. (65), and we have

moment of inertia in terms of radius of gyration.

$$I = MK'^2 + MD^2 \quad . \quad . \quad . \quad (66)''.$$

XI.

CENTRAL FORCES.

Central forces;

§ 168.—Conceive a body, whose weight is P, attached to a fixed point C by a rigid bar $A\ C$, and suppose it to have any velocity whatever in the direction AT, perpendicular to the bar. If the body were free, it would, in virtue of its inertia, move in the direction $A\ T$ with a constant velocity. But not being free, the bar will keep it at the same distance from C and cause it to describe the circumference of a circle about this point as a centre. There are, then, during this constrained motion of the body, two central efforts exerted in the direction of the bar, the one by the bar to draw the body from the tangential path $A\ T$, the other by the body

a body made to revolve about a fixed point by means of a bar;

Fig. 114.

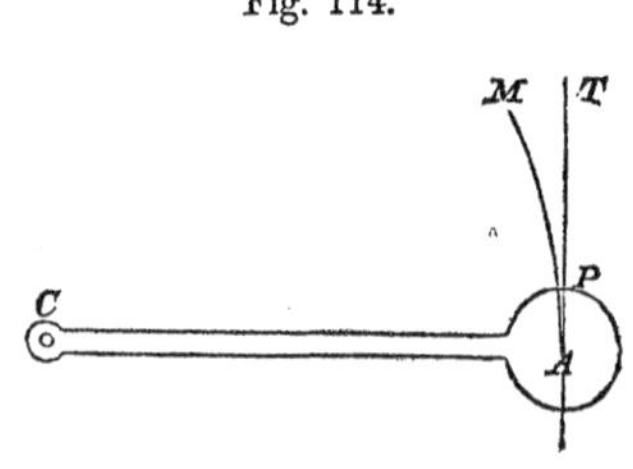

to stretch the bar out to that path. These forces are equal and directly opposed, because action and reaction are always equal and contrary. The first, or that which tends to draw the body within the tangent, is called the *centripetal force*, and the second, or that which tends to stretch the bar, the *centrifugal force*. *The centrifugal force is, then, the resistance which the inertia of a body in motion opposes to whatever deflects it from its rectilinear path.*

centripetal force; centrifugal force; centrifugal force arises from inertia;

We will first suppose that the dimensions of the body are so small as compared with its distance from the fixed centre, that it may be regarded as a material point, animated with a velocity V. For the circle which it describes, we may substitute a regular polygon $ABCDE$, of a great number of very small sides and having its angles in the circumference. This being supposed, it is first to be shown that the material point will describe each of the sides of this polygon with the same velocity, or that there will be no loss of velocity in passing from one side to another.

substitution of a polygon for the circle;

For this purpose, we remark, that if the body possess

Fig. 115.

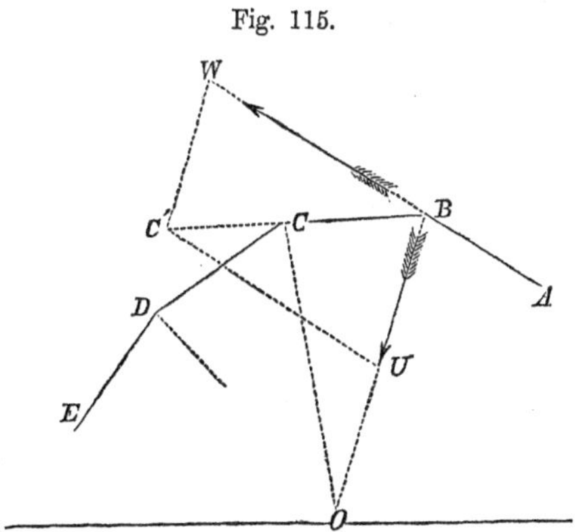

to prove there is no loss of velocity from the reaction of the curve;

the velocity V at the moment of its arriving at B, the beginning of the side BC, it will be animated, while describing this side, with two simultaneous velocities. One of these is the primitive velocity $V = BW$, in the prolongation of AB; the other BU, in the direction BO, is a velocity due to the action of the centripetal force while the body is passing from the side AB to the side BC of the polygon. But we have seen, § 106, that when a body receives two simultaneous velocities in different directions, its resultant velocity will be the same as if it

the body has two simultaneous velocities;

Fig. 115.

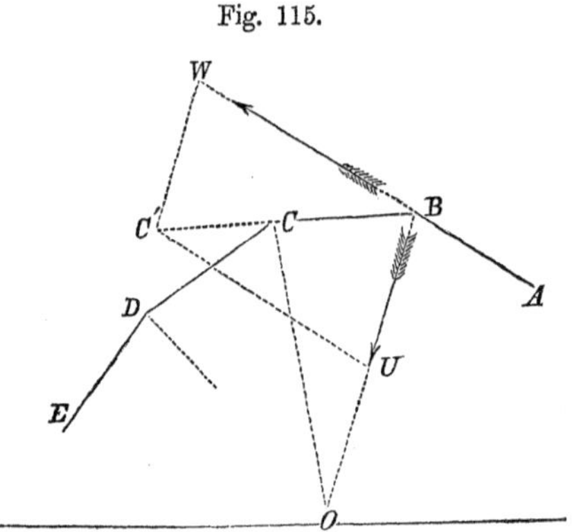

possessed them successively, and as though they were communicated one after the other in their respective directions. Thus the resultant velocity BC', with which the side BC is described, coincides in direction with this side, otherwise the body would take some other path, which is contrary to the hypothesis. BU and WC' are equal and parallel, from the parallelogram of velocities. The radius OB divides the angle ABC into the two equal angles ABO and OBC; the angle ABO is equal to the angle BWC', and the angle OBC is equal to the angle $BC'W$; hence the angles $BC'W$ and BWC' are equal, and the side BC' is equal to the side BW; in other words, the resultant velocity BC', with which the side BC is described, is equal to the velocity BW which the body had at the end of the side AB. *Whence it*

the resultant of which has the direction of the side next to be described;

no loss of velocity;

results, that the velocity communicated to a material point is not altered in its circular motion; a result easy to foresee, since the centripetal force, acting in a direction perpendicular to the direction of the motion, cannot work efficiently; it can neither accelerate nor retard the motion, and, therefore, can neither increase nor diminish the living force of the material point. velocity not altered by the deflecting cause;

Now observe, that $BU = WC'$, is the velocity generated by the centripetal force in its own direction during the time the material point is passing to the side BC. Denote this time by t, and the centrifugal, which is equal though opposed to the centripetal force, by F, and the mass of the point by M, then will the value of F, be given, Eq. (39), by the equation

$$F = M \cdot \frac{WC'}{t}.$$

Draw the radius CO; the triangles BOC and BWC' are similar, because the angle $OCB = OBC$ is equal to the angle $BC'W$, and the angle OBC is equal to the angle BWC'. Hence we have the proportion, to find value of centrifugal force;

$$BO : BC :: BW : WC';$$

or, denoting the radius BO by R, and replacing BW by its equal V,

$$R : BC :: V : WC';$$

whence

$$WC' = \frac{BC \times V}{R},$$

and this, in the value for F, gives

$$F = M \cdot \frac{BC}{t} \times \frac{V}{R};$$

but $B\ C$, the element of the space, divided by t, the element of the time, is equal to the velocity V, whence

the measure of the centrifugal force;

$$F = \frac{M \cdot V^2}{R} \quad . \quad . \quad . \quad . \quad . \quad (67).$$

Such is the expression for the *centrifugal* force. The numerator is the living force of the body, and the denominator is the radius of the circular arc which the body is describing for the instant; whence we conclude, that *the centrifugal force of a body of small dimensions, as compared with its distance from the centre about which it revolves, is equal to the living force impressed upon the body, divided by the radius of the circle described by its centre of gravity.*

equal to the living force impressed, divided by radius of curvature;

Suppose, for example, that the weight of the body is 100 pounds, that its centre describes a circle whose radius is 3 feet, with a velocity of 12 feet.

$$M = \frac{100}{32}; \quad V = 12; \quad V^2 = 144; \quad R = 3;$$

illustration;

$$F = \frac{100 \times 144}{32 \times 3} = 150 \text{ pounds};$$

the body, therefore, tends to stretch the bar with an effort of 150 pounds.

Denote by V_1 the angular velocity; then will

$$V^2 = V_1^2\, R^2,$$

and this, in Eq. (67), gives

expressed in terms of the angular velocity.

$$F = M\, V_1^2\, R \quad . \quad . \quad . \quad . \quad . \quad (68).$$

Extension to a body of any dimensions;

§ 169.—Let us next take the case of a thin layer of matter $D\ A\ B$, rotating about an axis O, perpendicular to its plane, with an angular velocity V_1. Taking any one of the elements of the layer whose mass is m, and de-

Fig. 116.

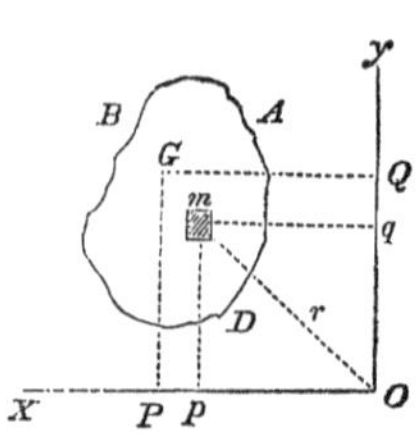

noting its distance from the axis O, by r, its living force will be

$$m r^2 V_1^2,$$

and its centrifugal force,

$$\frac{m r^2 V_1^2}{r} = m r V_1^2;$$

centrifugal force of an element;

which will act in the direction $O m$ of the radius of the circle described by m about the centre O. Through the point O, draw in the plane of the layer, any two rectangular axes, as $O x$ and $O y$. Resolve the centrifugal force into two components acting in the direction of these axes; these components and their resultant will be proportional to the sides and diagonal of the rectangle $p\, O q m$, and we shall have, denoting $O p$ by x, $O q$ by y, the component parallel to the axis $O x$ by X, and that parallel to $O y$ by Y,

$$r : x :: m r V_1^2 : X,$$

$$r : y :: m r V_1^2 : Y;$$

resolved into rectangular components;

whence,

$$X = m x V_1^2$$

$$Y = m y V_1^2;$$

and for any number of small masses m', m'', &c., by using the same notation with accents,

$$X' = m' x' V_1^2,$$

$$X'' = m'' x'' V_1^2,$$

$$\&c. = \&c.;$$

components parallel to the axis x, for other elements;

components parallel to the axis y;

$$Y' = m' y' V_1^2,$$

$$Y'' = m'' y'' V_1^2,$$

$$\&c. = \&c.;$$

by this process all the centrifugal forces have been reduced to two groups of forces acting upon the point O, in the direction of the axes Ox and Oy, and from the principle of parallel forces, each group will have for its resultant, denoted by X_1 and Y_1 respectively,

$$X_1 = V_1^2 (m x + m' x' + m'' x'' + \&c.),$$

$$Y_1 = V_1^2 (m y + m' y' + m'' y'' + \&c.);$$

that is,

resultants parallel to the axes x and y;

$$X_1 = V_1^2 M' x_{,},$$

$$Y_1 = V_1^2 M' y_{,};$$

in which M' denotes the entire mass of the layer, and $x_{,}$ and $y_{,}$ the co-ordinates OP and OQ of its centre of gravity G.

The resultant of the forces X_1 and Y_1 is the entire centrifugal force of the layer; and this denoted by F_1, is, from the principle of the parallelogram of forces,

$$F_1 = \sqrt{V_1^4 M'^2 y_{,}^2 + V_1^4 M'^2 x_{,}^2} = V_1^2 M' \sqrt{x_{,}^2 + y_{,}^2};$$

and making

$$\sqrt{x_{,}^2 + y_{,}^2} = OG = r_{,}$$

measure of the layer's centrifugal force;

$$F_1 = M' r_{,} V_1^2;$$

whence, *the centrifugal force of a thin layer of matter revolving about an axis perpendicular to its plane, is equal to the square of its angular velocity, multiplied by the product of its mass into the distance of its centre of gravity from the axis of rotation.* This force is applied to the centre of gravity, since it acts in the direction $O\ G$.

Now suppose any body, as $A\ B\ C$, to turn around the axis $L\ M$. Divide the body into thin layers whose planes are perpendicular to the axis. These layers will give rise to as many centrifugal forces acting at their centres of gravity, G, G', G'', &c. All these forces are perpendicular to the axis $L\ M$, without being parallel to each other. Sometimes they have a single resultant, sometimes they will reduce to two forces, and sometimes they will reduce to nothing, depending upon the form and density of the body, and the position of the axis. In the last case, viz.: that in which the forces reduce to nothing, there will be no pressure upon the axis.

centrifugal force of a body of any size;

Fig. 117.

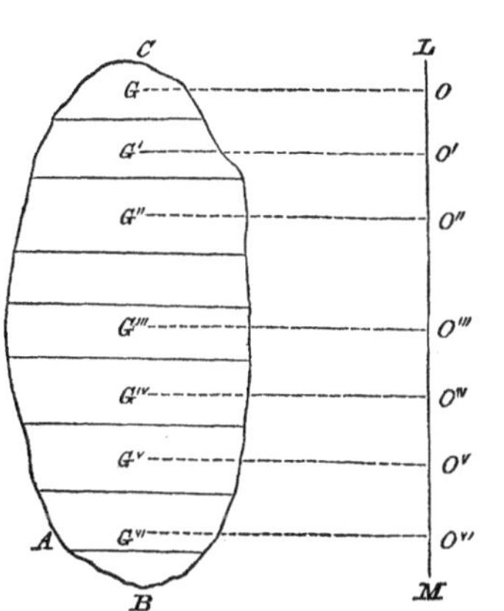

may reduce to a single force, two forces, or to zero;

in the last case no pressure on axis;

If the centres of gravity G, G', G'', &c., be all on the same straight line parallel to $L\ M$, the centrifugal forces will be parallel, will act in the same plane, at the same distance R from the axis of rotation, and their resultant, which becomes equal to their sum, will pass through the centre of gravity of the entire mass, and we shall have

$$F = V_1^2 R\,(M' + M'' + M''' + \&c.);$$

and making

$$M' + M'' + M''' + \&c. = M,$$

$$F = V_1^2 R \,.\, M.$$

the centres of gravity of the layers on same line parallel to the axis;

that is to say, *the centrifugal force of a body, whose sections perpendicular to the axis, have their centres of gravity in a straight line parallel to the axis, is the same as though the entire mass were concentrated at the common centre of gravity.* This simplification is peculiar to the sphere, the cylinder, and surfaces of revolution generally whose axes of figure are parallel to the axis of rotation.

the centrifugal force the same as though the body were reduced to centre of gravity;

examples.

Fig. 118.

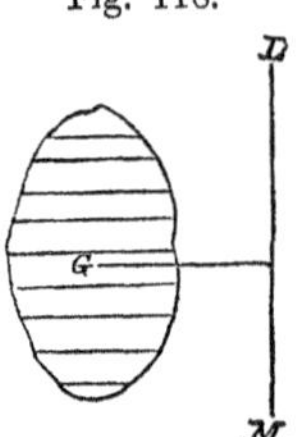

Illustration of the action of the centrifugal force;

§ 170.—The centrifugal force accounts for a multitude of interesting facts. When a horse is made to travel in the circumference of a circle, his centrifugal force will vary as his mass and the square of his velocity; when the latter is doubled, his centrifugal force is quadrupled; when trebled, it is made nine times as great, &c., so that it would soon become sufficient to overturn him or to cause him to recede from the centre *C*. It is to resist this effort that horses, under these circumstances, are seen to incline their bodies inward, and this inclination is determined by that of the resultant of his centrifugal force and weight, as the line of direction of this resultant must pierce the plane of his path somewhere within the polygon formed by joining his feet.

Fig. 119.

the horse travelling in a circle;

If, then, we lay off upon the vertical and horizontal drawn through his centre of gravity G, the distances GP and GF, to represent his weight and centrifugal force respectively, and construct the rectangle $PGFR$, the diagonal GR will give the inclination sought. Denoting the weight of the horse by P, his distance from the

his inclination;

centre by R, and his actual velocity by V, we have

$$F = \frac{P}{g} \cdot \frac{V^2}{R};$$

and consequently the pressure or

his oblique pressure on the ground;

$$GR = P\sqrt{1 + \frac{V^4}{g^2 R^2}}.$$

Finally, in order that the horse may not slip, the surface BA of his path, must be perpendicular to GR.

surface of his path·

When a horseman rapidly turns a corner, he leans his body towards the centre of the curve which he is describing, to bring the resultant of his weight and centrifugal force to pass between his points of support in the stirrups.

a horseman turning a corner;

When a wagon makes a quick turn, its centrifugal force tends to overthrow it towards the convex side of the curve it describes; and the risk of upsetting is directly proportional to its weight and the square of its velocity, and inversely proportional to the radius of the curve. This is why the exterior of the roadway is usually elevated in short turnings, and carriages diminish their speed when approaching them.

a wagon making a short and rapid turn;

Fig. 120.

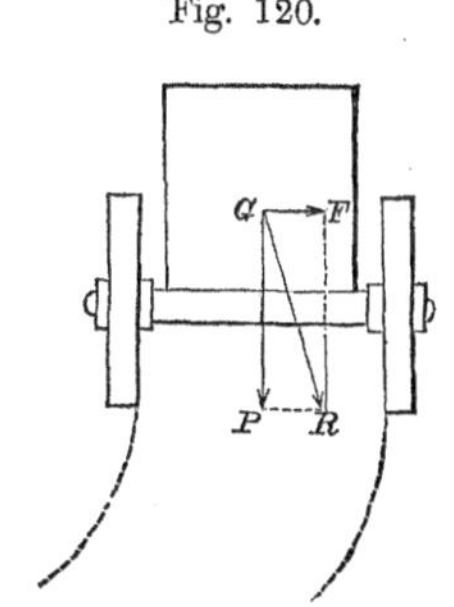

inclination of roadway;

The sling, the axe, the sabre, &c., exert upon the hand, when we give them a circular motion, a traction equal to the centrifugal force. The common wheel is usually composed of fellies A, A, &c., connected with the nave N, by means of radial arms, l, l,

other examples—the sling, axe, sabre, &c.;

Fig. 121.

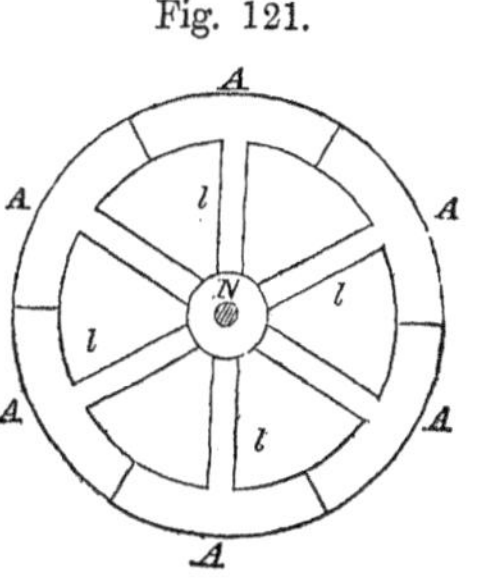

common carriage-wheel;

&c., and the centrifugal force is constantly acting when the wheel is in motion to draw these arms from their places, to enlarge the circumference, and thus to detach the fellies from each other; hence the tire not only protects the wheel from the wear and tear arising from the roughness of the road, but also counteracts the effect of the centrifugal force.

action upon the fellies of the common wheel.

§ 171.—We know that the earth revolves about its axis $A A'$ once in twenty-four hours, and that the circumferences of the parallels of latitude, that is to say, the circles perpendicular to the axis, have velocities which diminish from the equator to the poles. To find the law of this diminution, let P be the weight of a body on the surface of the earth in any parallel of which R' is the radius, its centrifugal force will, Eq. (68), be

Centrifugal force at earth's surface;

Fig. 122.

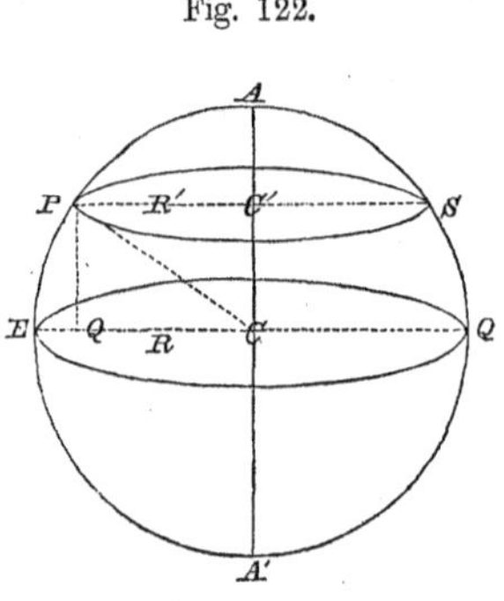

$$\frac{P}{g} \cdot V_1^2 R';$$

that of a body whose weight is P;

in which V_1 is the angular velocity of the earth. Substituting M for $\frac{P}{g}$, we have

$$F = M V_1^2 R'.$$

Denoting the equatorial radius $CE = CP$, by R, and the angle $CPC' = PCE$, which is the latitude of the place, by φ, we have in the triangle PCC',

$$R' = R \cos \varphi;$$

which substituted for R' above gives

$$F = M V_1^2 R \cos \varphi \quad . \quad . \quad . \quad . \quad (69).$$

Now, the only variable quantity in this expression, when the same mass is taken from one latitude to another, is φ; *whence we conclude that the centrifugal force varies as the cosine of the latitude.* law of variation of centrifugal force;

The centrifugal force is exerted in the direction of the radius R' of the parallel of latitude, and therefore in a direction oblique to the horizon TT'. Lay off on the prolongation of this radius, the distance PH, to represent this force, and resolve it into two components PN and PT, the one normal, the other tangent to the surface of the earth; the first will diminish the weight P by its entire value, being directly opposed to the force of gravity, the second will tend to urge the body towards the equator. the centrifugal force resolved into a vertical and horizontal component;

Fig. 123.

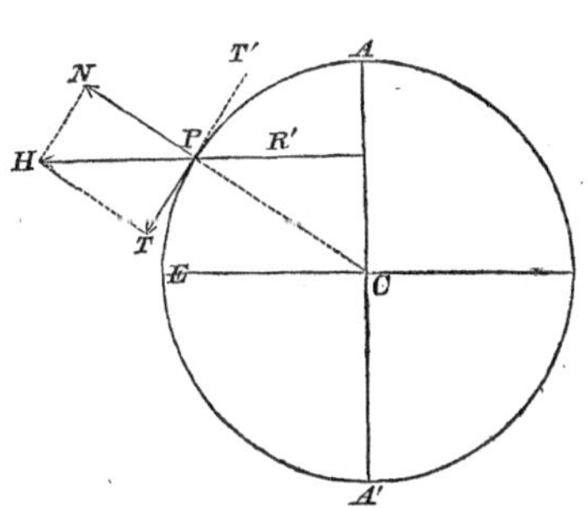

The angle HPN is equal to the angle PCE, which is the latitude, denoted by φ; whence the normal component

$$PN = PH \times \cos \varphi = F . \cos \varphi = M V_1^2 R \cos {}^2\varphi,$$ value of vertical component;

and

$$PT = PH \sin \varphi = F . \sin \varphi = M V_1^2 R . \sin \varphi \cos \varphi;$$ horizontal component;

but

$$\sin \varphi \,.\, \cos \varphi = \tfrac{1}{2} \sin 2\, \varphi;$$

therefore

its value;

$$P\,T = \tfrac{1}{2}\, M\, V_1^2\, R \sin 2\, \varphi$$

whence we conclude, *that the diminution of the weights of bodies arising from the centrifugal force at the earth's surface, varies as the square of the cosine of the latitude; and that all bodies are, in consequence of the centrifugal force, urged towards the equator by a force which varies as the sine of twice the latitude.*

effect upon the weights of bodies and figure of the earth;

Fig. 123.

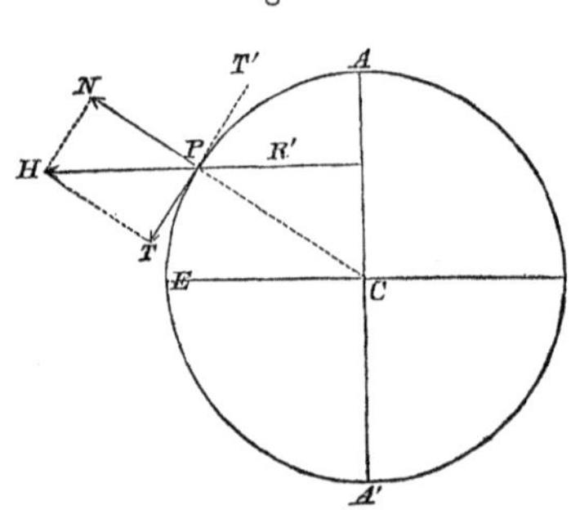

At the equator and poles this latter force is zero, and at the latitude of 45° it is a maximum, and equal to half of the entire centrifugal force at the equator.

At the equator the diminution of the force of gravity is a maximum, and equal to the entire centrifugal force; at the poles it is zero. The earth is not perfectly spherical, and all observations agree in demonstrating that it is protuberant at the equator and flattened at the poles, the difference between the equatorial and polar diameters being about twenty-six English miles. If we suppose the earth to have been at one time in a state of fluidity, or even approaching to it, its present figure is readily accounted for by the foregoing considerations.

cause of the present figure of the earth;

The force of gravity which varies, according to the Newtonian hypothesis, directly as the mass and inversely as the square of the distance from the centre of the earth, is, therefore, on account of a difference of distance and of the centrifugal force of the earth combined, less at the equator than at the poles. weight of the same body greatest at the poles and least at the equator;

To find the value of the centrifugal force at the equator, make, in Eq. (69), $M = 1$ and $\cos \varphi = 1$, which is equivalent to supposing a unit of mass on the equator, and we have

$$F = V_1^2 R.$$

centrifugal force at the equator;

The angular velocity is equal to the absolute velocity, divided by the equatorial radius of the earth. The absolute velocity is equal to the circumference of the equator in feet, divided by the number of solar seconds in one siderial day:

Diameter of earth in miles 7925	Log.	3.8989993
π 3.1416	Log.	0.4971507
Feet in one mile.......... 5280	Log.	3.7226340
Circumference of earth in feet	Log.	8.1187840
Length of a sid. day in Sol: seconds, 86400 × 0.997269,	Log.	4.9353259
Absolute velocity in feet........................	Log.	3.1834581
Radius of earth in feet	Log.	7.3206032
Angular velocity V_1..........................	Log.	5.8628549
		2
Square of angular velocity V_1^2	Log.	1.7257098
Radius of earth in feet.........................	Log.	7.3206032
Centrifugal force at equator.. $\overset{f}{0}.1112$		9.0463130

computed;

Thus the value of the centrifugal force at the equator is 0.1112 of one foot. its value;

By the aid of this value, it is very easy to find the angular velocity with which the earth should rotate, to make the centrifugal force of a body at the equator equal to its weight; for by the present rate of motion we find to find angular velocity sufficient to destroy weights at the equator;

$$\overset{f}{0}.1112 = V_1^2 R,$$

and by the new rate of motion

$$\overset{f}{32}.1937 = V_1'^2 R;$$

in which $\overset{f}{32}.1937$ is the force of gravity at the equator.

Dividing the second by the first, and we find

$$\frac{32.1937}{0.1112} = \frac{V_1'^2}{V_1^2} = 289, \text{ nearly};$$

whence

result;

$$V_1' = 17\, V_1;$$

that is to say, if the earth were to revolve seventeen times as fast as it does, bodies would possess no weight at the equator; and the weights of bodies at the various latitudes from the equator to the poles diminishing in the ratio of the squares of the cosines of latitude, the weights of all bodies, except at the poles, would be affected.

the weight of all bodies affected.

§ 172.—If we now suppose the body, instead of being connected with the point C by means of a rigid bar, to move about the same point in a circular groove, the effects, as regards the centrifugal force, will obviously be the same, since the body will be constrained, by the resistance of the groove, to remain at the same distance from the centre. If the plane of the groove be horizontal, the pressure of the body against the side will be constant and equal to the centrifugal force, that is to say, to

Motion in a circular groove,

when plane of groove is horizontal;

Fig. 124.

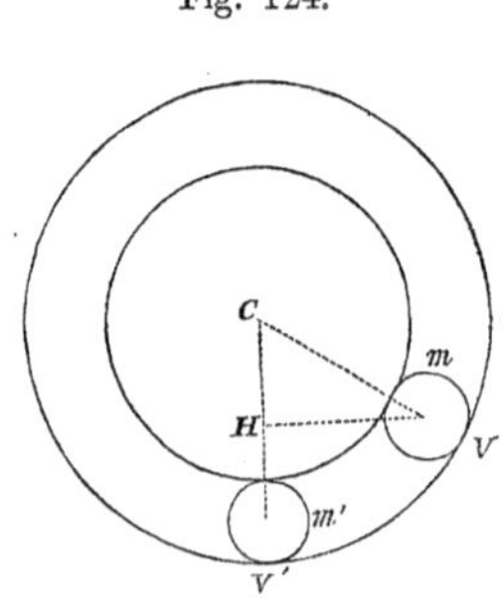

$$M \cdot \frac{V^2}{R}.$$

But if the plane of the groove be vertical, the weight of the body will also exert its influence; for the weight being resolved into two components, one tangent and the other normal to the curve at the place of the body, the latter will sometimes act with, and sometimes in opposition to the centrifugal force, while the former will sometimes increase and sometimes diminish the velocity; so that the pressure becomes greater or less than the centrifugal force depending upon these two circumstances. Knowing one of the velocities which the body may have, it is easy, by the principle of living forces, to find the others. Take the body at its lowest point m', and denote its velocity, supposed known, by V', and let it be required to find its velocity at any other point m, whose vertical height above m' is H. Denote the velocity at this latter point by V, then will the loss of living force in passing from m' to m be

when vertical, the effect of the body's weight;

from one velocity to find the others;

$$M V'^2 - M V^2;$$

and this being equal to double the quantity of action of the weight denoted by W, in the same interval, which quantity of work is $2\,WH$, we have,

$$M(V'^2 - V^2) = 2\,WH;$$

replacing M by its equal $\frac{W}{g}$, and reducing

$$V'^2 - V^2 = 2\,g\,H,$$

$$V = \sqrt{V'^2 - 2\,g\,H}.$$

value of velocity at any point;

Denoting by H', the height due to the velocity V', we have

$$V'^2 = 2\,g\,H';$$

which in the above equation gives

same in terms of difference of level of the points;

$$V = \sqrt{2\,g\,(H' - H)}.$$

Thus, the velocity of the body will be diminished by the action of its weight during its ascent, while, on the contrary, it will be increased during the descent, being always the same at points situated on the same horizontal line. The velocity will be greatest at the lowest and least at the highest point. During the descent, the body will acquire living force by absorbing the work of its weight, which living force will again be destroyed during the ascent because it is opposed to the weight.

velocity greatest at lowest point;

least at the highest.

gain and loss of living force.

Fig. 124.

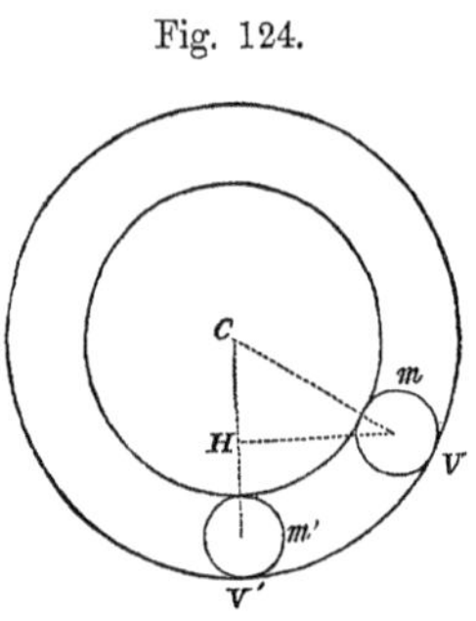

§ 173.—When a body, in virtue of the motive forces which act upon it, describes a curve in space, the effect is the same as though it passed over the arcs of the successive osculatory circles of which the curve is composed. If the positions of the centres C, C', C'', &c., of these successive circular arcs be known, as well as their radii $A\,C$, $A'\,C'$, $A''\,C''$, &c., the curve will be given

Centrifugal force of a body which describes any curve;

Fig. 125.

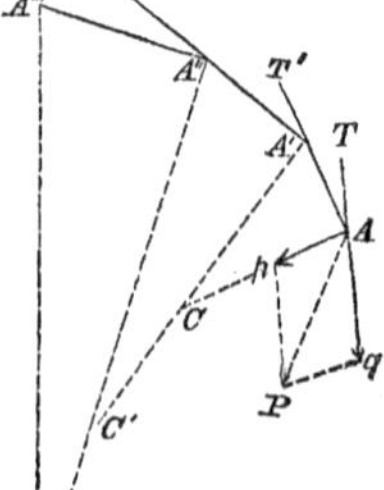

by the series of arcs $A\,A'$, $A'\,A''$, $A''\,A'''$, &c., described about these centres, and terminated by these radii. And it will be easy, from the consideration of the centrifugal and motive forces, to obtain for every point of the curve, the position of the centres and the magnitudes of the radii of the osculatory circles, and, consequently, to trace the path described by the body.

Let P denote the resultant of the motive forces which act upon the body at any particular point as A; M the mass of the body; V its velocity, of which the direction is $A\,T$; and r the radius $A\,C$; then will the centrifugal force be measured by

to trace the curve from the motive and centrifugal forces;

$$\frac{M\,V^2}{r}.$$

But the body, in describing the curve, does not abandon the small arc $A\,A'$, and must therefore be retained on it by a force equal and directly opposed to the centrifugal force; in other words, the motive force $A\,P$, being resolved into two components, one tangent and the other normal to the curve, this latter must be equal to the centrifugal force. Denote the normal component by p, then will

$$p = \frac{M\,V^2}{r} \quad . \quad . \quad . \quad . \quad . \quad (70);$$

value of the normal component of the motive force;

whence

$$r = \frac{M\,V^2}{p} \quad . \quad . \quad . \quad . \quad . \quad (71).$$

radius of curvature;

Such would be the radius of the initial arc $A\,A'$, provided the velocity V were constant during its description. This condition cannot, however, be fulfilled, since the tangential component of the motive force will either increase or diminish the velocity. It will be sufficient to make

$$V = \frac{n + n'}{2};$$

in which n and n' denote the velocities of the body at the beginning and ending of the arc. The former of these must be given, being the *initial* velocity; the latter must be found, and for this purpose we remark, that as the arc is described in a very short time, say the tenth of a second, the motive force, and therefore its tangential component, may be regarded as constant during this interval. Denoting the tangential component by q, and the time by t, we have, from the laws of uniformly varied motion, Eq. (11),

terminal velocity on the initial arc to be found;

Fig. 125.

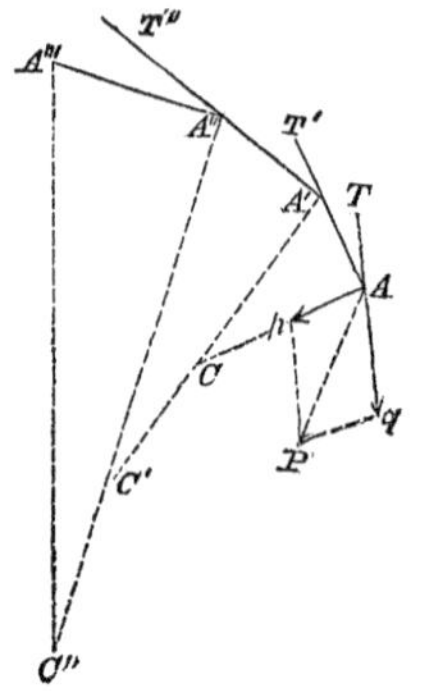

its value;

$$n' = n + \frac{q}{M} t,$$

and

value of mean velocity;

$$V = \frac{n + n'}{2} = n + \frac{q}{2M} t \quad . \quad . \quad (72);$$

which, in Eq. (71), gives

value of radius;

$$r = \frac{M\left(n + \frac{q}{2M} t\right)^2}{p} \quad . \quad . \quad . \quad (73).$$

This distance being laid off from the point A, upon the perpendicular to the tangent $A\,T$, will give the centre C.

The length of the arc, denoted by s, is found from Eq. (10), or

value of arc described;

$$s = n t + \tfrac{1}{2} \frac{q}{M} t^2 \quad . \quad . \quad . \quad . \quad (74).$$

The law of the motive force being known, the intensity of its action on the body at A' becomes known, and its component perpendicular to the tangent $A' T'$, denoted by p', will give

$$p' = \frac{M V'^2}{r'},$$

value of normal component of motive force;

or

$$r' = \frac{M V'^2}{p'};$$

in which r' is the radius of the arc $A' A''$, and V', the mean velocity with which it is described.

Denoting the new tangential component by q', we find, in the same way as before,

$$n'' = n' + \frac{q'}{M} t,$$

terminal velocity on second arc;

$$V' = \frac{n'' + n'}{2} = n' + \frac{q'}{2 M} t;$$

which in the equation above gives

$$r' = \frac{M \left(n' + \frac{q'}{2 M} t\right)^2}{p'};$$

radius of second arc;

and this being laid off, as before, upon the perpendicular to the tangent $A' T'$, will give the centre C'.

The length of the arc $A' A''$, denoted by s', will be found from

$$s' = n' t + \frac{q'}{2 M} t^2.$$

length of second arc;

Finding the value of the motive force at A'', its normal and tangential components p'' and q'', as well as the mean velocity V'', we obtain the value of the radius $C'' A''$, and the position of the centre C''; the tangential component and time will give us the length of the new

the same process for other arcs;

osculatory arc, and thus the description of the curve may be continued to the end.

application to the case of a bomb-shell thrown into the air;

To apply this general case to a particular example, take the instance of a bomb thrown into the air. The forces here are, that arising from the explosive action of the powder and which gives the initial velocity, the resistance of the air, and the weight of the bomb.

Fig. 126.

resistance of air;

Let A be the mouth of the piece, of which the axis coincides with the line $A\,T$. This line will be tangent to the path described by the bomb at the point A. Denote the weight of the bomb by W, the initial velocity by n, and the resistance of the air due to this velocity by f. The value of f may be taken from a table giving the resistances corresponding to different velocities and calibres. Through A draw $A\,H$ parallel to the horizon, and denote the angle $T A\,H$ by α; lay off upon the vertical through A, the distance $A\,W$ to represent the weight of the bomb, and resolve this weight into two components: one, $A\,c = p$, normal to the tangent $A\,T$; and the other, $A\,m = k$, in the direction of this line. The angle $W A\,c$ is equal to the angle $T A\,H = \alpha$; and hence,

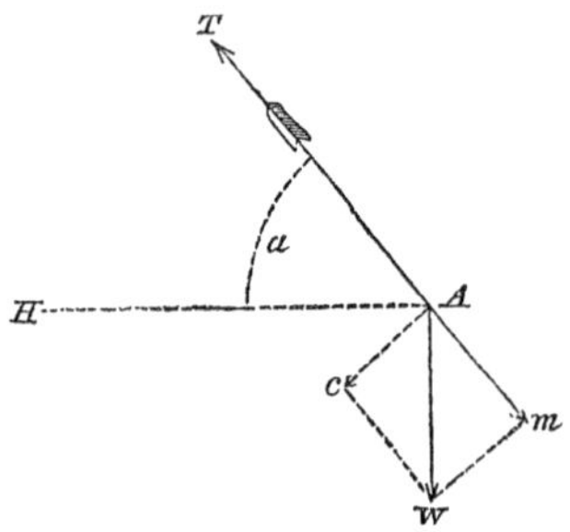

components of the weight of the bomb;

$$p = W \cos \alpha,$$

$$k = W \sin \alpha;$$

and since the resistance of the air is directly opposed to the motion, the force in the direction of the tangent, after the initial impulse, is retarding, and becomes

$$q = k + f = -\ (W \sin \alpha + f);$$ tangential component;

therefore

$$n' = n - \frac{W \sin \alpha + f}{M} \cdot t,$$ terminal velocity;

and

$$V = n - \frac{W \sin \alpha + f}{2\,M} \cdot t;$$ mean velocity;

this value and that of p, in Eq. (71), give

$$r = \frac{M\left(n - \dfrac{W \sin \alpha + f}{2\,M}\,t\right)^2}{W \cos \alpha};$$ radius of initial arc;

and writing, in Eq. (74), for q its value, we find

$$s = n\,t - \tfrac{1}{2}\,\frac{W \sin \alpha + f}{M} \cdot t^2.$$ length of initial arc;

Fig. 127.

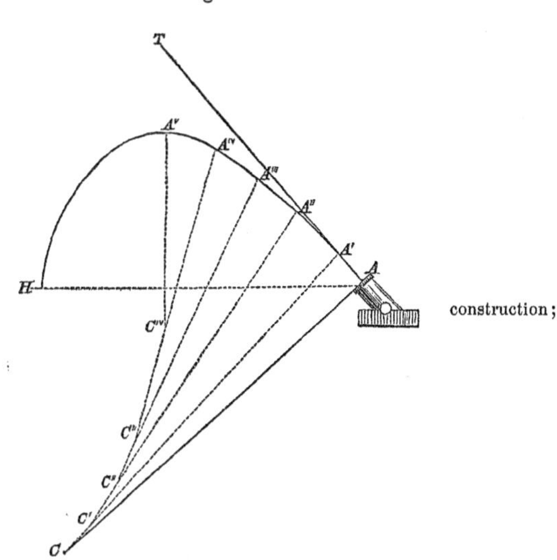

construction;

Through the point A, draw an indefinite perpendicular to the line $A\ T$, and lay off from A the distance $A\ C$, equal to r; with C as a centre, r as radius, describe the arc $A\ A'$ equal to s. This will give the initial arc.

The linear dimension of an arc at the unit's distance from C, is

length of arc at unit's distance from the centre;

$$\frac{s}{r};$$

and denoting the ratio of the circumference of the circle to its diameter by π, we have

$$2\pi : \frac{s}{r} :: 360° : z,$$

its value in arc;

$$z = \frac{360° \times s}{2\pi r};$$

in which z denotes the number of degrees in this arc, or the value of the angle $A\,C\,A'$. But this angle is equal to that made by the tangents $A\,T$ and $A'\,T'$ at the extremities of arc $A\,A'$, and the angle which the tangent at the beginning of the second arc, $A'\,A''$ makes with the horizon, or the angle $T''\,A'\,H'$, will be

angle of the tangents at the initial points of two consecutive arcs;

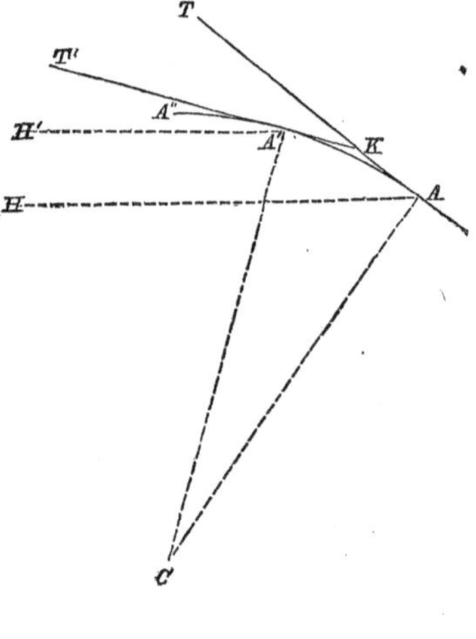

Fig. 128.

angle of projection at the beginning of the second arc;

$$\alpha - z = \alpha'.$$

Pursuing the same operation as before, we find

$$p' = W \cos \alpha',$$

$$k' = W \sin \alpha';$$

and taking from the tables the resistance f', corresponding to the new velocity n', we construct in the same way the second arc $A' A''$, &c., &c.

It is to be remarked, that as the angle denoted successively by α, α', &c., diminishes in passing from arc to arc, it will presently become equal to zero, at the summit, and afterward take the negative sign; in the first case, the tangential component of the weight of the bomb will be zero, its sign will then change, and instead of being a retarding, it will become an accelerating force. Hence, in this curve, three portions are to be distinguished, viz.: the ascending branch, the descending branch, and that immediately about the summit.

variation in the angles of projection;

three parts of the curve;

The resistance of the atmosphere to the motion of bodies in it, is found to vary as the square of the velocity of the moving body, and some idea of the intensity of this resistance may be formed from the fact, that a twenty-four pound shot, projected under an angle of 45°, in vacuo, with a velocity of 2000 feet a second, would have a range of 125000 feet, while the same ball, projected under the same circumstances in the atmosphere, would only attain to the range of 7300 feet; about one-seventeenth of the former.

range in atmosphere and in vacuo.

§ 174.—The laws of the centrifugal force may be illustrated experimentally by means of the *whirling-table.*

Fig. 124.

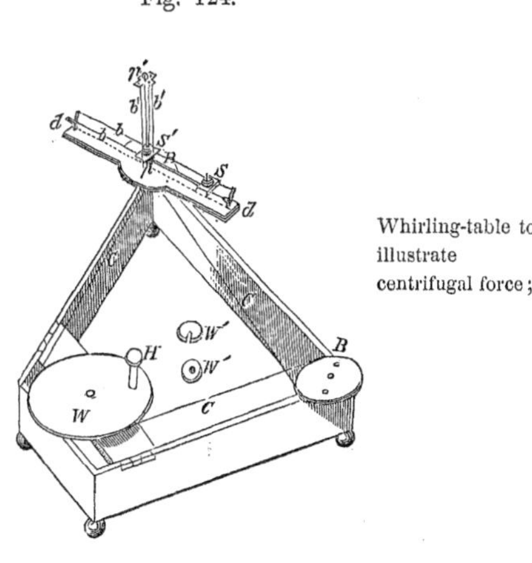

Whirling-table to illustrate centrifugal force;

This consists of a framework upon which are mounted two vertical axes. Upon the top of each axis is fastened a circular block B, B, having a groove cut in the circumference for the reception of an endless cord C, C, C, which also passes round a wheel W. This wheel is provided with a crank and handle H, for the purpose of communicating motion to the whole. The circular blocks are so made, that their circumferences, around which the cord passes, may be varied to change the velocity of rotation. A piece of wood $d\,d$, is mounted upon each of the circular blocks, by means of screws, to support two polished horizontal metallic bars b, b, along which a small stage S may slide with as little friction as possible. This stage is connected with another S', which slides freely on a pair of vertical bars b', b', by means of a piece of flexible catgut passing over the pulleys p, p', in such manner as to lift the stage S' in a vertical, when motion is communicated to S in a horizontal, direction.

arrangement of the parts of the table;

Fig. 124.

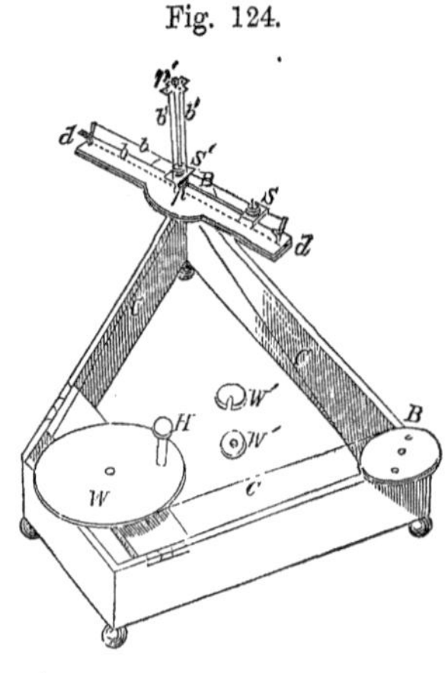

The stage S' is placed with its centre immediately over the axis of motion.

On the piece $d\,d$ is a graduated linear scale, having its zero in the axis, for the purpose of measuring the distance of the stage S from the centre of motion. A series of weights W', W', in the shape of small circular plates, complete this part of the apparatus. The weights, being perforated in the centre, are kept in place by a vertical pintle rising from the middle of each stage.

scale and moveable weights;

example first;

Example 1*st*. Load one of the stages S, with the weight 5, and place it over the division 8 of the scale; load the other stage S with the weight 2, and place it over the division 5; make the circumference of the first circular block double that of the second. The angular velocity of the first being V_1, that of the second will be $2\ V_1$. When motion is communicated, the centrifugal forces will, Eq. (68), be, respectively,

$$5 \times 8\ V_1^2 \text{ and } 2 \times 5 \times 4\ V_1^2,$$

or

$$40\ V_1^2 \text{ and } 40\ V_1^2;$$

that is to say, the centrifugal forces will always be equal to each other. Hence, if the stages S' be loaded equally, they will be drawn up simultaneously. result;

Example 2*d*. Retaining the same ratio as before between the angular velocities, viz., V_1 and 2 V_1, load one of the stages S with weight 6, and place it over the division 8 of the scale; load the other stage S with weight 3, and place it over the division 7. When rotation takes place, the centrifugal forces will be, respectively, example second;

$$6 \times 8\ V_1^2 = 48\ V_1^2,$$

$$3 \times 7 \times 4\ V_1^2 = 84\ V_1^2,$$

the ratio of which is

$$\frac{48}{84} = \frac{12}{21};$$

and hence, if the first stage S' be loaded with 12 weights, and the second with 21, they will rise together, and with a little care may be kept suspended by properly regulating the motion. result;

If the particles of which a body is composed may move among each other, that is, if the body be soft, a change may be effected by the action of this force in its figure.

Such a body of a spherical form, revolving about one of its diameters, acquires a flattened shape in the direction of this diameter or axis, because the parts that lie in the plane of the greatest circumference which can be drawn perpendicular to the axis, that is, in the plane of the body's equator, have the greatest centrifugal force, while those when a rotating body is soft, and of spherical figure, it acquires a flattened shape;

in the neighborhood of the poles have the least; the former will, therefore, recede from and the latter approach the centre. Hence the inference in regard to the causes of the flattened figure of the earth.

Example 3d. On the vertical axis $a\ b$, is an armillary sphere, composed of elastic wires, fitting round the axis by means of a ring, which holds them all together. By this contrivance it is possible for the elastic wires to assume an elliptical figure, having a shorter vertical diameter. Screw this apparatus into the middle of the circular block of the whirling table, and give to the whole a rotatory motion; the wires, instead of their original form represented by the dotted lines, will assume, in consequence of the centrifugal force, the figure shown in the dark lines.

experimental illustration.

Fig. 130.

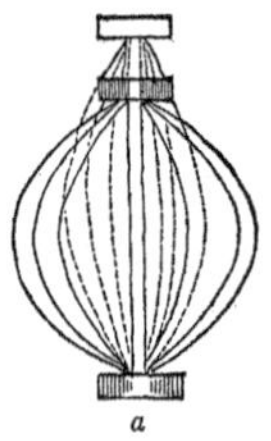

Principle of the areas;

§ 175.—When a body moves with uniform motion, it passes over equal spaces in equal times. Thus, suppose the body to start from A, and to move uniformly in the direction from A to B; the line $A\ B$ being divided into equal spaces $A\ m'$, $m'\ m''$, $m''\ m'''$, &c., these spaces will be described in equal times. If the several points of division be joined with any point as C, off the line, a series of triangles $A\ C\ m'$, $m'\ C\ m''$, $m''\ C\ m'''$, &c., will be formed, all having a common vertex and equal bases lying in the same straight line. The areas of these triangles will, therefore, be equal, and

Fig. 131.

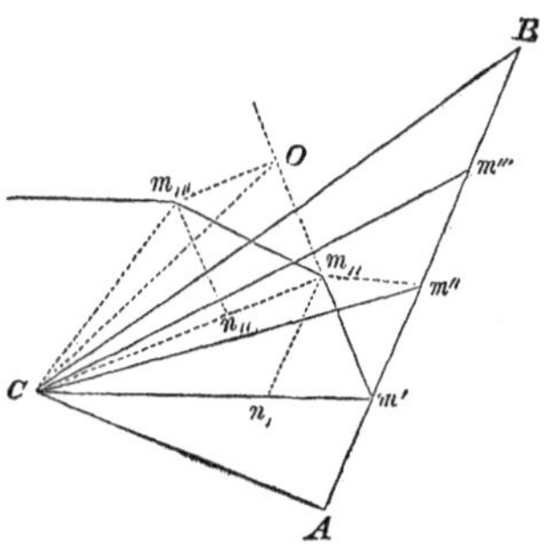

a body in motion under the action of the central force;

will have been described in equal times during the motion of the body by the line joining it with the point C.

If when the body arrives at m', it receive an impulse in the direction from m' to C, which would cause it, if moved from rest, to describe the path $m' n_{,}$ in the same time that it would have described $m' m''$ if unmolested, then will it describe, in the same time, the diagonal $m' m_{,,}$ of the parallelogram constructed upon $m' n_{,}$ and $m' m''$ as sides. The line $m'' m_{,,}$ being parallel to $m' C$, the triangles $C m' m''$ and $C m' m_{,,}$ will have the same base $C m'$, and equal altitudes; their areas will therefore be equal; hence the triangles $C A m'$ and $C m' m_{,,}$ will be equal. In like manner, if when the body arrives at $m_{,,}$, it receive another impulse directed towards C, which would cause it to describe $m_{,,} n_{,,}$, in the time it would have described $m_{,,} O = m' m_{,,}$ if undisturbed at $m_{,,}$, it will describe the diagonal $m_{,,} m_{,,,}$ of the parallelogram constructed upon $m_{,,} O$ and $m_{,,} n_{,,}$ as sides; the triangle $C m_{,,} m_{,,,}$ will be equal to the triangle $C m_{,,} O = C m' m_{,,} = C A m'$. These equal triangles are described in equal intervals of time by the line joining the moving body with the centre C. If now the impulses towards C be applied at intervals of time indefinitely small, the force may be considered incessant, the sides of the polygon $A m'$, $m' m_{,,}$, $m_{,,} m_{,,,}$, &c., will become indefinitely small, and the polygon itself will not differ from a curve. The line which joins the body and the centre C, is called the *radius vector;* and the incessant force acting in the direction of this line towards the centre, is called the *centripetal force.*

the forces first impulsive;

then incessant;

radius vector;

Whence we conclude, that when any body having received a motion, is acted upon by a centripetal force, of which the direction is oblique to that of the motion, its radius vector will describe equal areas in equal times.

areas described by radius vector proportional to the time of description;

And conversely, *if the radius vector of a body moving in a curve, be found to describe equal areas in equal times about a fixed point, the body must be urged towards this fixed point by a centripetal force,* for the equality of the triangles

conversely, the areas being equal in equal times, the force must tend to the fixed point;

$C m' m''$ and $C m' m_{,,,}$ $C m_{,,} O$ and $C m_{,,} m_{,,,,}$ &c., depends upon the lines $m'' m_{,,,}$ $O m_{,,,,}$ &c., being respectively parallel to $m' C$, $m_{,,} C$, &c., drawn from the positions in which the body receives the deflecting impulses to the centre C.

Fig. 131.

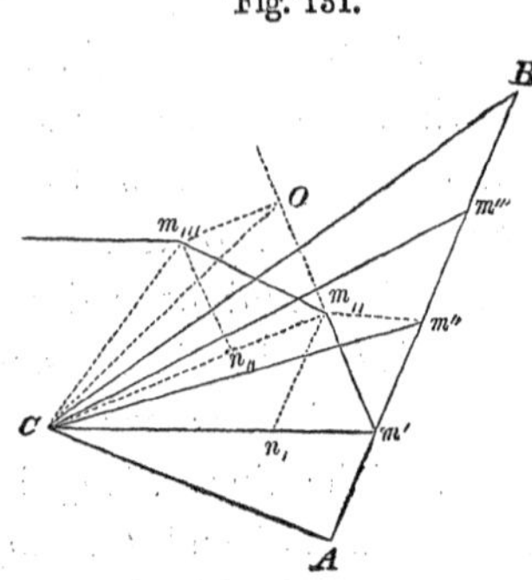

Denote the area by A, and the time in which it is described by t; the ratio of A to t, must, from what has just been shown, be constant. Denote this constant by a, and we shall have

ratio of areas to the times.

$$\frac{A}{t} = a,$$

or

$$A = a\,t \quad . \quad . \quad . \quad . \quad . \quad . \quad (74)';$$

and making t equal to unity, we find

$$A = a;$$

from which we conclude, that a denotes the area described in the unit of time.

Measure of the centripetal force;

§ 176.—Let a body describe the curve $A\,B$ under the action of a centripetal force directed to the centre C; and suppose m and m' to be two positions of the body very near to each other. Draw the tangent $m\,Q$ to the curve at the place

Fig. 132.

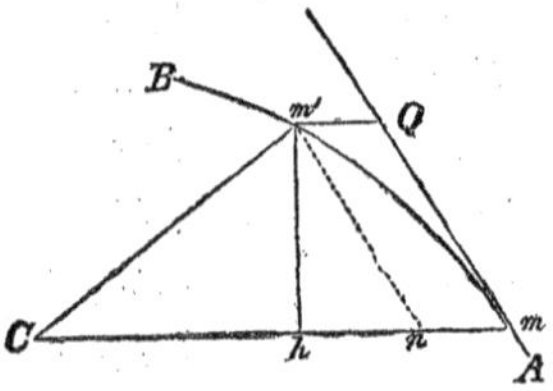

m, and draw $m'Q$ parallel to the radius vector Cm, and $m'n$ parallel to the tangent. If the centripetal force had ceased to act at m, the body would have described mQ in the time that it has actually described mm'. Again, if the body had been moved from rest at m by the centripetal force alone, it would have described the path $mn = m'Q$, in the same time; the path mn is, therefore, the path due to the action of the centripetal force. The places m and m' being very near each other, the centripetal force may be considered as constant during the passage of the body from the one to the other. Denote the velocity which the centripetal force can generate in the body at m, in a unit of time, by $v_{,}$, then, Eq. (7), will

components of the actual velocity;

$$mn = \tfrac{1}{2} v_{,} t^2,$$

whence

$$v_{,} = \frac{2\,mn}{t^2};$$

but, Eq. (74)′,

$$t = \frac{A}{a};$$

and substituting this for t, we find

$$v_{,} = \frac{2a^2 \times mn}{A^2}.$$

value of the acceleration due to the centripetal force;

Multiplying both members by the mass of the moving body, denoted by M, we have

$$Mv_{,} = \frac{2Ma^2 \times mn}{A^2}.$$

Draw from m', the line $m'h$ perpendicular to Cm, then, because A is the area of the triangle Cmm', will

$$A = \tfrac{1}{2}\,Cm \times m'h,$$

which in the above equation gives

the intensity of the centripetal force;

$$Mv_{,} = 4Ma^2 \times \frac{mn}{\overline{Cm}^2 \times \overline{m'h}^2} \quad . \quad . \quad (74)''.$$

versed sine;

altitude of the sector;

The distance mn is called the versed sine of the arc mm', and $m'h$ the altitude of the sector; the first member, or $Mv_{,}$, is the quantity of motion which the centripetal force can generate in a unit of time, and therefore measures its intensity; whence we conclude that, *the intensity of the centripetal force by which a body is made to describe a curve, is always equal to four times the mass of the body into the square of the area described by its radius vector in a unit of time, multiplied by the versed sine of the elementary arc and divided by the square of the radius vector into the square of the altitude of the sector.*

Fig. 132.

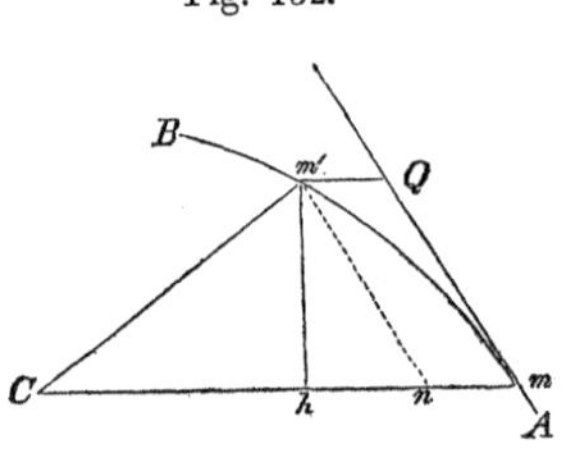

value of the intensity of the force in words.

XII.

MOTIONS OF THE HEAVENLY BODIES.

Phenomena of the heavenly bodies;

§ 177.—The phenomena of the heavenly bodies may be divided into three classes: the first, comprehending the motion of revolution round the sun; the second, the motion of rotation about their respective centres of inertia; and third, their figure and the oscillations of the fluids on their surfaces. It is only proposed to consider the force which produces the motion of revolution, and the orbits which the bodies would, if undisturbed, describe.

Observation has established three laws respecting the

motion of the planets, which, from their discoverer, are called KEPLER'S laws, viz.: laws of Kepler:

1st. *The planets move in plane curves, and their radii vectors describe round the centre of the sun, areas proportional to the times of their description.* 1st law;

2d. *The orbits of the planets are ellipses with the centre of the sun in one of the foci.* 2d law;

3d. *The squares of the times of revolution of the different planets are to one another as the cubes of their mean distances from the sun or semi-major axes of their orbits.* 3d law;

These laws relate only to a motion of translation, and must, therefore, be limited to the motion of the centres of gravity of the planets. only relate to motion of translation.

§ 178.—From the first of these laws, and the principle of areas proportional to the times, explained in § 175, it follows that, *the centripetal force which keeps the planets in their orbits is directed to the centre of the sun, and that this body is, therefore, the centre of the system.* Consequences of first law;

The consequence of the second law relates to the variation which takes place in the intensity of the centripetal force arising from a change in the body's place, and may be determined thus. Let m and m', be two consecutive places of the planet moving in an ellipse of which CA and CB are the semi-transverse and semi-conjugate axes, and having the sun, towards which the centripetal force is directed, in the focus S. Draw $m'n$ parallel to the tangent mQ, and produce it till it meets mC, drawn to the centre of the ellipse, in the point v; let fall the perpendicular $m'h$ upon the radius vector Sm; join the body at m with the other focus that of the second deduced;

Fig. 133.

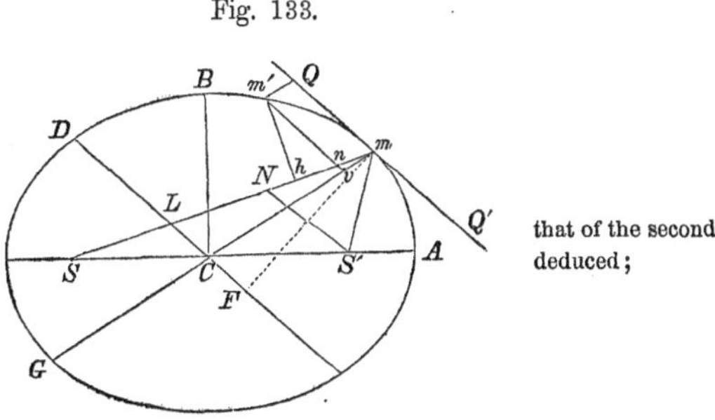

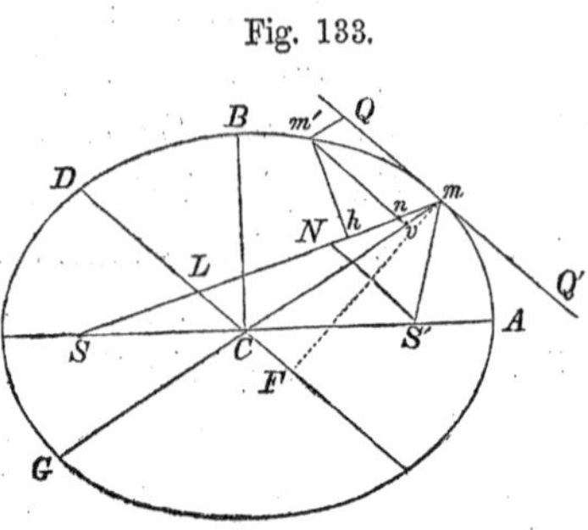

Fig. 133.

S'; draw $S'N$ and CD parallel to the tangent mQ, and produce mC to the curve at G.

Construction of the figure; The tangent QQ' makes equal angles, QmS and $Q'mS'$, with the line drawn from the place m to the foci, and because $S'N$ is parallel to this tangent, the triangle $mS'N$ is isosceles, making $S'm = Nm$; and because CD is parallel to $S'N$, and CS is equal to CS', the distance NL is equal to LS; hence $mL = \frac{mS + mS'}{2}$ is equal to the semi-transverse axis $CA = A$. Denote the semi-conjugate axis by B.

In the similar triangles mnv and mLC, we have,

$$mn : mv :: mL : mC;$$

whence, writing A for mL, we have

value of the versed sine;

$$mn = \frac{A \cdot mv}{mC}.$$

Again, drawing mF perpendicular to DC, we have, from the similar right-angled triangles mLF and $m'hn$,

$$\overline{m'h}^2 : \overline{m'n}^2 :: \overline{mF}^2 : \overline{mL}^2;$$

whence, writing A for mL, we have

value of the altitude of sector;

$$\overline{m'h}^2 = \frac{\overline{m'n}^2 \times \overline{mF}^2}{A^2};$$

and, dividing the last equation by this one, we have

$$\frac{m\,n}{\overline{m'\,h}^2} = A^3 \times \frac{m\,v}{m\,C} \times \frac{1}{\overline{m'\,n}^2 \times \overline{m\,F}^2}.$$ ratio of the versed sine to the square of altitude of sector;

The equation of the ellipse, referred to the conjugate diameters Cm and CD, gives, because the points n and v will sensibly coincide for consecutive places of the body,

$$\overline{m'\,n}^2 = \frac{\overline{C\,D}^2}{\overline{C\,m}^2} \times m v \times v\,G;$$

which, substituted for $\overline{m'\,n}^2$ above, we find

$$\frac{m\,n}{\overline{m'\,h}^2} = A^3 \times \frac{C\,m}{\overline{C\,D}^2 \times \overline{m\,F}^2 \times v\,G};$$ the same, in other terms;

and, because the rectangle of the semi-axes is equivalent to the parallelogram constructed upon the semi-conjugate diameters CD and Cm, we have

$$\overline{C\,D}^2 \times \overline{m\,F}^2 = A^2 \times B^2;$$

moreover, the points m and m' being contiguous, $G\,v$ will not differ sensibly from $2\,Cm$. Making these substitutions, the above equation reduces to

$$\frac{m\,n}{\overline{m'\,h}^2} = \frac{A}{2\,B^2};$$ its final value;

and, multiplying both members by $\frac{4\,M\,a^2}{\overline{C\,m}^2}$,

$$4\,M\,a^2 \times \frac{m\,n}{\overline{m'\,h}^2 \times \overline{C\,m}^2} = \frac{2\,M\,a^2\,A}{B^2} \times \frac{1}{\overline{C\,m}^2}.$$

The first member we have seen, Eq. (74)″, is the intensity of the centripetal force at m. Calling this force F and writing r for the radius vector Cm, we finally have

value of the force;

$$F = \frac{2\,M a^2 A}{B^2} \times \frac{1}{r^2}.$$

consequence of the second law;

Every thing being constant in the second member but r, it follows that, *the force which urges a planet towards the sun, varies inversely as the square of the planet's distance from that body.*

The consequence of the third law is not less important, and may be evolved thus. Multiply both members of the last equation by $\pi^2 A^2 B^2$, and we have

to find the consequence of the third law;

$$F \pi^2 A^2 B^2 = 2\,\pi^2 M a^2 A^3 \times \frac{1}{r^2};$$

divide both members of this equation by Fa^2, and there will result

$$\frac{\pi^2 A^2 B^2}{a^2} = \frac{2\,\pi^2 M}{F} \times A^3 \times \frac{1}{r^2}.$$

Now, $\pi A B$ is the area of the entire ellipse; a is the area described by its radius vector in a unit of time; hence $\frac{\pi A B}{a}$ is the number of units of time in one entire revolution of the planet, called the *periodic time.* Denote this by T, and substitute it for $\frac{\pi A B}{a}$, and we get

periodic time;

the value of its square;

$$T^2 = \frac{2\,\pi^2 M}{F} \cdot \frac{1}{r^2} \cdot A^3.$$

In like manner for any other planet, whose mass is M', mean distance A', radius vector r', periodic time T', and centripetal force F', we have

$$T'^2 = \frac{2\,\pi^2 M'}{F'} \cdot \frac{1}{r'^2} \cdot A'^3;$$

and dividing this equation by the one above

$$\frac{T'^2}{T^2} = \frac{M' F. r^2}{M F'. r'^2} \times \frac{A'^3}{A^3}.$$

ratio of the squares of periodic times;

But, by the *third law*,

$$\frac{T'^2}{T^2} = \frac{A'^3}{A^3};$$

whence

$$\frac{M' F r^2}{M F' r'^2} = 1,$$

or

$$\frac{F}{M} \times r^2 = \frac{F'}{M'} \times r'^2.$$

centripetal acceleration;

Now $\frac{F}{M}$ is the velocity which the centripetal force can generate in one unit of time, or, which is the same thing, it is the measure of the acceleration due to the force which acts upon the planet M; so, likewise, $\frac{F'}{M'}$ is the acceleration due to the centripetal force which acts upon the planet M'; and resolving the above equation into the proportion

$$\frac{F}{M} : \frac{F'}{M'} :: \frac{1}{r^2} : \frac{1}{r'^2},$$

consequence of the third law;

we see that the forces which urge two different planets towards the sun, are to each other in the inverse ratio of the squares of the distances; so that the same law which regulates the intensity of the force in a single orbit, also extends to different planets revolving in different orbits. If r be made equal to r', then will the accelerations due to the centripetal force be equal; that is to say, if all the

at same distance, the centripetal accelerations are equal;

planets were brought to the same distance from the sun, each unit of mass would be urged towards that body with the same intensity; and as the different planets might be inverted in respect to the order of their distances from the sun, without the relation of the periodic times as expressed by the third law being affected, it follows that the force which acts upon all the planets is absolutely the same in kind, and is only qualified, in intensity, by a change of distance. These considerations led Newton to adopt the celebrated hypothesis which laid the foundation of physical astronomy, viz.: *that all bodies attract each other with an energy which is directly proportional to their masses and inversely proportional to the squares of their distances from each other.*

Newtonian hypothesis of universal gravitation;

Starting from this hypothesis, it is easy to solve by a process not suited to an elementary work like this, the converse problem of that which led to the consequence of the second law, and to show, that a heavenly body may describe any one of the conic sections having the sun at one of the foci, depending upon the relation which subsists between its velocity and the energy with which the body and the sun attract each other. The orbit will be a *parabola*, an *ellipse*, or *hyperbola*, according as the square of the body's velocity is equal to, less, or greater than, twice the attractive force, multiplied by the distance from the sun.

consequences of this hypothesis;

the orbits might have been ellipses, parabolas, or hyperbolas.

§ 179.—Let $C\,m\,m'$ be the sector described in the unit of time: take the distance $C\,b$ equal to unity, and describe, with C as a centre and Cb as radius, the arc $b\,d = s_{,}$, which will measure the angular velocity. With C as a centre, and $C\,m' = r$ as radius, describe the arc $m'\,h'$; then will

The angular velocity;

Fig. 134.

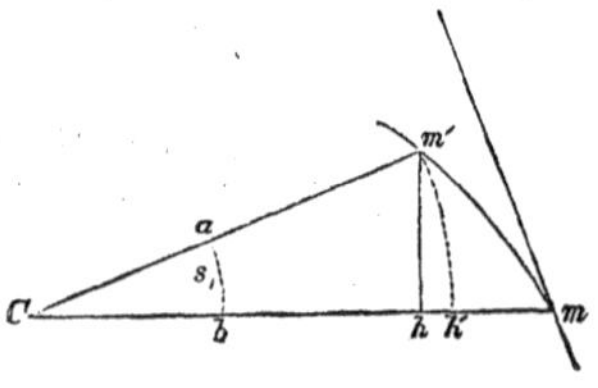

$$m'\,h' = r\,s_{,}.$$

Supposing the unit of time small, in which case m' will be very near to m, $m'h$ will be sensibly equal to $m'h'$, mC to $m'C$, and we have for the area of the sector Cmm',

$$\frac{1}{2} Cm \times m'h' = \frac{1}{2} r^2 s_{,} = a;$$

whence

$$s_{,} = \frac{2a}{r^2}; \qquad \text{its value;}$$

from which we find that, *the angular velocity of a planet about the sun, varies inversely as the square of its distance or radius vector.* law of its variation;

Supposing the planet to describe the ellipse $ABPD$, having the sun at the focus S, the extremities A and P of the transverse axis are called, the former the *Aphelion*, and the latter the *Perihelion*. The angular velocity of the planet is the least at aphelion and greatest at perihelion.

aphelion;

perihelion;

angular velocity greatest at perihelion and least at aphelion;

Fig. 135.

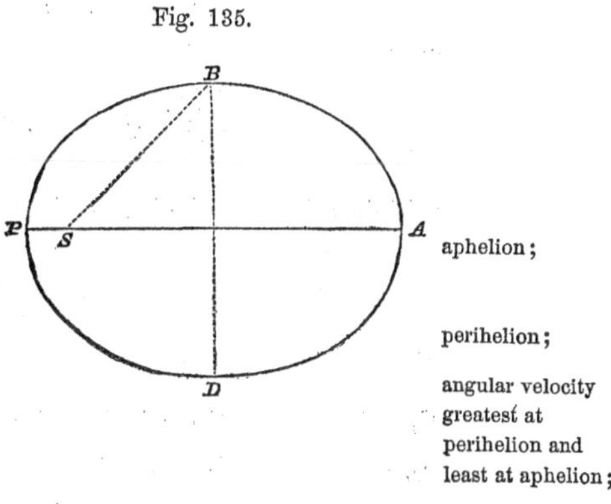

Again, denote the angle CmQ by α, and suppose the velocity of the body on the small arc mm' uniform, which we may do without sensible error, the length of mm' will measure the velocity of the planet at m, since it is described in a unit of time. Hence

absolute velocity;

Fig. 136.

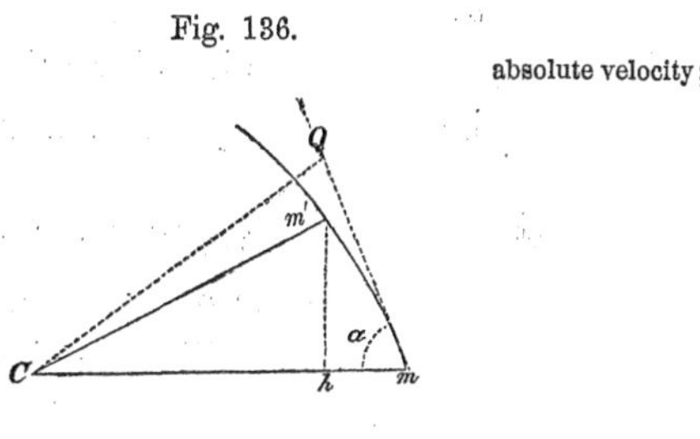

$$mm' \sin\alpha = m'h = V.\sin\alpha;$$

and the area of the triangle or sector Cmm' will be $\frac{1}{2} V . \sin \alpha \times r$; whence

$$\frac{V . r \sin \alpha}{2} = a,$$

or

value of the absolute velocity;

$$V = \frac{2a}{r . \sin \alpha}.$$

Draw the tangent $m Q$ to the curve at the point m, and from C let fall the perpendicular $C Q$, then in the right-angled triangle $C Q m$, will

Fig. 136.

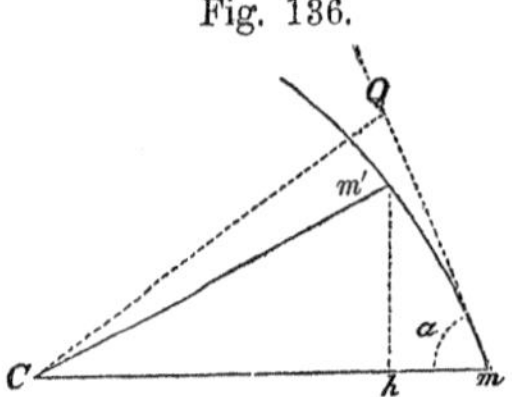

$$C Q = r . \sin \alpha = p,$$

which substituted above gives

the same in different terms;

$$V = \frac{2a}{p};$$

its law of variation;

that is to say, *the velocity of a planet in its orbit, varies inversely as the length of the perpendicular let fall from the centre of the sun upon the tangent drawn to the orbit at the body's place.*

greatest at perihelion and least at aphelion.

From this it follows that the velocity of the planet will be greatest at perihelion and least at aphelion.

Centripetal force directed to the centre of an elliptical orbit;

§ 180.—It will be found convenient when we come to discuss the nature of light, to know that when a body describes an ellipse under the action of a force directed towards the centre of that curve, the force will vary directly as the length of the radius vector, and that the periodic time will be the same for all ellipses, great and small.

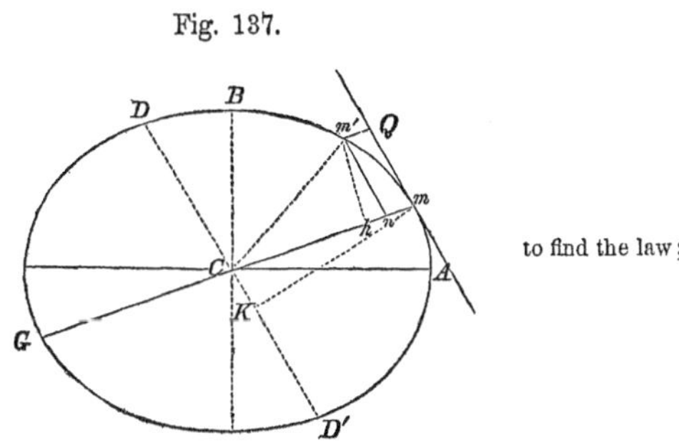

Fig. 137.

Let the body, under the action of a force directed to the centre C, describe the ellipse of which CA and CB are the semi-axes, denoted respectively by A and B; and suppose m and m' to be two of its consecutive places. Draw the tangent mQ at the point m, and parallel to this tangent draw the diameter DD', perpendicular to which, draw from m the line mK. From m' draw $m'n$ parallel to the tangent till it meets the radius vector Cm in n, and let fall upon the same radius vector the perpendicular $m'h$. to find the law;

The equation of the ellipse, referred to its conjugate diameters Cm and CD, gives

$$\overline{m'n}^2 = \frac{\overline{CD}^2}{\overline{Cm}^2} \times mn \times nG;$$

whence

$$mn = \frac{\overline{m'n}^2 \times \overline{Cm}^2}{\overline{CD}^2 \times nG}.$$

value of the versed sine;

Because $m'n$ and $m'h$ are respectively perpendicular to the lines mK and mC, the angles $hm'n$ and CmK are equal, and the angles at K and h being right angles, the triangles $m'nh$ and CmK are similar, and give the proportion

$$\overline{m'n}^2 : \overline{m'h}^2 :: \overline{Cm}^2 : \overline{mK}^2;$$

whence

$$\overline{m'h}^2 = \frac{\overline{m'n}^2 \times \overline{mK}^2}{\overline{Cm}^2};$$

value of the square of sector's altitude;

dividing the last equation by this one, we have

ratio of the versed sine to square of sector's altitude;

$$\frac{m\,n}{\overline{m'\,h}^2} = \frac{\overline{C\,m}^4}{\overline{C\,D}^2 \times \overline{m\,K}^2 \times n\,G}.$$

But the rectangle of the semi-axes is equivalent to the parallelogram described upon the semi-conjugate diameters, hence

$$\overline{C\,D}^2 \times \overline{m\,K}^2 = A^2 \times B^2;$$

moreover, $n\,G$ is sensibly equal to $2\,C\,m$; making these substitutions above, there will result

same in different terms;

$$\frac{m\,n}{\overline{m'\,h}^2} = \frac{\overline{C\,m}^3}{2\,A^2\,B^2};$$

multiplying both members by $4\,M a^2$, and dividing by $\overline{C m}^2$, we have, Eq. (74)″,

$$4\,M a^2 \times \frac{m\,n}{\overline{Cm}^2 \times \overline{m'\,h}^2} = F = \frac{2\,M a^2}{A^2\,B^2} \times C\,m,$$

in which M is the mass of the body. Finally, writing r for $C m$, we find

value of the centripetal force;

$$F = \frac{2\,M\,a^2}{A^2\,B^2} \times r;$$

the law of its variation;

that is to say, *the centripetal force which will cause a body to describe an ellipse when directed to the centre of that curve, varies directly as the radius vector.*

to find the periodic time;

Multiply both members of the last equation by $\pi^2\,A^2\,B^2$, and we have

$$F.\,\pi^2\,A^2\,B^2 = 2\,\pi^2\,M a^2 \times r.$$

Dividing both members of this equation by $F a^2$, and we have

$$\frac{\pi^2 A^2 B^2}{a^2} = 2\,\pi^2 \frac{M}{F} \cdot r = 2\,\pi^2 \times \frac{r}{\frac{F}{M}};$$

taking the square root, and recollecting that $\frac{\pi A B}{a}$, is the periodic time $= T$, we find

$$T = \pi \sqrt{\frac{2\,r}{\frac{F}{M}}}.$$

value of the periodic time;

The quotient $\frac{F}{M}$ is the measure of the acceleration due to the centripetal force, which we have just found to vary directly as the radius vector. This makes the radical expression constant; hence T must also be constant.

Whence we conclude, generally, that *when any number of bodies are solicited towards a fixed point by forces which vary directly as the distances of the bodies from that point, they will describe ellipses*, or circles, one of the varieties of the ellipse; *and that they will all perform their revolutions in the same time.*

conclusion.

XIII.

THE PENDULUM.

§ 181.—A body $M\,Q\,N$, suspended from a horizontal axis A, about which it may swing with freedom under the action of its own weight, is called, in general, a *compound pendulum*. When the body is reduced to a material heavy point, and the medium of connection with the axis is without weight, it is called a *simple pendulum.*

Fig. 138.

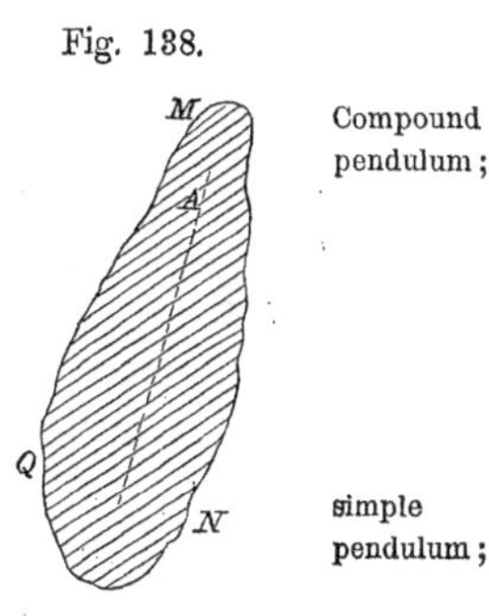

Compound pendulum;

simple pendulum;

Fig. 139.

has no real existence;

The simple pendulum is but a mere conception, and yet the expression for its length, which may easily be found in a manner soon to be explained, is of great practical importance.

When the pendulum is at rest, in such position that its centre of gravity G is below and on the vertical line passing through the axis A, it will be in a state of stable equilibrium, § 151; but as soon as it is deflected to one side, as indicated in the figure, and abandoned to itself, it will swing back and forth about the position of equilibrium, into which it will finally settle in consequence of the resistance of the air and friction on the axis. If these causes of resistance were removed, the pendulum would continue its motion indefinitely; but this cannot be accomplished in practice, and hence such figure and mode of suspension are resorted to as to give these impediments the least possible influence.

effect of friction and resistance of air;

Fig. 140.

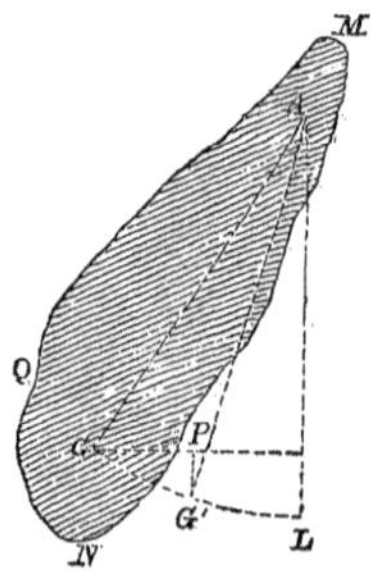

figure of pendulum and mode of suspension;

The pendulum is usually mounted upon a knife-edge A as an axis, resting upon a well-polished plate of metal, or other hard substance, B; and the figure of the pendulum is that of a flat bar C, supporting at its lower end a heavy lenticular-shaped mass D, called a *bob*.

knife-edge and bob;

Fig. 141.

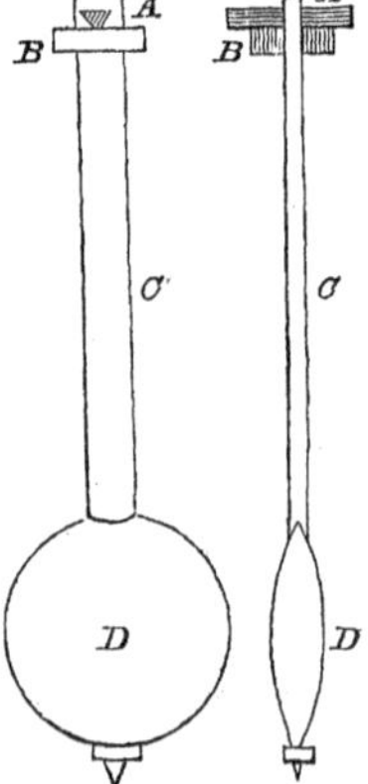

One entire swing of the pendulum, by which its centre of

gravity is carried from the extreme limit G of its path, on one side of the vertical $A L$, to G'' on the other, is called an *oscillation*. oscillation;

To find the time of a single oscillation, call the weight of the entire pendulum, W; its mass, M; its angular velocity at any instant, V_1; its moment of inertia with reference to the axis of suspension, I; the distance of its centre of gravity from the axis, D; the vertical distance $P G'$, through which the centre of gravity must descend from its highest point G to arrive at any point G', y. to find time of a single oscillation; notation;

Fig. 142.

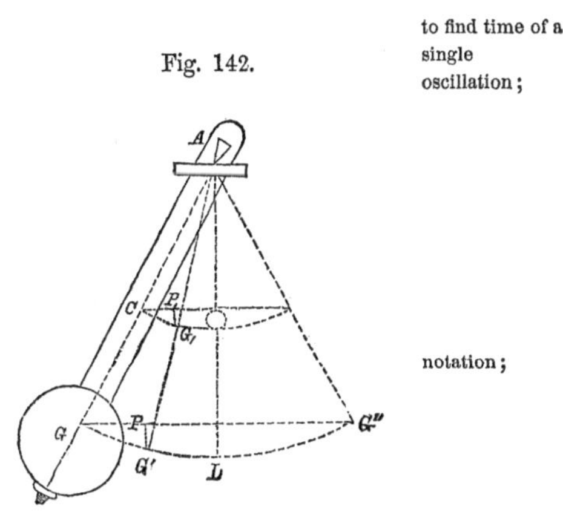

The living force of the pendulum when the centre of gravity reaches the point G' will, § 159, be

$$I \cdot V_1^2,$$

living force;

and the quantity of work of the weight will be

$$W y = M g y,$$

work of the weight;

and hence

$$I V_1^2 = 2 M g y.$$

The point C on the line $A G$ at the unit's distance from A, will, during the motion, describe an arc similar to $G G'$, and the vertical distance $G_{,} P_{,,}$ denoted by $y_{,,}$ through which this point will fall while G is passing to G', will be given by

$$y = D y_{,};$$

fall of the centre of gravity;

and this, in the above equation, gives

$$I \,.\, V_1^2 = 2 M g D y_{,};$$

whence

square of the angular velocity;

$$V_1^2 = \frac{M.D}{I} \cdot 2 g y_{,}.$$

Denoting by $s_{,}$ the small distance described by the point C during the very short interval t, succeeding the instant at which the angular velocity is V_1, we shall have

$$V_1 = \frac{s_{,}}{t},$$

which, in the preceding equation, gives

$$\frac{s_{,}^2}{t^2} = \frac{M \,.\, D}{I} \cdot 2 g y_{,};$$

whence

square of the time required to describe a very small arc;

$$t^2 = \frac{I}{M \,.\, D} \cdot \frac{s_{,}^2}{2 g y_{,}}.$$

Taking $A M$ equal to unity, let $C B C''$ be the arc described in one oscillation by the point M, and $M N$ the small arc $s_{,}$ described in the time t, immediately succeeding the instant at which the angular velocity is V_1. Draw $M E$ perpendicular to the vertical

to find the arc described in the small time;

Fig. 143.

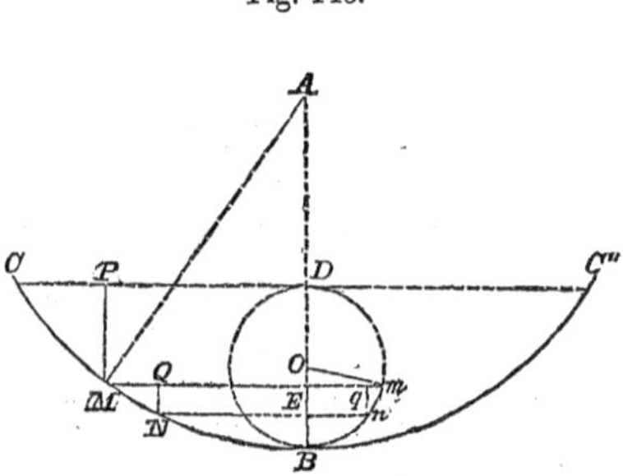

AB, and NQ perpendicular to ME: then, in the similar triangles AME and MNQ, we have

$$QN \; : \; EM \; :: \; MN \; : \; AM;$$

and because AM is unity, and MN is $s_{,}$,

$$s_{,} = \frac{QN}{EM}.$$ one value for the arc;

But from the property of the circle

$$EM = \sqrt{2AB.EB - \overline{EB}^2} = \sqrt{2EB - \overline{EB}^2},$$

and if we take the arc CBC'' very small, the versed sine EB will be a very small fraction, and its second power may be neglected in comparison with the first. Whence

$$EM = \sqrt{2EB};$$

which, in the value of $s_{,}$ above, gives

$$s_{,} = \frac{QN}{\sqrt{2EB}};$$ another value for the arc;

and this, in the value for t^2, gives

$$t^2 = \frac{I}{M.D} \cdot \frac{1}{4g} \cdot \frac{QN^2}{y_{,}BE}.$$ another value for the square of the time;

Upon BD as a diameter, describe a semi-circumference $DmnB$, and through the points M and N, the extremities of the arc $s_{,}$, draw the horizontal lines Mm and Nn, cutting this semi-circumference in the points m and n. Draw the radius Om, and the vertical nq. From the property of the circle we have

$$\overline{mE}^2 = BE \times ED = BE \times PM = BE \times y_{,};$$

Fig. 143.

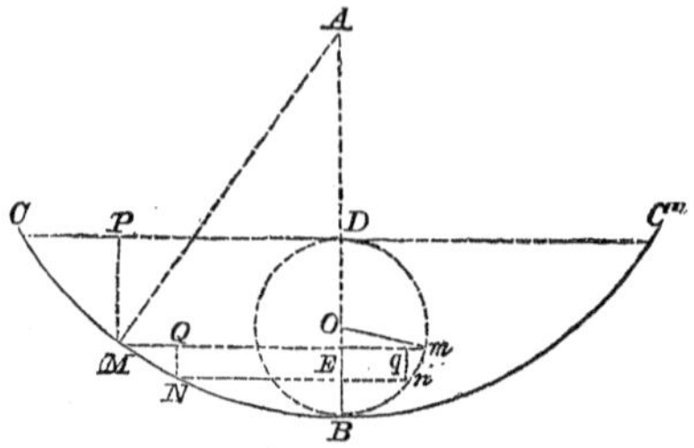

whence

vertical distance from last point;

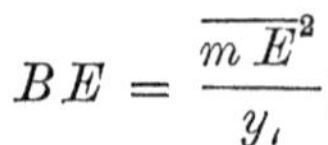

$$BE = \frac{\overline{mE}^2}{y_{\prime}};$$

which, substituted for BE in the equation above, gives

$$t^2 = \frac{I}{M.D} \cdot \frac{1}{4g} \cdot \frac{\overline{QN}^2}{\overline{mE}^2},$$

and, taking the square root,

value of the element of the time;

$$t = \tfrac{1}{2}\sqrt{\frac{I}{g \,.\, M \,.\, D}} \times \frac{QN}{mE}.$$

The two triangles $m\,O\,E$ and $m\,q\,n$ are similar, and give

$$qn = QN \;:\; mE \;::\; nm \;:\; Om;$$

whence

$$\frac{QN}{mE} = \frac{nm}{Om};$$

and this substituted above in the value of t, gives

$$t = \tfrac{1}{2}\sqrt{\frac{I}{g \,.\, M \,.\, D}} \times \frac{nm}{Om}.$$

proportional to the projection of arc on the circle whose diameter is versed sine of arc of oscillation;

Such is the value of the time required to describe the elementary arc MN, which we see is proportional to the arc mn, or to the projection of MN on the semi-circumference described upon DB as a diameter, every other quantity in the second member of the equation being constant; and hence, the time required to describe the whole arc CMB, which is obviously the sum of all the elementary times of describing the elementary arcs MN, &c., must be equal to

the time of making a semi-oscillation found;

$$\tfrac{1}{2}\sqrt{\frac{I}{g \,.\, M \,.\, D}} \times \frac{1}{Om},$$

into the sum of all the projections of MN, &c, on the semi-circumference DmB; but this sum is the semi-circumference itself; and denoting the time from C to B, or that of a semi-oscillation, by $\frac{1}{2}T$, we have

$$\tfrac{1}{2}T = \tfrac{1}{2}\sqrt{\frac{I}{g \,.\, M \,.\, D}} \times \frac{DmB}{Om};$$

its value;

but

$$\frac{DmB}{Om} = \pi = 3.1416,$$

the ratio of the circumference to the diameter; whence,

$$T = \pi\sqrt{\frac{I}{g \,.\, M \,.\, D}} \quad . \quad . \quad . \quad (75).$$

time of a single oscillation;

From this formula we see that the duration is independent of the amplitude of the oscillation, when this amplitude is small; and a pendulum slightly deflected from its vertical position and abandoned to itself, will oscillate in equal times whatever be the magnitude of the arc, provided it be inconsiderable. Such oscillations are said to be *Isochronal*.

isochronal oscillations;

If the number of oscillations performed in a given interval, say *ten* or *twenty minutes*, be counted, the duration of a single oscillation will be found by dividing the whole interval by this number.

time of a single oscillation found from observation;

Thus, let θ denote the time of observation, and N the number of oscillations, then will

$$T = \frac{\theta}{N} = \pi \sqrt{\frac{I}{g \,.\, M \,.\, D}};$$

and if the same pendulum be made to oscillate at some other location during the same interval θ, the force of gravity being different, the number N' of oscillations will be different; but we shall have, as before, g' being the new force of gravity,

the same for a second place;

$$\frac{\theta}{N'} = \pi \sqrt{\frac{I}{g' \,.\, M \,.\, D}}.$$

Squaring and dividing the first by the second, we find

$$\frac{N'^2}{N^2} = \frac{g'}{g} \quad . \quad . \quad . \quad . \quad . \quad (76);$$

forces of gravity are as squares of number of oscillations in same time.

that is to say, the intensities of the force of gravity, at different places, are to each other as the squares of the number of oscillations performed in the same time, by the same pendulum. Hence, if the intensity of gravity at one station be known, it will be easy to find it at others.

Simple pendulum;

§ 182.—Resuming the general value for I, Eq. (65), we have

$$I = I_1 + D^2 M;$$

which value of I, in Eq. (75), gives

$$T = \pi \sqrt{\frac{I_1 + D^2 M}{g \,.\, M \,.\, D}} \quad . \quad . \quad . \quad (77).$$

mass concentrated in a single point;

If, now, we suppose the entire mass of the pendulum to be concentrated into a single point, and this point connected with the axis by a medium without weight, we have

$$I_1 = \Sigma\, m\, r^2 = 0;$$

moment of inertia in reference to the centre of gravity;

since the centre of gravity must also go to that point, and $r = r' = r'' = \&c. = 0$; whence, writing l for the new value assumed by D, which now becomes the distance from the axis to the single heavy point, we have

time of oscillation of the simple pendulum;

$$T = \pi \sqrt{\frac{l}{g}} \quad . \quad . \quad . \quad . \quad (78);$$

which is the expression for the time of oscillation of a *simple pendulum* of which l is the length.

If the time of oscillation of the simple, be the same as that of the compound pendulum, we shall have, from Eqs. (75) and (78),

$$\pi \sqrt{\frac{I}{g \,.\, M \,.\, D}} = \pi \sqrt{\frac{l}{g}};$$

or

$$l = \frac{I}{M \,.\, D} = \frac{I_1 + M D^2}{M D} \quad . \quad . \quad (79);$$

equivalent simple pendulum;

in which case l is called the *equivalent simple pendulum;* that is to say, the length of a simple pendulum which will oscillate in the same time as a compound pendulum whose moment of inertia in reference to the axis of suspension is I, whose mass is M, and of which the axis of suspension is at a distance from the centre of gravity equal to D.

centre of oscillation;

The point situated on a line drawn through the centre of gravity of the pendulum, perpendicular to the axis of suspension, and at a distance from that axis equal to l, is called the *centre of oscillation;* and is that point of which the circumstances of oscillation would in nowise be altered were the entire pendulum concentrated into it, or were it disconnected from the other points of the pendulous mass, its connection with the axis being retained.

§ 183.—*A line drawn through the centre of oscillation, and parallel to the axis of suspension, is called the axis of oscillation. The axes of suspension and of oscillation are reciprocal.*

Axes of suspension and of oscillation are reciprocal;

Let D' denote the distance of the axis of oscillation from the centre of gravity; then will

$$l = D + D'.$$

Invert the pendulum and make the axis of oscillation the axis of suspension, take l' for the new equivalent simple pendulum, then will

new equivalent simple pendulum;

$$l' = \frac{I_1 + MD'^2}{M \, . \, D'};$$

but we have, from the foregoing equation,

$$D' = l - D;$$

and this, in the preceding value for l', gives

$$l' = \frac{I_1 + M(l - D)^2}{M \, . \, (l - D)}.$$

Again, from Eq. (79), we have

$$l - D = \frac{I_1}{MD};$$

substituting this in the above value for l', we finally get

the simple pendulum the same;

$$l' = \frac{I_1 + MD^2}{MD} = l;$$

that is to say, when the axis of oscillation is taken as the axis of suspension, the old axis of suspension becomes the new axis of oscillation. In other words, these axes are *reciprocal*. This furnishes an experimental method for finding the length of any equivalent simple pendulum, which is the more valuable in view of the great difficulty of computing the moment of inertia of a compound pendulum by the ordinary calculus, owing to the peculiar forms of that instrument rendered necessary by the circumstances under which it is employed. But before proceeding to the explanation of this method, it will be proper to premise, that the time of oscillation of a compound pendulum will be a minimum, when, in Eqs. (78) and (79),

conclusion; equivalent simple pendulum found from experiment;

$$\frac{I_1 + D^2 M}{M D} = \frac{\frac{I_1}{M} + D^2}{D} = l,$$

value of equivalent simple pendulum;

is the least possible; or replacing $\frac{I_1}{M}$ by its value K', deduced from Eq. (66)′ by making $D = 0$, the expression

$$\frac{K'^2 + D^2}{D},$$

must be the least possible.

But it may easily be shown, either by trial, or by a simple process of the calculus, that this expression is a minimum when

$$K' = D,$$

and consequently

$$l = 2K';$$

length of the shortest equivalent simple pendulum;

that is to say, the time of oscillation of a pendulum will be the least possible when the axis of suspension passes through the principal centre of gyration, and the length of the equivalent simple pendulum is twice the principal radius of gyration.

usual form of the compound pendulum;

Let A and A' be two acute parallel prismatic axes firmly connected with the pendulum, the acute edges being turned towards each other. The oscillation may be made to take place about either axis by simply inverting the pendulum. Also, let M be a sliding mass capable of being retained in any position by the clamp-screw H. For any assumed position of M, let the principal radius of gyration be $G\,C$; with G as a centre, $G\,C$ as radius, describe the circumference $C\,S\,S'$. From what has been explained, the time of oscillation about either axis will be shortened as it approaches, and lengthened as it recedes from this circumference, being a minimum, or least possible, when on it. By moving the mass M, the centre of gravity, and therefore the gyratory circle of which it is the centre, may be thrown towards either axis. The pendulum *bob* being made heavy, the centre of gravity may be brought so near one of the axes, say A', as to place the latter within the gyratory circumference, keeping the centre of this circumference between the axes, as indicated in the figure. In this position, it is obvious that any motion in the mass M would at the same time either shorten or lengthen the duration of the oscillation about both axes, but unequally, in consequence of their unequal distances from the gyratory circumference.

device to change the position of the centre of gravity;

Fig. 144.

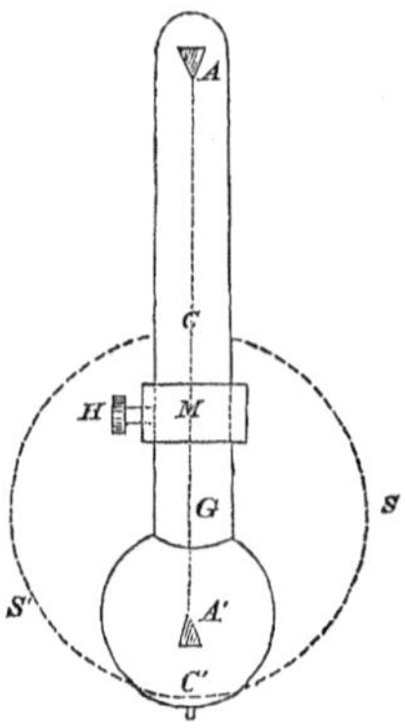

position of centre of gravity;

pendulum made to oscillate during same time;

The pendulum thus arranged, is made to vibrate about each axis in succession during equal intervals, say an hour or a day, and the number of oscillations carefully noted; if these numbers be the same, the distance between the axes is the length l of the equivalent simple pendulum;

if not, then the weight M must be moved towards that axis whose number is the least, and the trial repeated, till the numbers are made equal. The distance between the axes may be measured by a scale of equal parts. distance between the axes measured;

From this value of l, we may easily find that of the *simple second's pendulum;* that is to say, the simple pendulum which will perform its vibration in one second. Let N be the number of vibrations performed in one hour by the compound pendulum whose equivalent simple pendulum is l; the number performed in the same time by the second's pendulum, whose length we will denote by l', is of course 3600, being the number of seconds in 1 hour, and hence, from Eq. (78), simple second's pendulum:

$$\frac{1^h}{N} = T = \pi \sqrt{\frac{l}{g}},$$

$$\frac{1^h}{3600^s} = T' = \pi \sqrt{\frac{l'}{g}};$$

and because the force of gravity at the same station is constant, we find, after squaring and dividing the second equation by the first,

$$l' = \frac{l \, . \, N^2}{(3600^s)^2} \quad . \quad . \quad . \quad . \quad . \quad (80).$$ its length;

Such is, in outline, the beautiful process by which KATER determined the length of the simple second's pendulum at the Tower of London to be 39.13908 inches, or 3.26159 feet. value at London;

As the force of gravity at the same place is not supposed to change its intensity, this length of the simple second's pendulum must remain for ever invariable; and, on this account, the English have adopted it as the basis of their system of *weights* and *measures*. For this purpose, it was simply necessary to say that the $\frac{1}{3 \cdot 26159}$th part of the *simple second's pendulum at the Tower of London* shall basis of the English system of weights and measures;

English linear foot; the gallon; avoirdupois ounce;

be *one English foot*, and all linear dimensions at once result from the relation they bear to the foot; that the *gallon* shall contain $\frac{231}{1728}$th of a cubic foot, and all measures of *volume* are fixed by the relations which other volumes bear to the gallon; and finally, that a *cubic* foot of distilled water at the temperature of sixty degrees Fahr. shall weigh *one thousand ounces*, and all weights are fixed by the relation they bear to the ounce.

apparent force of gravity at London;

It is now easy to find the apparent force of gravity at London; that is to say, the force of gravity as affected by the centrifugal force and the oblateness of the earth. The time of oscillation being one second, and the length of the simple pendulum 3.26159 feet, Eq. (78) gives

$$1 = \pi \sqrt{\frac{\overset{ft.}{3.26159}}{g}};$$

whence

$$g = \pi^2 (3.26159) = (3.1416)^2 . (3.26159) = 32.1908 \text{ feet.}$$

From Eq. (78), we also find, by making T one second,

$$g = \pi^2 l,$$

and assuming

length of the simple second's pendulum, a function of the latitude;

$$l = x + y \cos 2 \psi,$$

we have

$$\frac{g}{\pi^2} = x + y \cos 2 \psi \quad . \quad . \quad . \quad . \quad (81).$$

Now starting with the value for g at London, and causing the same pendulum to vibrate at places whose latitudes are known, we obtain, from the relation given in Eq. (76), the corresponding values of g, or the force of gravity at these places; and these values and the corresponding latitudes being substituted successively in Eq. (81), give a series of equations involving but two unknown quantities, which may easily be found by the method of least squares.

force of gravity found at different places;

In this way it has been ascertained that

$$\pi^2 . x = 32.1803 \quad \text{and} \quad \pi^2 . y = - 0.0821;$$

whence, generally,

$$g = \overset{f}{32}.1803 - 0.0821 \cos 2 \psi \quad . \quad . \quad (81)';$$

Force of gravity in any latitude;

and substituting this value in Eq. (78), and making $T = 1$, we find

$$l = \overset{f}{3}.26058 - 0.008318 \cos 2 \psi \quad . \quad . \quad (82).$$

length of simple second's pendulum in any latitude;

Such is the length of the simple second's pendulum at any place of which the latitude is ψ.

If we make $\psi = 40° \, 42' \, 40''$, the latitude of the City-Hall of New York, we shall find

$$l = \overset{ft.}{3}.25938 = \overset{in.}{39}.11256.$$

length at City Hall of New York;

The principles which have just been explained, enable us to find the moment of inertia of any body turning about a fixed axis, with great accuracy, no matter what its figure, density, or the distribution of its matter. If the axis do not pass through its centre of gravity, the body will, when deflected from its position of equilibrium, oscillate, and become, in fact, a compound pendulum; and denoting the length of its equivalent simple pendulum by l, we have, Eq. (79),

moment of inertia found by means of simple pendulum;

$$M . D . l = I;$$

or, since

$$M = \frac{W}{g},$$

its value;

$$\frac{W}{g} . D . l = I \quad . \quad . \quad . \quad . \quad (83);$$

in which W denotes the weight of the body.

simple second's pendulum known from latitude;

Knowing the latitude of the place, the length l' of the simple second's pendulum is known from Eq. (82); and counting the number N of oscillations performed by the body in one hour, Eq. (80), gives

the body's equivalent simple pendulum;

$$l = \frac{l' . (3600)^2}{N^2}.$$

To find the value of D, which is the distance of the centre of gravity from the axis, attach a spring or other balance to any point of the body, say its lower end, and bring the centre of gravity to a horizontal plane through the axis, which position will be indicated by the maximum reading of the balance. Denoting by a the distance from the axis C to the point of support R, and by b the maximum indication of the balance, we have, from the principles of moments,

distance from centre of gravity to axis found;

Fig. 145.

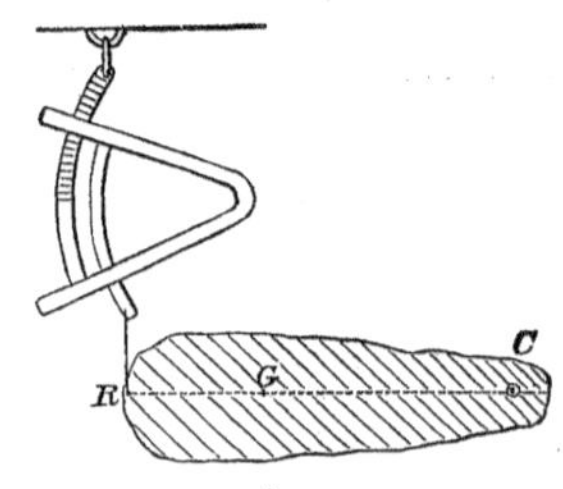

$$b\,a = W D.$$

The distance a may be measured by a scale of equal parts. Substituting the values of $W D$ and l in the expression for the moment of inertia, Eq. (83), we get

value of the moment of inertia;

$$\frac{b . a . l' . (3600)^2}{g . N^2} = I. \quad . \quad . \quad . \quad (84).$$

If the axis pass through the centre of gravity, as, for example, in the *fly-wheel*, take Eq. (79),

the moment of inertia found when the axis passes through the centre of gravity;

$$l = \frac{I_1 + MD^2}{MD};$$

whence

$$I_1 = M.D.l - MD^2 \quad . \quad . \quad . \quad (85).$$

Fig. 146.

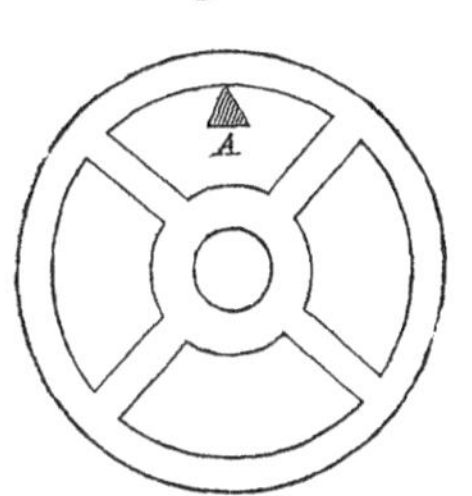

example of the fly-wheel;

Mount the body upon a parallel axis A, not passing through the centre of gravity, and cause it to vibrate for an hour as before; from the number of these vibrations and the length of the simple second's pendulum, the value of l may be found as before; M is known, being the weight W divided by g; and D may be found by direct measurement, or by the aid of the spring balance, as already indicated; whence I_1 becomes known.

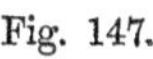
Fig. 147.

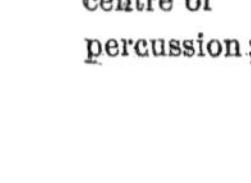
centre of percussion;

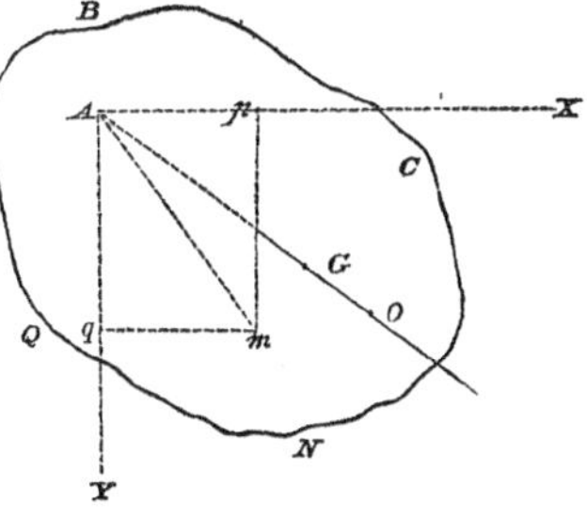

§ 184.—When a body, $B\,Q\,N\,C$ receives a motion of rotation about an axis A, which is here supposed perpendicular to the plane of the paper, each elementary mass m, will develop a force of inertia whose direction is perpendicular to the shortest line connecting it with the axis, and whose intensity will be measured by

$$m \, . \, r \, . \, \frac{V_1}{t},$$

inertia exerted by an elementary mass during an elementary time;

notation;

in which r is the distance of m from the axis, and V_1 the elementary amount of angular velocity generated in the very small portion of time denoted by t.

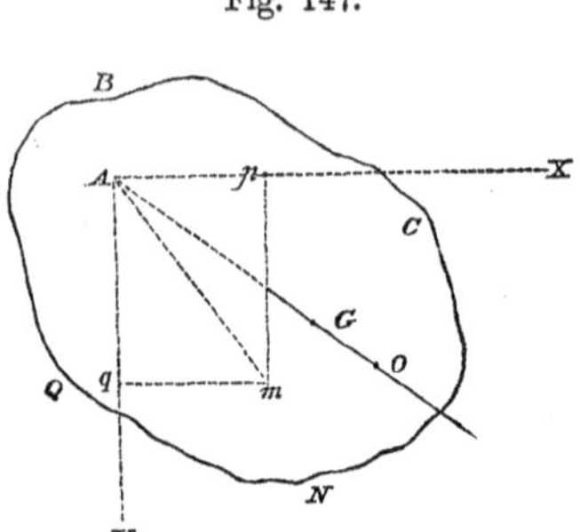

Fig. 147.

co-ordinate planes;

Through the axis A, draw two planes at right angles to each other, and let their traces on the paper be $A\,X$ and $A\,Y$. Denoting the co-ordinates $A\,p$ and $A\,q$ of m, referred to these planes, by x and y, respectively, we shall have

$$\cos m\,A\,p = \frac{x}{r},$$

$$\cos m\,A\,q = \frac{y}{r}.$$

Resolve the force of inertia, above given, into two components in the direction of these planes. The component parallel to the plane of which the trace is $A\,y$, will be

component of the inertia parallel to the plane $A\,y$;

$$m\,r \cdot \frac{V_1}{t} \cdot \frac{x}{r} = m\,.\,x\,\frac{V_1}{t},$$

and that parallel to the plane whose trace is $A\,x$, will be

that parallel to the plane $A\,x$;

$$m\,r \cdot \frac{V_1}{t} \cdot \frac{y}{r} = m\,y \cdot \frac{V_1}{t};$$

and for other elementary masses m', m'', &c., of which the co-ordinates are $x'\,y'$, $x''\,y''$, &c., we shall have the components

the same for other elementary masses;

$$m'\,x' \cdot \frac{V_1}{t}, \qquad m''\,x''\,\frac{V_1}{t}, \text{ \&c.},$$

$$m'\,y' \cdot \frac{V_1}{t}, \qquad m''\,y''\,\frac{V_1}{t}, \text{ \&c.};$$

the resultant of the components parallel to the plane $A\,y$, will be

$$\frac{V_1}{t}(m\,x + m'\,x' + m''\,x'' + \&c.) = \frac{V_1}{t}M x_{,},$$ resultant of the components parallel to $A\,y$,

and of the components parallel to the plane $A\,x$,

$$\frac{V_1}{t}(m\,y + m'\,y' + m''\,y'' + \&c.) = \frac{V_1}{t}M y_{,};$$ resultant of those parallel to $A\,x$;

in which M denotes the entire mass of the rotating body, and $x_{,}$ and $y_{,}$ the co-ordinates of its centre of gravity. And the intensity of the general resultant will, from the parallelogram of forces, be

$$\frac{V_1}{t}M\sqrt{x_{,}^2 + y_{,}^2} = \frac{V_1}{t}M\,.\,D;$$ resultant of the whole;

in which D represents the distance of the centre of gravity G, of the whole mass, from the axis. The direction of this resultant will be perpendicular to $A\,G$, drawn through the centre of gravity perpendicular to the axis, as will readily appear by reference to its components parallel to the planes $A\,y$ and $A\,x$ found above. its direction;

The moment of this force, with reference to the axis, will therefore be its intensity multiplied into some distance as $A\,O = L$, on this line,

$$\frac{V_1}{t}M\,.\,D\,.\,L.$$ its moment;

But, Eq. (63), the sum of the moments of all the forces of inertia actually exerted, in reference to the axis A, is equal to the product of the entire moment of inertia I, multiplied by the ratio $\frac{V_1}{t}$, therefore

$$\frac{V_1}{t}\,.\,M\,.\,D\,.\,L = I\,.\,\frac{V_1}{t},$$

or

$$L = \frac{I}{M \,.\, D} \quad . \quad . \quad . \quad . \quad . \quad (86);$$

point at which the resultant inertia of a rotating mass is exerted;

whence we conclude that, *the point at which the resultant inertia of a rotating mass is exerted, is on a line drawn through its centre of gravity perpendicular to the axis, and at a distance from the axis equal to the moment of inertia divided by the product of the mass into the distance of the centre of gravity from the axis.*

Fig. 148.

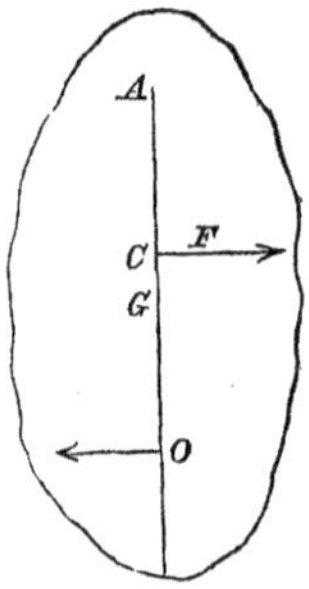

This being understood, suppose a force F applied at the point C in a direction perpendicular to the line $A\ O$, and immediately opposed to the direction of the motion; this force would obviously tend to bend the line $A\ O$, the point A being retained by the axis, and the point O being urged onward by the inertia concentrated at it. If the force be suddenly applied, the axis must receive a shock, and to estimate its intensity S, denote by X the distance $A\ C$; then, from the principles of parallel forces already explained, we have

shock experienced by the axis when the body is struck;

$$L \;:\; L - X \;::\; F \;:\; S;$$

whence

$$S = F \cdot \frac{L - X}{L} = F\left(1 - \frac{X}{L}\right) \quad . \quad . \quad (87);$$

or, substituting the value of L, Eq. (86),

its intensity;

$$S = F\left(1 - \frac{MD}{I} \cdot X\right). \quad . \quad . \quad (88).$$

If we suppose the body at rest, and desire to apply the

force F so as to communicate no shock, we make

$$S = 0,$$

the blow applied so as to communicate no shock to the axis;

a condition that can only be satisfied by making

$$1 - \frac{MD}{I} \times X = 0;$$

whence

$$X = \frac{I}{MD} = L = AO.$$

distance from the axis at which it must be applied;

There being no shock to the axis, it can oppose no resistance to the motion of rotation, and hence we infer that this latter will be the same as though the body were perfectly free. The point O is, on this account, called the centre of percussion, which may be defined, *that point of a body retained by a fixed axis, at which it may be struck in a direction perpendicular to the plane of the centre of gravity and axis without communicating any shock to the axis.*

centre of percussion defined;

The centre of percussion may be found experimentally thus:—lay the axis C upon a support $A\,A$, and permit the body to fall upon a moveable edge B, resting on a horizontal plane; when this edge is placed in such position that the axis C will not move when the body falls upon it, the centre of percussion will be immediately above the point struck. Since the distance of the centre of percussion from the axis is equal to

Fig. 149.

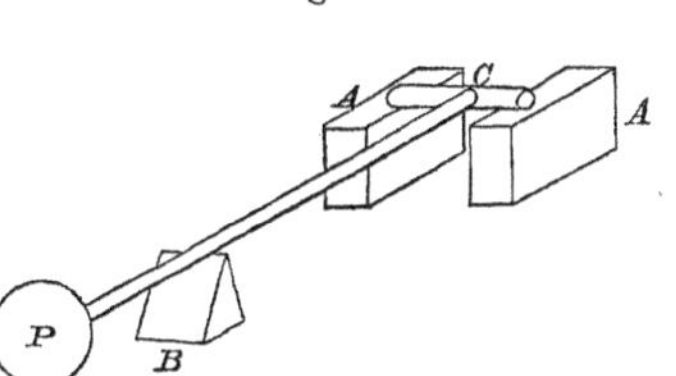

centre of percussion found experimentally;

$$\frac{I}{MD},$$

to put a pendulum in motion, the force should be applied to centre of oscillation.

it must be at the centre of oscillation. To move a pendulum without communicating action to its axis, the force must be applied at the centre of oscillation.

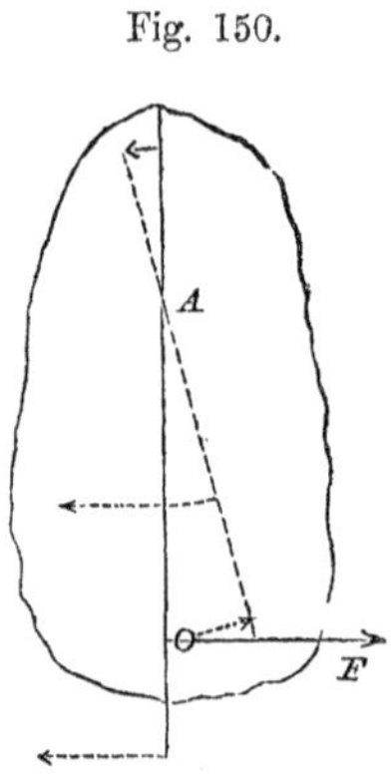

Fig. 150.

§ 185.—Resuming Eq. (87), we see that the shock upon the axis A will be positive, that is to say, will act in the direction of the impressed force F, as long as X is less than L: when X is equal to L, there will be no shock; when X is greater than L, there will again be a shock, but with a negative sign, which indicates that it will be exerted in a direction opposite to that of the impressed force. Now these shocks in opposite directions, with a neutral point A, can only arise from an effort of the particles, which are situated on opposite sides of the axis, to move in contrary directions when the body is struck at the centre of oscillation; and as the effect upon the neutral point A is the same in this latter case, whether the body be retained by an axis or a force, it follows that every free body, when struck, in general, begins to move for the instant, but only an instant, about a single point. This point is called the centre of *spontaneous rotation.* If the blow be impressed at any point, as O, the centre of spontaneous rotation will be upon the axis corresponding to the point O as a centre of oscillation, and hence its distance from the latter will be given by

The shock may be positive, nothing, or negative;

centre of spontaneous rotation;

distance of centre of spontaneous rotation from axis;

$$L = \frac{I}{MD} \quad . \quad . \quad . \quad . \quad . \quad (89);$$

and since the centre of oscillation and axis of suspension are reciprocal, I will denote the moment of inertia taken with reference to an axis through the point O, and D the distance of the latter from the centre of gravity.

relation of centre of spontaneous rotation to centre of oscillation;

Referring to Eq. (88), if the axis be supposed to pass through the centre of gravity, D will be equal to zero, and

$$S = F;$$

the entire shock always communicated to centre of gravity;

that is to say, no matter where the force F be applied, its entire effect will be communicated to the centre of gravity, which is a confirmation of the result given in § 146.

if direction of impact pass through centre of gravity, the body will not rotate.

If the line of direction of the force pass through the centre of gravity, D, in Eq. (89), will be zero, and the distance of the centre of spontaneous rotation will be at an infinite distance from the point of impact; in other words the body will not rotate, which is another result of § 146.

Collision of a body having a motion of translation against another retained by a fixed axis;

§ 186.—Let Q be a body suspended from an axis A perpendicular to the plane of the figure. This body being at rest, suppose it to be struck at the point T by another body P, moving in the direction $T\,L$ at right angles to the surface of contact, and in a plane perpendicular to the axis A. Denote by m and w the mass and weight of the impinging body, and by V its velocity before the impact. At the instant of meeting there will be developed a force of compression F, which will act equally upon each body along the line $T\,L$, but in opposite directions. The pressure upon both bodies, which is nothing when they begin to touch each other, will aug-

Fig. 151.

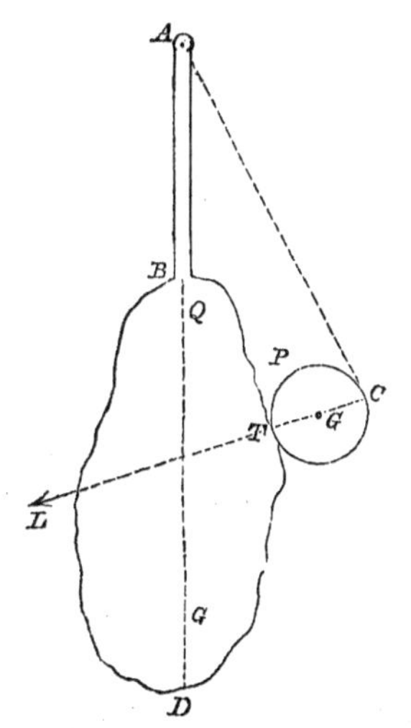

the action and reaction variable; ment by degrees as they approach to the state of greatest compression; so that F, although always representing a number of pounds weight, is, nevertheless, not a fixed, but a variable quantity. We may disregard for a moment the body Q, and suppose the force F applied to the body P, considered as free; the force will deprive this body of a series of small degrees of velocity denoted by v, each in the small time t, so that its measure at any instant will, Eq. (39), be given by

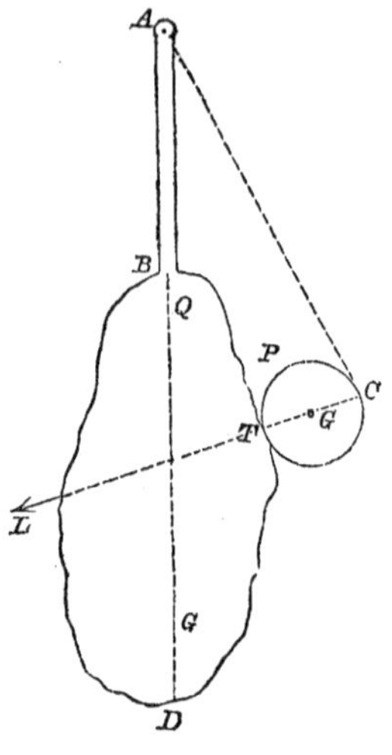

Fig. 151.

measure of the force of reaction;

$$F = \frac{m v}{t}.$$

But the force F also acts upon the body Q, and turns it about the axis A, generating in it, during the same interval of time t, an angular velocity $v_{\prime}$; and the forces of inertia thence arising, must be in equilibrio with the force F; in other words, the sum of the moments of the first in reference to the axis A, must be equal to the product of the force F into the perpendicular $A\,C$, drawn from the axis to the line of direction $T\,L$. Hence, Eq. (63),

moment of action equal to moment of reaction;

$$F \,.\, A\,C = I \,.\, \frac{v_{\prime}}{t};$$

and substituting the value of F above, and dividing by $A\,C$, which we will represent by the single letter p,

$$\frac{m \,.\, v}{t} = \frac{I \,.\, v_{\prime}}{p \,.\, t};$$

or, finally,

$$p \,.\, m \,.\, v = I \,.\, v_{,}.$$

result for a single instant of time;

Denote by v', v'', v''', &c., the small degrees of velocity lost by the body P, during the second, third, fourth, &c., intervals of time t, supposed to be always of the same length; and by $v_{,}'$, $v_{,}''$, $v_{,}'''$, &c., the angular velocities acquired by the body Q during the same intervals; we shall have

$$p \,.\, m \,.\, v' = I v_{,}',$$

$$p \,.\, m \quad v'' = I v_{,}'',$$

$$\&c. \quad = \quad \&c.;$$

the same for other instants of time;

by taking the sum of the whole,

$$p\,(v + v' + v'' + v''' + \&c.)\, m = I\,(v_{,} + v_{,}' + v_{,}'' + \&c.);$$

the sum of the whole;

and denoting by U the whole velocity lost by the body P, and by V_1 the whole angular velocity gained by the body Q during the entire action, we shall have

$$U = v + v' + v'' + v''' + \&c.,$$

velocity lost;

$$V_1 = v_{,} + v_{,}' + v_{,}'' + v_{,}''' + \&c.;$$

angular velocity gained;

whence, by substituting above,

$$p \,.\, m \,.\, U = I V_1 \quad . \quad . \quad . \quad . \quad (90).$$

result for the entire duration of the impact;

If the bodies be not elastic, it will only be necessary to consider the impact from the instant in which they first come in contact, to that in which the body P has lost its excess of velocity over that part of Q into which it becomes imbedded; for, as soon as the body P has taken the

If the bodies be not elastic, they will ultimately constitute a single one;

angular velocity of the other about the axis, there will be no effort to regain lost figure, and the two bodies will turn about A as though they constituted but a single one.

But the angular velocity of Q about A being V_1, that of P will be $p\ V_1$, and we shall have

$$U = V - p\,V_1;$$

substituting this value of U in Eq. (90), we find

$$pm\,(V - pV_1) = I\,V_1;$$

whence

angular velocity generated by the impact.

$$V_1 = \frac{p\,.\,m\,.\,V}{m\,.\,p^2 + I} \quad . \quad . \quad . \quad . \quad (91);$$

which gives the angular velocity of the body struck, after the impact, in terms of its moment of inertia, the mass and velocity of the impinging body, and the distance from the axis to the path described by its centre of gravity.

Application to the balistic pendulum;

§ 187.—In artillery, the initial velocity of projectiles is ascertained by means of the *balistic pendulum*, which consists of a mass of matter suspended from a horizontal axis in the shape of a knife-edge, after the manner of the compound pendulum. The bob is either made of some unelastic substance, as wood, or of metal provided with a large cavity

its construction;

Fig. 152.

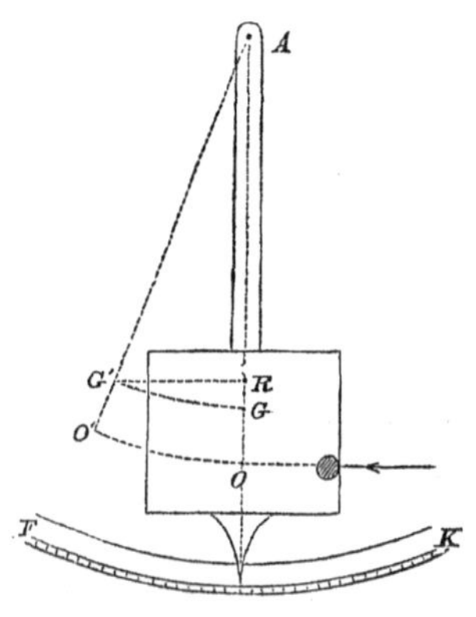

filled with some soft matter, as dirt, which receives the projectile and retains the shape impressed upon it by the blow.

Denote by V and m, the initial velocity and mass of the ball; V_1 the angular velocity of the balistic pendulum after the blow, I and M its moment of inertia and mass. Also let r represent the distance of the centre of oscillation of the pendulum from the axis A. That no motion may be lost by the resistance of the axis arising from a shock, the ball must be received in the direction of a line passing through this centre and perpendicular to the line $A\ O$. This condition being satisfied, we have

notation;

the pendulum must be struck at centre of oscillation;

$$p = r,$$

and Eq. (91) becomes

$$V_1 = \frac{r\,m\,V}{m\,r^2 + I};$$

from which we find

$$V = \frac{(m\,r^2 + I)\,V_1}{m\,r} \quad . \quad . \quad . \quad (92);$$

value for the velocity of projectile;

the velocity V becomes known, therefore, when V_1 is known, since all the other quantities may be easily found by the methods already explained. To find V_1, denote by H the greatest height to which the centre of gravity of the pendulum is elevated by virtue of this angular velocity; then, since the moment of inertia of the ball is $m\,r^2$, we have, from the principle of the living force,

$$(I + m\,r^2)\,V_1^2 = 2\,(M + m)\,g\,H;$$

equation of living force;

whence

$$\frac{(I + m\,r^2)\,V_1^2}{(M + m)\,g} = 2\,H.$$

Denoting by T the time of a single oscillation of the pendulum after it receives the ball, we have, Eq. (75),

time of a single oscillation of balistic pendulum;

$$T = \pi \sqrt{\frac{I + m r^2}{(M + m) D \cdot g}},$$

D being the distance from the axis to the centre of gravity; whence,

$$\frac{I + m r^2}{(M + m) g} = \frac{D T^2}{\pi^2};$$

and this value, substituted in the equation of the living force, gives

$$\frac{D T^2}{\pi^2} V_1^2 = 2 H;$$

whence

angular velocity of the pendulum;

$$V_1 = \frac{\pi}{T} \cdot \sqrt{\frac{2 H}{D}};$$

also

moment of inertia of the whole;

$$I + m r^2 = \frac{(M + m) g \cdot D \cdot T^2}{\pi^2};$$

and because, Eq. (78),

time of oscillation of the equivalent simple pendulum;

$$T = \pi \sqrt{\frac{r}{g}},$$

we find

length of this pendulum;

$$r = \frac{T^2 g}{\pi^2}.$$

Substituting these values of V_1, $I + m r^2$ and r in Eq. (92), we find

$$V = \frac{\pi}{T} \sqrt{2 H D} \cdot \frac{M + m}{m};$$

or, replacing the masses by the weight divided by the force of gravity,

$$V = \frac{\pi}{T} \sqrt{2 H . D} \times \frac{W + w}{w};$$ simpler value for velocity of projectile;

in which W and w denote the weights of the pendulum and ball respectively.

Observe that H is the height to which the centre of gravity rises in describing the arc of a circle of which D is the radius. Let $G G' K$ be half of the circumference of which this arc is a part, G and G' the initial and terminal positions of the centre of gravity during the ascent; draw $G' R$ perpendicular to $K G$. Then, because $A G = D$, and $G R = H$, we have, from the property of the circle,

Fig. 153.

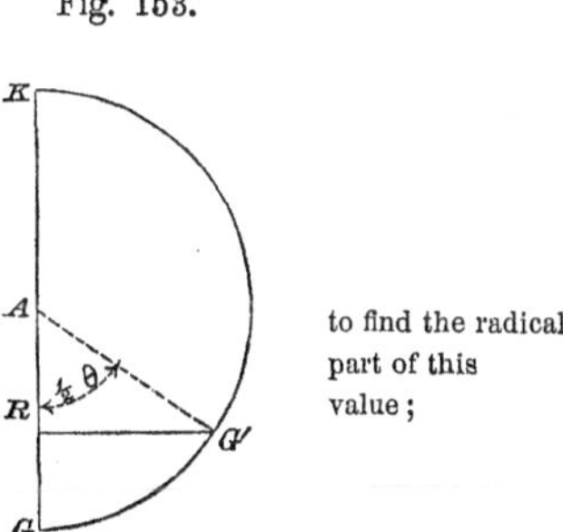

to find the radical part of this value;

$$R G' = \sqrt{H (2 D - H)};$$

and if the pendulum be made large, so that the arc $G G'$ shall be very small, which is usually the case, H may be neglected in comparison with $2 D$, and therefore

$$R G' = \sqrt{2 H . D};$$ value of radical part found;

$\sqrt{2 H D}$ is half the chord of the arc described by the centre of gravity in one entire oscillation. Denoting this chord by C, and substituting above, we have

velocity of projectile in terms of the chord of the arc of vibration;

$$V = \tfrac{1}{2} \cdot \frac{\pi}{T} \cdot C \cdot \frac{W + w}{w}.$$

From this equation, we may find the initial velocity V; and for this purpose, it will only be necessary to have the duration of a single oscillation, and the amplitude of the arc described by the centre of gravity of the pendulum. The process for finding the time has been explained. To find the arc, it will be sufficient to attach to the lower extremity of the pendulum a pointer, and to fix on a permanent stand below, a circular graduated groove, whose centre of curvature is at A; the groove being filled with some soft substance, as tallow, the pointer will mark on it the extent of the oscillation. Knowing thus the arc, denoted by θ, and the value of D, found as already described, § 184, we have

to find the arc of vibration;

$$R\,G' = \tfrac{1}{2}\,C = D\,.\,\sin\tfrac{1}{2}\,\theta\,;$$

whence

its value found;

$$C = 2\,D\,.\,\sin\tfrac{1}{2}\,\theta\,;$$

and finally

final value of velocity.

$$V = \frac{\pi}{T}\cdot D\cdot\frac{W + w}{w}\sin\tfrac{1}{2}\,\theta\quad .\quad .\quad (93).$$

SIMPLE MACHINES.

A machine defined.

§ 188.—A *machine* is any device by which the action of a force is received at one set of points and transmitted to another set, where it may either balance or overcome the action of one or more opposing forces and perform its effective work. The force impressed is usually called *the power*, and that overcome, *the resistance.* We proceed to discuss the simple machines, so named because some one or more of them enter as elements into the composition of all machinery.

XIV.

FUNICULAR MACHINE.

§ 189.—This consists of an assemblage of *cords* or bars; the former united by knots, and the latter by joints or hinges. The cords are supposed, for simplification, perfectly flexible, the bars perfectly rigid, and both inextensible, without weight, and devoid of inertia. The weight and inertia of the several parts of every machine, are usually small when compared with the intensity of the power and resistance; and when this is not the case, they may be estimated and taken into the account by the methods already explained. The hypothesis of inextensibility is also admissible, because when a cord or bar is extended or the latter compressed under the action of one or of several forces, the maximum change of dimensions is soon attained, after which the figure remains unaltered during the subsequent action.

Funicular machine; weight and inertia small as compared with the power and the resistance; inextensibility admissible;

Let the extremities of the straight cord AB be solicited by several forces. Each force may be resolved into two components, one in the direction of the cord, the other at right angles to it. Since the cord is perfectly flexible, if it be in equilibrio, the perpendicular components at each end must destroy each other, otherwise they would produce flexure. The components in the direction of the cord must reduce to two forces, which are equal in intensity and immediately opposed. They must also act to stretch the cord, for compression would only bend it, and

Fig. 154.

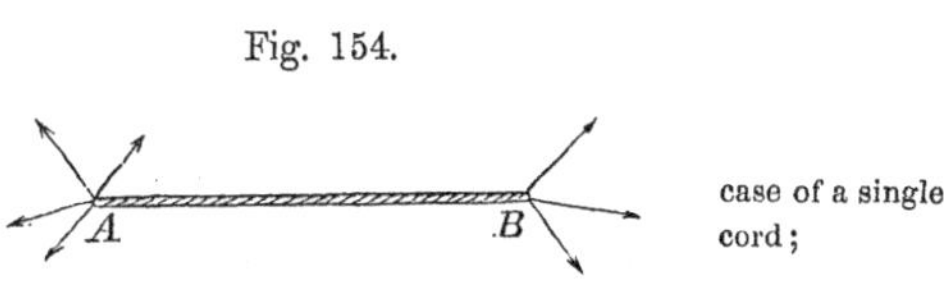

case of a single cord; conditions of equilibrium; forces must act to stretch the cord;

the action of one force could not be transmitted to the point of application of the other.

in the case of a bar, the forces may also act to compress it;

If instead of a cord we suppose a bar, the conditions of equilibrium will be the same, only that the bar being inflexible, the forces in the direction of its length may act either to stretch or to compress it. By recalling what was said of the physical constitution of bodies, we may regard the molecular forces as so many springs which, as soon as an effort is made to disturb the particles from their positions of rest, are extended or compressed everywhere equally by the equal and contrary forces which act at the ends of the cord or bar. Hence the *tension*, that is, the effort by which any two consecutive elements are urged to approach each other or to separate, in the direction of the cord or bar, must be equal throughout, and equal to one of the equal forces in question, except when the cord or bar is vertical; in which case, the tension at any point is increased by the weight of all the particles below it.

action of the molecular springs;

the tension the same throughout, except when vertical;

When a cord or bar is subjected to a force of traction, it stretches, and may even break. If it be equally strong throughout, the rupture ought to take place simultaneously at all its points, and yet this is never found to be the case in practice, and it is because bars and cords are not homogeneous, and break at the weakest point. When two pieces of cord of the same kind, are of the same length, no reason can be assigned why one should break rather than the other under the same resistance; but when of unequal length, the chance of rupture is greater for the longer; and this is the reason why cords and ropes, which to all external appearances are the same in kind, are generally found to be weaker as they are longer.

cords never equally strong throughout;

in practice, cords and bars are weaker as they are longer.

§ 190.—We have seen that when forces which act upon the extremities of a cord are in equilibrio, the resultant of those acting at one end, must be equal and directly opposed to that of those acting at the other; and that their common line of direction must coincide with that of the cord. The work of these resultants must be equal, and hence we conclude that the work of the forces which act at one end of a cord is equal to the work of those which act at the other. The work of each resultant must also be equal to that of the tension of the cord at any one of its points, as C; and to find the value of this work, it is only necessary to multiply this tension by the path described by the point C in the direction of the tension. *Thus the quantity of work of several forces applied to one end of a cord, is equal to the quantity of work of its tension.* In the example of the common device for ringing large bells, it is usual to attach to one end A of a rope, which connects with the machinery of the bell, several cords C, upon each of which a man may pull. It would be difficult to estimate the work performed by each man, because his effort, as well in intensity as direction, varies at each instant; but there is a general tension exerted upon the main rope, and the quantity of work of this tension is equal to the sum of the effective quantities of work of the several men. The effort of each man is resolved into two components, one in the direction of the main rope $A B$, the other perpendicular to

The work of the forces which act at the ends of a cord must be equal;

Fig. 155.

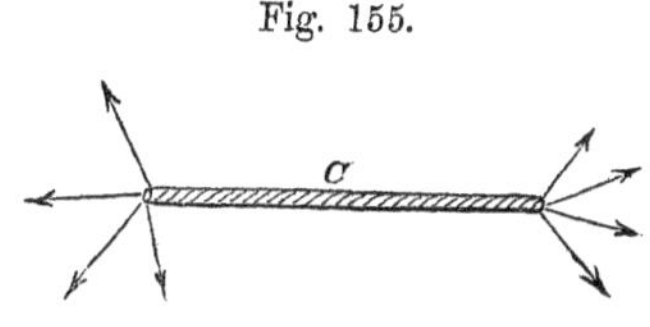

quantity of work of the forces applied to one end of a cord is equal to that of the tension;

Fig. 156.

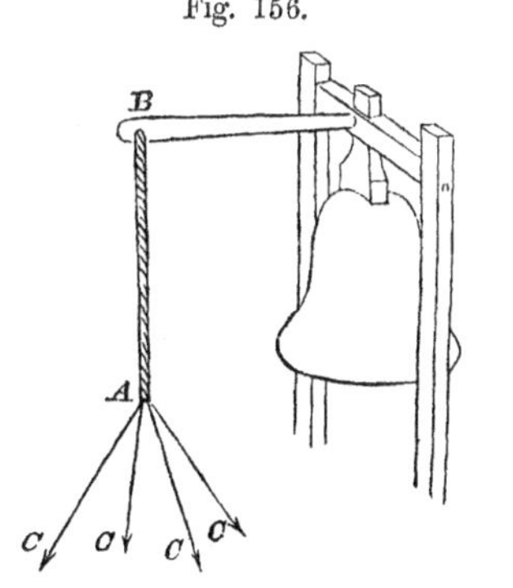

example of the bell-ropes;

it. The perpendicular components must be in equilibrio, while the parallel components are alone effective in producing useful work. The perpendicular components only produce fatigue, and exhaust uselessly the strength of the men. And, although the total quantity of work is transmitted to the main rope, yet the disposition of inclined cords is a source of real loss, which is the greater in proportion as the inclination is greater. It is for this reason that a rigid hoop *m n* is so introduced as to separate the cords, and give the portions to which the efforts are immediately applied parallel directions.

effect of components perpendicular to the main rope of the bell;

effect of a hoop.

Fig. 157.

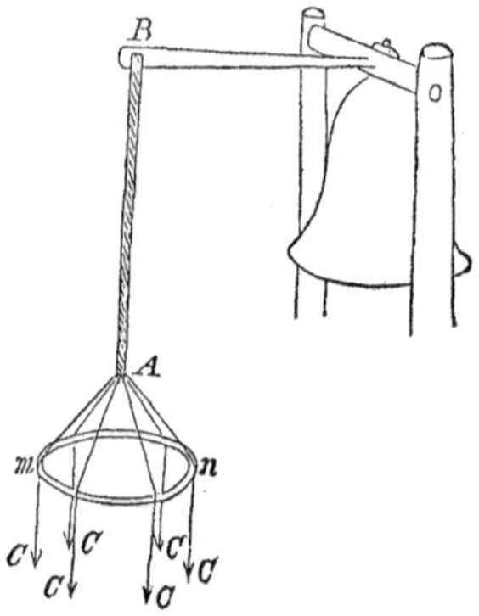

§ 191.—When several forces act upon cords which meet in a point and are united by a knot, the tension of any one is equal to the resultant of the efforts exerted upon the others, and the equilibrium requires that this same tension shall be equal and directly opposed to the force which solicits the cord in question. Hence, when forces are applied to cords which meet in a knot, the condition of their equilibrium requires that the effort of any one shall be equal and directly opposed to the resultant of all the others.

Equilibrium of several cords meeting in a point;

When a force P is applied to a point D, which may slide along a cord whose ends A and B are fixed, the equilibrium of the point D requires that the direction of the force P shall bisect the angle $A D B$ formed by the portions of the cord separated by the bend at D; for

equilibrium of a sliding knot;

the force P must be equal and directly opposed to the resultant of the tensions on DA and DB; but the whole cord ADB being continuous, these tensions must be equal, since the tension is the same throughout; if, therefore, the distance DC be laid off on PD produced, proportional to the intensity P, and from C, the lines Cm and Cn be drawn parallel to DB and DA respectively, the figure $CmDn$ will be a rhombus, because Dm and Dn, which represent the tensions, must be equal.

Fig. 158.

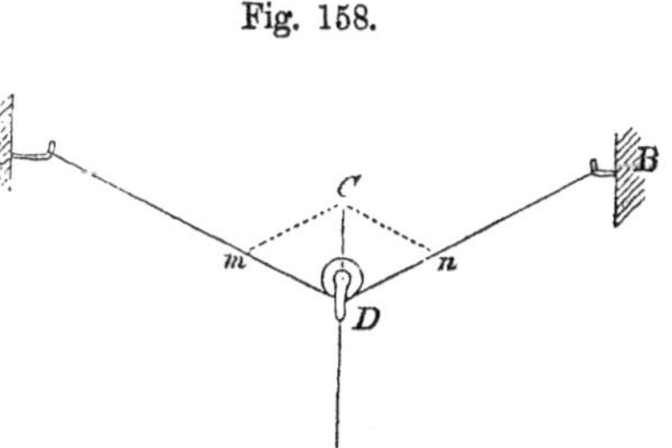

direction of the force applied to the knot, must bisect the angle of the two parts of the cord;

An example of this mode of action is furnished by the manner of suspending a common lantern L from a small pulley D, of which the groove receives the cord ADB, whose ends are fastened to hooks at A and B. The weight of the lantern will cause the pulley to move till the direction of the weight bisects the angle made by the branches of the cord; the pulley will then come to rest and remain in a state of stable equilibrium. The equilibrium will be stable because, being a heavy system, the centre of gravity is the lowest possible; and to show this, it will be sufficient to remark that the length of the entire cord being constant, the point D will, when in motion, describe an ellipse of which A and B are the

Fig. 159.

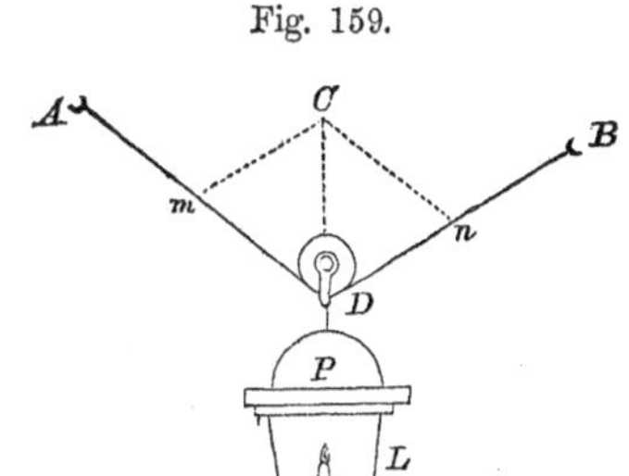

example in the mode of suspending the common lantern;

the pulley will be in stable equilibrium when at the lowest point;

foci, and as the direction PC, of the weight of the lantern, bisects the angle ADB, it will be perpendicular to the tangent to the curve at D, which must therefore be horizontal, and no point of the curve can lie below it.

position of the horizontal tangent;

Fig. 159.

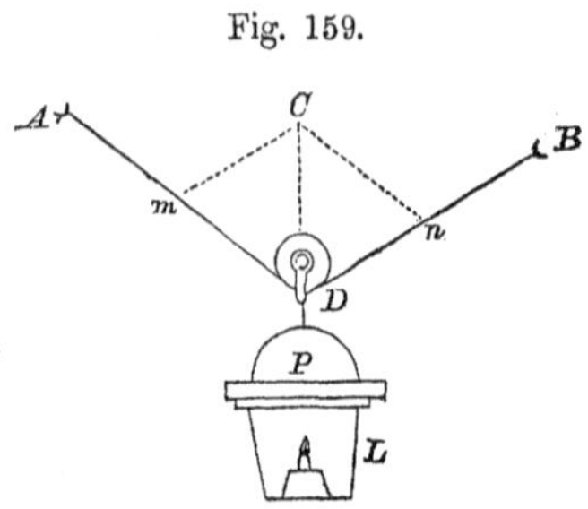

If the pulley be removed and the lantern be attached by a knot arbitrarily to some point as D, the freedom of motion will be destroyed, the tension will no longer be the same throughout, and the conditions of equilibrium will be those of forces applied to three cords meeting at a single point. Produce the vertical PD, and lay off DC to represent the weight of the lantern. Denote its weight by W; the tension on DA by a, and that on DB by b; the angle ADB by φ, and ADC by θ; then, drawing Cn and Cm, parallel respectively to DA and DB, we have, from the parallelogram of forces,

when the pulley is replaced by a knot;

tension will not be the same throughout;

the conditions of the equilibrium will be the same as those of three oblique forces;

Fig. 160.

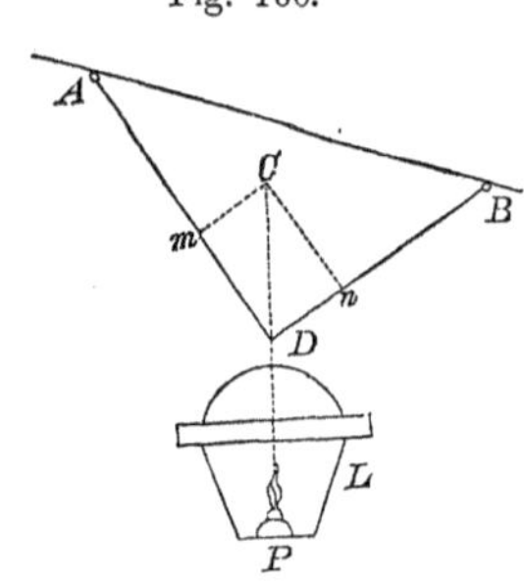

$$W : a :: \sin\varphi : \sin(\varphi - \theta),$$

$$W : b :: \sin\varphi : \sin\theta;$$

whence

tension on one branch;

$$a = \frac{W . \sin(\varphi - \theta)}{\sin\varphi} \quad . \quad . \quad . \quad (94),$$

$$b = \frac{W \cdot \sin \theta}{\sin \varphi} \quad . \quad . \quad . \quad . \quad . \quad . \quad (95).$$ tension on the other;

If θ be less than $\varphi - \theta$, a will be greater than b; that is to say, *the tension will be the greater upon that branch with which the direction of the weight makes the least angle.* branch most inclined has the greatest tension;

If the cord ADB be drawn into a straight horizontal line, φ will become equal to 180°, the sine of which is zero, and the tensions a and b will become infinite; in other words, there is no force sufficiently great to bring the whole cord to a horizontal position. no force sufficient to make the cord horizontal.

§ 192.—Let us now consider a polygon $ABCD$, composed of an assemblage of cords or bars, and acted upon at the angular points by the forces P, Q, R, S. Moreover, let N and N' be two forces drawing on the points A and D, in the directions AA' and DD', respectively; these latter forces will represent the efforts exerted at the two extremities where the polygon is attached to fixed supports. The conditions of equilibrium about each of the several angles are the same as in the preceding case, and the figure formed by the sides, in turning about the angular points to satisfy them, is called a *funicular polygon.* This figure must be such that the equilibrium will subsist at each angle. If, therefore, any one of the forces, as R, be resolved into two components in the directions of the sides DC and BC, adjacent to its point of application, these components will To find conditions of equilibrium of the funicular polygon; equilibrium must subsist at each angle;

Fig. 161.

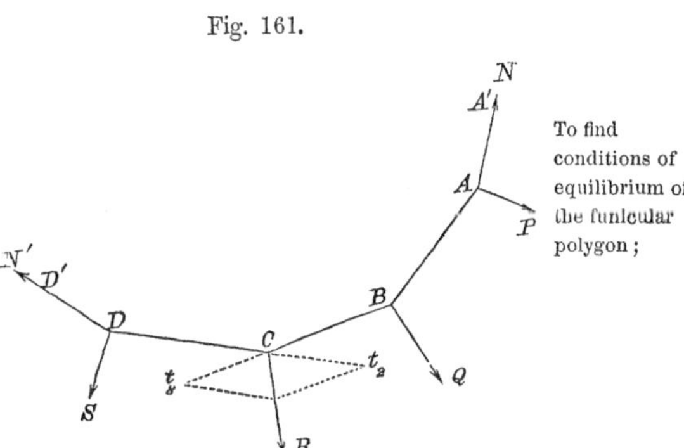

be equal and directly opposed to the tensions of the sides. The equilibrium is entirely independent of the length of the sides, and will subsist when these are reduced to zero, in which case, all the forces and tensions will be transferred parallel to their primitive directions to the same point; and as each side is drawn by two equal and contrary tensions, these latter will disappear or destroy each other, so that the conditions of equilibrium of several forces applied to a funicular polygon is, *that these forces shall remain in equilibrio when transferred parallel to their primitive directions and applied to a single point.*

is independent of length of sides; and will subsist when the sides are zero;

conditions of equilibrium in words.

§ 193.—If all the forces P, Q, R, &c., be weights, and the polygon in equilibrio, since the force R will be in the plane of the sides $B\,C$ and $C\,D$, adjacent to the angle C; the force Q equally in the plane of the sides $B\,C$ and $A\,B$; the sides $A\,B$, $B\,C$, and $C\,D$, will be in the plane of the parallel forces Q and R. In the same way it may be shown that the entire polygon and the forces applied to it are in the same plane. If the polygon be a collection of heavy bars, each side will be solicited by its own weight in addition to the weights applied to the angles. Denote by w the weight of the bar $A\,B$; this weight must pass through the centre of gravity of $A\,B$. Resolve it into two

When the forces are parallel, the polygon and direction of forces are in same plane;

the polygon a collection of heavy bars;

Fig. 162.

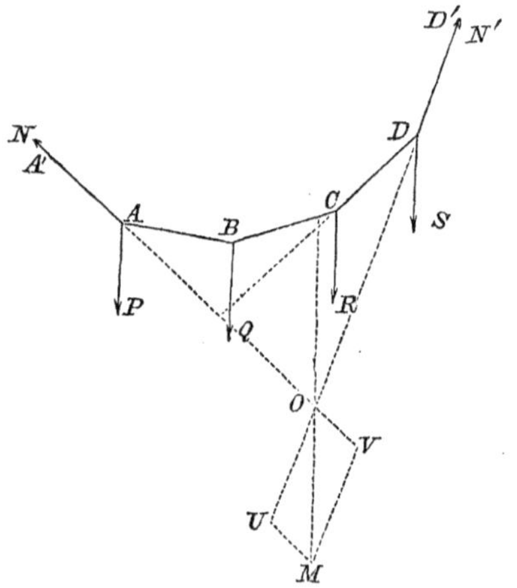

components acting at the extremities of the bar. If the bar have the same cross section throughout and be of homogeneous density, the components at A and B will be $\frac{1}{2}w$. In like manner, if w' be the weight of the side BC, the components at B and C will be $\frac{1}{2}w'$, and so on for the other sides. Thus the angles B and C will be acted upon by the weights $\frac{1}{2}(w + w')$ and $\frac{1}{2}(w' + w'')$ respectively, that is, by the half sum of the weights of the adjacent sides. The extreme ends will each be acted upon by half the weight of the adjacent side; and thus we have but to consider the polygon as without weight and solicited by forces applied to its angular points. Since all the weights P, Q, R, S, and the weights w, w', w'', &c., are maintained in equilibrio by the reaction N and N' of the fixed points, which are equal to the tensions of the sides $A'A$ and DD' respectively, the resultant of these tensions must be equal and directly opposed to that of all the weights. If, therefore, the lines AA' and DD' be produced, their intersection O will give one point through which the resultant of the weights P, Q, R, S, and that of the polygon, will pass; and this resultant being vertical, if the distance OM be laid off, by any scale of equal parts, so as to contain as many linear units as there are pounds in $P + Q + R + S + w + w' + w''$, &c., and two lines MU and MV be drawn through M parallel respectively to AA' and DD', the distances OV and OU will give, by the same scale, the tensions at A' and D', or the values of N and N'.

Weight of sides resolved into parallel components;

Resultant of extreme tensions, equal and opposed to that of all the weights;

Value of extreme tensions found.

If the polygon be only subjected to the action of its own weight, the line OM may be drawn vertically through its centre of gravity.

§ 194.—It is often of great practical importance to know the tensions on the sides of a funicular polygon subjected to the action of weights, in order to proportion the dimensions of its several parts.

Method of finding the tensions of the sides;

Let $ABCDE$ be a polygon in equilibrio, under the action of the weights P, Q, R, S, T, including the weights of the sides, and the extreme forces N and N', of which the directions are $A A'$ and $E E'$, respectively. Denote the tension of the side $A B$ by t_1, that of $B C$ by t_2, that of $C D$ by t_3, &c. Since the equilibrium subsists about each angle, as A for example, the force N which acts from A to A', is equal and directly opposed to the resultant of the two forces P and t_1; and if $A n$ be taken on the prolongation of $A' A$ to represent N, the parallelogram $A p n o$, constructed on $A n$ as a diagonal, will give $A p$ for the weight P, and $p n$ for the value of the tension t_1. This being understood, draw the *horizontal* line $a' e$, upon which lay off the distances $a' a$, $a b$, $b c$, $c d$, $d e$, proportional to the weights P, Q, R, S, and T. From the point a' draw $a' S$ perpendicular to $A A'$, and proportional in length to the tension N, and join S with the several points a, b, c, d, and e; then will $a S$, $b S$, $c S$, $d S$, and $e S$, represent, respectively, the tensions t_1, t_2, t_3, t_4, and N'. For the two triangles $A p n$ and $a' S a$ are similar, because $a' S$ and $a' a$ are respectively perpendicular to $A n$ and $A p$; hence the angles $S a' a$ and $p A n$ are equal; moreover, the sides about these equal angles are proportional by construction and we, therefore, have

funicular polygon in equilibrio under the action of weights;

determination of a single tension;

general construction for finding the tensions;

Fig. 163.

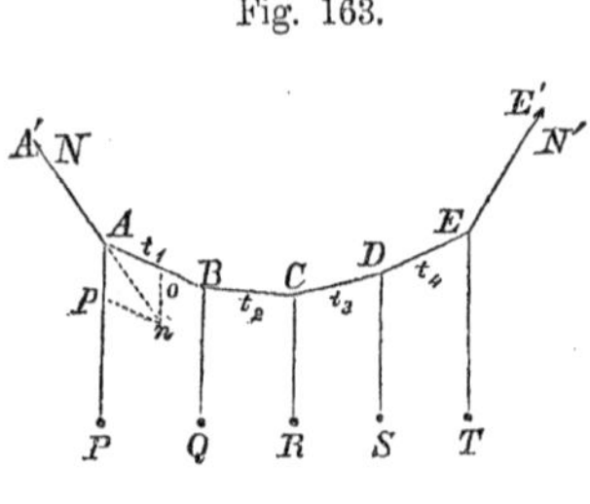

Fig. 164.

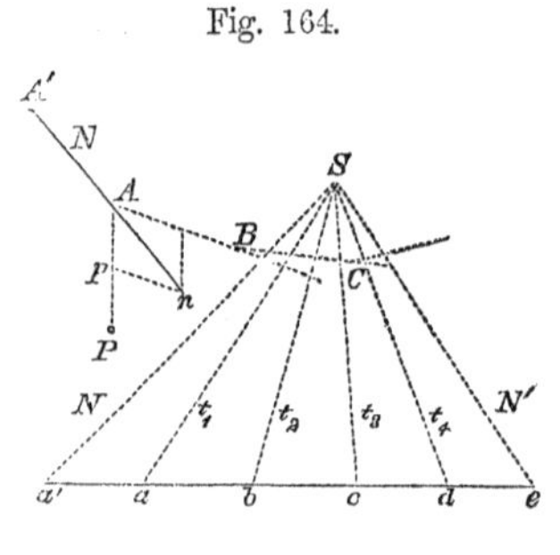

$$A n = N \ : \ p n = t_1 \ :: \ a' S \ : \ S a;$$

and if $a'\,S$ represent the tension N, $S\,a$ must represent the tension t_1. For the same reason, $a\,b$ being proportional to Q, the third side $b\,S$, of the triangle $a\,S\,b$, will be proportional to t_2, since the three forces t_1, Q, and t_2, are in equilibrio about the point B. Finally, since $a\,a'$ and $a'\,S$ are perpendicular to the directions $A\,p$ and $A\,n$ of the forces P and N, $a\,S$ will be perpendicular to the side $A\,B$ of which it measures the tension t_1. It will be the same of $B\,C$ and $b\,S$, and so on. Therefore, when a funicular polygon is in equilibrio under the action of weights, if a series of distances be taken on a horizontal line proportional to these weights, the lines drawn through the points of division perpendicular to the corresponding sides of the polygon will meet in a point, and the lengths of these perpendiculars, included between the common point of intersection and the horizontal line, will measure the tensions of the sides of the polygon. The point S is called the *point of tensions*.

demonstration;

lines which represent the tensions, perpendicular to the sides of the funicular polygon;

point of tensions.

§ 195.—The sides of the polygon may be very short and only subjected to the action of their own weight, which would be the case with a heavy chain $A\,C\,B$ suspended from its extremities. The polygon of equilibrium then becomes a curve, called the *catenary*. This curve is employed to give form to arches and domes. The use of the catenary for such purposes may be illustrated by conceiving a series of equal spherical balls held together by mutual attrac-

The catenary;

its use in the arts;

Fig. 165.

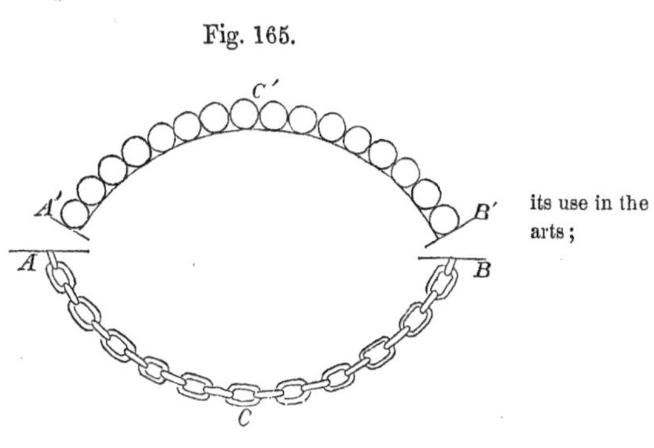

illustration by a string of balls;

tions, but with perfect freedom to slide the one over the other. Such a collection of balls would resemble a string of beads, and if supported at the ends would, under the action of their own weight, assume the form of the catenary, or rather funicular polygon, of which the sides would be the chords of the spheres joining the points of contact. If the whole arrangement be reversed, and the balls, instead of being suspended, be supported upon the ends as fixed points, after the manner indicated in $A' C' B'$, the figure will remain unchanged and the balls will still be in equilibrio; for, the action of the weights will be the same as before, and the reciprocal action of the balls upon each other will simply be changed from a force of extension to one of compression. If we now suppose the points of contact to be extended into tangent planes, and the spaces between filled up with solid matter, as wood, stone, or metal, we shall have a perfect system of voussoirs or arch-solids in equilibrio under the action of their own weight, requiring no aid from friction or any other principle of sup-

sides of the polygon, the chords of the balls;

Fig. 165.

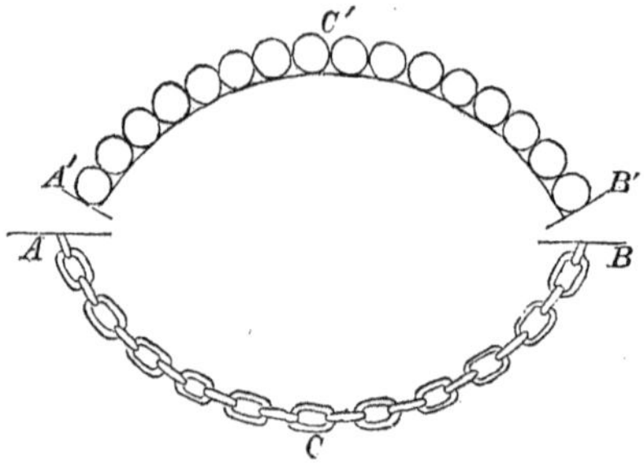

the string of balls reversed;

points of contact extended to tangent planes;

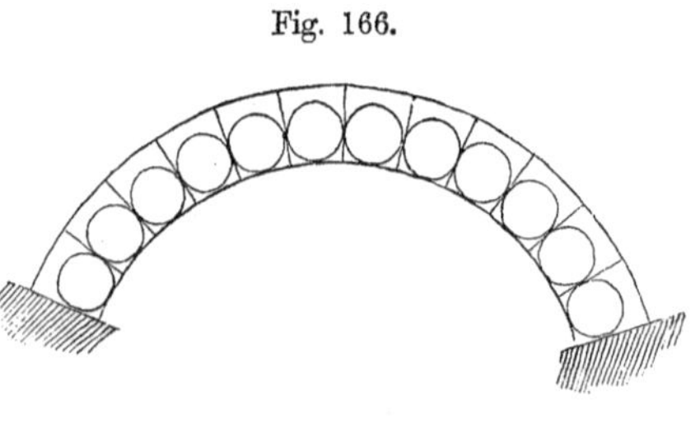

Fig. 166.

arch-stones or voussoirs;

port. The tangent planes or joints of the voussoirs will be normal to the curve. The catenary is also employed in suspension-bridges supported upon two or more parallel chains stretched across a river. In the construction of such catenaries it is important to determine the tension at the ends, in order to secure an adequate resistance at those points.

position of the joints;

also used in suspension-bridges.

§ 196.—The catenary $A\,C\,B$, suspended from two points A and B, is nothing more, as we have seen, than a heavy polygon in equilibrio, and whose sides are indefinitely small; so that, if upon a horizontal line, a length $A'\,B'$ be taken proportional to its weight, and this length be divided into a number of equal parts, there will exist a certain point S such, that all the right lines drawn from it to the points of division, will be perpendicular to the small successive sides or elements of the catenary, and that the lengths $S\,A'$, $S\,F'$, $S\,C'$, &c., of these lines, are proportional to the tensions of the same elements. Of all the tensions, the least is given by the line $S\,C'$, drawn perpendicular to the horizontal line $A'\,B'$. But the element of the catenary to which this tension corresponds being itself horizontal, it will occupy the lowest point of the curve. This length becoming greater

General properties of the catenary;

Fig. 167.

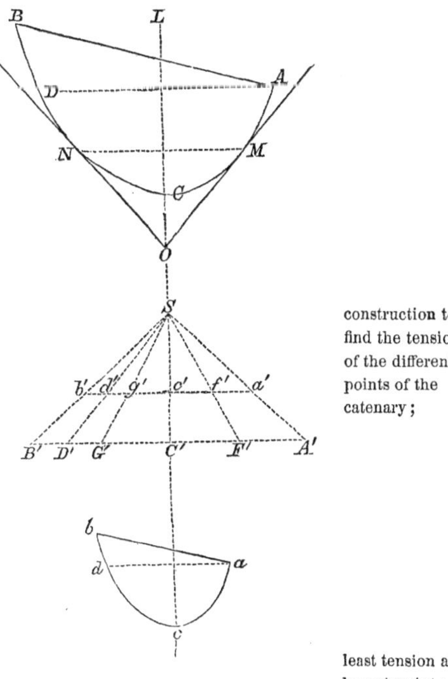

construction to find the tension of the different points of the catenary;

least tension at lowest point;

and greater in proportion as the oblique lines SF', &c., recede from the perpendicular SC', the tensions of the elements of the catenary will increase in proportion as they are at a greater distance from the lowest point. Whence it follows, that the tension is the greatest possible at the extremities A and B. Two equal tensions SF' and SG', appertain to two elements equally distant from the lowest point C: moreover, these elements form equal angles with the vertical LC passing through this point; hence, these elements, M and N, are situated on the same horizontal line MN, and the chord MN, as well as all similar chords, will be divided equally by this vertical line. The catenary is, therefore, a symmetrical curve in reference to a vertical line passing through its lowest point. It follows, also, that when the extremities or attached points A and B are on the same horizontal line, the extreme tensions are equal, and that the point of meeting which determines the tensions is upon the perpendicular drawn through the middle of the horizontal line $A'B'$, which is proportional to the weight of the catenary. A and D being, for example, the two points of suspension, and $A'D'$ being the length proportional to the weight of the catenary ACD, SC', perpendicular to $A'D'$ and passing through the

tension greater as the element is at a greater distance from the lowest point;

elements of equal tensions form equal angles with the vertical through lowest point;

catenary symmetrical in reference to this line;

on same level the extreme tensions are equal; position of the point of tensions.

Fig. 167.

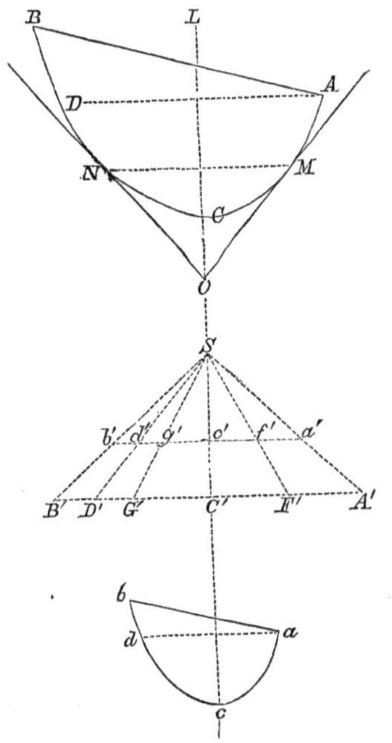

point S, will divide $A'D'$ into two equal parts $A'C'$ and $C'D'$.

§ 197.—Two catenaries, (last figure), ACB and acb are similar when the points of suspension A and B of the one, and a and b of the other, are situated upon parallel right lines, and when their lengths ACB and acb are proportional to the distances AB and ab, between their points of suspension. If the equilibrium subsists in the catenary ACB, this equilibrium will not be disturbed if the length of its elements and its other dimensions be proportionally diminished indefinitely, § 192. Therefore, when ACB is reduced to the size acb, the equilibrium will not only exist, but there will be no one of its parts which will not be parallel and proportional to the corresponding part of the original. But since the elements of the smaller catenary acb are parallel to those of the larger ACB, all the tensions of the former are comprised within the angle $A'SB'$, which contains the different tensions of the latter. We have, then, but to find in this angle, the position of a line $a'b'$ parallel to $A'B'$, which represents the weight of the smaller catenary, as $A'B'$ represents the weight of the larger, and the slightest consideration will show that the two tensions Sf' and SF' situated upon the same line converging to S will appertain to parallel elements of the two curves. These are called *homogeneous tensions*. But because $A'B'$ and $a'b'$ are parallel, we have the proportion

Similar catenaries;

equilibrium independent of size;

tensions of one catenary found from those of a similar one;

homogeneous tensions;

$$Sf' : SF' :: a'b' : A'B';$$

whence we conclude that, *in two similar catenaries, the tensions of elements similarly situated are to each other as the weights of the catenaries.*

tensions of elements similarly situated are as the weights of the entire curves.

To construct the catenary from its weight, length, and the point of tensions;

§ 198.—Let $A' B'$ be a horizontal line proportional to the weight of the catenary, S the point of tensions. Divide the line $A' B'$, and the length of the catenary into the same and a great number of equal parts; those of the catenary may be regarded as its elements, and those of $A' B'$ their corresponding weights. Draw the lines $S A$, $S 1'$, $S 2'$, $S 3' \ldots S B'$; these will be perpendicular to the different elements of the catenary. From any point A, on $S A$, draw A 1 perpendicular to $S A$ and equal to an element of the catenary; from the point 1 draw 1–2 perpendicular to $S 1'$ and equal to an element; again 2–3 perpendicular to $S 2'$, and equal to an element, and so on to the end. The polygon A–1–2–3 . . . B, will approximate to the required catenary the nearer in proportion as the number of divisions is greater.

Fig. 168.

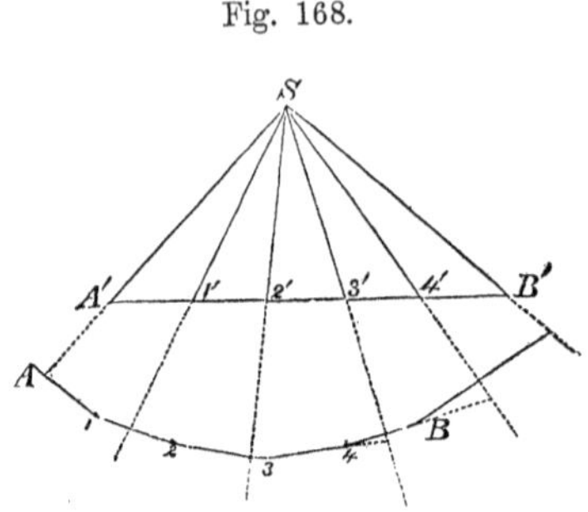

to draw a tangent to any point of the catenary.

The point of tensions S gives the means of drawing a tangent to the catenary at any point. Let E be the given point, and let $A' e$ represent the weight of the portion $A E$ of the catenary; through e and S draw the indefinite line $e G$, and from E draw $E G$ perpendicular to $e S$, $E G$ will be the tangent line.

Fig. 169.

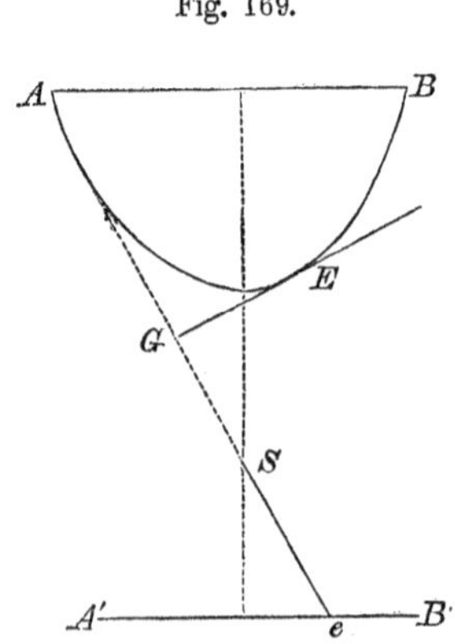

§ 199.—The point of tensions in the catenary depends upon the intensity and directions of the extreme tensions. For $A'B'$ being the horizontal line proportional to the weight of the entire catenary, if from the extremities A' and B' arcs be described with radii proportional to the extreme tensions, their intersection S will give the point of meeting.

Determination of the point of tensions;

Fig. 170.

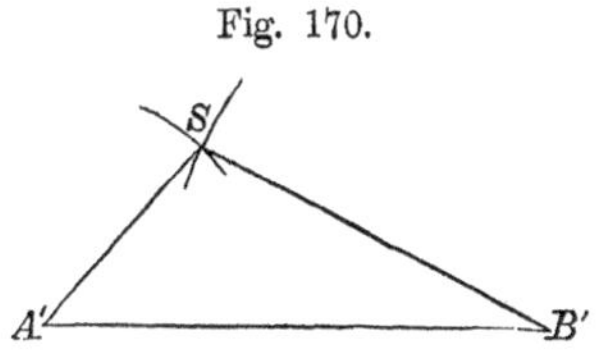

The process for finding the extreme tensions must of course depend upon the data given. Let us first suppose the catenary $A\,CB$ to be given and traced out. It is evident from the conditions of equilibrium, that the vertical $O\,L$ drawn through the intersection O of the extreme tangents $A\,O$ and $B\,O$, will pass through the centre of gravity of the catenary. If, therefore, a distance $O\,G$ be taken on this line to represent the entire weight of the catenary, and the parallelogram $O\,B'\,G\,A'$ be constructed upon the tangents, the sides $O\,A'$ and $O\,B'$ will represent the tensions at A and B respectively.

to find the extreme tensions from the curve traced;

Fig. 171.

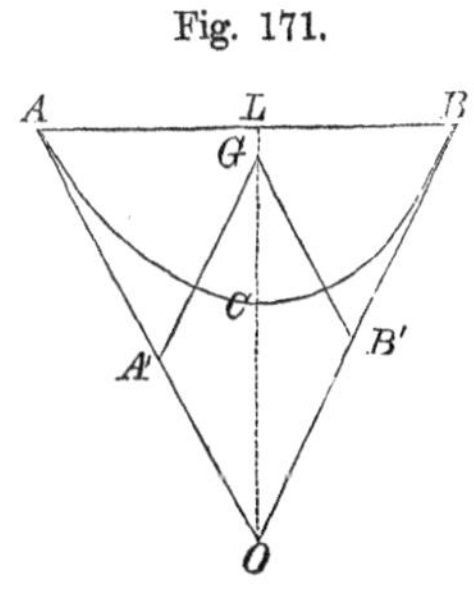

But if only the two points A and B of suspension, the weight, and entire length of the catenary be given, the process for finding the extreme tensions is as follows, viz.: Take a small chain and suspend it against a ver-

to find the extreme tensions from the points of support, the weight, and length of the curve;

Fig. 172.

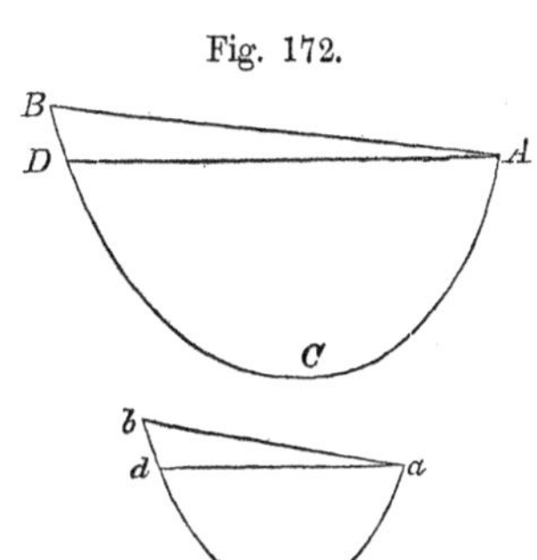

figure found by means of a small chain;

tical plane from two points *a* and *b*, situated upon a right line parallel to *A B*, and whose distance apart shall be to the distance from *A* to *B*, as the length of the smaller chain is to the length of the longer. The smaller chain being thus suspended, measure by means of a spring balance the tension exerted at the points *a* and *b*. The tensions on the points *A* and *B* produced by the larger chain, will be equal to the tensions at *a* and *b*, multiplied by the number of times which the weight of the larger chain contains that of the smaller. § 197.

Fig. 172.

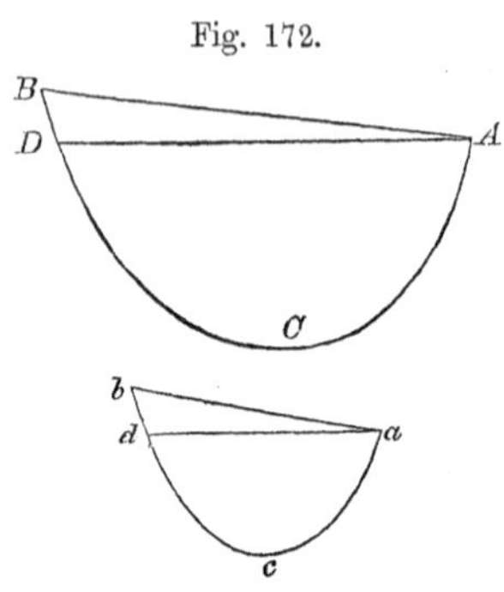

the tensions found by construction;

Instead of measuring with a spring balance the tensions at the ends of the catenary, we may proceed as follows: Draw through the lowest point of suspension *a*, a horizontal line cutting the opposite branch of the small chain in the point *d*. Upon a horizontal line take the distance *a′ b′* to represent the weight of the entire chain, and lay off the distance *a′ d′* proportional to the length *a c d*. The portion *a c d* of the catenary would be in equilibrio if the point *d* were fixed and the remainder *d b* removed; the point of tensions for *a c d*, and therefore for *a c b*, will, from what has already been explained, be found

Fig. 173.

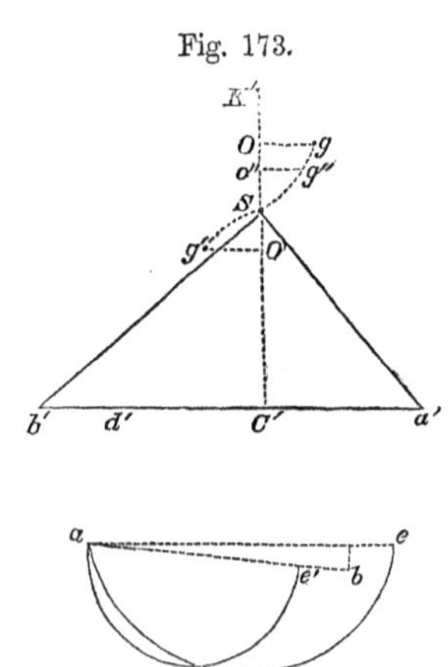

somewhere on the perpendicular $C' K'$ drawn to the middle of $a' d'$; assume it at O, and by means of this point and the line $a' b'$, construct a catenary after the manner of § 198, and let $a\,e$ be the resulting distance between its points of support. Through O draw a perpendicular to $C' K'$, and lay off upon it from the point O, the distance $O g = a e - a b$, to the right when $a\,e$ is greater than $a\,b$, and to the left when the reverse is the case. Assume another point as O' below O, and do the same as before; we shall find a new point g', say to the left of $C' K'$; repeat the process with points between O and O' several times, and pass through the points g, g', g'', &c., thus determined, a curve; its intersection S with $C' K'$ will be the true point of tensions. The distances $S a'$ and $S b'$ will represent the extreme tensions.

Construction of an approximate curve.

§ 200.—We have seen that in the catenary the tensions at the different points are different, and that the smallest tension is at the lowest point. This is still true when the catenary becomes a vertical chain loaded with a weight. For the lowest link supports only the attached weight Q; the link C' only supports the weight Q and link C, and so on to the topmost link, which supports all below it; so that if the chain were proportioned to the tension of its different parts, it would be made stronger above than below.

Fig. 174.

The smallest and greatest tension of a vertical chain.

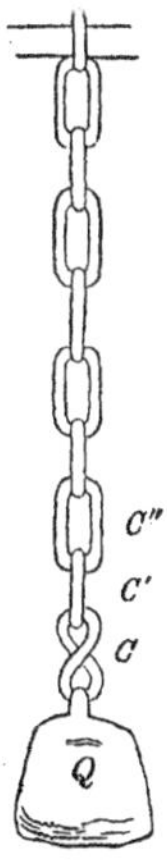

§ 201.—The point S being the point of meeting of the tensions, and $A' B'$ a horizontal line representing the weight of the catenary, we have seen that the tension at

Fig. 175.

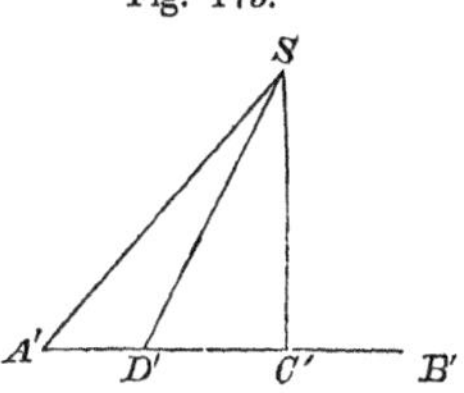

Direct measure of the tension on any point of the catenary;

D is represented by the length $D' S$, and that at C, the lowest point, by $S C'$, perpendicular to $A' B'$, the lengths $A' D'$ and $D' C'$ representing, respectively, the weights of the portions $A D$ and $D C$ of the curve; that is to say, the tension at any point D, is represented by the hypothenuse of a right-angled triangle, of which one side represents the tension at the lowest point of the curve, and the other the weight of that portion of the catenary included between the lowest point and the point whose tension is to be found. Hence, *the tension at any point of the curve, estimated in a horizontal direction, is constant and equal to the entire tension at the lowest point; and estimated in the vertical direction, is equal to the weight of that portion of the catenary included between this point and the lowest point.*

construction; the tension at any point is the hypothenuse of a right-angled triangle; tension in horizontal direction; in vertical direction;

Fig. 175.

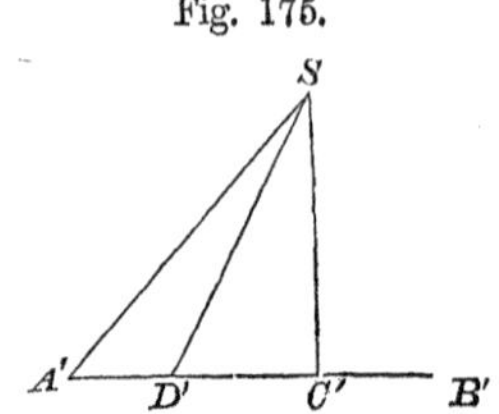

Fig. 176.

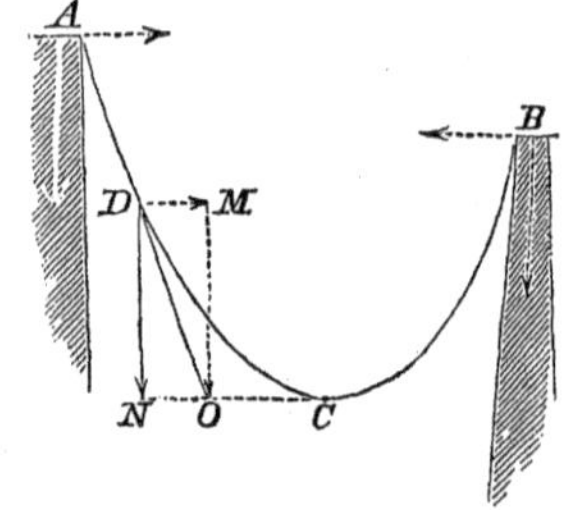

The horizontal tensions at A and B are therefore the same, although they may be situated on very different levels. If the catenary be suspended from the tops of piers, the vertical components will promote their stability by pressing them down, while the horizontal components will tend to overturn them.

effect of these tensions on piers.

§ 202.—It is comparatively easy to compute the extreme tensions of the catenary when the versed sine of its arc is small. Let $A C B$ be a catenary, of which $C D$, the distance of the lowest point below the horizontal line BA, is very small. The curve being in equilibrio, the equi- To find extreme tensions when the versed sine of the curve is small;

Fig. 177.

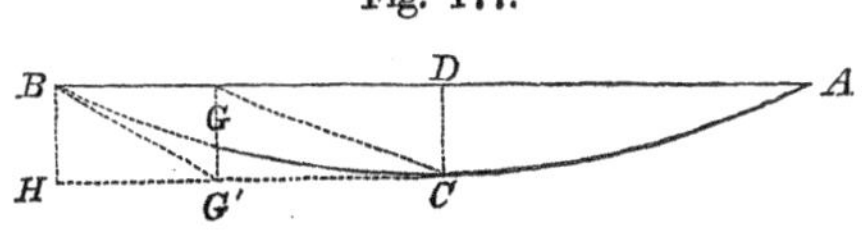

librium of the part BC will not be disturbed by taking the point C as fixed, and regarding it and the point B as the points of suspension. But because of the smallness of DC, the curvature must be very small, and the centre of gravity of BC may, without sensible error, be regarded as at the middle point G. The tangents CH and BG', at the points of suspension, will intersect at G' on a vertical line drawn through the point G. Denote by T, the tension at B; by T_0, the tension at C; and by p, the weight of the portion BC. notation;

Because the three forces p, T, and T_0, are in equilibrio about the point G', we have

$$p : T_0 :: BH : HG',$$

$$p : T :: BH : BG';$$

whence

$$T_0 = p \cdot \frac{HG'}{BH}, \quad \text{tension at lowest point;}$$

$$T = p \cdot \frac{BG'}{BH}. \quad \text{tension at highest point;}$$

Observe that BH is the versed sine, which denote by f; and, because BGC may be regarded a right line, HG' is half the semi-space BD, which semi-space denote by l. Then, since the triangle $BG'H$ is right angled,

$$BG' = \sqrt{\overline{BH}^2 + \overline{G'H}^2} = \sqrt{f^2 + \frac{l^2}{4}}.$$

Substituting these quantities in the above equations, we find

horizontal tension or thrust;

$$T_0 = \frac{p\,l}{2f},$$

tension at highest point.

$$T = \frac{p}{f} \cdot \sqrt{f^2 + \frac{l^2}{4}} = p\sqrt{1 + \frac{l^2}{4f^2}}.$$

The first expresses the tension at the lowest point, which we have seen is equal to the horizontal thrust at the points of suspension. The second gives the entire tension at the same points, which must be known in order to adjust the dimensions of the chain.

Application to suspension-bridge;

§ 203.—To conclude the subject of the catenary, and show the application of the preceding principles, take the case of a bridge suspended from two parallel chains extended from one bank of a river to the other.

Fig. 178.

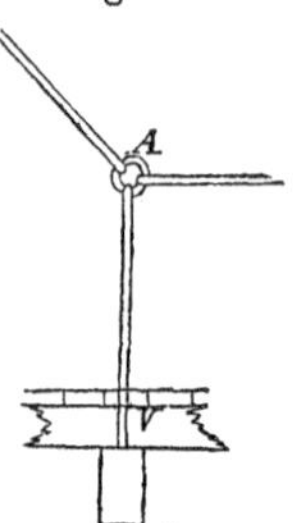

suspending rods; joists; sections;

To the different points, A, B, C, &c., of the catenaries, or rather to the angles of the funicular polygons thus formed, are attached vertical *suspending rods*, which are united at the bottom in pairs by transverse pieces called *sleepers;* these receive a set of longitudinal *joists*, which, in their turn, support the floor plank. The distances between the suspending pieces in longitudinal direction are supposed equal. These equal portions of the roadway included between two consecutive sleepers, are called *sections*. Each sleeper is loaded with half the section which precedes and half that which follows it; that is to say, with the weight of an entire section. This weight is known, and determines the cross section of the suspending rods. The weight of the suspenders being small compared with that of the roadway, may be neglected, and thus the weight of the bridge will be equally distributed.

each pair of suspending rods supports the weight of one section;

Fig. 179.

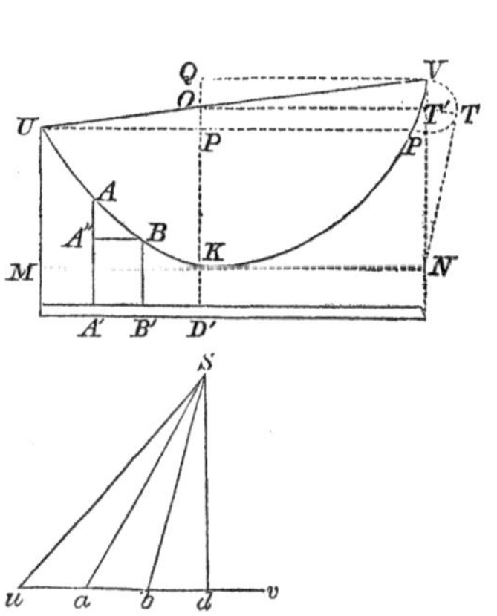

Draw a horizontal right line, and take $u\,v$ proportional to the weight of the bridge; let S be a point such that $S\,u$ shall be perpendicular to the side $U\,A$, and proportional to its tension. Take upon $u\,v$, tho portions $u\,a$, $a\,b$, &c., proportional to the weights supported at the angles A, B, &c.; the converging lines $a\,S$, $b\,S$, &c., will be proportional to the tensions on the sides $A\,B$, &c., and the perpendicular $S\,d$, to the tension on the horizontal side of the polygon. First, find the difference of level between any two consecutive angles, as A and B. Draw the horizontal line $B\,A''$, and the two triangles $A\,A''B$ and $S\,u\,d$, will be similar and give

tensions on the sides of the funicular polygon;

$$A\,A'' \;:\; A''\,B \;::\; u\,d \;:\; S\,d;$$

whence

$$A\,A'' = \frac{A''\,B}{S\,d}\,u\,d.$$

difference of level between two consecutive angles;

Because of the equality of distances between the suspending rods, $A''\,B$ will be constant. Moreover, $u\,d$ and $S\,d$ being proportional respectively to the weight of the portion $A'\,D'$, and the tension t_0 upon the horizontal side, if we denote by ω the weight of a unit of length of the bridge,

ratio of weight of half the bridge to the horizontal tension;

$$\frac{u\,d}{S\,d} = \frac{\omega\,A'\,D'}{t_0};$$

which in the preceding gives

$$A\,A'' = \frac{\omega\,A''\,B\,.\,A'\,D'}{t_0};$$

but $\omega\,A''\,B$ is the weight of a section of the bridge. Denoting this by p, we have

$$A\,A'' = \frac{p}{t_0}\,.\,A'\,D';$$

and denoting the constant ratio of the weight p to the tension t_0 at the lowest point by k,

value of the difference of level of two consecutive angles;

$$A\,A'' = k\,.\,A'\,D';$$

from which we conclude, that the difference of level of two consecutive angles, is equal to the constant ratio k, multiplied by the horizontal distance of the higher of the two angles from the lowest angle of the funicular polygon. Denoting by l the constant length of a section, and beginning at the lowest angle K, the horizontal distances will be successively l, $2\,l$, $3\,l$ . . . $n\,l$, for the 1st, 2d, 3d, . . . nth, angle to the right and left. Thus the difference of level between the lowest angle K and the next in order C, is $k\,l$; between C and B, $2\,k\,l$; between B and A, $3\,k\,l$, &c. The heights of the angles C, B, A, &c., above the lowest point K, will be respectively $k\,l$, $k\,l + 2\,k\,l$, $k\,l + 2\,k\,l + 3\,k\,l$, $k\,l + 2\,k\,l + 3\,k\,l + 4\,k\,l$, and, in general,

difference of level of the angles of the polygon above the lowest angle;

Fig. 180.

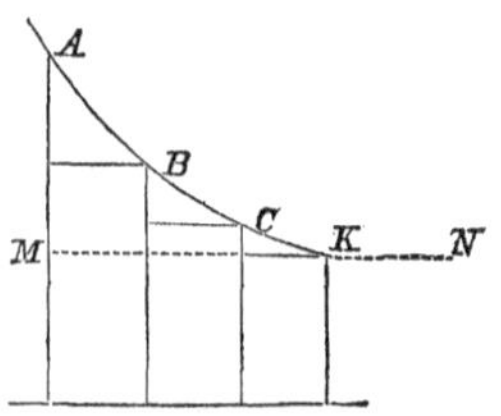

if there be n sections between the lowest angle and that under consideration, the height of the latter above the former will be given by the expression

$$k\,l\,(1 + 2 + 3 + 4 \ldots . + n) = k\,l\,.\,n\,\frac{n + 1}{2}.$$

height of the nth angle above the lowest one;

In this expression, if we make successively $n = 1$, $n = 2$, $n = 3$, $n = 4$, &c., we have $k\,l$, $3\,k\,l$, $6\,k\,l$, $10\,k\,l$, &c., for the heights of 1st, 2d, 3d, 4th, &c., angles above the horizontal side of the funicular polygon.

the locus of the angles is a parabola;

The locus of all these angles is a parabola, for if $y = KP = MU$ denote the height of one of these angles above the lowest point K, n being the number of its place from the latter, we have

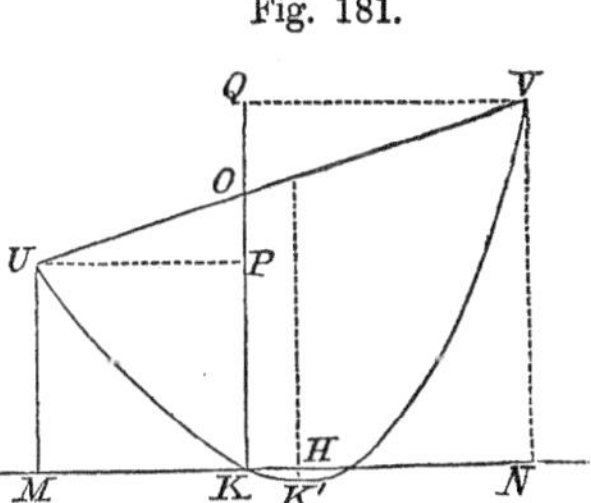

Fig. 181.

$$y = k\,l\,.\,n\,\frac{n + 1}{2} \quad . \quad . \quad . \quad . \quad (96);$$

and making

$$n\,l = x = KM,$$

$$y = k\,.\left(\frac{n + 1}{2}\right)x = \frac{k}{2}\,(x + l)\,\frac{x}{l};$$

or

$$y = \frac{k\,x^2}{2\,l} + \frac{k}{2}\,x;$$

equation of the locus of the angles;

this is the equation of a parabola, of which the vertex is

to the right of the point K, and at a distance from it equal to

$$x = -\frac{l}{2};$$

place of the vertex of the locus curve;

it is below the horizontal side by the distance

$$HK' = y = -\frac{k\,l}{8};$$

a quantity so small that it may be neglected in practice.

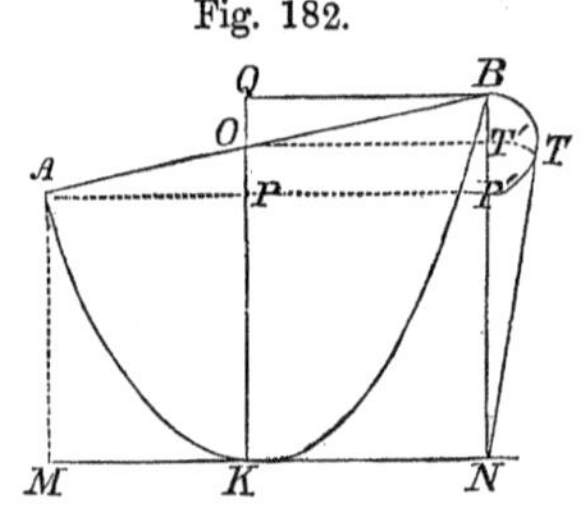

Fig. 182.

Moreover, from the property of the parabola, the squares of the ordinates are to each other as the abscisses; that is to say,

to find the point in which the vertical through the vertex cuts the line of supports;

$$\overline{AP}^2 : \overline{BQ}^2 :: MA : NB;$$

and from the similar triangles obtained by joining A and B,

$$\overline{AP}^2 : \overline{BQ}^2 :: \overline{PO}^2 : \overline{QO}^2;$$

whence

$$\overline{PO}^2 : \overline{QO}^2 :: MA : NB;$$

or

$$\overline{PO}^2 \times NB = \overline{QO}^2 \times MA;$$

but

$$PO = OK - KP = OK - MA,$$

$$QO = QK - OK = NB - KO;$$

which, substituted above, give

$$(OK - MA)^2 \times NB = (NB - KO)^2 . MA;$$

developing the squares and reducing, we get

$$\overline{KO}^2 = MA \times NB.$$

That is to say, the distance KO, at which the vertical line drawn through the vertex of the curve cuts the chord joining any two of its points, is a mean proportional between the heights of these points above the vertex. This property furnishes an easy method of finding the lowest point K on the level MN. For this purpose, join the points of suspension U and V, by the cord UV; draw the horizontal line UP through the lower point U, and produce it till it cuts the vertical VN in P'. Upon the distance $P'V$ describe the semicircle VTP', and from the point N draw the tangent NT; with N as a centre and NT as a radius, describe the arc TT'' till it cuts VN in T'', and through the point T'' draw a horizontal line; this line will cut the cord UV in the point O, through which draw a vertical line OP, and its intersection with the horizontal side will give the lowest point K. Taking this point as the extremity of the horizontal side, and laying off on the line MN the equal lengths of the sections; the points of division will correspond to the vertical ordinates kl, $3\,kl$, $6\,kl$, . . . $n . \frac{n+1}{2}\,kl$. This last appertaining to the point U, whose height h is given,

distance of this point above the vertex;

Fig. 183.

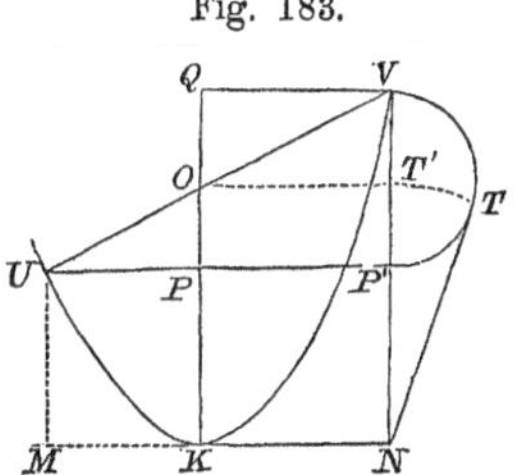

construction for finding the position of the lowest point;

and the abscisses of the angular points;

we have

$$\frac{n \,.\, (n + 1)}{2} k \, l = h;$$

whence we have

ratio of the weight of a section to horizontal tension found;

$$k = \frac{2\,h}{n\,(n + 1)\,l}; \quad . \quad . \quad . \quad . \quad (97);$$

and hence the lengths of the several suspenders $k\,l$, $3\,k\,l$, &c., are known.

from which the lengths of the suspenders are known;

We have seen that

$$\frac{p}{t_0} = k = \frac{2\,h}{n\,(n + 1)\,l},$$

and therefore

and horizontal tension found;

$$t_0 = \frac{n\,(n + 1) \,.\, l \,.\, p}{2\,h} \quad . \quad . \quad . \quad (98);$$

the tension on the horizontal side is, therefore, also known. The tension on the side next in order to the horizontal side is

tension on side next in order;

$$\sqrt{t_0^2 + p^2},$$

that of the second in order

that on the second in order;

$$\sqrt{t_0^2 + (2\,p)^2},$$

that of the third

that on third;

$$\sqrt{t_0^2 + (3\,p)^2},$$

and so on to

that on the *n*th in order;

$$\sqrt{t_0^2 + (n\,p)^2},$$

which is the tension on the nth side from the horizontal one.

If the points U and V be on the same level, it is obvious that the curve or polygon becomes symmetrical in reference to the vertical $O\,K$, in which case it is only necessary to find the lengths of the suspenders for one half the bridge. Having given the points of suspension, their horizontal distance apart, and the level of the lowest side of the funicular polygon, it is easy to determine the dimensions of every part of the bridge.

Data necessary to make known the dimensions of the bridge.

XV.

OF BODIES RESTING UPON EACH OTHER, AND UPON INCLINED PLANES.

§ 204.—When two bodies touch and compress each other, there is immediately a depression or yielding in a direction perpendicular to the surfaces at the point of contact, which indicates that the reaction of the two bodies takes place in the same direction; that is to say, in the direction of the normal common to both surfaces. Let us suppose one of the two bodies as A to be solicited by forces of which the resultant shall coincide with this normal, and that the other body A' is fixed; it is plain that the reaction of the latter body will destroy this resultant, and that the body A will remain at rest. But the equilibrium will also subsist if the body A' be replaced by a force equal to the reaction which it exerts on the body A, while this latter body is perfectly free to move and acted upon by this new force in conjunction with the given forces. This property of all bodies, by which they resist the re-

Action and reaction of bodies forced into apparent contact;

Fig. 184.

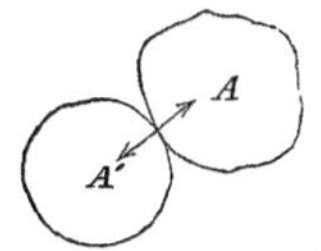

action and reaction of two bodies;

ciprocal action of each other in directions normal to both surfaces at the common point of contact, extends to the general case of a single body pressing upon two or more bodies at the same time. The reaction of these last are so many real forces which may be substituted for the resisting bodies at the several points of contact, and in virtue of this substitution, the conditions of equilibrium of the first body will be the same as though it were free to move in any direction whatever. Let us examine the circumstances of the simple case of a body resting upon a plane, and having first but one point of contact, then two, three, &c.

the principle of the reaction of two bodies extends to several.

Illustration;

§ 205.—Let us consider a sphere subjected to the action of its own weight, and resting upon a level plane $A\ B$ with a single point of contact m. Since the reaction takes place in the direction of the perpendicular to the plane through the point of contact, and must be in equilibrio with the weight W of the sphere, the centre of gravity G must be upon a vertical line, in order that the weight and reaction may destroy each other. In like manner, when a body rests upon any plane whatever, and is solicited by forces, no matter how directed, their resultant must be perpendicular to the plane, and pass through the point of contact; for if the resultant were oblique, it might be resolved into two components, one normal, and the other parallel to the plane; the first would be destroyed by the reaction of the plane, while the latter would put the body in motion. In order, therefore, that a body, supported against a plane, and having a single point of contact with it, shall be in equilibrio, it is necessary, 1st, *that the resultant of the forces which act upon it be perpendicular to the plane;* and 2d, *that this resultant pass through the point of contact.*

the bodies having but one point of contact;

Fig. 185.

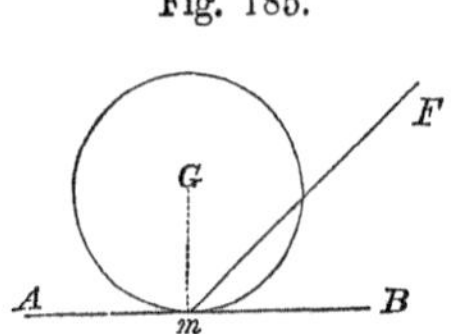

conditions which will keep a body at rest against a plane.

§ 206.—But when the body has two points of contact, A and B, with the plane, it is not necessary that the resultant of the forces shall pass through either. It will be sufficient if it meet the line $A B$ in any point between A and B, and be perpendicular to the plane. For the reaction of these points of support being both perpendicular to the plane, their resultant, which is parallel to them, will also be perpendicular to it: this resultant and that of the forces acting upon the body must be in equilibrio; they must, therefore, be equal and directly opposed; in other words, the resultant of the forces acting upon the body must admit of being resolved into two components, respectively equal and directly opposed to the resistances at the points of support. But these latter act in the same direction, so also must the former, and hence their resultant will have its point of application between A and B; and this resultant being parallel to its components, will be perpendicular to the plane.

Fig. 186.

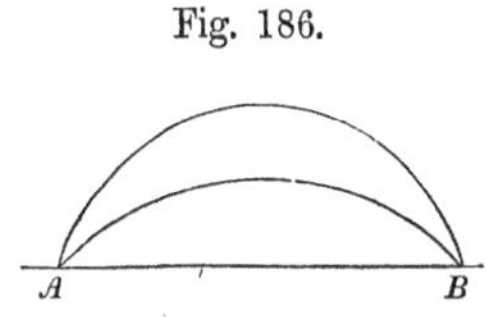

If the body have two points of contact the resultant need not pass through either;

it must be normal to the plane, and intersect the line joining the points of contact;

If the body be laid on a horizontal plane, the equilibrium will subsist whenever the vertical drawn through the centre of gravity intersects the line joining the points of support somewhere between them.

when the plane is horizontal.

§ 207.—Now let us suppose three or more points of contact. The *resistances* of these points are perpendicular to the plane, and cannot maintain the forces which act upon the body in equilibrio unless the resultant of the latter may be decomposed into components which are respectively equal and directly opposed to these resistances; this resultant must, therefore, be perpendicular to the plane, and as its components must act in the same direction, its point of application will, from the principles of parallel forces, be within the polygon formed, by joining the points of contact. If the line of direction of the resultant, pierce

Case of three or more points;

resultant still normal, and within the polygon of contact;

if the resultant pierce the plane without the polygon of contact, the body will overturn;

the plane in a point m, exterior to the polygon which connects the points of support, the body will tend to overturn around the edge $a\,b$ of this polygon nearest to m; if the line of contact be a curve, the body will overturn about the tangent nearest to m. The effort by which the body will be urged to overturn is measured by the intensity of the resultant of the forces, into the shortest distance from its line of direction to that about which the motion of rotation takes place.

effort by which the body is urged to overturn.

Fig. 187.

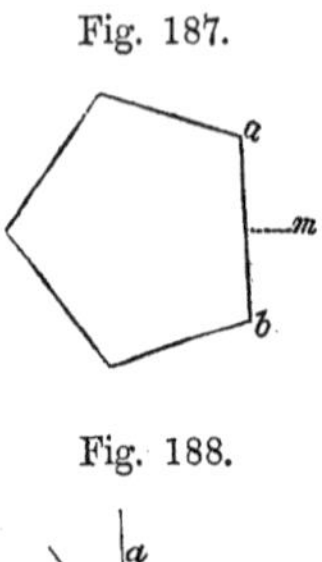

Fig. 188.

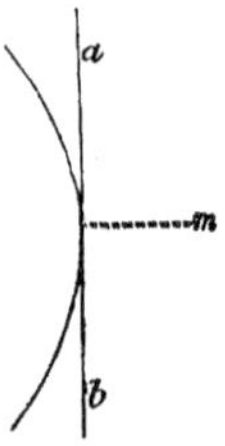

§ 208.—The conditions of equilibrium of a heavy sphere, resting upon a horizontal plane, have already been considered. Let us apply the same principles to other examples, and take first the case of a heavy body resting upon a table having but three feet. If the feet be upon a horizontal plane and in the same right line, and the vertical line through the centre of gravity be not in the vertical plane passing through this line, the table will overturn towards the side on which the centre of gravity is situated, and with an effort equal to the product of the weight into the distance $A\,g$ of the projection of the

Examples;

table having but three feet;

when the feet are in same right line;

will overturn unless the weight pass through this line;

Fig. 189.

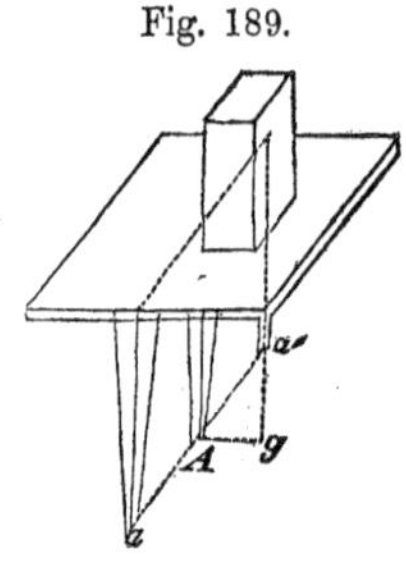

Fig. 190.

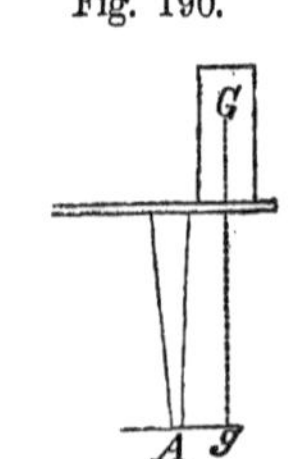

centre of gravity from the line $a a'$ of rotation. This product is called the *moment of stability*. If the distance $A g$ is zero, the weight will pass through the line of support, and there will be an equilibrium; but it will be unstable, since the centre of gravity will be at the highest point.

moment of stability;

If the three feet be not in the same right line, and the weight pass within the triangle formed by joining the feet, the table will be in equilibrio. But if the line of direction of the weight pass without the triangle of the feet, the table will overturn about the nearest edge $a b$. In the first case, the equilibrium is stable, because no derangement can take place about the line of either two of the feet without causing the centre of gravity to ascend. And, generally, if the table have any number of feet, there will be stable equilibrium whenever the line of direction of the weight passes within the polygon formed by joining them.

if the feet be not in same right line;

stable equilibrium;

in case of any number of feet the resultant must pass within the polygon;

Fig. 191

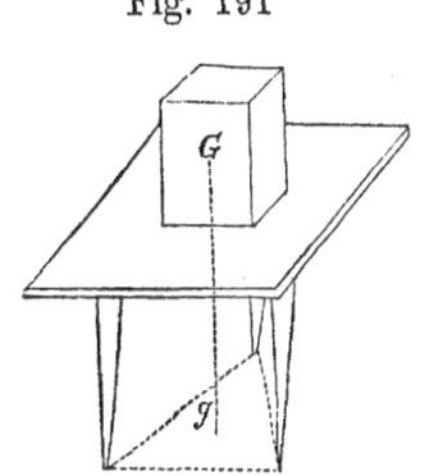

Fig. 192.

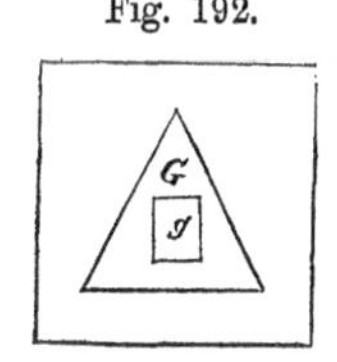

Fig. 193.

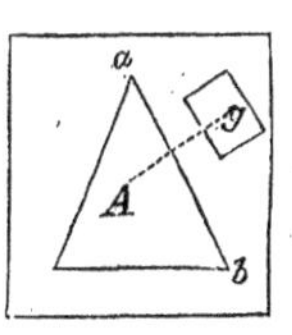

Fig. 194.

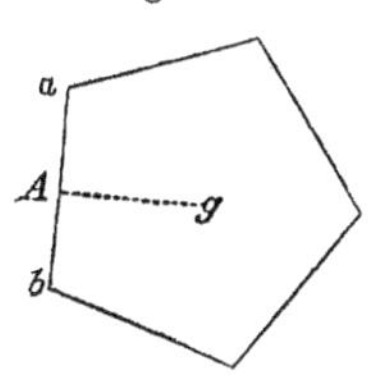

The effort with which the table or any other body will resist a cause which tends to upset it, is measured by the product of its weight into the shortest distance $A g$ from the line of direction of the weight to the line $a b$ about which the motion is to take place;

effort by which a body resists a cause to overturn it;

moment of stability of a heavy body; and this effort will be smaller in proportion as the distance $A g$ is less. For this reason, the *moment of stability* of a heavy body is the smallest moment of its weight taken with reference to the different lines of its polygonal base.

the same principles apply to solids resting on plane faces; The conditions are the same if the body rest upon a plane face bounded by a polygon or curve. The equilibrium will exist when the line of direction of the weight passes within the base. Such, for example, is the case with the cube resting upon a level plane; also with a right prism, whatever its height, only that its stability diminishes as the height increases; for, in proportion as the centre of gravity G is more and more elevated, the angle $G A B$ becomes less and less, and the centre of gravity will not have to be raised so much above its position of rest when the body is overturned about the edge $a a'$, as it would if the angle $G A B$ were greater, or the centre of gravity lower. In proportion as the centre of gravity is placed higher and higher above the same base, the body will approach more and more to the condition of unstable equilibrium.

example of the cube and right prism;

stability diminishes as the centre of gravity is higher;

Fig. 195.

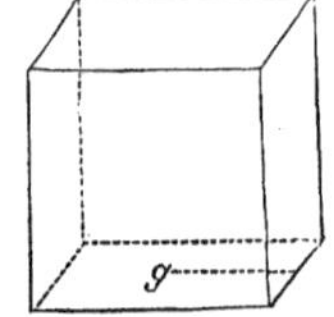

Fig. 196.

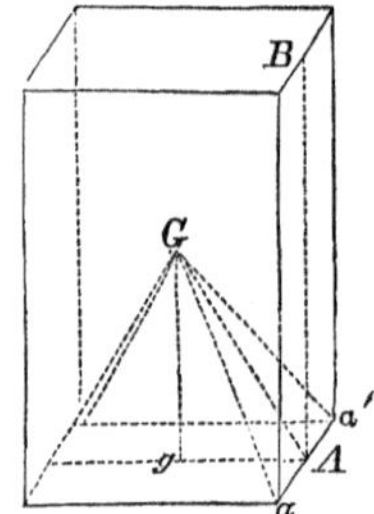

inclined prism; An inclined prism will preserve its equilibrium as long as the direction of its weight falls within its base. The difficulty of overturning it will be less in proportion as the

Fig. 197.

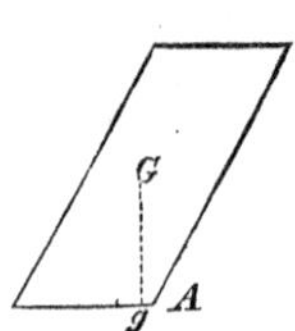

distance $A\,g$ becomes smaller. When g falls without the base, the prism will overturn of itself. The Tower of Pisa, though considerably inclined, preserves its equilibrium because the line of direction of its weight passes within its base. A pile of dominos or bricks, in which each one projects beyond that immediately below it, will preserve its equilibrium till the line of direction of the weight of the entire pile falls without the domino or brick at the bottom, when it will overturn. We see, therefore, that the natural stability of bodies increases as their bases increase, and the heights of their centres of gravity decrease; and that it is the greatest possible when the centre of gravity is at the centre of figure of the base. This is the reason why walls are usually made of elements like brick, cut-stone, &c., placed with their faces vertical, and laid upon large bases, called foundations.

will overturn when weight falls without the base;

Tower of Pisa;

Fig. 198

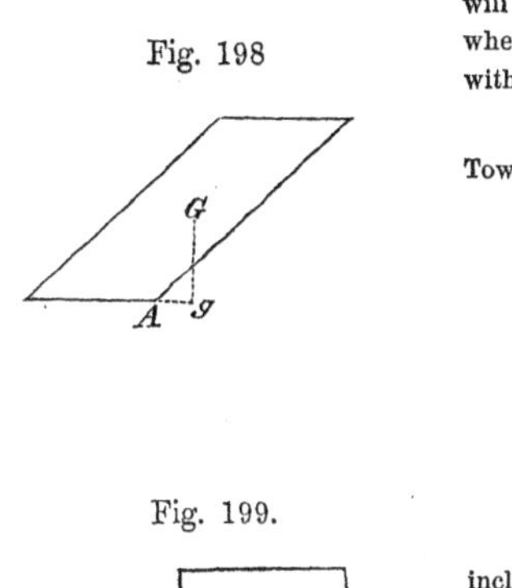

Fig. 199.

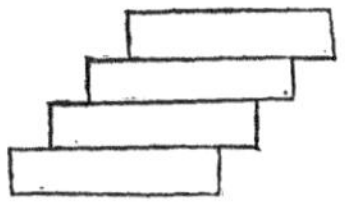

inclined pile of brick;

stability increases as the base increases and as the centre of gravity is lower;

If the heavy bodies are solicited by other forces than their weights, the resultant of the whole, weight included, must act in the direction of a line passing within the base. The resultant of the extraneous forces may unite with the weight and increase the stability of the body. Thus an inclined prism, the direction $G\,g$ of whose weight falls without the base $A\,B$, would, if abandoned to itself, overturn; whereas, if it were acted upon by a force in the direction $G\,E$, of such intensity as to give, with the weight, a resultant which intersects

heavy bodies solicited by other forces than their weights;

Fig. 200.

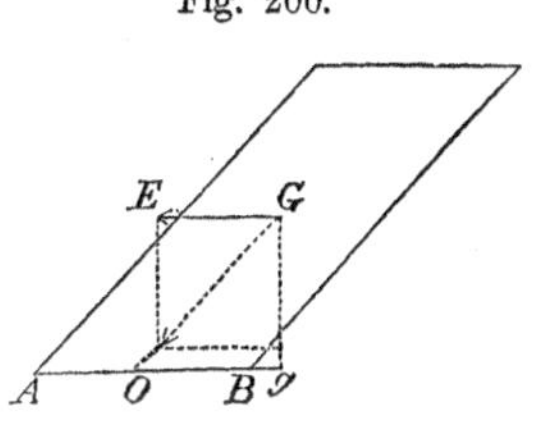

these may act to increase the stability;

equilibrium stable;

the base at O, it would be supported, and the equilibrium would be stable. Reciprocally, the weight W of the prism is opposed to the force $GE = F$, when the latter acts to turn the solid about the edge A. The measure of this opposing effort is

Fig. 200.

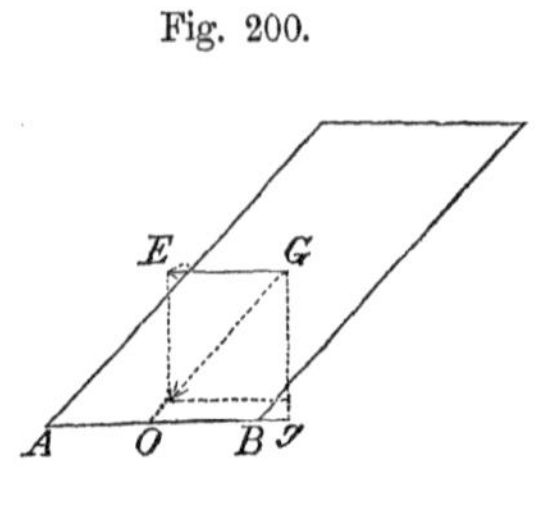

moment of stability;

$$W \, . \, Ag;$$

and in this view, we see that the moment of the natural stability will increase as $A\,g$ increases.

illustration of the foregoing in the construction of sustaining walls;

In walls destined to support an embankment of earth or a head of water, in order to resist the thrust with greater effect, the lower exterior edge A is thrown as far as convenience will permit from the vertical line $G\,g$ of the weight. This is done either by an exterior slope $B\,A$, or by masses of masonry C, called counterforts, attached to the back of the wall. It will be sufficient, in general, for the stability of the wall, if the resultant of its weight W and the pressure against it, intersects the base $A\,D$. The moment of natural stability of such structures is always equal to the product of the weight into the distance $A\,g$; and therefore the figure of the cross-section of the wall may be varied at pleasure without injury to the sta-

principle of counterforts;

moment of natural stability;

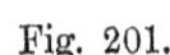
Fig. 201.

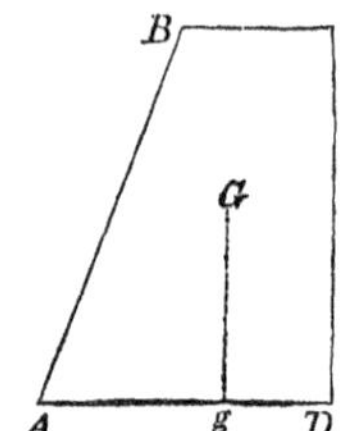

Fig. 202.

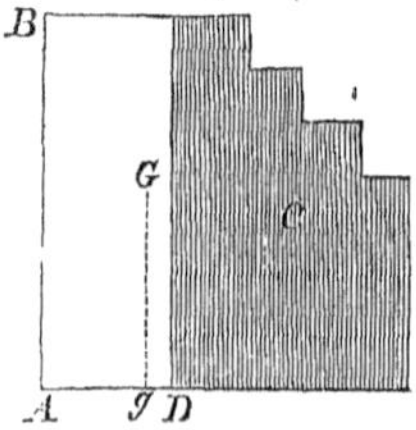

bility, provided this product remain the same. Hence the external slope may be suppressed, if the thickness of the wall be so increased that its augmented weight shall compensate for the diminution in $A g$.

external slope and weight;

If the ground upon which the wall rests be compressible, it will not be sufficient that the resultant of the weight and pressure pass within the base; it must also pass through its centre of figure; otherwise there would be more pressure on one side of this point than on the other, and the wall would incline in that direction.

Fig. 203.

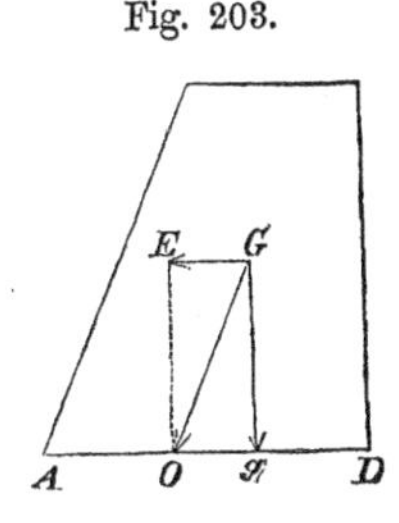

when the ground is compressible, the resultant should intersect middle of the base;

If the load of a two-wheel cart be such that the direction of its weight does not intersect the axle-tree, it will tend to overturn on the side of the weight, and will either exert a pressure upon the horse or an effort to lift him from the ground, according as the weight passes in front or in rear of the axle-tree. If the centre of gravity of the load be immediately above the axle-tree on a level road, then, when the cart is ascending a slope, the weight will pass behind, and the tendency of the load will be to lift the horse; while, on the contrary, when the cart is

Fig. 204.

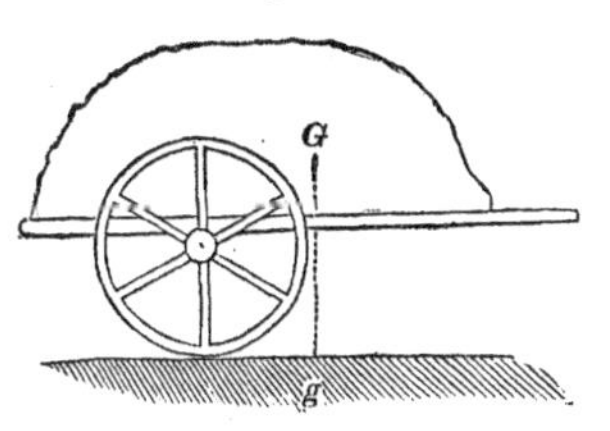

case of a loaded cart on a level;

Fig. 205.

on an inclined road ascending the tendency is to lift the horse;

Fig. 206.

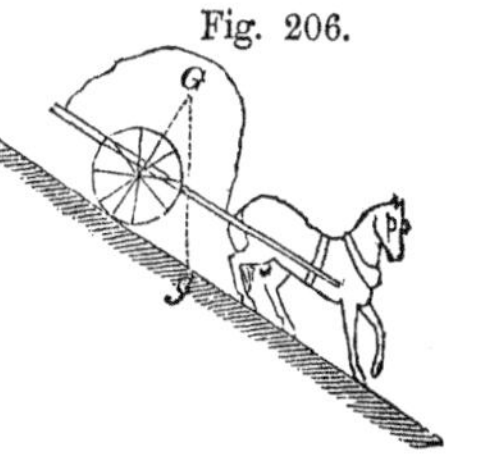

descending;

the tendency is to press upon the horse.

descending a slope, the tendency of the load will be to throw a pressure upon him. If the centre of gravity be on the axle-tree, the horse will experience no effort of the kind referred to.

Inclined plane;

§ 209.—Let AB represent the section of an inclined plane in the direction of its greatest declivity. Although the plane be indefinitely prolonged, it will be sufficiently defined by the relation of the base AC to the height CB, corresponding to a given length AB.

defined by ratio of height to base;

Fig. 207.

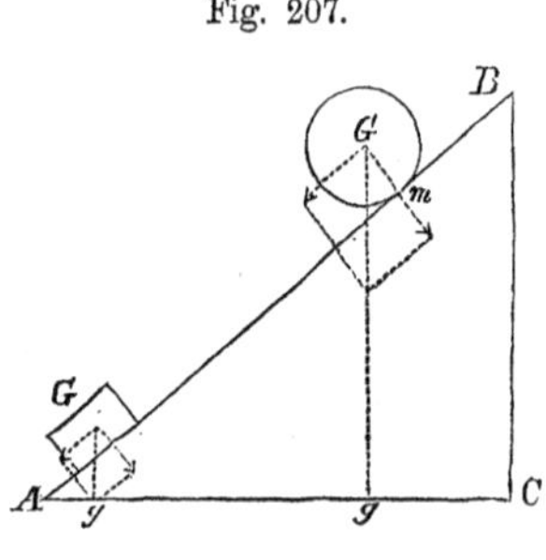

Conceive a heavy body resting upon this plane, and of which G is the centre of gravity. The equilibrium of this body requires, 1st, that its weight shall intersect the plane within the polygon formed by joining the points of contact; 2d, that the weight shall be perpendicular to the plane. This last condition cannot be satisfied for any but a horizontal plane, since the weight is always vertical. If the weight be replaced by its two components, one perpendicular and the other parallel to the plane, the former will be destroyed by the resistance of the plane, while the latter will cause the body to move in the direction of its length BA. If the direction of the weight meet the plane within the polygon of contact, the parallel component will cause the body to slide, otherwise it will cause it to roll. This last will happen in the case of a spherical ball, since the weight will not meet the plane in the single point of contact m.

a body on an inclined plane;

the body may slide or roll;

Let a force P be applied in the direction GS, next figure, to prevent the body from moving down the plane. Since the body must be in equilibrio under the action of its weight W and the force P, these must have a resultant, and this resultant must be perpendicular to the plane and intersect

conditions of equilibrium;

it within the polygon of contact, or in the case of the sphere, at the point m. The force P must, therefore, be applied in a vertical plane which passes through the centre of gravity, and which is, at the same time, perpendicular to the inclined plane. plane in which the force must be applied;

Lay off on the vertical through the centre of gravity G, the distance $G\,G'$ to represent the weight W, through the same point draw $G\,M$ perpendicular to the inclined plane, and through G', the line $G'\,M$ parallel to the direction of the force P; from the point M draw $M\,Q$ parallel to $G\,G'$; the distance $G\,Q$ will represent the intensity of the force P, and $G\,M$ that of the resultant, R, of W and P. From the principle of the parallelogram of forces, we have intensity of the force found by construction;

Fig. 208.

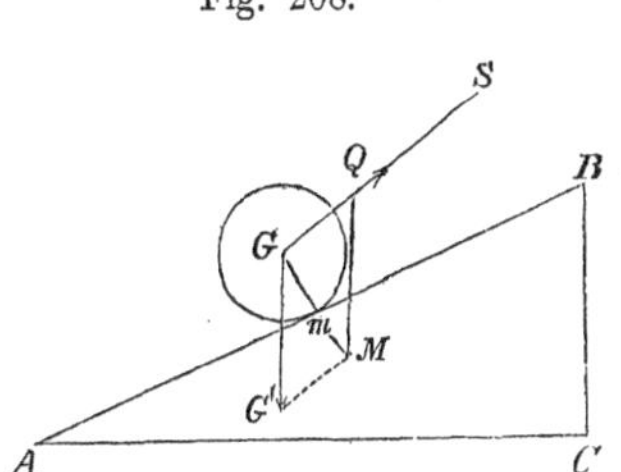

$$W : R : P :: \sin Q\,G\,M : \sin G'\,G\,Q : \sin G'G\,M;$$ intensity of the force found analytically;

but $G\,G'$ and $G\,M$ being respectively perpendicular to $A\,C$ and $A\,B$, the angle A is equal to the angle $G'\,G\,M$, and we have

$$\sin G'\,G\,M = \sin B\,A\,C = \frac{B\,C}{A\,B};$$

and this substituted in the foregoing proportion gives, after reduction,

$$W : R : P :: A\,B \,.\, \sin Q\,G\,M : A\,B \,.\, \sin G'\,G\,Q : B\,C;$$

from which we find

$$P = W \cdot \frac{B\,C}{A\,B \,.\, \sin Q\,G\,M} \quad . \quad . \quad (99);$$ value of the force;

$$R = W \cdot \frac{\sin G'\,G\,Q}{\sin Q\,G\,M} \quad . \quad . \quad . \quad (100).$$ value of the pressure against the plane;

power applied parallel to the plane;

If the power P be applied parallel to the plane, the angle $Q\,G\,M = 90°$; and the angle $G'\,G\,Q$ becomes the supplement of the angle $A\,B\,C$; whence we have

Fig. 209.

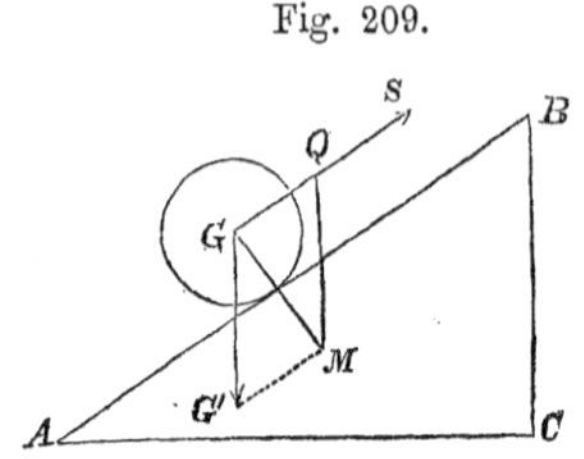

$$\sin Q\,G\,M = \sin 90° = 1;$$

$$\sin G'\,G\,Q = \sin A\,B\,C = \frac{A\,C}{A\,B};$$

which, in the above equations, give

value of force;

$$P = W \cdot \frac{B\,C}{A\,B},$$

value of the pressure against the plane;

$$R = W \cdot \frac{A\,C}{A\,B}.$$

relation of power, weight, and resistance of plane;

That is to say, *when the power is applied parallel to the plane*, 1st, *the power will be to the weight as the height of the plane is to its length;* 2d, *the resistance of the plane will be to the weight as the base of the plane is to its length.*

power applied parallel to the base;

If the power be applied parallel to the base of the plane, the angle $Q\,G\,M$ becomes equal to the angle $A\,B\,C$, because $G\,Q$ and $G\,M$ are respectively perpendicular to $B\,C$ and $A\,B$; and the angle $G'\,G\,Q$ becomes $90°$, whence

Fig. 210.

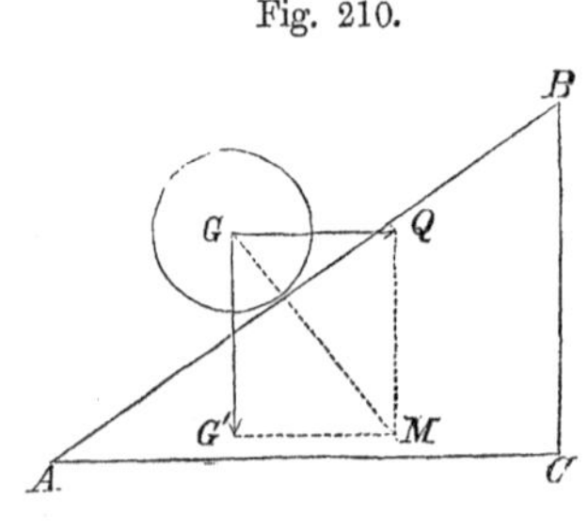

relation of the angles;

$$\sin Q\,G\,M = \sin A\,B\,C = \frac{A\,C}{A\,B},$$

$$\sin G'\,G\,Q = 1;$$

which, in Eqs. (99) and (100), give

$$P = W \cdot \frac{BC}{AC},$$ value of power;

$$R = W \cdot \frac{AB}{AC}.$$ pressure on plane;

That is to say, *when the power is applied parallel to the base of the plane,* 1st, *the power will be to the weight as the height of the plane is to its base;* 2d, *the resistance of the plane will be to the weight as the length of the plane is to its base.* relation of power, weight, and resistance;

In the application of the power parallel to the plane, the power will always be less than the weight. When applied parallel to the base, the power will be less than the weight, while the inclination of the plane is less than 45°. When the inclination is 45°, the power and weight will be equal. When the inclination exceeds 45°, the power will be greater than the weight. limits within which the power will be less than weight.

§ 210.—Let us now consider the *motion* of a heavy body on the inclined plane. The body being acted upon by its weight $G\,G'$ alone, this may be resolved into two components, the one $G\,M$, perpendicular, the other $G\,N$, parallel to the plane. The first will be totally destroyed by the resistance of the plane, while the second will be effective in giving motion. Denote the weight of the body by W, the height $B\,C$ of the plane by h, and its length $A\,B$ by l; then, from the similarity of the triangles $A\,B\,C$ and $G\,G'\,N$, will Motion of a heavy body on an inclined plane; to find the component of the weight parallel to the plane;

Fig. 211.

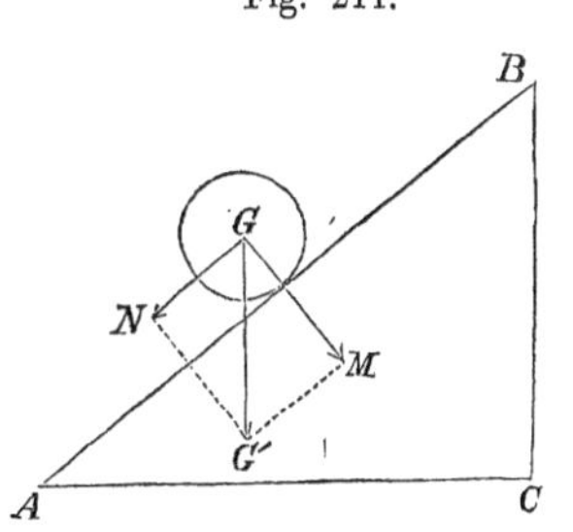

$$W : GN :: l : h;$$

whence

its value;

$$GN = \frac{h}{l} \cdot W;$$

and because the inclination of the plane is the same throughout, the ratio $\frac{h}{l}$ will be constant, from the top to the bottom; whence we see that the motion of the same body down the plane, is that arising from the action of a constant force. It will, therefore, be uniformly varied, and the circumstances of motion will be given by the laws of constant forces.

the motion is that arising from the action of a constant force;

it will be uniformly varied;

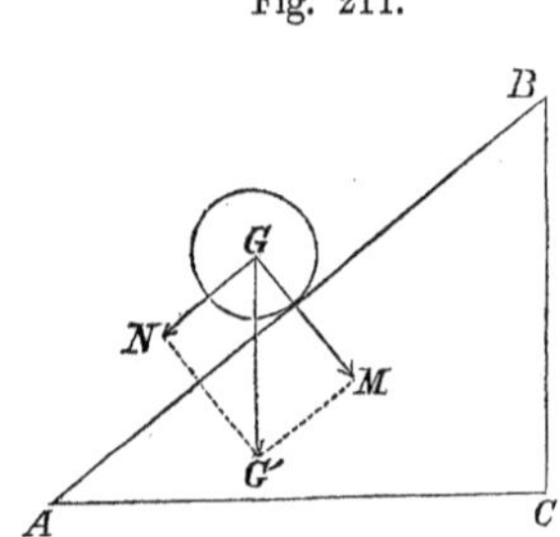

Fig. 211.

Substituting Mg for W, we have

$$GN = \frac{h}{l} Mg;$$

and making M equal to unity, and denoting by g' the corresponding value of the component GN, we find

component of the force of gravity in direction of the plane;

$$g' = \frac{h}{l} \cdot g.$$

Such is the intensity of the force of gravity in the direction of the inclined plane. This may be varied at pleasure by changing the ratio $\frac{h}{l}$; in other words, by altering the inclination of the plane. Now, since the velocities impressed during the first unit of time on the same body, moved from rest, are proportional to the forces producing them, the motion may be made as slow as we please by diminishing $\frac{h}{l}$. It was in this way that Galileo discovered

the motion may be regulated by varying the inclination of the plane;

the laws which regulate the fall of heavy bodies. These being the same as for bodies moving on an inclined plane, it was easy so to regulate the inclination of the plane as to enable him to note and compare the spaces described, times elapsed, and velocities acquired, with each other.

In this way Galileo discovered the laws of falling bodies;

If the body be mounted upon wheels, as in the case of the loaded cart referred to in § 209, it will be urged to *roll* along the inclined plane by an effort of which the measure is

Fig. 212.

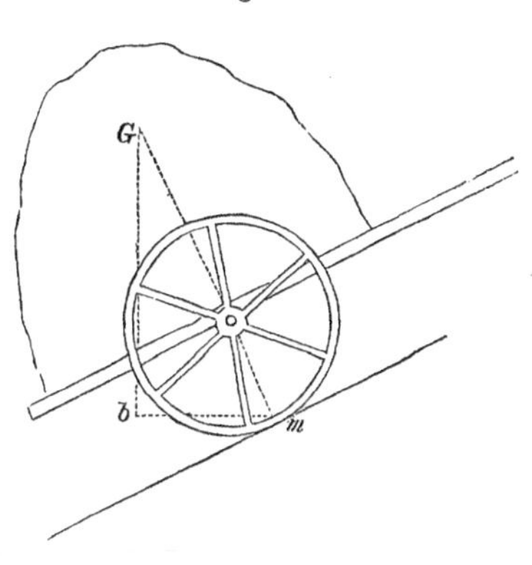

when the body is mounted on wheels it will roll;

$$W \cdot D;$$

in which W denotes the weight of the cart and its load, and D the perpendicular distance $m\,b$ from the point of contact m, to the line of direction $G\,b$ of the weight W.

example of the loaded cart;

moment of the effort by which rotation is produced.

XVI.

FRICTION AND ADHESION.

§ 211.—When two bodies are pressed together, experience shows that a certain effort is always required to cause one to roll or slide along the other. This arises almost entirely from the inequalities in the surfaces of contact interlocking with each other, thus rendering it necessary, when motion takes place, either to break them off, compress them, or force the bodies to separate far enough to allow them to pass each other. This cause of resistance to motion is called *friction*, of which we distin-

Friction;

manifested when two bodies are pressed together and one is moved over the other;

guish two kinds, according as it accompanies a sliding or rolling motion. The first is denominated *sliding*, and the second *rolling friction*. They are governed by the same laws; the former is much greater in amount than the latter under given circumstances, and being of more importance in machines, will principally occupy our attention.

sliding and rolling friction;

The intensity of friction, in any given case, is measured by the force exerted in the direction of the surface of contact, which will place the bodies in a condition to resist, during a change of state, in respect to motion or rest, only by their inertia.

the measure of its intensity.

§ 212.—The friction between two bodies may be measured directly by means of the spring balance. For this purpose, let the surface CD of one of the bodies M, be made perfectly level, so that the other body M', when laid upon it, may press with its entire weight. To some point, as E, of the body M', attach a cord with a spring balance in the manner indicated in the figure, and apply to the latter a force F of such intensity as to produce in the body M' a uniform motion. The motion being uniform, the accelerating and retarding forces must be equal and contrary; that is to say, the friction must be equal and contrary to the force F, of which the intensity is indicated by the balance.

Intensity measured by spring balance;

Fig. 213.

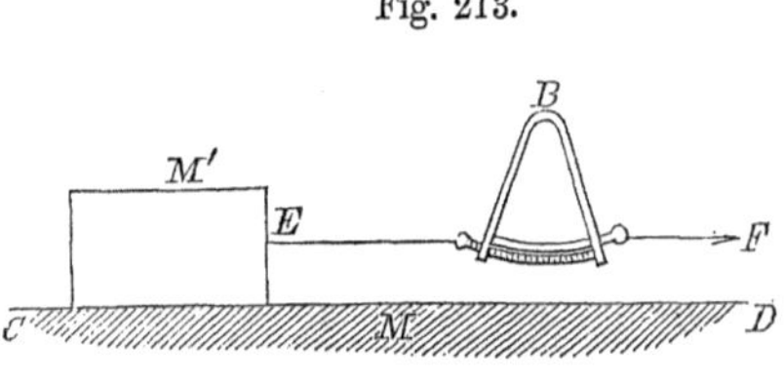

the indication of the balance, when the motion is uniform is the measure;

The experiments on friction which seem most entitled to confidence, are those performed at Metz by M. Morin, under the orders of the French government, in the years 1831, 1832, and 1833. They were made by the aid of a contrivance, first suggested by M. Poncellet, which is one of the most beautiful and valuable contributions that

the most valuable experiments are those of M. Morin;

theory has ever made to practical mechanics. Its details are given in a work by M. Morin, entitled "*Nouvelles Expériences sur le Frottement.*" Paris, 1833. where these experiments may be found;

The following conclusions have been drawn from these experiments, viz.:

The friction of two surfaces which have been for a considerable time in contact and at rest, is not only different in amount, but also in nature from the friction of surfaces in continuous motion; especially in this, that the friction of quiescence is subjected to causes of variation and uncertainty from which the friction during motion is exempt. This variation does not appear to depend upon the *extent* of the surface of contact; for, with different pressures, the ratio of the friction to the pressure varied greatly, although the surfaces of contact were the same. conclusions from these experiments;

The slightest jar or shock, producing the most imperceptible movement of the surfaces of contact, causes the friction of quiescence to pass to that which accompanies *motion.* As every machine may be regarded as being subject to slight shocks, producing imperceptible motions in the surfaces of contact, the kind of friction to be employed in all questions of equilibrium, as well as of motions of machines, should obviously be this last mentioned, or that which accompanies continuous motion. in machinery, the friction which accompanies motion to be considered;

The LAWS of friction which accompanies continuous motion are remarkably *uniform* and *definite.* These laws are: the laws of this friction are uniform and definite;

1st. Friction accompanying continuous motion of two surfaces, between which no unguent is interposed, bears a constant proportion to the force by which those surfaces are pressed together, whatever be the intensity of the force. first law;

2d. Friction is wholly independent of the *extent* of the surfaces in contact. second law;

3d. Where *unguents* are interposed, a distinction is to be made between the case in which the surfaces are simply *unctuous* and in intimate contact with each other, and that in which the surfaces are wholly *separated* from one another third law;

by an *interposed stratum of the unguent.* The friction in these two cases is not the same in amount under the same pressure, although the law of the independence of extent of surface obtains in each. When the pressure is increased sufficiently to *press out* the unguent so as to bring the unctuous surfaces in contact, the latter of these cases passes into the first; and this fact may give rise to an *apparent* exception to the law of the independence of the extent of surface, since a diminution of the surface of contact may so concentrate a given pressure as to remove the unguent from between the surfaces. The exception is however but apparent, and occurs at the passage from one of the cases above-named to the other. To this extent, the law of independence of the extent of surface is, therefore, to be received with restriction.

influence of unguents; an apparent exception to second law;

There are then three conditions in respect to friction, under which the surfaces of bodies in contact may be considered to exist, viz.: 1st, that in which no unguent is present; 2d, that in which the surfaces are simply *unctuous;* 3d, that in which there is an interposed stratum of the unguent. Throughout each of these states the friction which accompanies motion is always proportional to the pressure, but for the same pressure in each, very different in amount.

three conditions of the surfaces in respect to friction;

4th. The friction, which accompanies motion, is always independent of the *velocity* with which the bodies move; and this, whether the surfaces be without unguents or lubricated with water, oils, grease, glutinous liquids, syrups, pitch, &c., &c.

fourth law;

The variety of the circumstances under which these laws obtain, and the accuracy with which the phenomena of motion accord with them, may be inferred from a single example taken from the first set of Morin's experiments upon the friction of surfaces of oak, whose fibres were parallel to the direction of the motion. The surfaces of contact were made to vary in extent from 1 to 84; the forces which pressed them together from 88 to 2205

remarkable instance of the uniformity of these laws;

pounds; and the velocities from the slowest perceptible motion to 9.8 feet a second, causing them to be at one time accelerated, at another, uniform, and at another, retarded; yet, throughout all this wide range of variation, in no instance did the ratio of the pressure to the friction differ from its mean value of 0.478 by more than $\frac{1}{24}$ of this same fraction.

result;

Denote the constant ratio of the normal pressure P, to the entire friction F, by f; then will the first law of friction be expressed by the following equation,

first law expressed by an equation;

$$\frac{F}{P} = f \quad . \quad . \quad . \quad . \quad . \quad (101);$$

whence

$$F = f \, . \, P.$$

This constant ratio f is called the *coefficient of friction*, because, when multiplied by the total normal pressure, the product gives the entire friction.

coefficient of friction;

Assuming the first law of friction, the coefficient of friction may easily be obtained by means of the inclined plane. Let W denote the weight of any body placed upon the inclined plane AB. Resolve this weight GG' into two components, one GM perpendicular to the plane, and the other parallel to it. Because the angles $G'GM$ and BAC are equal, the first of these components will be

its value found by means of the inclined plane;

Fig. 214.

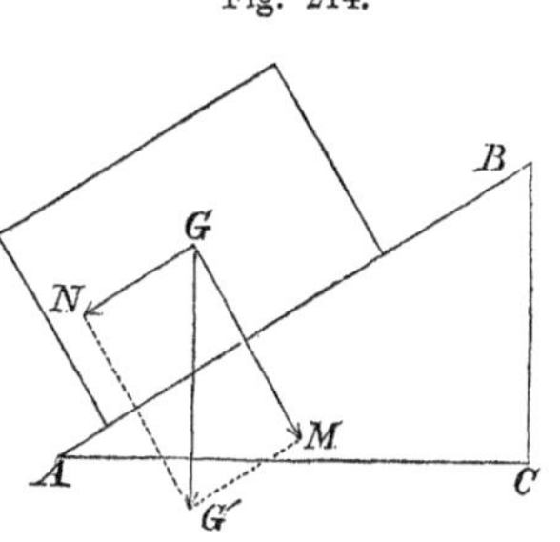

$$W \, . \cos A,$$

component of the weight perpendicular to the plane;

and the second,

that parallel to the plane;

$$W \cdot \sin A,$$

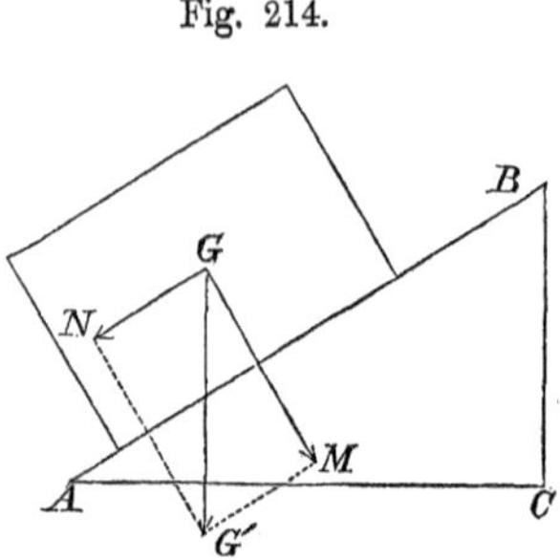

Fig. 214.

in which A denotes the angle $B A C$.

The first of these components determines the total pressure upon the plane, and the friction due to this pressure, will be

the friction on the plane;

$$f \cdot W \cos A.$$

The second component urges the body to move down the plane. If the inclination of the plane be gradually increased till the body move with uniform velocity, the total friction and this component must be equal and opposed; hence

friction and parallel component equal;

$$f \cdot W \cdot \cos A = W \cdot \sin A;$$

whence

value of the coefficient of friction;

$$f = \frac{\sin A}{\cos A} = \tan A.$$

We, therefore, conclude, that the *unit* or *coefficient* of friction between any two surfaces, is equal to the tangent of the angle which one of the surfaces must make with the horizon in order that the other may slide over it with a uniform velocity, the body to which the moving surface belongs being acted upon by its own weight alone. This angle is called the *angle of friction* or *limiting angle of resistance*.

angle of friction; limiting angle of resistance;

The values of the *unit* of friction and of the *limiting angles* for many of the various substances employed in the art of construction, are given in the following tables:

TABLE I.

Experiments on Friction, without Unguents. By M. Morin.

The surfaces of friction were varied from .03336 to 2.7987 square feet, the pressures from 88 lbs. to 2205 lbs., and the velocities from a scarcely perceptible motion to 9.84 feet per second. The surfaces of wood were planed, and those of metal filed and polished with the greatest care, and carefully wiped after every experiment. The presence of unguents was especially guarded against.

SURFACES OF CONTACT.	Friction of Motion.*		Friction of Quiescence.†	
	Coefficient of Friction.	Limiting Angle of Resistance.	Coefficient of Friction.	Limiting Angle of Resistance.
Oak upon oak, the direction of the fibres being parallel to the motion	0.478	25° 33′	0.625	32° 1′
Oak upon oak, the directions of the fibres of the moving surface being perpendicular to those of the quiescent surface and to the direction of the motion‡ - - - - - - - -	0.324	17 58	0.540	28 23
Oak upon oak, the fibres of both surfaces being perpendicular to the direction of the motion - - - -	0.336	18 35		
Oak upon oak, the fibres of the moving surface being perpendicular to the surface of contact, and those of the surface at rest parallel to the direction of the motion - - -	0.192	10 52	0.271	15 10
Oak upon oak, the fibres of both surfaces being perpendicular to the surface of contact, or the pieces end to end - - - - - - - - - -	- -	- -	0.43	23 17.
Elm upon oak, the direction of the fibres being parallel to the motion	0.432	23 22	0.694	34 46
Oak upon elm, ditto§ - - - - -	0.246	13 50	0.376	20 37
Elm upon oak, the fibres of the moving surface (the elm) being perpendicular to those of the quiescent surface (the oak) and to the direction of the motion - - - - - -	0.450	24 16	0.570	29 41

* The friction in this case varies but very slightly from the mean.

† The friction in this case varies considerably from the mean. In all the experiments the surfaces had been 15 minutes in contact.

‡ The dimensions of the surfaces of contact were in this experiment .947 square feet, and the results were nearly uniform. When the dimensions were diminished to .043, a tearing of the fibre became apparent in the case of motion, and there were symptoms of the combustion of the wood; from these circumstances there resulted an irregularity in the friction, indicative of excessive pressure.

§ It is worthy of remark that the friction of oak upon elm is but five-ninths of that of elm upon oak.

TABLE I.—*continued.*

SURFACES OF CONTACT.	FRICTION OF MOTION.		FRICTION OF QUIESCENCE.	
	Coefficient of Friction.	Limiting Angle of Resistance.	Coefficient of Friction.	Limiting Angle of Resistance.
Ash upon oak, the fibres of both surfaces being parallel to the direction of the motion	0.400	21° 49′	0.570	29° 41′
Fir upon oak, the fibres of both surfaces being parallel to the direction of the motion	0.355	19 33	0.520	27 29
Beech upon oak, ditto	0.360	19 48	0.53	27 56
Wild pear-tree upon oak, ditto	0.370	20 19	0.440	23 45
Service-tree upon oak, ditto	0.400	21 49	0.570	29 41
Wrought iron upon oak, ditto*	0.619	31 47	0.619	31 47
Ditto, the surfaces being greased and well wetted	0.256	14 22	0.649	33 0
Wrought iron upon elm	0.252	14 9	- -	- -
Wrought iron upon cast iron, the fibres of the iron being parallel to the motion	0.194	10 59	0.194	10 59
Wrought iron upon wrought iron, the fibres of both surfaces being parallel to the motion	0.138	7 52	0.137	7 49
Cast iron upon oak, ditto	0.490	26 7		
Ditto, the surfaces being greased and wetted	- -	- -	0.646	32 52
Cast iron upon elm	0.195	11 3		
Cast iron upon cast iron	0.152	8 39	0.162	9 13
Ditto, water being interposed between the surfaces	0.314	17 26		
Cast iron upon brass	0.147	8 22		
Oak upon cast iron, the fibres of the wood being perpendicular to the direction of the motion	0.372	20 25		
Hornbeam upon cast iron—fibres parallel to motion	0.394	21 31		
Wild pear-tree upon cast iron—fibres parallel to the motion	0.436	23 34		
Steel upon cast iron	0.202	11 26		
Steel upon brass	0.152	8 39		
Yellow copper upon cast iron	0.189	10 49		
Ditto oak	0.617	31 41	0.617	31 41
Brass upon cast iron	0.217	12 15		
Brass upon wrought iron, the fibres of the iron being parallel to the motion	0.161	9 9		
Wrought iron upon brass	0.172	9 46		
Brass upon brass	0.201	11 22		

* In the experiments in which one of the surfaces was of metal, small particles of the metal began, after a time, to be apparent upon the wood, giving it a polished metallic appearance; these were at every experiment wiped off; they indicated a wearing of the metal. The friction of motion and that of quiescence, in these experiments, coincided. The results were remarkably uniform.

TABLE I.—*continued.*

SURFACES OF CONTACT.	FRICTION OF MOTION.		FRICTION OF QUIESCENCE.	
	Coefficient of Friction.	Limiting Angle of Resistance.	Coefficient of Friction.	Limiting Angle of Resistance.
Black leather (curried) upon oak* -	0.265	14° 51′	0.74	36° 31′
Ox hide (such as that used for soles and for the stuffing of pistons) upon oak, rough - - - - - - - - -	0.52	27 29	0.605	31 11
Ditto ditto ditto, smooth	0.335	18 31	0.43	23 17
Leather as above, polished and hardened by hammering - - - - -	0.296	16 30	- -	- -
Hempen girth, or pulley-band, (sangle de chanvre,) upon oak, the fibres of the wood and the direction of the cord being *parallel* to the motion -	0.52	27 29	0.64	32 38
Hempen matting, woven with small cords, ditto - - - - - - - - -	0.32	17 45	0.50	26 34
Old cordage 1½ inch in diameter, ditto†	0.52	27 29	0.79	38 19
Calcareous oolitic stone, used in building, of a moderately hard quality, called stone of Jaumont—upon the same stone - - - - - - - - -	0.64	32 38	0.74	36 31
Hard calcareous stone of Brouck, of a light gray color, susceptible of taking a fine polish, (the muschelkalk,) moving upon the same stone	0.38	20 49	0.70	35 0
The soft stone mentioned above, upon the hard - - - - - - - - - -	0.65	33 2	0.75	36 53
The hard stone mentioned above, upon the soft - - - - - - - - -	0.67	33 50	0.75	36 53
Common brick upon the stone of Jaumont - - - - - - - - - - - -	0.65	33 2	0.65	33 2
Oak upon ditto, the fibres of the wood being perpendicular to the surface of the stone - - - - - - - -	0.38	20 49	0.63	32 13
Wrought iron upon ditto, ditto - -	0.69	34 37	0.49	26 7
Common brick upon the stone of Brouck	0.60	30 58	0.67	33 50
Oak as before (endwise) upon ditto -	0.38	20 49	0.64	32 38
Iron, ditto ditto - -	0.24	13 30	0.42	22 47

* The friction of motion was very nearly the same whether the surface of contact was the inside or the outside of the skin.—The *constancy* of the coefficient of the friction of motion was equally apparent in the rough and the smooth skins.

† All the above experiments, except that with curried black leather, presented the phenomenon of a change in the polish of the surfaces of friction—a state of their surfaces necessary to, and dependent upon, their motion upon one another.

TABLE II.

Experiments on the Friction of Unctuous Surfaces. By M. Morin.

In these experiments the surfaces, after having been smeared with an unguent, were wiped, so that no interposing layer of the unguent prevented their intimate contact.

Surfaces of Contact.	Friction of Motion.		Friction of Quiescence.	
	Coefficient of Friction.	Limiting Angle of Resistance.	Coefficient of Friction.	Limiting Angle of Resistance.
Oak upon oak, the fibres being parallel to the motion - - - - - -	0.108	6° 10′	0.390	21° 19′
Ditto, the fibres of the moving body being perpendicular to the motion	0.143	8 9	0.314	17 26
Oak upon elm, fibres parallel - - -	0.136	7 45		
Elm upon oak, ditto - - - - - -	0.119	6 48	0.420	22 47
Beech upon oak, ditto - - - - -	0.330	18 16		
Elm upon elm, ditto - - - - - -	0.140	7 59		
Wrought iron upon elm, ditto - -	0.138	7 52		
Ditto upon wrought iron, ditto - -	0.177	10 3		
Ditto upon cast iron, ditto - - - -	- -	- -	0.118	6 44
Cast iron upon wrought iron, ditto -	0.143	8 9		
Wrought iron upon brass, ditto - -	0.160	9 6		
Brass upon wrought iron - - - -	0.166	9 26		
Cast iron upon oak, ditto - - - -	0.107	6 7	0.100	5 43
Ditto upon elm, ditto, the unguent being tallow - - - - - - - -	0.125	7 8		
Ditto, ditto, the unguent being hog's lard and black lead - - - - -	0.137	7 49		
Elm upon cast iron, fibres parallel -	0.135	7 42	0.098	5 36
Cast iron upon cast iron - - - -	0.144	8 12		
Ditto upon brass - - - - - - -	0.132	7 32		
Brass upon cast iron - - - - - -	0.107	6 7		
Ditto upon brass - - - - - - -	0.134	7 38	0.164	9 19
Copper upon oak - - - - - - -	0.100	5 43		
Yellow copper upon cast iron - -	0.115	6 34		
Leather (ox hide) well tanned upon cast iron, wetted - - - - - -	0.229	12 54	0.267	14 57
Ditto upon brass, wetted - - - -	0.244	13 43		

The distinction between the friction of surfaces to which no unguent is present, those which are merely unctuous, and those between which a uniform stratum of the unguent is interposed, appears first to have been remarked by M. Morin; it has suggested to him what appears to be the true explanation of the difference between his results and those of Coulomb. He conceives, that in the experiments of this celebrated engineer, the requisite precautions had not been taken to exclude unguents from the surfaces of contact. The slightest unctuosity, such as might present itself accidentally, unless expressly guarded against—such, for instance, as might have been left by the hands of the workman who had given the last polish to the surfaces of contact—is sufficient materially to affect the coefficient of friction.

cause of the discrepancy between the results of Morin and of Coulomb;

Thus, for instance, surfaces of oak having been rubbed with hard dry soap, and then thoroughly wiped, so as to show no traces whatever of the unguent, were found by its presence to have lost $\frac{2}{3}$ds of their friction, the coefficient having passed from 0.478 to 0.164.

example illustrative of this;

This effect of the unguent upon the friction of the surfaces may be traced to the fact, that their motion upon one another without unguents was always found to be attended by a wearing of both the surfaces; small particles of a dark color continually separated from them, which it was found from time to time necessary to remove, and which manifestly influenced the friction: now with the presence of an unguent the formation of these particles, and the consequent wear of the surfaces, completely ceased. Instead of a new surface of contact being continually presented by the wear, the same surface remained, receiving by the motion continually a more perfect polish.

effect of friction upon surfaces without unguents;

TABLE III.

Experiments on Friction with Unguents interposed. By M. Morin.

The extent of the surfaces in these experiments bore such a relation to the pressure, as to cause them to be separated from one another throughout by an interposed stratum of the unguent.

SURFACES OF CONTACT.	Friction of Motion.	Friction of Quiescence.	UNGUENTS.
	Coefficient of Friction.	Coefficient of Friction.	
Oak upon oak, fibres parallel	0.164	0.440	Dry soap.
Ditto ditto - - - -	0.075	0.164	Tallow.
Ditto ditto - - - -	0.067	- -	Hogs' lard.
Ditto, fibres perpendicular	0.083	0.254	Tallow.
Ditto ditto - - - -	0.072	- -	Hogs' lard.
Ditto ditto - - - -	0.250	- -	Water.
Ditto upon elm, fibres parallel	0.136	- -	Dry soap.
Ditto ditto - - - -	0.073	0.178	Tallow.
Ditto ditto - - - -	0.066	- -	Hogs' lard.
Ditto upon cast iron, ditto -	0.080	- -	Tallow.
Ditto upon wrought iron, ditto	0.098	- -	Tallow.
Beech upon oak, ditto - - -	0.055	- -	Tallow.
Elm upon oak, ditto - - -	0.137	0.411	Dry soap.
Ditto ditto - - - -	0.070	0.142	Tallow.
Ditto ditto - - - -	0.060	- -	Hogs' lard.
Ditto upon elm, ditto - -	0.139	0.217	Dry soap.
Ditto upon cast iron, ditto -	0.066	- -	Tallow.
Wrought iron upon oak, ditto	0.256	0.649	Greased, and saturated with water.
Ditto ditto ditto -	0.214	- -	Dry soap.
Ditto ditto ditto -	0.085	0.108	Tallow.
Ditto upon elm, ditto -	0.078	- -	Tallow.
Ditto ditto ditto -	0.076	- -	Hogs' lard.
Ditto ditto ditto -	0.055	- -	Olive oil.
Ditto upon cast iron, ditto -	0.103	- -	Tallow.
Ditto ditto ditto -	0.076	- -	Hogs' lard.
Ditto ditto ditto -	0.066	0.100	Olive oil.
Ditto upon wrought iron, ditto	0.082	- -	Tallow.
Ditto ditto ditto -	0.081	- -	Hogs' lard.
Ditto ditto ditto -	0.070	0.115	Olive oil.
Wrought iron upon brass, fibres parallel - - - - -	0.103	- -	Tallow.
Ditto ditto ditto -	0.075	- -	Hogs' lard.
Ditto ditto ditto -	0.078	- -	Olive oil.
Cast iron upon oak, ditto -	0.189	- -	Dry soap.
Ditto ditto ditto -	0.218	0.646	Greased, and saturated with water.
Ditto ditto ditto -	0.078	0.100	Tallow.
Ditto ditto ditto -	0.075	- -	Hogs' lard.
Ditto ditto ditto -	0.075	0.100	Olive oil.
Ditto upon elm, ditto -	0.077	- -	Tallow.

TABLE III.—*Continued.*

SURFACES OF CONTACT.	FRICTION OF MOTION. Coefficient of Friction.	FRICTION OF QUIESCENCE. Coefficient of Friction.	UNGUENTS.
Cast iron upon elm—fibres parallel - - - - - - - -	0.061	- -	Olive oil.
Ditto ditto ditto -	0.091	- -	Hogs' lard and plumbago.
Ditto, ditto upon wrought iron	- -	0.100	Tallow.
Cast iron upon cast iron - -	0.314	- -	Water.
Ditto ditto - - - -	0.197	- -	Soap.
Ditto ditto - - - -	0.100	0.100	Tallow.
Ditto ditto - - - -	0.070	0.100	Hogs' lard.
Ditto ditto - - - -	0.064	- -	Olive oil.
Ditto ditto - - - -	0.055	- -	Lard and plumbago.
Ditto upon brass - - - -	0.103	- -	Tallow.
Ditto ditto - - - -	0.075	- -	Hogs' lard.
Ditto ditto - - - -	0.078	- -	Olive oil.
Copper upon oak, fibres parallel	0.069	0.100	Tallow.
Yellow copper upon cast iron	0.072	0.103	Tallow.
Ditto ditto -	0.068	- -	Hogs' lard.
Ditto ditto - - - -	0.066	- -	Olive oil.
Brass upon cast iron - - -	0.086	0.106	Tallow.
Ditto ditto - -	0.077	- -	Olive oil.
Ditto upon wrought iron -	0.081	- -	Tallow.
Ditto ditto - - - -	0.089	- -	Lard and plumbago.
Ditto ditto - - - -	0.072	- -	Olive oil.
Ditto upon brass - - - -	0.058	- -	Olive oil.
Steel upon cast iron - - -	0.105	0.108	Tallow.
Ditto ditto - - - -	0.081	- -	Hogs' lard.
Ditto ditto - - - -	0.079	- -	Olive oil.
Ditto upon wrought iron -	0.093	- -	Tallow.
Ditto ditto - - - -	0.076	- -	Hogs' lard.
Ditto upon brass - - - -	0.056	- -	Tallow.
Ditto ditto - - - -	0.053	- -	Olive oil.
Ditto ditto - - - -	0.067	- -	Lard and plumbago.
Tanned ox hide upon cast iron	0.365	- -	Greased, and saturated with water.
Ditto ditto - - - -	0.159	- -	Tallow.
Ditto ditto - - - -	0.133	0.122	Olive oil.
Ditto upon brass - - - -	0.241	- -	Tallow.
Ditto ditto - - - -	0.191	- -	Olive oil.
Ditto upon oak - - - -	0.29	0.79	Water.
Hempen fibres not twisted, moving upon oak, the fibres of the hemp being placed in a direction perpendicular to the direction of the motion, and those of the oak parallel to it - - - - - - - - -	0.332	0.869	Greased, and saturated with water.

TABLE III.—*continued.*

SURFACES OF CONTACT.	FRICTION OF MOTION. Coefficient of Friction.	FRICTION OF QUIESCENCE. Coefficient of Friction.	UNGUENTS.
The same as above, moving upon cast iron - - - - -	0.194	- -	Tallow.
Ditto ditto - - - -	0.153	- -	Olive oil.
Soft calcareous stone of Jaumont upon the same, with a layer of mortar, of sand, and lime interposed, after from 10 to 15 minutes' contact	- -	0.74	

A comparison of the results enumerated in the above table leads to the following remarkable conclusion, easily fixing itself in the memory, *that with the unguents hogs' lard and olive oil interposed in a continuous stratum between them, surfaces of wood on metal, wood on wood, metal on wood, and metal on metal, when in motion, have all of them very nearly the same coefficient of friction, the value of that coefficient being in all cases included between* 0.07 *and* 0.08, *and the limiting angle of resistance therefore between* 4° *and* 4° 35′.

conclusions in regard to olive oil and lard;

For the unguent tallow the coefficient is the same as the above in every case, except in that of metals upon metals; this unguent seems less suited to metallic surfaces than the others, and gives for the mean value of its coefficient 0.10, *and for its limiting angle of resistance* 5° 43′.

tallow not so well suited to metal.

Adhesion;

§ 213.—Besides friction, there is another cause of resistance to the motion of bodies when moving over one another. The same forces which hold the elements of bodies together, also tend to keep the bodies themselves together, when brought into sensible contact. The effort by which two bodies are thus united, is called the *force of Adhesion.*

Familiar illustrations of the existence of this force are furnished by the pertinacity with which sealing-wax, wafers, ink, chalk, and black-lead cleave to paper, dust to articles of dress, paint to the surface of wood, whitewash to the walls of buildings, and the like. Illustrations of the force of adhesion;

The intensity of this force, arising as it does from the affinity of the elements of matter for each other, must vary with the number of attracting elements, and therefore with the *extent of the surface of contact.* its intensity depends upon the extent of the surface of contact;

This law is best verified, and the actual amount of adhesion between different substances determined, by means of a delicate spring-balance. For this purpose, the surfaces of solids are reduced to polished planes, and pressed together to exclude the air, and the efforts necessary to separate them noted by means of this instrument. The experiment being often repeated with the same substances, having different extent of surfaces in contact, it is found that the area of the surface divided by the effort necessary to produce the separation gives a constant ratio. Thus, let S denote the area of the surfaces of contact expressed in square feet, square inches, or any other superficial unit; A, the effort required to separate them, and a the constant ratio in question, then will measured by the spring balance; mode of operation;

Fig. 215.

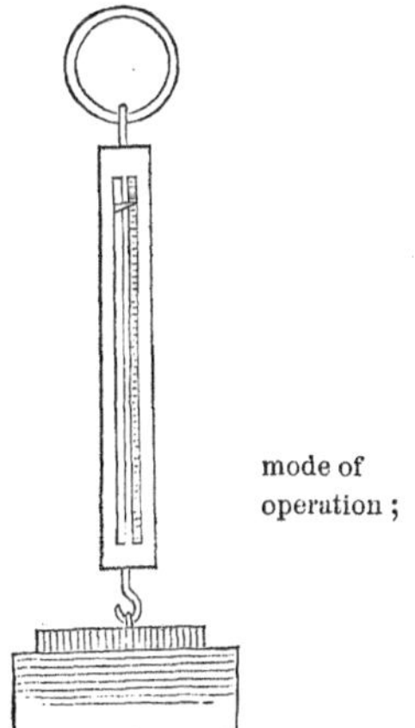

$$\frac{A}{S} = a,$$

or,

$$A = a \cdot S.$$

The constant a is called the *unit* or *coefficient of adhesion*, and obviously expresses the value of adhesion on each unit of surface, for making coefficient of adhesion;

$$S = 1,$$

we have

$$A = a.$$

adhesion between solids and liquids;

To find the adhesion between solids and liquids, suspend the solid from the balance, with its polished surface downward and in a horizontal position; note the weight of the solid, then bring it in contact with the horizontal surface of the fluid and note the indication of the balance when the separation takes place, on drawing the balance up; the difference between this indication and that of the weight will give the adhesion; and this divided by the extent of surface, will give, as before, the coefficient a. But in this experiment two opposite conditions must be carefully noted, else the cohesion of the elements of the liquid for each other may be mistaken for the adhesion of the solid for the fluid. If the solid on being removed take with it a layer of the fluid; in other words, if the solid has been wet by the fluid, then the attraction of the elements of the solid for those of the liquid is stronger than that of the elements of the liquid for each other, and a will be the unit of adhesion of two surfaces of the fluid. If, on the contrary, the solid on leaving the fluid be perfectly dry, the elements of the fluid will attract each other more powerfully than they will those of the solid, and a will denote the unit of adhesion of the solid for the liquid.

mode of ascertaining its amount in any case;

Fig. 216.

i

precaution to be observed;

attraction of fluid elements for each other and for those of solids;

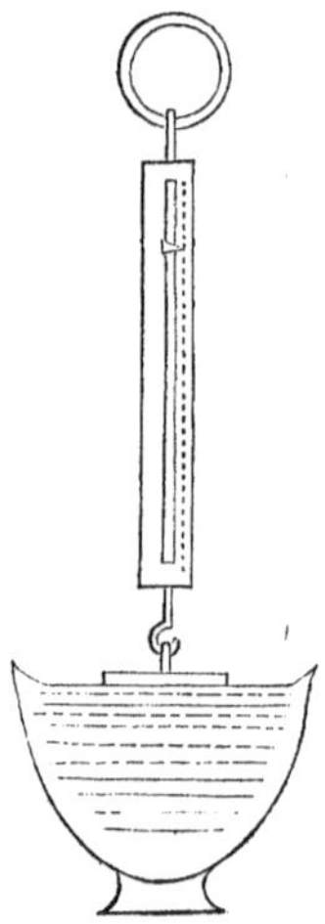

diversity in the action of bodies in this respect;

It is easy to multiply instances of this diversity in the action of solids and fluids upon each other. A drop of water or spirits of wine, placed upon a wooden table or piece of glass, loses its globular form and spreads itself over the surface of the solid; a drop of mercury will not do so. Immerse the finger in water, it becomes wet; in quicksilver, it remains dry. A tallow-candle or a feather from any species of water-fowl remains dry though dipped in water. Gold, silver, tin, lead, &c., become moist on being immersed in quicksilver, but iron and platinum do not. Quicksilver when poured into a gauze bag will not run through; water will: place the gauze containing the quicksilver in contact with water, and the metal will also flow through.

illustration of this diversity;

Solids which become wet on being immersed in a fluid, lose this property if covered with any matter not similarly affected by that particular fluid. A drop of water placed upon a wooden table or piece of glass, smeared with oil or tallow, will not spread, but retain its globular shape and roll off, if the surface be sufficiently inclined. Pour water from a clean common glass tumbler nearly full, and it will run along the exterior surface; smear the rim with hogs' lard or tallow, and the fluid will flow clear of the tumbler. The living force with which the elements of the water in contact with the glass tend to leave the tumbler by the pressure from behind, is, in a great measure, overcome by the attraction between the glass and water, and they are thus made to flow along the surface, while the viscosity of the water, or the attraction of the fluid particles for each other, drags the remote elements after them; and thus the water, under the combined action of its living force, adhesion for the glass and viscosity, becomes spread out into a sheet of which the plane is normal to the surface of the tumbler. When the tumbler is smeared with grease, the adhesion is so much reduced as to offer but feeble opposition to the living force with which the water reaches the edge of the tumbler, it will, therefore, pass the edge after the manner of a projectile. Quicksilver poured out of a glass or wooden vessel will, in like manner, flow clear of the outer surface; but the contrary will happen if a tin vessel be used.

effect of covering surfaces with lard, oil, &c.;

illustrated in the flow of water from a tumbler;

explanation;

case of quicksilver poured from different kinds of vessels;

effect of interposing a fluid between surfaces in contact;

The adhesion of solids is *apparently* increased by introducing a liquid between them. The fluid fills up the existing inequalities of the surfaces, and thus, by increasing the number of points of contact, increases the adhesion by an amount equal either to that of the fluid particles for each other, or to that of the fluid for the solid for which it has the least affinity, depending upon whether the solids are wetted or not by the interposed fluid. This is strikingly exemplified by means of common window-glass, blocks of wood, metallic plates, and the like.

it is difficult to find the adhesion between the rubbing surfaces of machinery;

It is difficult to ascertain the precise value of the force of adhesion between the rubbing surfaces of machinery, apart from that of friction. But this is attended with little practical inconvenience, as long as a machine is in motion. The experiments of which the results are given in the table of § 212, and which are applicable to machinery, were made under considerable pressures, such as those with which the parts of the larger machines are accustomed to move upon one another. Under such pressures, the adhesion of unguents to the surfaces of contact, and the opposition to motion presented by their viscosity, are causes whose influence may be safely disregarded as compared with that of friction. In the cases of lighter machinery, however, such as watches, clocks, and the like, these considerations rise into importance, and cannot be neglected.

this adhesion may be disregarded;

except in watches and the like.

Friction on a plane;

§ 214.—Let any body M, rest with one of its faces in contact with the inclined plane $A\,B$. Denote its weight by W, and suppose it to be solicited by a force F in the direction $G\,Q$, making with the inclined plane the angle $Q\,G\,q'$, which denote by φ. Denote the inclination $B\,A\,C$ of the plane to the horizon by α. Resolve the weight $W = G\,G'$ into two components, $G p$ and $G p'$, one perpendicular and the other parallel to the plane. The angle $G'\,G p$ being equal to the angle $B\,A\,C$, the first of these components will be,

normal component of the weight;

$$W \,.\, \cos \alpha\,;$$

Fig. 217.

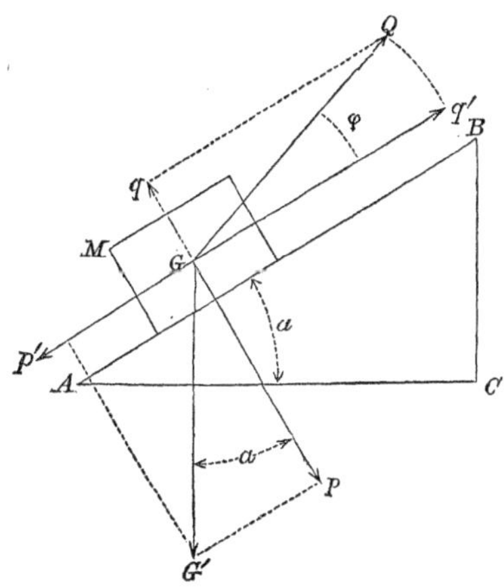

and the second,

$$W \,.\, \sin \alpha.$$

parallel component of the weight;

In like manner, resolve the force $F = G\,Q$, into two components $G\,q$ and $G\,q'$, the first normal and the second parallel to the plane. The first of these will be,

$$F \,.\, \sin \varphi;$$

normal component of the force;

and the second

$$F \,.\, \cos \varphi.$$

its parallel component;

The total pressure upon the plane will be

$$W \,.\, \cos \alpha \;-\; F \,.\, \sin \varphi;$$

pressure upon the plane;

and the friction thence arising

$$f\,(W \,.\, \cos \alpha \;-\; F \,.\, \sin \varphi);$$

corresponding friction;

in which f denotes the coefficient of friction. The force which solicits the body in the direction of the plane will be,

whole force in direction of the plane;

$$F \,.\, \cos \varphi \; - \; W \,.\, \sin \alpha.$$

This will tend to accelerate the body; the friction will tend to retard it. When they are in equilibrio, the body will either have a uniform motion or be just on the eve of motion; which condition will therefore be expressed by

$$F \,.\, \cos \varphi \; - \; W \sin \alpha \; = f \,(W \cos \alpha \; - \; F \,.\, \sin \varphi);$$

whence

force necessary to hold the body in equilibrio, or to keep it in uniform motion up the plane;

$$F = \frac{W\,(f \cos \alpha \; + \; \sin \alpha)}{\cos \varphi \; + \; f \,.\, \sin \varphi} \quad . \; . \; . \quad (102).$$

Here the force F will be the smallest possible, or will be applied under the most advantageous circumstances, when the denominator is the greatest possible, since all the quantities in the numerator are constant. To ascertain the relation between the quantities of the denominator to satisfy this condition, draw $G\,Q$ making with the plane $A\,B$ the angle $Q\,G\,B$ equal to φ; from G lay off the distance $G\,b$ equal to unity, and draw $b\,c$ perpendicular to $A\,B$; then will

Fig. 218.

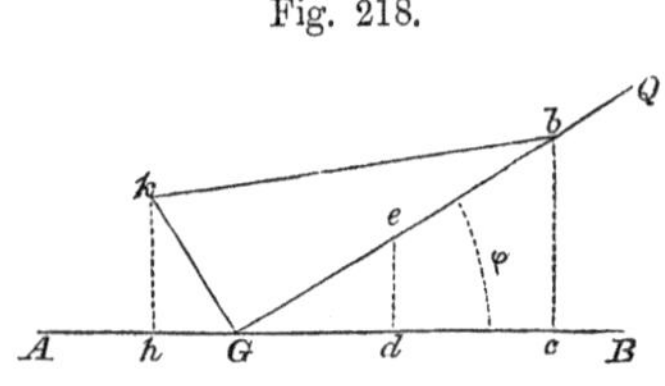

to find under what angle to the plane this force may be applied to greatest advantage;

$$G\,c \; = \; \cos \varphi,$$

$$b\,c \; = \; \sin \varphi.$$

Take the distance $G\,e$ equal to f, and we have

$$e\,d \; = f \sin \varphi.$$

Make $G h$ equal to $e d$, and there will result

$$h c = \cos \varphi + f . \sin \varphi,$$ value of the denominator;

which is the value of the denominator in Eq. (102). Draw $G k$ perpendicular to $G Q$, and erect at h a perpendicular to $A B$, then, because the angle $k G h$ is the complement of $B G Q = \varphi$, will

$$k h = G h \cot \varphi;$$

or, substituting the value of $G h$, as given above,

$$k h = f . \sin \varphi . \cot \varphi = f \cos \varphi.$$

Join k and b, and it will be obvious that $h c$ is the projection of the line $k b$ on $A B$, and that this projection will be the greatest possible when $k b$ is parallel to $A B$; that is, when $k h$ and $b c$ are equal; which condition is expressed by the equation,

$$f \cos \varphi = \sin \varphi,$$

or

$$f = \frac{\sin \varphi}{\cos \varphi} = \tan \varphi;$$ the value of the tangent of this angle;

that is to say, *the power will be applied to the greatest advantage, when its direction makes with the inclined plane an angle of which the tangent is equal to the coefficient of the friction between the plane and the body on it.* conclusion;

If the plane be horizontal, the angle α will be zero, and Eq. (102) reduces to

$$F = \frac{W f}{\cos \varphi + f \sin \varphi}.$$ value of the force when plane is horizontal;

when in equilibrium on eve of motion down the plane;

Finally, if the body is to be retained in equilibrio on the eve of motion *up* the plane, the condition for this purpose is given by Eq. (102) as it stands, but if the equilibrium is maintained on the eve of motion *down* the plane, the friction will act in aid of the force F, and the equation becomes

the value of the force;

$$F' = \frac{W(\sin\alpha - f\cos\alpha)}{\cos\varphi - f\sin\varphi} \quad . \quad . \quad (103);$$

whence it follows, that there are an indefinite number of different values for the force between F and F' which will maintain the body in equilibrio on the plane. If the body be in motion up the plane, the force whose intensity is F will make it uniform; if in motion down the plane, the force whose value is F' will make it uniform. The importance of this will be perceived when we come to treat of the screw.

infinity of forces that will maintain the equilibrium.

Quantity of work on the inclined plane;

§ 215.—The inclined plane is one of the most useful machines employed in the arts, and facilitates the transportation of the heaviest burdens to considerable elevations. To build a stone wall, for instance, to any height, the labor of many men would be required to elevate the necessary materials in a vertical direction, whereas that of a few accomplishes the same end over a ramp or inclined plane whose slope is sufficiently gentle to admit the easy passage of men, horses, carts, &c. Burdens are conveyed up inclined planes by applying the power parallel to its length, and the force for this purpose is given by Eq. (102), after making the angle φ equal

usual direction of the power;

Fig. 219.

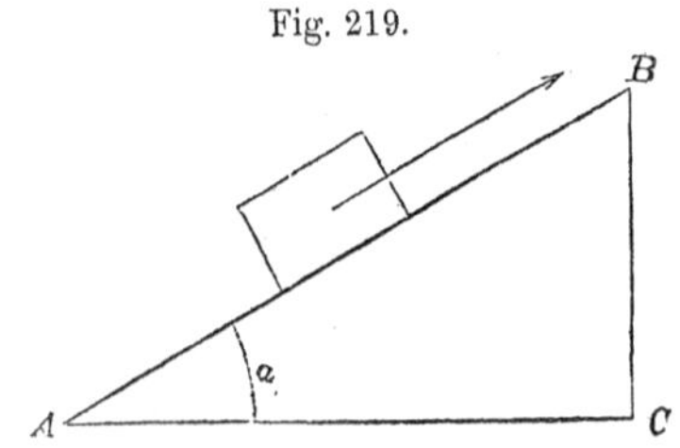

to zero, that is by

$$F = W(\sin \alpha + f \cos \alpha).$$

its value;

Multiplying both members by AB, the distance through which F is exerted, we have,

$$F \times AB = W[AB \sin \alpha + f \,.\, AB \cos \alpha];$$

which reduces to

$$F \times AB = W \,.\, BC + f \,.\, W \,.\, AC \,.\,. \quad (104).$$

its quantity of work;

The first member is the quantity of work performed by the power in moving the burden from the bottom to the top of the plane; and this, we see, is equal to the quantity of work which the weight of the burden would have performed if raised vertically through the same height, increased by the quantity of work which the friction due to a pressure equal to the entire weight, would have exerted through a distance equal to the horizontal projection of the plane.

this value expressed in words;

If the burden be rolled, in which case the friction may be disregarded, or if it be transported in any way to avoid the friction, f would be zero, and we should have

$$F \,.\, AB = W \,.\, BC.$$

value when the body is rolled;

That is to say, the work in the direction of the plane is equal to the work in the vertical direction. What, then, is gained by the use of the plane? Why nothing more than the ability, which it gives, of putting in motion by a feeble power, applied in the direction of its length, a burden which the same power could not move vertically upward.

advantage of the plane;

Resuming Eq. (104), we shall find that what is true of an inclined *plane* is equally true of a curved surface, such as that of a common road or railroad over an undulating piece of ground. For, portions of the road, as $A\,b$, $b\,b'$, $b'\,b''$, &c., may be taken so short as to differ insensibly from a plane, in which case we shall have, by denoting the intensities of the forces on these several elementary planes by F', F'', F''', &c.

all equally true of inclined curved surfaces;

Fig. 220.

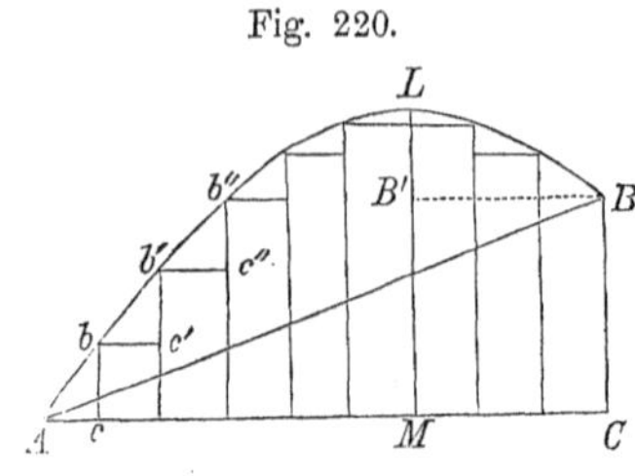

forces on elementary portions of the surface;

$$F' \times A\,b = W\,.\,b\,c + f\,.\,W\,.\,A\,c,$$

$$F'' \times b\,b' = W\,.\,b'\,c' + f\,.\,W\,.\,b\,c',$$

$$F''' \times b'\,b'' = W\,.\,b''\,c'' + f\,.\,W\,.\,b'\,c'',$$

$$\&c., \quad = \quad \&c., \quad + \quad \&c.$$

Adding these equations together, and denoting the first member, which will be the total amount of work in the direction of the surface, by Q', we have

total quantity of work on entire surface;

$$Q' = W\,[b\,c + b'\,c' + b''\,c'' + \&c.] + f\,W\,[A\,c + b\,c' + b'\,c'' + \&c.]\,;$$

and supposing the burden to reach the highest point L, we shall have

$$b\,c + b'\,c' + b''\,c'' + \&c. = L\,M,$$

$$A\,c + b\,c' + b'\,c'' + \&c. = A\,M;$$

which, in the above equation, give

$$Q' = W \,.\, LM + f \,.\, W \,.\, AM \,.\,.\,. \quad (105).$$ quantity of work in the ascent;

After passing the highest point L, the weight acts in favor of the force applied in the direction of the plane, and the first terms of the second members will all change their signs; and denoting the quantity of work in the direction of the plane from L to B by Q'', we shall have, by the same process,

$$Q'' = - \; W \,.\, LB' + f \,.\, W \,.\, B'B \,.\,. \quad (106);$$ quantity in the descent;

adding this to Eq. (105), and denoting the total quantity of work in the direction of the planes from A to B by Q, we find

$$Q = Q' + Q'' = W\,[ML - LB'] + f\,W\,[AM + BB'],$$

or

$$Q = W \times BC + f \,.\, W \,.\, AC \,.\,. \quad (107).$$ quantity in the ascent and descent;

Now it is to be remarked, that every trace of the path actually described by the burden whose weight is W, has disappeared from this value for the quantity of work; this latter is, therefore, wholly independent of this path, and for the same burden, only depends upon the difference of level from A to B, and the horizontal distance AC between these points; so that, the work would be the same as though the load had been transported from A to B along one continuous plane. Nothing is said here of the resistance of the atmosphere, which, like the friction, would be a cause of opposition to the motion.

quantity of work the same as though the path had been straight.

§ 216.—We are now prepared to measure the tension of a cord arising from the action of its own weight. For

Tension of cords;

this purpose take the cord $PAMF$, resting upon any surface of which A is the highest point, and consider the part AF which tends by its weight to move in the direction from A to F. Omit the consideration of friction for the present, and the question will consist in this, viz.: to find a force which, acting in the direction of its length, will keep the cord in equilibrio. This force must be equal and directly opposed to the tension on the part AF. Designate by W, the weight of a unit of length of the cord; then considering the element whose length is MN, its weight will be

the tension of a cord arising from its own weight;

Fig. 221.

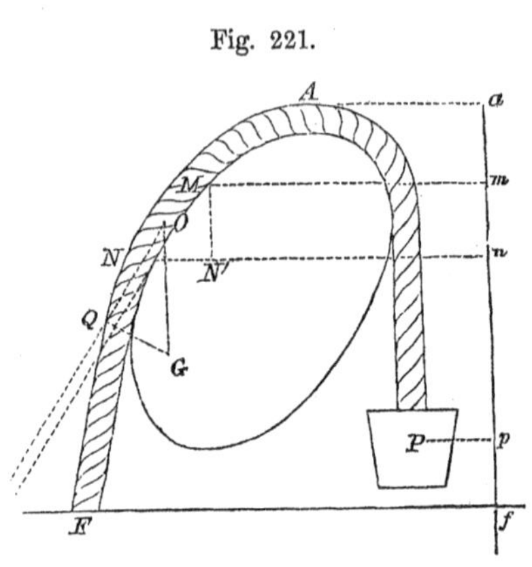

weight of a given portion;

$$W . MN.$$

Through the centre of gravity O of this element, draw the vertical OG to represent this weight, which resolve into two components GQ and QO, the one perpendicular and the other parallel to the cord. The first will be destroyed by the reaction of the surface; the second will act to move the cord in the direction of its length, and will determine its tension. Draw MN' perpendicular and NN' parallel to the horizon; then will the triangles GQO and MNN' be similar, both being right-angled triangles, and the angle QGO of the one, equal to the angle MNN' of the other, because the side GQ is perpendicular to MN, and OG to NN'; hence the proportion,

this weight resolved into components;

$$QO : OG :: MN' : MN;$$

whence

$$Q\,O = \frac{O\,G \times MN'}{MN}.$$

component of weight parallel to the cord;

Denote the tension by t, which will be equal to $Q\,O$; $O\,G$ represents the weight, equal to $W \times MN$; and projecting the points A, M, N, F, P, upon the vertical by the horizontal lines $A\,a$, Mm, Nn, Ff, and Pp, we have MN' equal to $m\,n$, and the last equation becomes,

$$t = \frac{W \times MN \times mn}{MN} = W \times m\,n.$$

value of the tension for a single element;

The second member is the weight of a portion of the cord equal in length to the vertical projection mn of the element MN. Now the length $A\,F$ is composed of a number of elements, each one of which produces, in like manner, a tension equal to the weight of a portion of the cord of the same length as its vertical projection. The tension on each element is transmitted in the direction of the cord to the elements above. Hence, the entire tension at any point of the cord, is measured by the weight of a portion equal in length to the vertical projection of all the cord below it. Thus, if F be the end of the cord, the tension at A will be measured by the weight of a portion of the cord equal to $a\,f$, provided no motion take place. In like manner, the tension at A, arising from the weight of $A\,P$, will be measured by the weight of a portion equal to $a\,p$, so that if the cord have no fixed point it will move in the direction of the lower end F, under the action of a force equal to

tension at any point is measured by the weight of the vertical projection of all the cord below it;

$$W\,(a\,f\,-\,a\,p).$$

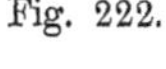

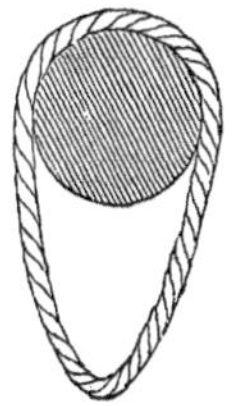

Fig. 222.

If the ends of the cord be upon the same level, or if the cord be endless, it will be in equilibrio.

an endless cord is in equilibrio.

§ 217.—We shall now take into consideration the friction of a cord when sliding around any body, say a fixed cylindrical beam in a horizontal position. Let the cord support at one end a weight W, and be subjected to the action of a force F applied at the other end. If the force communicate motion, it must not only raise the weight W, but must also overcome the friction between the cord and solid. If the surface were perfectly polished, the friction would be zero, and the force F would be equal to the weight W, in the case of an equilibrium. Divide the enveloping portion of the cord, a, t_1, t_2, t_3, &c., into an indefinite number of very small and equal parts, and draw through the points of division, t_1, t_2, t_3, &c., tangents to the cord; these tangents will intersect, two and two, at the points b, b', b'', &c., and the extreme ones will coincide with the straight portions of the cord to which the force and weight are applied. The points of division being extremely close, the arcs will be sensibly confounded with their chords $a\,t_1$, $t_1\,t_2$, $t_2\,t_3$, &c. The tension of the cord on the tangent $a\,b$, with which the cord sensibly coincides, is obviously equal to W, if we neglect the weight of the cord. Let t_1 be the tension which acts at t_1 on the second tangent $b\,b'$; this tension must overcome the weight W and the friction on the arc $t_1\,a$, comprised between the points of contact. Denote by p the pressure exerted by this element upon the cylinder, and by f the coefficient of friction, then will

Friction of a cord sliding around a fixed cylindrical beam;

construction of the figure and notation;

to find the tension on a single element of the cord;

Fig. 223.

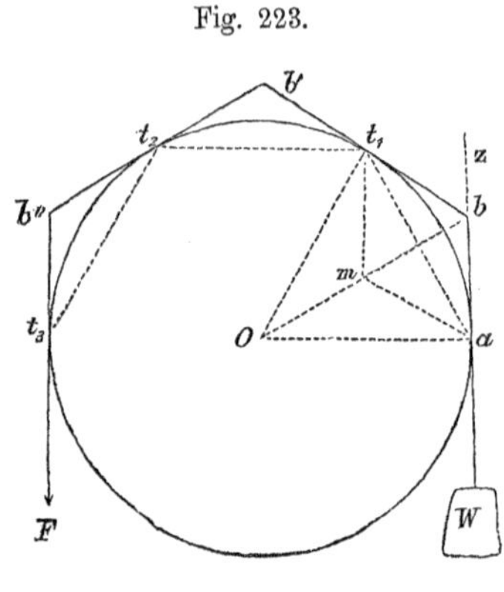

$$t_1 = W + fp.$$

its value;

To find the pressure, we will still disregard the weight of the cord, and remark that the two tangents $a\,b$ and $b\,t_1$ are equal. Moreover, if we construct the rhombus $a\,m\,t_1\,b$, and consider $a\,b$ as proportional to the weight W, this same side will represent the tension of the cord from a to t_1. The diagonal $b\,m$, will be normal to the chord $a\,t_1$, and therefore to the surface of the cylinder, and being the resultant of the tensions at a and t_1 will be the pressure arising from the tension, and consequently equal to p. The triangles $a\,O\,t_1$ and $m\,a\,b$ are similar, because they are both isosceles, and the angle O of the one is equal to $m\,a\,b$ of the other; hence

to find the normal pressure arising from the tension;

$$m\,b \;:\; a\,t_1 \;::\; a\,b \;:\; O\,a;$$

$m\,b$ represents the pressure p; $a\,t_1$ may be taken equal to the arc of which it is chord, which denote by s; $a\,b$ represents the weight W; and $O\,a$ is the radius of the cylinder, which denote by R, and the proportion may be written

$$p \;:\; s \;::\; W \;:\; R;$$

whence

$$p = \frac{s\,.\,W}{R};$$

value of this normal pressure;

and this, substituted in the value of t_1, gives

$$t_1 = W\left(1 + \frac{f\,.\,s}{R}\right).$$

value of the tension on first element nearest the resistance;

Denoting by t_2, the tension along the third tangent $b''\,t_2$, and at the third point of division t_2, this tension must overcome the tension t_1 and friction produced by the elementary arc $t_2\,t_1$, equal in length to $a\,t_1$, or s. In a word, t_2 will be circumstanced in respect to t_1 as t_1 was in regard to W. Hence

to find tension on next element in order;

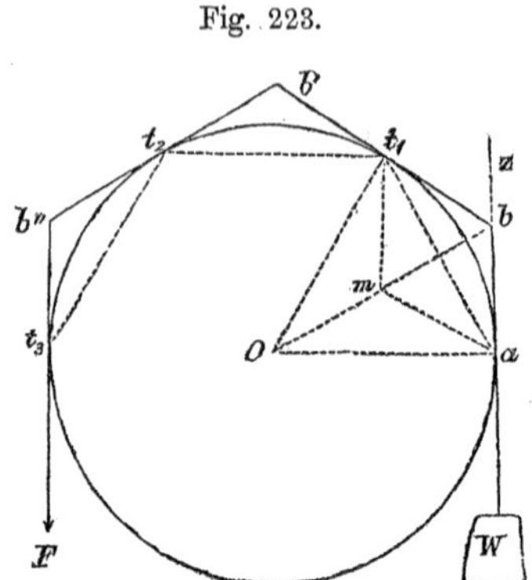

Fig. 223.

its value;

$$t_2 = t_1\left(1 + \frac{f\,.\,s}{R}\right);$$

and if t_3, t_4, t_5, . . . t_n be the tensions on the consecutive tangents, and at the points t_3, t_4, t_5, . . . t_n in order around the beam, we shall have

values for the successive tensions in order around the beam;

$$t_3 = t_2\left(1 + \frac{f\,s}{R}\right),$$

$$t_4 = t_3\left(1 + \frac{f\,s}{R}\right),$$

$$t_n = t_{n-1}\left(1 + \frac{f\,s}{R}\right).$$

Multiplying these equations together and dividing out the common factor, we have

value of the tension on the last element of contact;

$$t_n = W\left(1 + \frac{f\,s}{R}\right)^n.$$

The tension t_n, being the last in order, brings us to the straight portion of the cord to which F is applied, and, therefore, t_n must be equal to F; whence

$$F = W.\left(1 + \frac{fs}{R}\right)^n.$$

relation between the power and the resistance;

Developing this by the rules for the binomial theorem, we have

$$F = W[1 + n \cdot \frac{fs}{R} + \frac{n(n-1)}{1 \cdot 2} \frac{f^2 s^2}{R^2}$$
$$+ \frac{n(n-1)(n-2)}{1 \cdot 2 \cdot 3} \cdot \frac{f^3 s^3}{R^3} + \&c.]$$

this value developed;

It must be remembered that s was taken indefinitely small, and therefore for any definite extent of contac between the cord and cylinder, n must be indefinitely great; hence the numbers 1, 2, 3, 4, &c., connected with n by the sign minus, may be neglected in comparison with n; this gives

$$F = W[1 + \frac{nfs}{R} + \frac{n^2 f^2 s^2}{2 \cdot R^2} + \frac{n^3 f^3 s^3}{2 \cdot 3 \cdot R^3} + \&c.]:$$

under a different form;

but ns is equal to the entire arc enveloped. Denote this by S, and the above becomes

$$F = W[1 + \frac{fS}{R} + \frac{f^2 S^2}{2 \cdot R^2} + \frac{f^3 S^3}{2 \cdot 3 \cdot R^3} + \&c.]:$$

the quantity within the brackets is the development of the function $e^{\frac{fs}{R}}$; whence

$$F = W \times e^{\frac{fs}{R}} \quad . \quad . \quad . \quad . \quad (108),$$

final relation between the power and the resistance;

in which $e = 2.71825$, the base of the Nap. system of logarithms.

example to illustrate;

Suppose the cord to be wound around the cylinder three times, and $f = \frac{1}{3}$, then will

$$S = 3\pi \,.\, 2R = 6 \times 3.1416 \,.\, R = 18.849\, R,$$

and

$$F = W \times e^{\frac{1}{3} \times 18.849} = W \times (2.71825)^{6.2832};$$

or

$$F = W \,.\, 535.3;$$

that is to say, one man at the end W could resist the combined effort of 535 men, of the same strength as himself, to put the cord in motion when wound three times around the cylinder. This explains why it is that a single man, by a few turns of her hawser around a dock-post, is enabled to prevent the progress of a steamboat although her machinery may be in motion. Here friction comes in aid of the power, and there are numerous instances of this; indeed, without friction many of the most useful contrivances and constructions would be useless. It is by the aid of friction that the capstan is enabled to do its work; the friction between the rails of a railroad and the wheels of the locomotive enables the latter to put itself and its train of cars in motion. But for the friction between the feet of draft animals and the ground, they could perform no work; nor, indeed, could any animal walk or even stand with safety, if they were deprived of the aid of this principle.

importance of friction;

its absolute necessity.

XVII.

THE WEDGE.

§ 218.—Thus far we have only considered the cases of a body pressing against a single surface. The same body may also act against two or more surfaces at the same time. Such, for example, is the case with the *Wedge*, which consists of an acute right triangular prism $A\,B\,C$, usually employed in the operation of separating and splitting. The acute dihedral angle $A\,C\,b$, is called the *edge;* the opposite plane face $A\,b$, the *back;* and the planes $A\,c$ and $C\,b$, which terminate in the edge, the *faces*. The more common application of the wedge consists in driving it, by a blow upon its back, into any substance which we wish to split or divide into parts, in such manner that after each advance it shall be supported against the faces of the opening till the work is accomplished.

The wedge; description and use; definitions; common application of the wedge;

Fig. 224.

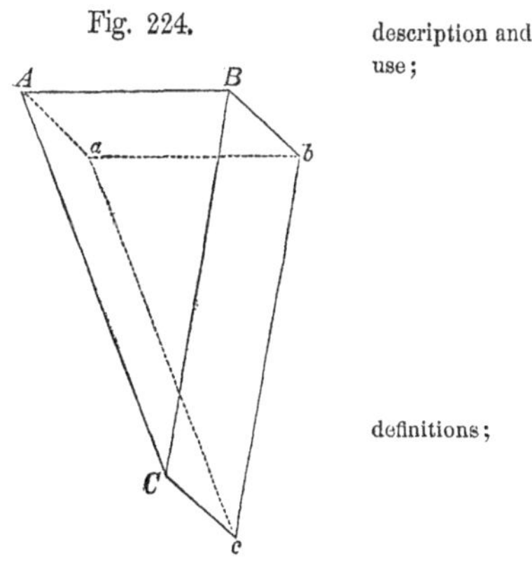

§ 219.—The blow by which the wedge is driven forward will be supposed perpendicular to its back, for if it were oblique, it would only tend to impart a rotary motion, and give rise to complications which it would be unprofitable to consider. And to make the case conform still further to practice, we will suppose the wedge to be isosceles.

the blow upon the wedge should be perpendicular to the back;

to find the resultant of the reactions on the faces;

The wedge ACB being inserted in the opening ahb, and in contact with its jaws at a and b, we know that the resistance of the latter will be perpendicular to the faces of the wedge. Through the points a and b, draw the lines aq and bp normal to the faces AC and BC; from their point of intersection O, lay off the distances Oq and Op equal, respectively, to the resistances at a and b. Denote the first by Q, and the second by P. Completing the parallelogram $Oqmp$, Om will represent the resultant of the resistances Q and P. Denote this resultant by R', and the angle ACB, of the wedge, by θ, which, in the quadrilateral $aObC$, will be equal to the supplement of the angle $aOb = pOq$, the angle made by the directions of Q and P. From the parallelogram of forces we have,

construction and notation;

Fig. 225.

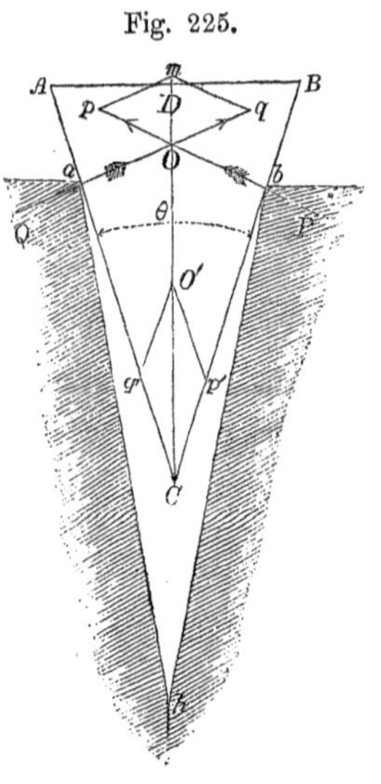

$$R'^2 = P^2 + Q^2 + 2PQ \cos pOq = P^2 + Q^2 - 2PQ \cos \theta;$$

or

value of the resultant;

$$R' = \sqrt{P^2 + Q^2 - 2PQ \cos \theta}.$$

to find resultant of frictions;

The resistance Q will produce a friction on the face AC equal to fQ, and the resistance P will produce on the face BC, the friction fP; these act in the directions of the faces of the wedge. Produce them till they meet in C, and lay off the distances Cq' and Cp' to represent their intensities, and complete the parallelogram $Cq'O'p'$; CO' will represent the resultant of the frictions. Denote this by R'', and we have, from the parallelogram of forces,

$$R''^2 = f^2 Q^2 + f^2 P^2 + 2 f^2 P Q \cos\theta;$$

or

$$R'' = f \sqrt{P^2 + Q^2 + 2 P Q \cos\theta}.$$

value of the resultant of the frictions;

The wedge being isosceles, the resistances P and Q will be equal, their directions being normal to the faces will intersect on the line CD, which bisects the angle $C = \theta$; and their resultant will coincide with this line. In like manner the frictions will be equal, and their resultant will coincide with the same line. Making Q and P equal, we have, from the above equation,

the wedge being isosceles;

$$R' = P\sqrt{2(1 - \cos\theta)},$$

$$R'' = fP\sqrt{2(1 + \cos\theta)}.$$

these values result;

But

$$1 - \cos\theta = 2\sin^2 \tfrac{1}{2}\theta,$$

$$1 + \cos\theta = 2\cos^2 \tfrac{1}{2}\theta;$$

whence we obtain, by substituting and reducing,

$$R' = 2P \, . \sin \tfrac{1}{2}\theta,$$

$$R'' = 2f \, . \, P \, . \cos \tfrac{1}{2}\theta;$$

or these;

and further,

$$\sin \tfrac{1}{2}\theta = \tfrac{1}{2}\frac{AB}{AC},$$

$$\cos \tfrac{1}{2}\theta = \frac{CD}{AC};$$

circular functions in terms of elements of the wedge;

therefore,

final value of these resultants;

$$R' = P \cdot \frac{A\,B}{A\,C},$$

$$R'' = 2\,f \cdot P \cdot \frac{C\,D}{A\,C}.$$

Denote by F the intensity of the blow on the back of the wedge. If this blow be just sufficient to produce an equilibrium bordering on motion forward, call it F'; the friction will oppose it, and we must have,

value of the blow when the wedge is on the eve of moving forward;

$$F' = R' + R'' = P \cdot \frac{A\,B}{A\,C} + 2\,f \cdot P \cdot \frac{C\,D}{A\,C} \quad . \quad . \quad (109).$$

If, on the contrary, the blow be just sufficient to prevent the wedge from flying back, call it F''; the friction will aid it, and we must have,

value, when on the eve of moving back;

$$F'' = P \cdot \frac{A\,B}{A\,C} - 2\,f \cdot P \cdot \frac{C\,D}{A\,C} \quad . \quad . \quad . \quad (110).$$

The wedge will not move under the action of any force whose intensity is between F' and F''. Any force less than F'', will allow it to fly back; any force greater than F' will drive it forward. The range through which the force may vary without producing motion, is obviously,

limits within which the blow may vary to produce no motion;

$$F' - F'' = 4\,f P \cdot \frac{C\,D}{A\,C} \quad . \quad . \quad . \quad . \quad . \quad (111);$$

which becomes greater and greater, in proportion as CD and $A\,C$ become more nearly equal; that is to say, in proportion as the wedge becomes more and more acute.

The ordinary mode of employing the wedge requires that it shall retain of itself whatever position it may be driven to. This makes it necessary that, Eq. (110),

$$P \cdot \frac{A\,B}{A\,C} = 2f \cdot P \cdot \frac{C\,D}{A\,C}, \quad \text{or} \quad P \frac{A\,B}{A\,C} < 2f \cdot P \cdot \frac{C\,D}{A\,C};$$

conditions that the wedge may retain the place to which it is driven;

or, omitting the common factors and dividing both members of the equation and inequality by $2\,CD$,

$$\frac{\frac{1}{2} A\,B}{C\,D} = f, \quad \text{or} \quad \frac{\frac{1}{2} A\,B}{C\,D} < f;$$

but $\frac{\frac{1}{2} A\,B}{CD}$ is the tangent of the angle $A\,C\,D$; hence we conclude, that the wedge will retain its place when its semi-angle does not exceed that whose tangent is the coefficient of friction between the surface of the wedge and the surface of the opening which it is intended to enlarge.

conclusion;

Resuming Eq. (110), and supposing the last term of the second member greater than the first term, F'' becomes negative, and will represent the intensity of the force necessary to withdraw the wedge; which will obviously be the greatest possible when $A\,B$ is the least possible. This explains why it is that nails retain with such pertinacity their places when driven into wood, &c.

why nails retain their places.

§ 220.—One of the most important uses of the wedge, is in its application to what is called the *Wedge Press*. This, in its simplest form, consists of a truncated wedge $A\,B\,C$, which, by a blow upon its back, is made to slide between two blocks, B' and B''; one of

Application to the wedge press;

Fig. 226.

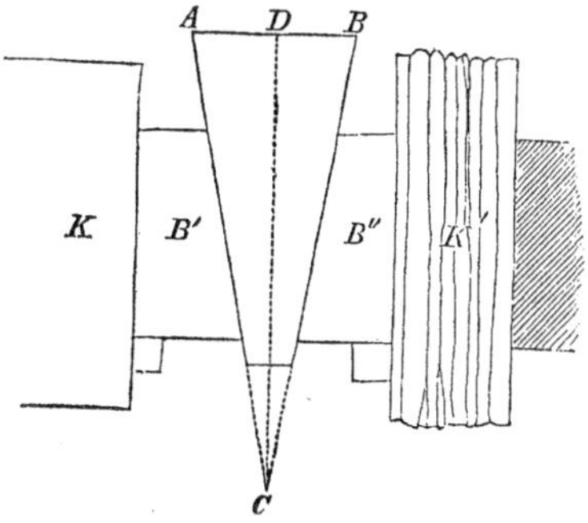

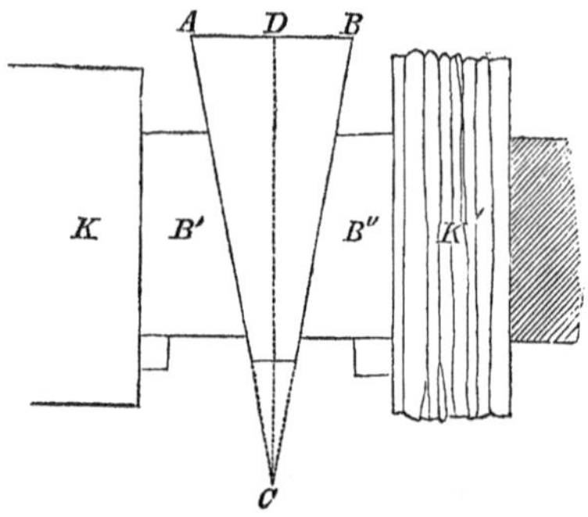

these blocks rests against a fixed support K, and the other against some yielding substance K' to be pressed. This machine is frequently employed to pack goods—wool, cotton, skins, and the like; and, to express the vegetable oils and juices from seeds, fruit, &c. The quantity of work performed by the power will obviously be the product of the intensity of the force F into the distance, in a direction perpendicular to the back, through which the wedge has been driven. Call this distance x, by which multiply both members of Eq. (109), and writing F for F' we have

description,

uses, &c.;

quantity of work of the power;

Fig. 226.

Fig. 227.

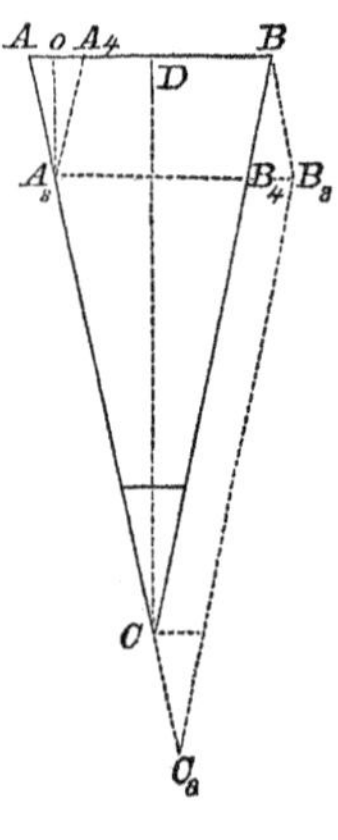

its value;

$$Fx = P \cdot x \cdot \frac{A\,B}{A\,C} + 2f \cdot P \cdot x \cdot \frac{C\,D}{A\,C} \;.\,.\; (112).$$

To obtain from this expression for the quantity of work of the power, a relation which will enable us to compare the intensity F with the reaction of the substance K', let ABC be the primitive position of the wedge, and $A_3 B_3 C_3$ any subsequent position; letting fall from A_3 the perpendicular $A_3 o$ on the back; $A_3 o$ will be equal to x. Moreover, $B_4 B_3$ is the distance through which the whole wedge has been moved towards the yielding substance K', and will, therefore, be equal to the distance through which the reaction of the latter has been exerted. Call this distance s, and the intensity of the reaction S. Draw $A_3 A_4$ parallel to CB; then, in the triangles $A A_3 A_4$ and $B_3 B_4 B$, the sides $A A_3$, and $A_3 A_4$ are respectively equal and parallel to BB_3 and BB_4, and, consequently, $A A_4$ will be equal to $B_4 B_3 = s$. The two triangles $A_3 A A_4$, and ABC, are similar, and give the proportion

to find relation between the power and resistance;

notation and construction;

$$A_3 o \; : \; A A_4 \; :: \; CD \; : \; AB;$$

or

$$x \; : \; s \; :: \; CD \; : \; AB;$$

whence

$$x = s \cdot \frac{CD}{AB}.$$

relation of the elementary paths of power and resistance;

If there were no friction, there would be no obstruction to the free transmission of the effect of the force F to the substance to be compressed. But, making f zero, we have, Eq. (112),

$$Fx = P \cdot x \cdot \frac{AB}{AC};$$

work of the power without friction;

and, from the principle of virtual velocities,

$$Fx = P \cdot x \cdot \frac{AB}{AC} = Ss;$$

whence

$$Px = Ss \times \frac{A\,C}{A\,B}.$$

Now substituting this value for Px, and the above value for x, in Eq. (112), and it reduces to

relation of quantities of work, of power, resistance, and friction;

$$F \cdot s \cdot \frac{CD}{AB} = Ss + 2f \cdot S \cdot s \cdot \frac{CD}{AB} \quad . \quad . \quad (113).$$

The first term of the second member is, obviously, the *effective* quantity of work done, being the reaction of the yielding body multiplied into the distance through which this reaction has been exerted, or through which the body has been compressed. This, we see, is less than the quantity of work of the power F, by the quantity

quantity of work absorbed and lost;

$$2f \,.\, S \,.\, s \times \frac{CD}{AB};$$

which has been totally absorbed, and therefore lost, in consequence of the friction. This loss is often very great, and to illustrate, suppose the reaction S to be 1000 pounds, and that the back of the wedge AB is $\frac{1}{20}$ of its length CD; then will

illustration by an example;

$$20 \,.\, F \,.\, s = \overset{lbs.}{1000} \,.\, s + 40 \,.\, f \,.\, s \,.\, \overset{lbs.}{1000};$$

and, taking $f = \frac{1}{10}$,

$$20 \,.\, F \,.\, s = \overset{lbs.}{1000} \,.\, s + \overset{lbs.}{4000}\, s.$$

numerical loss; Assigning any particular value to s we please, it appears that the useful effect is only $1000s$, while the loss from friction is $4000s$, and that the work performed by the force F is $5000\,s$. Dividing the above equation by $20\,s$, we get

$$F = \overset{lbs.}{50} + \overset{lbs.}{200} = \overset{lbs.}{250};$$ numerical value of the power;

which is much less than 1000, the value of the reaction S. Hence we see that the advantage of the wedge press consists in this, viz.: by its aid the work may be executed with comparatively a feeble power. The machine is, however, defective, on account of the large amount of work absorbed by its friction. defect of the machine.

§ 221.—As before remarked, the wedge is driven forward by a blow on its back. This mode of employing force is an additional source of loss of work. When a hammer strikes the wedge, two periods are to be distinguished, viz.: the first corresponds to the duration of the shock, that is to say, from the instant the hammer touches the wedge to that in which the greatest compression of the wedge and hammer takes place; the second follows immediately and includes the interval during which the reaction of the body to be pressed gives rise to the resistance called S, and to the frictions due to the pressures P and Q. While the wedge is acquiring motion under the blow, during the first period, its inertia acts as a resistance; in the second period, the inertia becomes a power to overcome the resistance S. The blow develops at each instant, between the hammer and wedge, real pressures, which are measurable in pounds; and these pressures are greater, for the same effect, in proportion as the duration of the shock or blow is shorter. The wedge will, in the first period, have a motion from the action of these pressures in consequence of its lateral compression; the inertia due to this motion being opposed by the lateral parts of the machine will give rise to friction, which friction, together with the inertia exerted by the wedge in Effect of the blow in the use of the wedge; explanation of the effect;

acquiring an increased velocity under the continued action of the hammer, will be in equilibrio with these pressures.

loss of living force;

The work of these frictions, during the first period, will be absorbed by the machine, and therefore lost to the inertia or living force of the wedge when this living force becomes, in the second period, a power to overcome the resistance *S*. The quantity of action, or half the living force, preserved by the wedge at the close of the first period, and with which it enters upon the work to be performed during the second, will be given by the rule furnished in Eq. (113), and from which it appears, that this quantity of action will be equal to the quantity of action of the hammer on the back of the wedge during the first period, diminished by that consumed by the friction due to the wedge's inertia within the same period.

its measure;

But this is not all. A part of the work of the hammer is consumed by the permanent change of figure of the wedge arising from the violence of the action. Thus we see, that a considerable portion of the living force with which the hammer begins its work, is lost by change of figure, and by friction due to the sudden development of inertia; neither of which would take place under a force of gradual and ordinary pressure.

loss of living force from permanent change of figure.

§ 222.—Notwithstanding the disadvantages arising from the great and wasteful consumption of work which accompanies the employment of the wedge, this machine is in universal use. It has not, however, always the

The wedge in universal use;

Fig. 228.

its different figures;

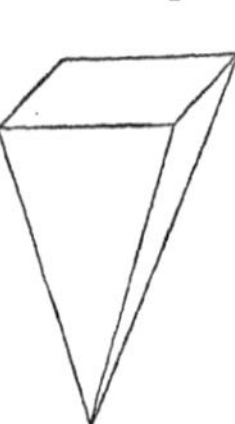

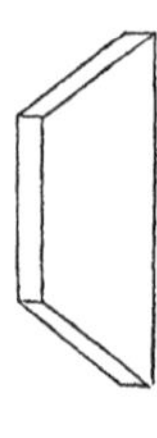

prismatic figure. It sometimes has the form of a pyramid with three, four, or more edges; in which case, the relations between the power and resistance when in equilibrio are altogether analogous to those of the prism; the power applied to the back is equal and directly opposed to the resultant of the resistances and the frictions against the faces. The wedge may also have the form of a truncated pyramid or prism. Often it is nothing more than a cone at the extremity of a cylinder. It may be a pyramid; a truncated pyramid, or even a cone;

Examples in tools. Almost all the tools employed in the arts have some relation to the wedge; such as the different kinds of knives, axes, shears, scissors, files, chisels, saws, hoes, ploughs, &c., &c. All wedges, of whatever kind and however employed, are destined to act by their pointed ends, and the shape of this should be regulated with special reference to the object in view. If too acute, it will break off; if too obtuse, it will not penetrate; and the angle adopted is generally the result of a compromise between these difficulties, determined by the nature of the material of which the wedge is made and that of the substance to be worked. If the substance to be worked be hard, as cold iron, copper, &c., the basil angle abc should be large; this angle in the chisel of a carpenter's plane, which is only intended for wood, is about 30°; it is made still more acute in knives employed to cut the softer substances, meat, bread, and the like. examples in tools; rules in regard to the degree of acuteness.

Fig. 229.

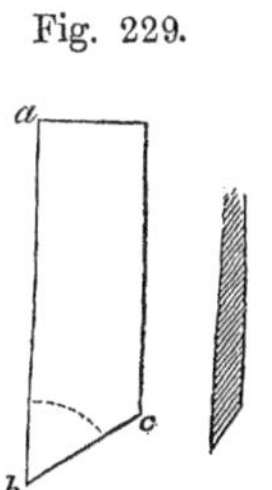

§ 223.—All rotating pieces, such as wheels supported upon other pieces, give rise by their motion to friction. This is an important element in all computations relating to the performance of machinery, and cannot safely be neglected. It seems to be different according as the rotating pieces are kept in place by *trunnions* or by Friction of rotating pieces; trunnions;

pivots; *pivots.* By *trunnions* are meant cylindrical projections $a\, a$ from the ends of the arbor $A\, B$ of a wheel; they are usually made as small in diameter as may be found consistent with the requisite strength, and are so placed that their axes coincide with that of the arbor which is perpendicular to the plane of the wheel.

definition and description of a trunnion;

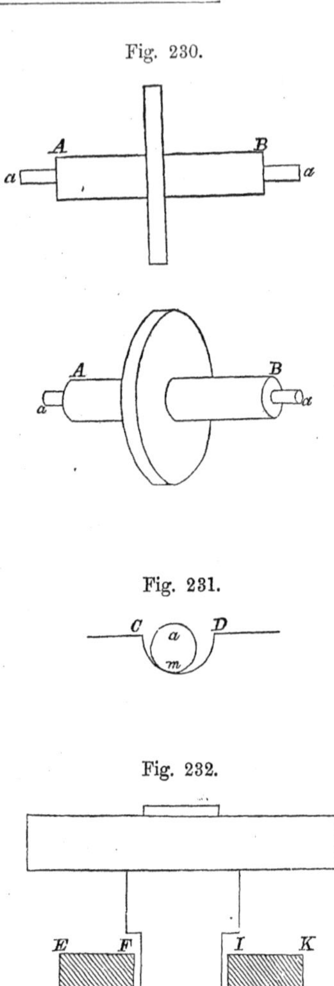

boxes; The trunnions rest on the concave surfaces of cylindrical boxes $C\, D$, with which they usually have a small surface of contact m, the linear elements of both being parallel.

definition and description of pivots and their sockets; *Pivots* are shaped like the trunnions, but support the weight of the wheel and its arbor upon their circular end, which rests against the bottom of cylindrical sockets, $E\, F\, G\, H\, I\, K$. If the forces which give motion to the wheel press its pivot against the cylindrical surface of the socket, the friction will partake of the nature of that due to the trunnion as well as the pivot; but this is usually prevented by special arrangements in the mounting of the wheel. Of the two frictions here referred to, one takes place between the end of the pivot and the circular bottom of the socket, and is in all respects similar to that of two surfaces sliding over each other. The friction due to the motion of the trunnion has been found by Coulomb to be much less than that of the pivot; and there is also less adhesion on account of the smallness of the surface of contact. A table of the coefficient of frictions which accompany the motion of trunnions will be given in its proper place.

nature of the friction on the pivot;

friction on trunnions less than on pivots.

§ 224.—It is not sufficient in case of rotary motion, to know the ratio of the friction to the pressure; we must also know how the friction arising from the peculiar arrangements of the rubbing parts as just indicated, acts with respect to the other forces. We shall first take the case of the pivot turning around its axis. Let N denote the force, in the direction of the axis, by which the pivot is pressed against the bottom of the socket. This force may be regarded as passing through the centre of the circular end of the pivot, and as the resultant of the partial pressures exerted upon all the elementary surfaces of which this circle is composed. Denote by A the area of the entire circle, then will the pressure sustained by each unit of surface be

To find the value of the friction of the pivot against the bottom of its socket;

Fig. 233.

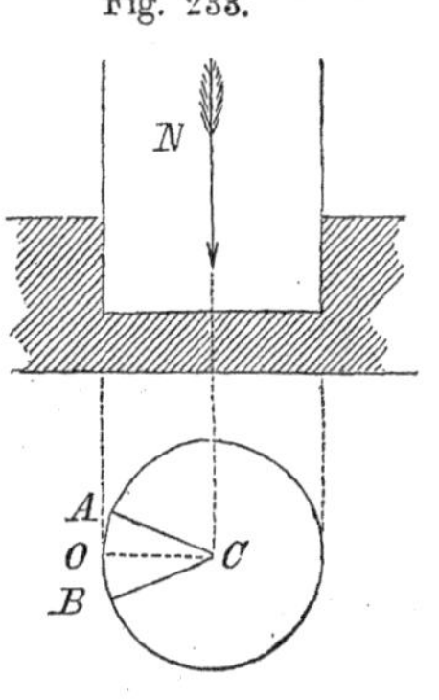

$$\frac{N}{A};$$

pressure on unit of surface;

and the pressure on any small portion of the surface

denoted by a, will obviously be

pressure on a single elementary surface;

$$\frac{a \cdot N}{A};$$

and the friction on the same will be

the corresponding friction;

$$\frac{f \cdot a \cdot N}{A}.$$

This friction may be regarded as applied to the centre of the elementary surface a; it is opposed to the motion, and the direction of its action is tangent to the circle described by the centre of the element. Denote the radius of this circle by r, then will the moment of the friction be

moment of this friction;

$$f \cdot \frac{a \cdot N}{A} \cdot r.$$

If we now consider all the elementary surfaces within the sector $A\,C\,B$, of which the angle at C is very small, we may regard the frictions on these elements as parallel to each other, and perpendicular to the radius $O\,C$, which bisects the angle $A\,C\,B$; in virtue of their parallelism, their resultant will be equal to their sum; and, because of their equality on equal elementary surfaces, the line of direction of this resultant will pass through the centre of gravity of the sector $A\,C\,B$. But this sector being very acute, will not differ from an isosceles triangle, of which

to find the moment of the friction on a single sector;

Fig. 233.

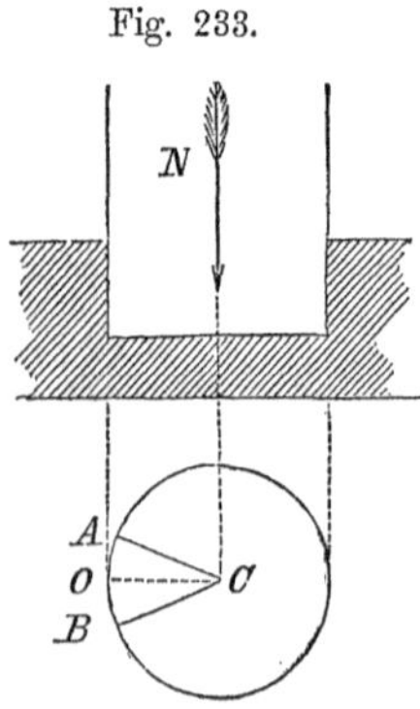

the equal sides, and perpendicular drawn to the base from the vertex C, will be sensibly equal to each other and to R, the radius CA of the circle: whence the distance of the resultant friction on the sector from the centre C will be $\frac{2}{3}R$. Substituting the small sector for a, and $\frac{2}{3}R$ for r, in the foregoing expression, and we have, for the moment of the friction on the sector,

$$f \cdot \frac{N}{A} \cdot \text{sect}. ACB \times \tfrac{2}{3}R;$$

the value of this moment;

and the same is true of any other sector. If the moments be taken for all the sectors which make up the circle, and these be added together, we shall have the moment of the entire friction. The quantity $f \cdot \frac{N}{A} \cdot \frac{2}{3}R$, is constant, and hence the sum of these moments will be

sum of all the similar moments;

$$f \cdot \frac{N}{A} \cdot \tfrac{2}{3}R \times (\text{sum of the sectors } ACB);$$

but the sum of all the sectors is equal to the area of the circle, or A; whence the moment of the friction on the entire base of the pivot is

$$f . N . \tfrac{2}{3}R \quad . \quad . \quad . \quad . \quad . \quad (114);$$

moment of the friction on the entire end of the pivot;

whence we conclude, that, in the friction of a pivot, *we may regard the whole friction due to the pressure as acting in a single point, and at a distance from the centre of motion equal to two thirds of the radius of the base of the pivot.* This distance is called the *mean lever* of friction.

mean lever of friction;

It may happen, that the extremity of the pivot, instead of rubbing upon an entire circle, is only in contact with a ring or surface comprised between two concentric circles. This occurs when the arbor of a wheel is urged

when the friction is on an annular surface;

find friction against a ring;

in the direction of its length by a force N against a shoulder $d\,c\,b\,a$. Denoting, as before, the area of the ring which sustains the pressure by A, the moment of the friction on the elementary sector $A\,B\,C$ is, as before found,

Fig. 234.

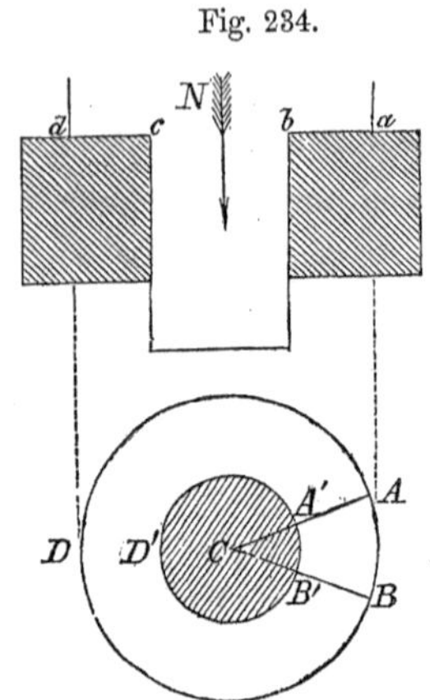

moment on a sector of the larger circle;

$$f \cdot \frac{N}{A} \cdot \tfrac{2}{3} R \times \text{sector } A\,C\,B;$$

in which R denotes the radius of the larger circle. Again, the moment of the friction on the sector $A'\,C\,B'$ is

that on the smaller;

$$f \cdot \frac{N}{A} \cdot \tfrac{2}{3} R' \times \text{sector } A'\,C\,B';$$

and the difference,

that on an element of the ring;

$$\tfrac{2}{3} f \cdot \frac{N}{A} \left[R \times \text{sector } A\,C\,B - R' \text{ sector } A'CB'\right],$$

will be the moment of the friction on the surface, $A'B'B\,A$. Taking the moments for the remaining elementary surfaces which make up the ring, and adding them together, observing that the sums of the sectors make up the areas of the circles to which they respectively belong, we find, for the moment of the friction on the whole ring,

moment for the entire ring;

$$\tfrac{2}{3} f \cdot \frac{N}{A} \left(R \times \text{area of circle } R - R' \times \text{area of circle } R'\right].$$

But the area of the circle whose radius is R, is

$$\pi R^2;$$ area of one circle;

that of the circle whose radius is R', is

$$\pi R'^2;$$ that of the other;

and the area A of the ring, is

$$\pi (R^2 - R'^2).$$ area of the ring;

Substituting these values in the above expression, we find

$$\tfrac{2}{3} f . N \times \frac{R^3 - R'^3}{R^2 - R'^2}.$$ moment of friction on the ring;

Finally, denote by l the breadth of the ring, that is, the distance $A'A$; by r, its mean radius or distance from C to a point half way between A' and A, and we shall have

$$R = r + \tfrac{1}{2} l,$$

$$R' = r - \tfrac{1}{2} l;$$

substituting these values above and reducing, we have

$$f \cdot N \times \left[r + \tfrac{1}{12} \cdot \frac{l^2}{r}\right] \quad . \quad . \quad (115);$$ same in different form;

and making

$$r + \frac{l^2}{12\, r} = r_{\prime},$$ mean lever arm;

we obtain, for the moment of the friction on the entire ring,

$$f . N . r_{\prime} \quad . \quad . \quad . \quad . \quad . \quad . \quad (115)'.$$

The quantity $r_{\prime}$ is called the *mean lever* of friction for a ring. Since the whole friction $f N$, may be considered as applied at a point whose distance from the centre is $\tfrac{2}{3} R$, or

$r_{,} = r + \frac{l^2}{12\,r}$, according as the friction is exerted over an entire circle or over a ring; and since the path described by this point lies always in the direction in which the friction acts, the quantity of work consumed by it will be equal to the product of its intensity $f N$ into this path. Designating the length of the arc described at the unit's distance from C by $s_{,}$, the path in question will be either

work consumed by friction;

$$\tfrac{2}{3} R s_{,}, \quad \text{or} \quad r_{,} s_{,};$$

and the quantity of work either

its value for an entire circle;

$$\tfrac{2}{3} R \,.\, s_{,} \,.\, f \,.\, N$$

for an entire circle, or

its value for a ring;

$$f \,.\, N \left(r + \frac{l^2}{12\, r} \right) s_{,}$$

for a ring. Let Q denote the quantity of work consumed by friction in the unit of time, and n the number of revolutions performed by the pivot in the same time; then will

$$s_{,} = 2\pi \times n;$$

and we shall have

work consumed by friction in a unit of time;

$$Q = \tfrac{4}{3}\pi \,.\, R \,.\, f \,.\, N \,.\, n \quad . \quad . \quad . \quad (116)$$

for the circle, and

for a ring;

$$Q = 2\pi \,.\, f \,.\, N \,.\left(r + \frac{l^2}{12\, r} \right) \cdot n \quad . \quad . \quad (117)$$

for a ring; in which $\pi = 3.1416$.

The coefficient of friction f, when employed in either of the foregoing cases, must be taken from the tables in § 212.

From these expressions, it appears that the quantity of work consumed by friction, in a given time, augments with the radius of the pivot, or mean radius of the ring; and as this work is always opposed to the motion, there is an advantage in reducing these radii as much as possible, consistently with the strength of the pivot. With this view, the pivots are sometimes made in the form of a truncated cone, and often with a convex ellipsoidal or spherical termination, and the socket having a corresponding shape, it will only be necessary to consider the small circle of contact which arises from the compression of the material.

work consumed by friction proportional to mean lever;

Fig. 235.

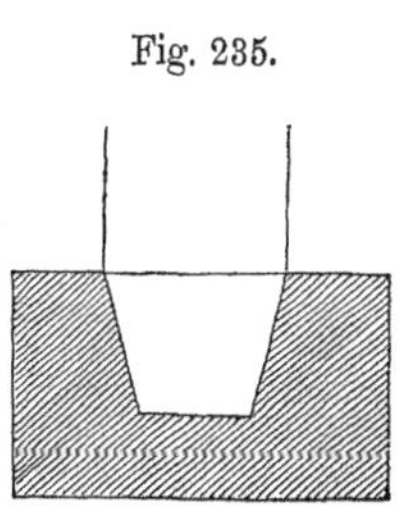

pivots should be as small as possible;

conical and spherical terminations;

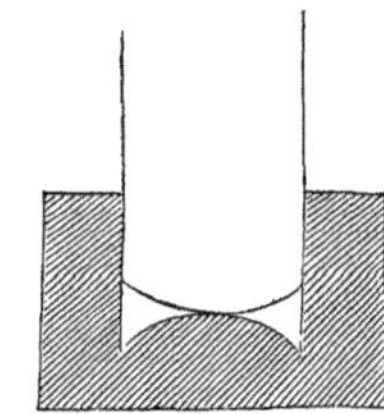

Referring to the expressions (114) and (115)′, we see, that to obtain the moment of friction, in the case of the pivot, either for an entire circle or ring, we *multiply the coefficient of friction, as given in the table of* § 212, *by the pressure, and this product by the mean lever.* And referring to Eqs. (116) and (117) we find, that *the quantity of work is obtained by multiplying the moment of friction into the path described by a point at the unit's distance from the centre of motion.*

the moment of the friction;

quantity of work of the friction;

Example. Required the moment of the friction on a pivot of cast iron, working in a socket of brass, and which supports a weight of 1784 pounds, the diameter of the circular end of the pivot being 6 inches. Here

example of the circular base;

$$R = \tfrac{6}{2} = \overset{in.}{3} = \overset{ft.}{0.25},$$

$$N = \overset{lbs.}{1784},$$

$$f = 0.147;$$

numerical value of elements;

which, substituted in expression (114), gives

value of the moment;

$$0.147 \times \overset{lbs.}{1784} \times \tfrac{2}{3} \times \overset{ft}{0.25} = 43.708.$$

And to obtain the quantity of work in one unit of time, say a minute, there being 20 revolutions in this unit, we make $n = 20$, and $\pi = 3.1416$ in Eq. (116), and find,

work consumed in unit of time;

$$Q = \tfrac{4}{3} \times 3.1416 \times 0.25 \times 0.147 \times 1784 \times 20$$
$$= 5492.80;$$

that is to say, during each unit of time, there is a quantity of work lost which would be sufficient to raise a weight of 5492.80 pounds, through a vertical distance of one foot.

example of the annular base;

Example. Required the moment of friction, when the pivot supports a weight of 2046 pounds, and works upon a shoulder whose exterior and interior diameters are respectively 6 and 4 inches; the pivot and socket being of cast iron, with water interposed.

numerical value of the elements;

$$l = \frac{6 - 4}{2} = 1 \text{ inch},$$

$$r = 2 + 0.5 = 2.5 \text{ inches},$$

$$r_{\prime} = 2.5 + \frac{(1)^2}{12 \times 2.5} = \overset{in.}{2.5333} = \overset{ft.}{0.2111},$$

$$N = 2046 \text{ pounds},$$

$$f = 0.314;$$

which, substituted in expression (115), gives for the moment of friction,

value of the moment;

$$0.314 \times \overset{lbs.}{2046} \times \overset{ft.}{0.2111} = 135.62.$$

The quantity of work consumed in one minute, there

being supposed 10 revolutions in that unit, will be found by making in Eq. (117), as before, $\pi = 3.1416$ and $n = 10$,

work consumed in unit of time;

$$Q = 2 \times 3.1416 \times 0.314 \times 2046 \times 0.211 \times 10$$
$$= 8517.24;$$

that is to say, friction will, in one unit of time, consume a quantity of work which would raise 8517.24 pounds through a vertical distance of one foot. The quantity of work consumed in any given time would result from multiplying the work above found, by the time reduced to minutes.

rule for finding the work in any time.

§ 225.—The friction on trunnions and axles, which we now proceed to consider, gives a considerably less coefficient than that which accompanies the kinds of motion referred to in the tables of § 212. This will appear from the following table, which is the result of careful experiment, viz.:—

Friction on trunnions and axles;

TABLE IV.

FRICTION OF TRUNNIONS IN THEIR BOXES.

KINDS OF MATERIALS.	STATE OF SURFACES.	Ratio of friction to pressure when the unguent is renewed.	
		By the ordinary method.	Or, continuously.
Trunnions of cast iron and boxes of cast iron.	Unguents of olive oil, hogs' lard, and tallow - - -	0.07 to 0.08	0.054
	The same unguents moistened with water - - -	0.08	0.054
	Unguent of asphaltum -	0.054	0.054
	Unctuous - - - - -	0.14	- -
	Unctuous and moistened with water - - - -	0.14	- -
Trunnions of cast iron and boxes of brass.	Unguents of olive oil, hogs' lard, and tallow - - -	0.07 to 0.08	0.054
	Unctuous - - - - -	0.16	- -
	Unctuous and moistened with water - - - -	0.16	- -
	Very slightly unctuous* -	0.19	- -

* The surfaces began to move about.

TABLE IV.—*continued.*

KINDS OF MATERIALS.	STATE OF SURFACES.	Ratio of friction to pressure when the unguent is renewed,	
		By the ordinary method.	Or, continuously.
Trunnions of cast iron and boxes of lignum-vitæ.	Without unguents*	0.18	- -
	Unguents of olive oil and hogs' lard	- -	0.090
	Unctuous with oil and hogs' lard	0.10	- -
	Unctuous with a mixture of hogs' lard and plumbago	0.14	- -
Trunnions of wrought iron and boxes of cast iron.	Unguents of olive oil, tallow, and hogs' lard	0.07 to 0.08	0.054
Trunnions of wrought iron and boxes of brass.	Unguents of olive oil, hogs' lard, and tallow	0.07 to 0.08	0.054
	Old unguents hardened	0.09	- -
	Unctuous and moistened with water	0.19	- -
	Very slightly unctuous†	0.25	- -
Trunnions of wrought iron and boxes of lignum-vitæ.	Unguents of oil or hogs' lard	0.11	- -
	Unctuous	0.19	- -
Trunnions of brass and boxes of brass.	Unguent of oil	0.10	- -
	Unguent of hogs' lard	0.09	- -
Trunnions of brass and boxes of cast iron.	Unguents of tallow or of olive oil	- -	0.045 to 0.052
Trunnions of lignum-vitæ and boxes of cast iron.	Unguents of hogs' lard	0.12	- -
	Unctuous	0.15	- -
Trunnions of lignum-vitæ and boxes of lignum-vitæ.	Unguent of hogs' lard	- -	0.07

to investigate the friction of trunnions;

Let us now examine the part performed by friction in connection with the forces which give motion. We have seen that the contact of the trunnion with its box is along a linear element, common to the surfaces of both. A section perpendicular to its length would cut from the trunnion and its box, two circles tangent to each other internally. The trunnion being acted on only by

* The wood being a little unctuous.

† The surfaces began to move about.

its weight, would, when at rest, give this tangential point at o, the lowest point of the section $p\,o\,q$ of the box. If the trunnion be put in motion by the application of a force, it would turn around the point of contact and roll indefinitely along the surface of the box, if the latter were level; but this not being the case, it will ascend along the inclined surface $o\,p$ to some point as m, where the inclination of the tangent $u\,m\,v$ is such, that the friction is just sufficient to prevent the trunnion from sliding. Here let the trunnion be in equilibrio. But the equilibrium requires that the resultant of all the forces which act, friction included, shall pass through the point m and be normal to the surface of the trunnion at that point. The friction is applied at the point m; hence the resultant N of all the other forces must pass through m in some direction as $m\,d$; the friction acts in the direction of the tangent; and hence, in order that the resultant of the friction and the force N shall be normal to the surface, the tangential component of the latter must, when the other component is normal, be equal and directly opposed to the friction.

Fig. 236.

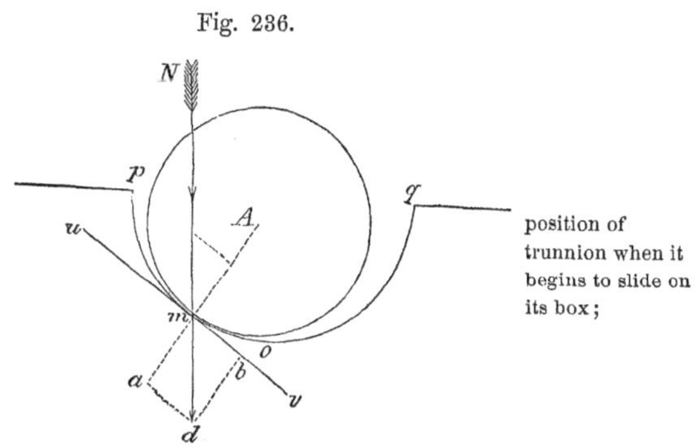

position of trunnion when it begins to slide on its box;

point in which the resultant intersects the surface of box;

tangential component of resultant, equal and opposed to friction;

Take upon the direction of the force N, the distance $m\,d$ to represent its intensity, and form the rectangle $a\,d\,b\,m$, of which the side $m\,b$ shall coincide with the tangent, then, denoting the angle $d\,m\,a$ by φ, will the component of N perpendicular to the tangent be

$$N \,.\, \cos \varphi;$$

normal component;

and the friction due to this pressure will be

friction due to this component;

$$f \,.\, N \,.\, \cos \varphi.$$

The component of N, in the direction of the tangent, will be

tangential component;

$$N \,.\, \sin \varphi;$$

and as this must be equal to the friction, we have

$$f \,.\, N \,.\, \cos. \varphi = N \,.\, \sin \varphi \quad . \quad . \quad (118);$$

whence

value of the unit of friction;

$$f = \tan \varphi;$$

that is to say, *the ratio of the friction to the pressure on the trunnion, is equal to the tangent of the angle which the direction of the resultant N of all the forces except the friction, makes with the normal to the surface of the trunnion at the point of contact.* This gives an easy method of finding the point of contact. For this purpose, we have but to draw through the centre A, a line $A\,Z$, parallel to the direction of N, and through A the line $A\,n$, making with $A\,Z$ an angle of which the tangent is f; the point m, in which this line cuts the circular section of the trunnion will be the point of contact.

to find the point of contact;

Fig. 237.

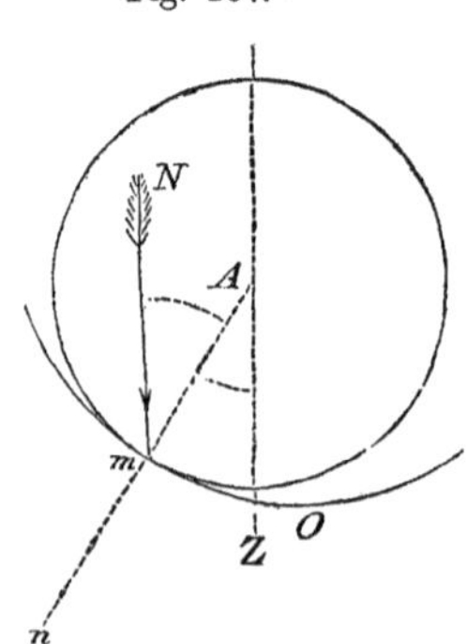

Because $m\,a\,d\,b$, last figure, is a rectangle, we have

$$N^2 = N^2 \cos^2 \varphi + N^2 \sin^2 \varphi;$$

and, substituting for $N^2 \sin^2 \varphi$ its equal $f^2 N^2 \cos^2 \varphi$, we have

to find the value of the total friction;

$$N^2 = N^2 \cos^2 \varphi + f^2 N^2 \cos^2 \varphi = N^2 \cos^2 \varphi (1 + f^2);$$

whence

$$N \cos \varphi = N \times \frac{1}{\sqrt{1 + f^2}};$$

and multiplying both members by f,

$$f \,.\, N \,.\, \cos \varphi = N \cdot \frac{f}{\sqrt{1 + f^2}} \quad . \quad . \quad (119);$$

its value;

but the first member is the total friction; whence we conclude, that *to find the friction upon a trunnion, we have but to multiply the resultant of the forces which act upon it, by the unit of friction, found in Table* IV, *and divide this product by the square root of the square of this same unit increased by unity.*

rule;

This friction acting at the extremity of the radius R of the trunnion and in the direction of the tangent, its moment will be

$$N \cdot \frac{f}{\sqrt{1 + f^2}} \times R \quad . \quad . \quad . \quad (120).$$

moment of the total friction;

And the path described by the point of application of the friction being denoted by $Rs_{,}$, the quantity of work of the friction will be

$$N \,.\, R \,.\, s_{,} \times \frac{f}{\sqrt{1 + f^2}} \quad . \quad . \quad (121);$$

quantity of work of friction;

in which $s_{,}$ denotes the path described by a point at the unit's distance from the centre of the trunnion. Denoting, as in the case of the pivot, the number of revolutions performed by the trunnion in a unit of time, say a minute, by n; the quantity of work performed by friction in this time by $Q_{,}$; and making $\pi = 3.1416$, we have

$$s_{,} = 2\pi . n;$$

and

quantity of work in a unit of time;

$$Q_{,} = 2\pi \cdot R . n \cdot N \cdot \frac{f}{\sqrt{1 + f^2}} \quad . \quad . \quad (122).$$

When the trunnion remains fixed and does not form part of the rotating body, the latter will turn about the trunnion, which then takes the name axle, having the centre of motion at A, the centre of the eye of the wheel; in this case, the lever of friction becomes the radius of the eye of the wheel. As the quantity of work consumed by friction is the greater, Eq. (122), in proportion as this radius is greater, and as the radius of the eye of the wheel must be greater than that of the axle, the trunnion has the advantage, in this respect, over the axle.

axle; lever of friction; trunnion better than the axle in regard to friction;

Fig. 238.

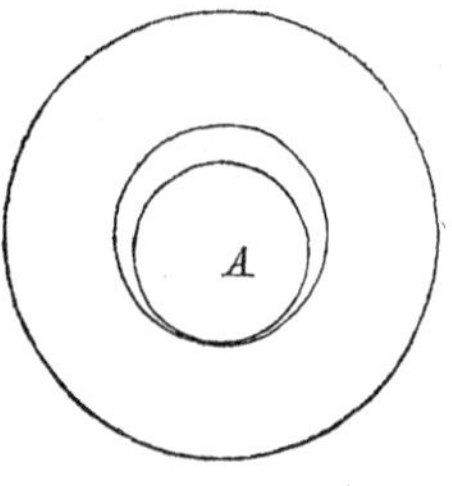

The value of the quantity of work consumed by friction is wholly independent of the length of the trunnion or axle, and no advantage is therefore gained by making it shorter or longer.

friction same for long and short trunnions and axles.

§ 226.—If we examine Eq. (122), we find that, all other things being equal, the value of the work consumed by friction will depend upon the radius R of the trunnion, and that as the latter diminishes, in the same proportion will this consumption diminish. The trunnion should, therefore, be made as small as possible, and of the hardest and

Trunnions should be small as possible;

strongest material, as steel. The consumption of work by friction may also be diminished by lessening the force N; but with these two exceptions there is no way of avoiding the effects of friction for given materials. When the trunnion is employed to support a piece which simply oscillates through an arc, as in the case of the pendulum and weighing-balance, the knife-edge may be used to great advantage, for, in that case, the radius R is reduced to the smallest conceivable length, and the work of friction to almost nothing.

consumption of work by friction lessened by diminishing the pressure;

Fig. 239.

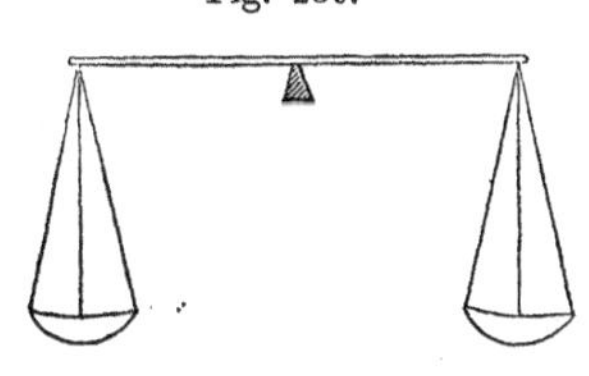

advantages of the knife-edge as an axle.

§ 227.—There is another species of friction yet to be mentioned, viz: that which arises from the rolling of one body over another. As the surfaces of contact are in this case applied to each other, and separated in a direction perpendicular to that of the motion, there would, at first view, appear to be no friction, nor would there if the surfaces were perfect—that is to say, free from all irregularities. But there can, in practice, be no such surface; when bodies are brought in contact in the manner here referred to, the slight protuberances on the surface of one will enter into the corresponding cavities on that of the other, after the manner of so many wedges, and cannot be again withdrawn without giving

Friction of rolling motion;

Fig. 240.

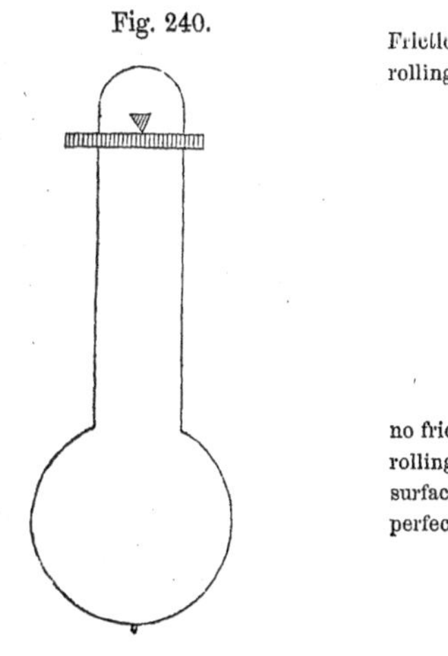

no friction in rolling motion if surfaces were perfect;

but in actual practice there is friction;

this kind of friction, as well as adhesion, may be neglected;

rise to an amount of friction due to their dimension, depth of insertion, and nature of material. Here adhesion assumes a value which is appreciable, as compared with this friction, but both together are found, in practice, to be exceedingly small, and generally, adhesion in rolling, as well as in sliding friction, may, without much error, be neglected. In general this friction, called *friction of the second kind*, is less in proportion as the diameter of the rolling body is less. A wheel of two feet diameter, loaded with a weight of 100 pounds, and rolling over a piece of level and smooth ground, only gives rise to a friction of 0.03 of the pressure—that is to say, to only three pounds. The wheels of carriages meet often with considerable resistance when rolling over compressible or rough ground, but this is because the carriage must be raised over the inclined planes formed in front by the sinking of the wheels, or over obstacles which project above the common surface. The little resistance to motion arising from friction of the second kind, is well illustrated by the comparative facility with which heavy blocks of stone are often transported upon rollers over considerable distances. A roadway is first usually made by placing straight pieces of timber along the ground to prevent the rollers from sinking into it; the stone is then mounted upon the rollers, which are placed upon these pieces at right angles to their length, and drawn in the direction of the road by the application of any convenient power. As fast as a roller is detached from behind, it is brought forward and interposed, in time to prevent the stone from tipping forward in consequence of its centre of gravity getting in advance of the leading roller. The quantity of work necessary to con-

friction of the second kind;

illustration;

illustration of its practical substitution for sliding friction.

Fig. 241.

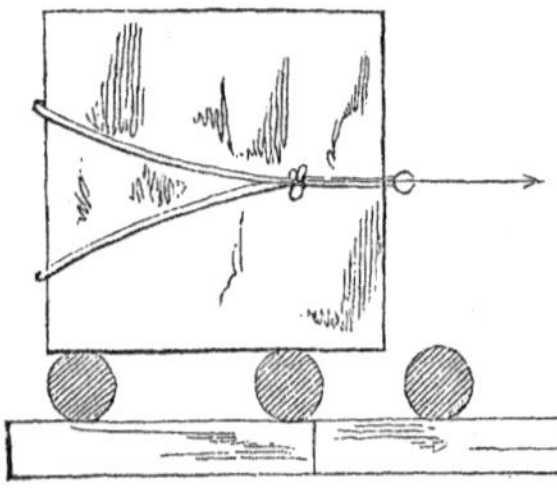

vey a stone over any considerable distance, in this way, is incomparably less than if it were to rest with its face against the ground.

§ 228.—The different kinds of friction may be so combined as to diminish both its intensity and the quantity of its work. Thus, let a pair of wheels CD be mounted upon an axle, and suppose a force $F_{\prime}$ applied to the latter, parallel to a level plane AB, to put it in motion. Denote the weight of the axle and its load by W, that of the wheels by w. Suppose, for a moment, that the wheels are firmly connected with the axle and that they cannot rotate, but, when put in motion, must slide along AB; the force $F_{\prime}$ requisite to impart motion and keep it uniform, will be given by the equation

Employment of rolling to diminish sliding friction;

Fig. 242.

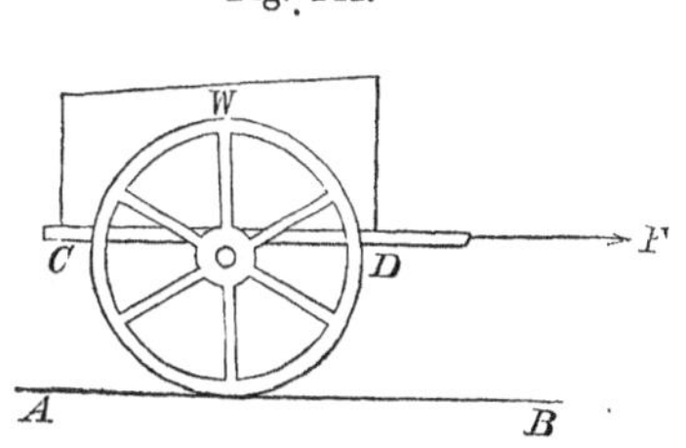

case of a common cart;

$$F_{\prime} = (W + w)f \quad . \quad . \quad . \quad (123);$$

force necessary to keep it in uniform motion when sliding;

in which f is the coefficient of sliding friction between the wheels and plane AB. Next, suppose the wheels capable of turning about the axle, and the force requisite to keep the motion uniform to be denoted by $F_{\prime\prime}$. This force, acting to communicate motion, will give rise to friction between the circumference of the wheel and the plane AB, and also between the axle and the inner surface of the eye: the latter will yield first, and the whole will move forward, the wheels having a rotary as well as a progressive motion. The friction at the axle will, Eq. (119), be

$$N \cdot \frac{f_{\prime}}{\sqrt{1 + f_{\prime}^2}};$$

friction on axle when rotating;

in which $f_{\prime}$ denotes the coefficient of friction at the axle. The weight W is thrown upon the axle and acts vertically; the force $F_{\prime\prime}$ applied also to the axle, acts horizontally, and hence

normal pressure on axle;

$$N = \sqrt{W^2 + F_{\prime\prime}^2};$$

and the friction at the axle becomes

friction on the axle;

$$\sqrt{W^2 + F_{\prime\prime}^2} \times \frac{f_{\prime}}{\sqrt{1 + f_{\prime}^2}}.$$

Denote the radius of the wheel by R, and that of its eye by r, and the space described by a point at the unit's distance from the centre of motion by $s_{\prime}$; then will the quantity of work of the friction be

quantity of work of friction;

$$\sqrt{W^2 + F_{\prime\prime}^2} \times \frac{f_{\prime}}{\sqrt{1 + f_{\prime}^2}} \times r \,.\, s_{\prime}.$$

The path described by the point of application of the power, and in the direction of the power, will be equal to the development of the arc of the circumference of the wheel corresponding to the arc $s_{\prime\prime}$, that is to say, to $Rs_{\prime\prime}$, and hence the quantity of work of the power will be

work of the power;

$$F_{\prime\prime}\, R\, s_{\prime};$$

whence we have

$$F_{\prime\prime}\, R\, s_{\prime} = \sqrt{W^2 + F_{\prime\prime}^2} \times \frac{f_{\prime}}{\sqrt{1 + f_{\prime}^2}} \times r \,.\, s_{\prime};$$

from which we find

power to keep the cart in uniform motion when the wheels are turning;

$$F_{\prime\prime} = W \cdot f_{\prime} \cdot \frac{r}{R} \cdot \frac{1}{\sqrt{(1 + f_{\prime}^2)\left(1 - \frac{f_{\prime}^2}{1 + f_{\prime}^2} \times \frac{r^2}{R^2}\right)}}.$$

The value of $f_{\prime}$ being a small fraction, as will appear from Table IV., the fraction

$$\frac{1}{\sqrt{(1 + f_{\prime}^2)\left(1 - \frac{f_{\prime}^2}{1 + f_{\prime}^2} \cdot \frac{r^2}{R^2}\right)}}$$

this factor equal to unity;

will differ but slightly from unity, and hence may be replaced by unity, which will give

$$F_{\prime\prime} = W \cdot f_{\prime} \cdot \frac{r}{R} \quad . \quad . \quad . \quad (124).$$

practical relation of power to load;

Dividing this by Eq. (123), we find

$$\frac{F_{\prime\prime}}{F_{\prime}} = \frac{W . f_{\prime}}{(W + w) f} \times \frac{r}{R} \quad . \quad . \quad (125).$$

relation of the powers to produce sliding and rolling motion;

Here W is less than $W + w$; $f_{\prime}$ is, by the tables, less than f, and r is usually very much less than R, so that the second member must be a small fraction, and $F_{\prime}$, consequently, much greater than $F_{\prime\prime}$. This is the theory of carriage-wheels of every kind, of castors, rollers for smoothing ground, and the like.

theory of the carriage-wheel, &c.;

Example. Suppose a carriage with four wheels, whose joint weight is 50 pounds, to be loaded with 2040 pounds, the weight of the axle-trees and body being together equal to 320 pounds. Let the wheels be of cast iron, the axles of wrought iron, the radius of the eye half an inch, that of the wheel one foot and a half, and suppose an unguent of tallow and the carriage placed upon a rail-track of wrought iron. Here

example—carriage on railway;

numerical value of the elements;

$$W = 2040 + 320 = \overset{lbs.}{2360},$$

$$w = \dots\dots \overset{lbs.}{50},$$

$$W + w = \dots\dots \overset{lbs.}{2410},$$

$$f_{,} = \text{Table IV.} \dots 0.08,$$

$$f = \text{Table I.} \dots 0.194,$$

$$r = \dots\dots \overset{ft.}{0.042}, \text{ nearly},$$

$$R = \dots\dots \overset{ft.}{1.5};$$

and Eq. (125),

ratio of the forces;

$$\frac{F_{,,}}{F_{,}} = \frac{2360}{2410} \times \frac{0.08}{0.194} \times \frac{0.042}{1.5} = 0.0113;$$

or

value of one in terms of the other;

$$F_{,} = 88.49 \,.\, F_{,,};$$

conclusion;

that is to say, the force requisite to put the carriage in motion when its wheels are free to rotate, is only about one eighty-eighth part of that which would be necessary to drag it, were its wheels locked.

runners substituted for wheels on ice;

If we examine Eq. (125), we shall find that by taking f equal to zero, the force $F_{,,}$ will be vastly greater than $F_{,}$, and the wheels will not turn. Now, although this extreme case can never occur in practice, yet, when a carriage is placed upon ice, we approximate to it; and this is why runners are usually substituted for wheels, under such circumstances. The same equation explains why it is that so much more advantage arises from large wheels than small ones.

Multiplying both members of Eq. (124), by $R s_{,}$, we find

equality of the work of friction and of power;

$$F_{,,} R s_{,} = W \,.\, f_{,} \,.\, r \,.\, s_{,};$$

the second member is obviously the work performed by friction, as the first is that performed by the power $F_{\prime\prime}$. Denoting by n the number of revolutions performed by the wheels, we have

$$s_{\prime} = 2\pi . n;$$

path described by point at unit's distance from axis of axle;

which, in the above equation, gives

$$F_{\prime\prime} R s_{\prime} = W . f_{\prime} \times 2\pi r . n.$$

Denote the distance travelled by d, then will

$$n = \frac{d}{2\pi R},$$

number of revolutions of the wheel;

and

$$F_{\prime\prime} R s_{\prime} = W \cdot f_{\prime} \cdot \frac{r}{R} \cdot d \quad . \quad . \quad (126).$$

work of power and of friction;

If we make d equal to one mile $= 5280$ feet, and take the dimensions and other elements the same as in the last example, we shall find

example;

$$F_{\prime\prime} R \cdot s_{\prime} = \frac{2360 \times 0.08 \times 0.042 \times 5280}{1.5} = 27912, \text{ nearly};$$

in words, the work expended in moving the carriage one mile, or the work consumed by its friction, is equivalent to that which would raise 27912 pounds through a vertical height of one foot.

When a trunnion is destined to support considerable weight, its dimensions must be proportionably large; but as the radius of the trunnion increases, the effect of friction will increase in the same ratio. To avoid the inconvenience

method by which friction may be still further reduced;

that would arise from this, when great freedom of motion is desirable, we may have recourse to the following device. Conceive the trunnion A to rest upon the circumference of two equal wheels, supported upon smaller trunnions C, C', whose distance apart is slightly greater than the radius of the wheels. Resolve, by the parallelogram of forces, the pressure upon the larger trunnion into two components, normal to the circumferences of the wheels; these will be transmitted to the smaller trunnions C and C', where they will be supported. Denote these components by N and N'. If the wheels could not turn, the friction between their circumferences and the larger trunnion would be fN and $f'N'$; and the quantity of work consumed by this friction would be

device for diminishing friction;

Fig. 243.

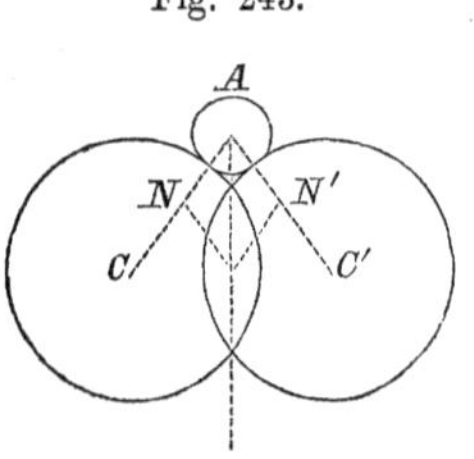

work of friction when wheels cannot turn;

$$(fN + f'N')\,R_{,}\,s_{,};$$

in which $R_{,}$ denotes the radius of the larger trunnion, and $s_{,}$ the arc described by a point at the unit's distance from its axis. If, on the contrary, the wheels may turn, the frictions on the trunnions C and C' will yield before that at the circumference of their wheels, and from what has just been shown, Eq. (124), the frictions there become

friction on the smaller trunnions;

$$Nf \cdot \frac{r}{R}, \quad \text{and} \quad N'f' \cdot \frac{r'}{R'};$$

in which r and r' denote the radii of the smaller trunnions, and R and R' the radii of their corresponding wheels; and thus the quantity of work of friction will become

work of friction;

$$\left(Nf \cdot \frac{r}{R} + N'f' \cdot \frac{r'}{R'}\right) \cdot R_{,}\,s_{,};$$

a quantity obviously much less than that obtained above. comparison of results;

If the wheels and their trunnions be of the same size, and the trunnions as well as their boxes be of the same material, the above expression becomes

$$f_{,} \cdot (N + N') \cdot \frac{r}{R} \cdot R_{,} s_{,};$$

work of friction, when wheels same size and of same material;

the value of this expression may be made as small as we please, indeed inappreciable, in a practical point of view, by selecting surfaces and unguents for which f is the least possible, and making r very small. A beautiful application of this principle is exhibited in Atwood's machine, which will be referred to hereafter. used in Atwood's machine.

XVIII.

STIFFNESS OF CORDAGE.

§ 229.—Let us now consider a wheel turning freely about an axle or trunnion, and having in its circumference a groove to receive a cord or rope. A weight W, being suspended from one end of the rope while a force F is applied to the other extremity to draw it up, the latter will experience a resistance in consequence of the rigidity of the rope, Resistance from stiffness of cordage;

Fig. 244.

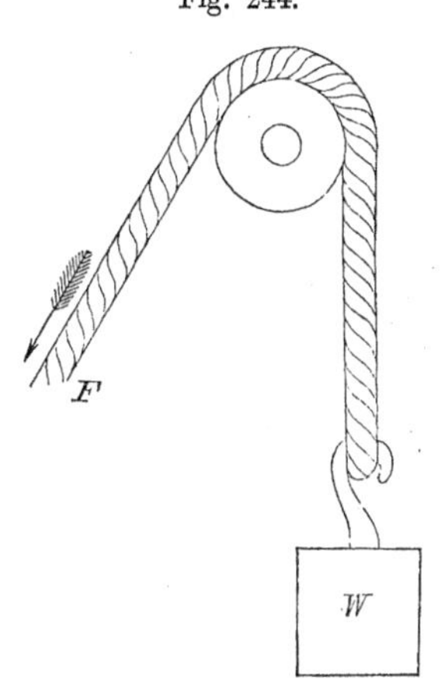

which opposes every effort to bend it around the wheel. This resistance must, of necessity, consume a portion of the work of the force F.

measure of the rigidity of cordage;

The measure of the resistance due to the rigidity of cordage has been made the subject of experiment by Coulomb; and, according to him, it results that for the same cord and same wheel, this measure is composed of two parts, of which one remains constant, while the other varies with the weight W, and is directly proportional to it; so that, designating the constant part by K, and the ratio of the variable part to the weight W by I, the measure will be given by the expression

Fig. 244

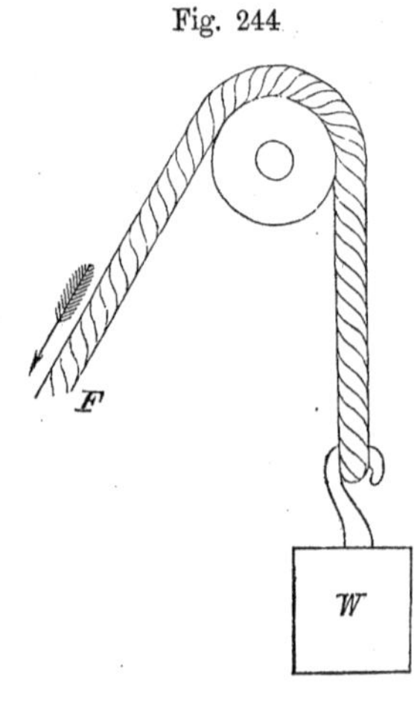

the value of this measure;

$$K + I \, . \, W;$$

in which K represents the stiffness arising from the natural torsion or tension of the threads, and I the stiffness of the same cord due to a tension resulting from one unit of weight; for, making $W = 1$, the above becomes

$$K + I.$$

Coulomb also found that on changing the wheel, the stiffness varied in the inverse ratio of its diameter; so that if

stiffness on a wheel whose diameter is unity;

$$K + I \, . \, W$$

be the measure of the stiffness for a wheel of one foot diameter, then will

stiffness on a wheel of any diameter;

$$\frac{K + I \, . \, W}{2R}$$

be the measure when the wheel has a diameter of $2R$. A

table giving the values of K and I for all ropes and cords employed in practice, when wound around a wheel of one foot diameter, and subjected to a tension arising from a unit of weight, would, therefore, enable us to find the stiffness answering to any other wheel and weight whatever. data from which to find the numerical measure of stiffness;

But as it would be impossible to anticipate all the different sizes of ropes used under the various circumstances of practice, Coulomb also ascertained the law which connects the stiffness with the diameter of the cross-section of the rope. To express this law in all cases, he found it necessary to distinguish 1st, *new white rope*, either dry or moist; 2d, *white ropes partly worn*, either dry or moist; 3d, *tarred ropes;* 4th, *packthread.* The stiffness of the first class he found nearly proportional to the square of the diameter of the cross-section; that of the second, to the square root of the cube of this diameter, nearly; that of the third, to the number of yarns in the rope; and that of the fourth, to the diameter of the cross-section. So that, if S denote the resistance due to the stiffness of any given rope; d the ratio of its diameter to that of the table; and n the ratio of the number of yarns in any tarred rope to that of the table, we shall have for these data abridged; different kinds of rope; the laws which govern them;

New white rope, dry or moist.

$$S = d^2 \cdot \frac{K + I \cdot W}{2R} \quad . \quad . \quad . \quad (127).$$

that for new white rope;

Half worn white rope, dry or moist.

$$S = d^{\frac{3}{2}} \cdot \frac{K + I \cdot W}{2R} \quad . \quad . \quad . \quad (128).$$

old white rope;

Tarred rope.

$$S = n \cdot \frac{K + I \cdot W}{2R} \quad . \quad . \quad . \quad (129).$$

tarred rope;

Packthread.

$$S = d \cdot \frac{K + I \cdot W}{2R} \quad . \quad . \quad . \quad (130).$$

packthread;

TABLES

OF WEIGHTS NECESSARY TO BEND DIFFERENT ROPES AROUND A WHEEL ONE FOOT IN DIAMETER.

No. 1. White Ropes—new and dry.

Stiffness proportional to the square of the diameter.

for new white ropes, dry;

Diameter of rope in inches.	Natural stiffness, or value of *K*.	Stiffness for load of 1 lb., or value of *I*.
	lbs.	*lbs.*
0.39	0.4024	0.0079877
0.79	1.6097	0.0319501
1.57	6.4389	0.1278019
3.15	25.7553	0.5112019

No. 2. White Ropes—new and moistened with water.

Stiffness proportional to square of diameter.

for new white ropes, moist;

Diameter of rope in inches.	Natural stiffness, or value of *K*.	Stiffness for load of 1 lb., or value of *I*.
	lbs.	*lbs.*
0.39	0.8048	0.0079877
0.79	3.2194	0.0319501
1.57	12.8772	0.1278019
3.15	51.5111	0.5112019

Squares of the ratios of diameter, or values of d^2.

Ratios d.	Squares d^2.
1.00	1.00
1.10	1.21
1.20	1.44
1.30	1.69
1.40	1.96
1.50	2.25
1.60	2.56
1.70	2.89
1.80	3.24
1.90	3.61
2.00	4.00

No. 3. White Ropes—half worn and dry.

Stiffness proportional to the square root of the cube of the diameter.

old white ropes, dry;

Diameter of rope in inches.	Natural stiffness, or value of *K*.	Stiffness for load of 1 lb., or value of *I*.
	lbs.	*lbs.*
0.39	0.40243	0.0079877
0.79	1.13801	0.0525889
1.57	3.21844	0.0638794
3.15	9.10150	0.1806573

No. 4. White Ropes—half worn and moistened with water.

Stiffness proportional to the square root of the cube of the diameter.

old white ropes, moistened;

Diameter of rope in inches.	Natural stiffness, or value of *K*.	Stiffness for load of 1 lb., or value of *I*.
	lbs.	*lbs.*
0.39	0.8048	0.0079877
0.79	2.2761	0.0525889
1.57	6.4324	0.0638794
3.15	18.2037	0.1806573

Square roots of the cubes of the ratios of diameters, or values of $d^{\frac{3}{2}}$.

Ratios or d.	Power $\frac{3}{2}$, or $d^{\frac{3}{2}}$.
1.00	1.000
1.10	1.154
1.20	1.315
1.30	1.482
1.40	1.657
1.50	1.837
1.60	2.024
1.70	2.217
1.80	2.415
1.90	2.619
2.00	2.828

No. 5. TARRED ROPES.

Stiffness proportional to the number of yarns.

[These ropes are usually made of three strands twisted around each other, each strand being composed of a certain number of yarns, also twisted about each other in the same manner.]

No. of yarns.	Weight of 1 foot in length of rope.	Natural stiffness, or value of K.	Stiffness for load of 1 lb., or value of I.
	lbs.	*lbs.*	*lbs.*
6	0.0211	0.1534	0.0085198
15	0.0497	0.7664	0.0198796
30	1.0137	2.5297	0.0411799

for tarred ropes;

For packthread, it will always be sufficient to use the tabular values given above, corresponding to the least tabular diameters, and substitute them in Eq. (130). An example or two will be sufficient to illustrate the use of these tables.

Example 1*st.* Required the resistance due to the stiffness of a new dry white rope, whose diameter is 1.18 inches, when loaded with a weight of 882 pounds, and wound about a wheel 1.64 feet in diameter.

examples to illustrate the use of Table No. 1;

Seek in Table No. 1 the diameter nearest that of the given rope; it is 0.79; hence

$$d = \frac{1.18}{0.79} = 1.5 \text{ nearly};$$

and from the table at the side,

$$d^2 = 2.25.$$

elements obtained from the tables;

From Table No. 1, opposite 0.79, we find

$$K = 1.6097,$$

$$I = 0.03195;$$

which, together with the weight $W = 882$ lbs., and $2\,R = \overset{ft.}{1.64}$, substituted in Eq. (127), give result;

$$S = 2.25 \cdot \frac{\overset{lb.}{1.6097} + \overset{lb.}{0.03195} \times 882}{1.64} = \overset{lbs.}{40.817},$$

which is the true resistance due to the stiffness of the rope in question.

example to illustrate Table No. 4;

Example 2*d*. What is the resistance due to the stiffness of a white rope, half worn and moistened with water, having a diameter equal to 1.97 inches, wound about a wheel 0.82 of a foot in diameter, and loaded with a weight of 2205 pounds?

The tabular diameter in Table No. 4, next below 1.97, is 1.57, and hence

$$d = \frac{1.97}{1.57} = 1.3 \text{ nearly};$$

the square root of the cube of which is, by the table at the side,

data from the table;

$$d^{\frac{3}{2}} = 1.482.$$

In Table No. 4 we find, opposite 1.57,

$$K = 6.4324,$$

$$I = 0.06387;$$

which values, together with $W = 2205$ lbs., and $2\,R = \overset{ft.}{0.82}$, in Eq. (128), give

result;

$$S = 1.482 \times \frac{\overset{lbs.}{6.4324} + \overset{lb.}{0.06387} \times 2205}{0.82} = \overset{lbs.}{266.109},$$

which is the required resistance.

Example 3*d*. What is the resistance due to the stiffness of a tarred rope of 22 yarns, when subjected to the action of a weight equal to 4212 pounds, and wound about a wheel 1.3 feet diameter, the weight of one running foot of the rope being about 0.6 of a pound? example to illustrate Table No. 5;

By referring to Table No 5, we find the tabular number of yarns next below 22 to be 15, and hence

$$n = \frac{22}{15} = 1.466 \text{ nearly.}$$

In the same table, opposite W, we find

$$K = 0.7664,$$

$$I = 0.019879;$$

data obtained from the table;

which, together with $W = 4212$, and $2\,R = \overset{ft.}{1.3}$, in Eq. (129), give

$$S = 1.466\ \frac{0.7664 + 0.019879 \times 4212}{1.3} = \overset{lbs.}{95.188}.$$ result;

Example 4th. Required the resistance due to the stiffness of a new white packthread, whose diameter is 0.196 inches, when moistened or wet with water, wound about a wheel 0.5 of a foot in diameter, and loaded with a weight of 275 pounds. example to illustrate the case of packthread;

The lowest tabular diameter is 0.39 of an inch, and hence

$$d = \frac{0.196}{0.390} = 0.5 \text{ nearly.}$$

In Table No. 2 we find, opposite 0.39,

$$K = \overset{lb.}{0.8048},$$

$$I = 0.00798;$$

data from Table No. 2;

which, with $W = 275$, and $2R = 0.5$, we find, after substituting in Eq. (130),

result.

$$S = 0.5 \frac{0.8048 + 0.00798 \times 275}{0.5} = \overset{lbs.}{2.999}.$$

Work due to stiffness of cordage;

§ 230.—The resistance just found is expressed in pounds, and is the amount of weight which would be necessary to bend any given rope around a vertical wheel, so that the portion $A\ E$, between the first point of contact A, and the point E, where the rope is attached to the weight, shall be perfectly straight. The entire process of bending takes place at this first or tangential point A; for, if motion be communicated to the wheel in the direction indicated by the arrow-head, the rope, supposed not to slide, will, at this point, take and retain the constant curvature of the wheel, till it passes from the latter on the side of the power F. When, therefore, by the motion of the wheel, the point m of the rope, now at the tangential point, passes to m', the working point of the force S will have described in its own direction the distance $A\ D$. Denoting the arc described by a point at the unit's distance from the centre of the wheel by $s_{,}$, and the radius of the wheel by R, we shall have

the bending takes place at the first point of contact;

Fig. 245.

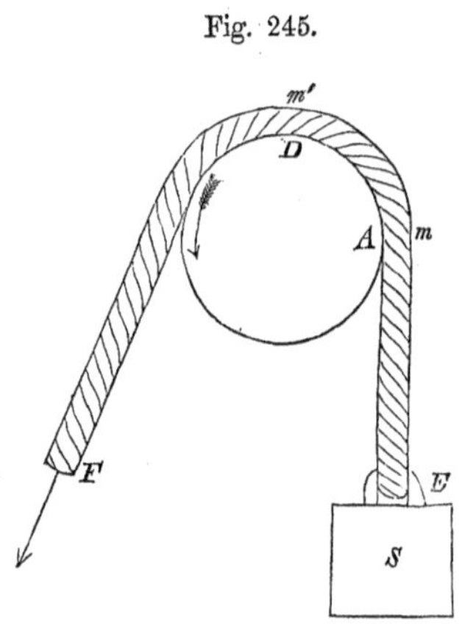

path described by the working point;

$$A\ D = R s_{,};$$

and representing the quantity of work of the force S, by

L, we get

$$L = S \,.\, R s_, ;$$

replacing S by its value in Eqs. (127) to (130),

$$L = R s_, d_, \frac{K + I \,.\, W}{2 R} \quad . \quad . \quad (131);$$ work of the stiffness;

in which $d_,$ represents the quantity d^2, $d^{\frac{3}{2}}$, n, or d, in Eqs. (127), (128), (129), or (130), according to the nature of the rope.

Example. Taking the 2d example of § 229, and supposing a portion of the rope, equal to 20 feet in length, to have been brought in contact with the wheel, after the motion begins, we shall have examples;

$$L = 20 \times 266{,}109 = \overset{lbs.}{5322.18};$$ result;

that is, the quantity of work consumed by the resistance due to the stiffness of the rope, while the latter is moving over a distance of 20 feet, would be sufficient to raise a weight of 5322.18 pounds through a vertical height of one foot. in words.

XIX.

WHEEL AND PULLEY.

§ 231.—A plane wheel, free to turn about its trunnion or axle, supported in a fixed box, may be moved in either direction by two forces F and Q, which act in its Wheel and pulley;

Fig. 246.

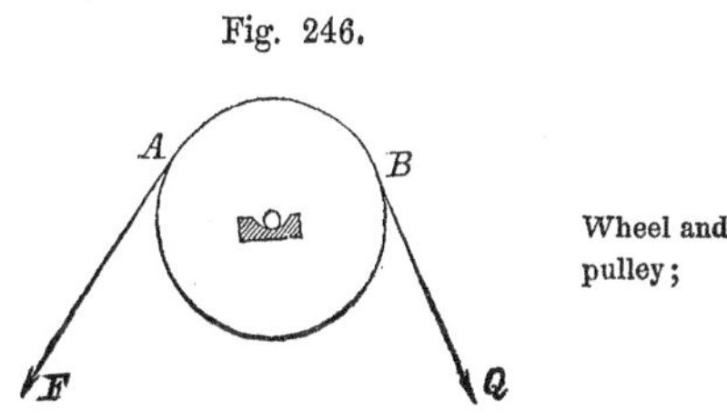

plane, and tangent to its circumference at A and B. These forces, acting in the same plane, perpendicular to the axle, and tending to turn the wheel in opposite directions, will be in equilibrio when the elementary quantity of work developed by each is the same with contrary signs. But the points of application A and B, belonging to the same circumference, the paths which they simultaneously describe will be equal; and since the product of these paths by the forces F and Q must be equal, it follows that *whenever the forces are in equilibrio, they must also be equal.*

Fig. 246.

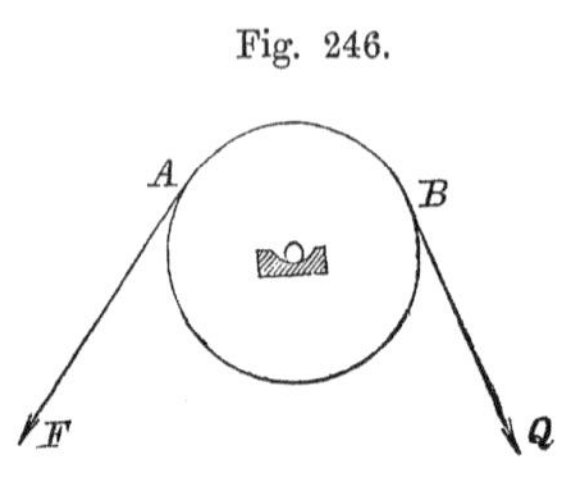

equilibrium of two forces acting upon the circumference of a wheel;

the forces must be equal;

This supposes the wheel free to turn, without obstruction of any kind. But if we consider the friction at the trunnion or axle, then, supposing the equilibrium still to exist, but the wheel on the eve of motion in the direction of the force F, the elementary quantity of work of the latter must be equal to that of the resistance Q, increased by that of the friction; in which case F and Q will not be equal; and denoting the radius of the wheel by R, that of its trunnion or eye by r, and the resultant of F and Q by N, we shall have

when friction is taken into the account;

relation of work of power, resistance, and friction;

$$F R s_{,} = Q R s_{,} + f N . r s_{,} \quad . \quad . \quad (132);$$

in which f is the coefficient of friction at the axle or trunnion, and $s_{,}$ the arc described by a point at the unit's distance from the axis during motion. Dividing by $R\ s_{,,}$ we find

relation of these forces;

$$F = Q + f N \cdot \frac{r}{R} \quad . \quad . \quad . \quad . \quad (133).$$

From which we might conclude the value of F, but that N is unknown, being the resultant of F and Q.

Now, two cases may arise, viz.: either the value of r may be very small in comparison with R, or it may not. In the first case, any error committed in the determination of N would but slightly affect the value of F, since only the small fractional portion $\frac{r}{R}$ of N is taken. We may, therefore, be content with an approximate value for N. two cases may arise;

To obtain this, we first omit the consideration of friction, which will make $f = 0$, in the above equation, which then reduces to to find the resultant of the power and resistance;

$$F = Q.$$

Fig. 247.

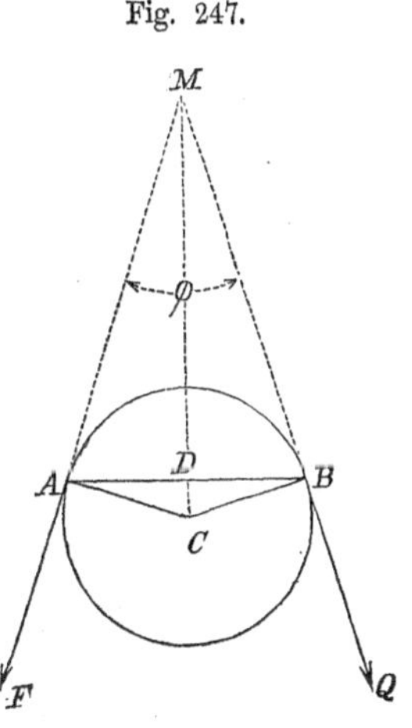

Denote by φ the angle AMB, which the two forces F and Q make with each other; then, from the parallelogram of forces, will

$$N = \sqrt{F^2 + Q^2 + 2FQ \cdot \cos \varphi};$$ its value;

and, because F and Q are, in this case, equal,

$$N = Q\sqrt{2 + 2\cos\varphi};$$ or this;

but

$$2 + 2\cos\varphi = 4\cos^2 \tfrac{1}{2}\varphi;$$

whence

$$N = Q \times 2\cos\tfrac{1}{2}\varphi;$$ and finally this;

but joining A and B by a right line, as also M and C, we have the angle $A\,M\,C$, equal to $\frac{1}{2}\,\varphi$, and

$$\cos \tfrac{1}{2}\,\varphi = \sin M\,C\,A = \frac{A\,D}{A\,C} = \frac{A\,D}{R};$$

which, substituted above, gives

value of resultant under a more convenient form;

$$N = Q \cdot \frac{A\,B}{R},$$

since $2\,A\,D = A\,B$. That is to say, the resultant N is obtained by multiplying the resistance Q, by the chord of the arc between the tangential points, and dividing the product by the radius of the wheel. This value of N, substituted in Eqs. (132) and (133), gives

rule;

quantity of work;

$$F\,R\,s_{,} = Q\,R\,s_{,} + f \,.\, r\,s_{,}\,Q\,\frac{A\,B}{R} \quad . \quad . \quad (134),$$

value of the force;

$$F \quad = Q + f \cdot \frac{r}{R}\,Q \times \frac{A\,B}{R} \quad . \quad . \quad (135);$$

the first of which will give the quantity of work of the power, and the latter the relation of the power F to the resistance Q, necessary to produce an equilibrium. The first shows that the work of the power is equal to the work of the resistance, increased by that consumed by friction.

inferences;

We now come to the second case, viz.: that in which r is not very small

Fig. 247

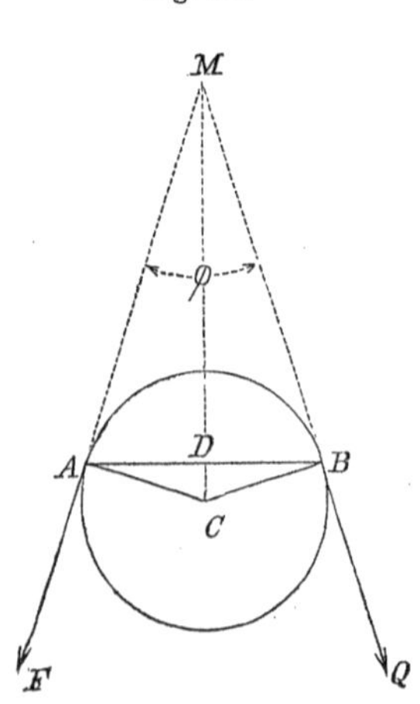

in comparison with R. And first we remark, that F is always greater than Q, and that the resultant obtained under the hypothesis of F being equal to Q, is, therefore, too small. Calling N_1 this latter resultant, we have

case in which trunnion is not so small;

$$N_1 = Q \cdot \frac{A\,B}{R} \quad . \quad . \quad . \quad . \quad (136);$$

first approximation to the resultant;

and this value, substituted in Eq. (133) for N, gives

$$F = Q + \frac{r}{R} \cdot f \cdot N_1 = F_1 . \quad . \quad (137).$$

first approximation to the power;

Now if N_1 be too small, it is obvious that F_1 will also be too small. But this value of F_1 is greater than Q, and if we find the resultant of two forces each equal to F_1, or make

$$N = \frac{F_1\,A\,B}{R} = N_2 \quad . \quad . \quad . \quad (138);$$

second approximation to resultant;

it is obvious that N_2 will be too great, and so of the value

$$F = Q + \frac{r}{R} \cdot f \cdot N_2 = F_2 .$$

second approximation to the power;

Thus tne true value of F is greater than F_1, and less than F_2, and as these two values will not differ much, we may take the true value of F to be an arithmetical mean between them, that is,

$$F = \frac{F_1 + F_2}{2},$$

mean of the approximations;

or

$$F = Q + \frac{r}{R} \cdot f \cdot \frac{N_1 + N_2}{2};$$

value of the power;

and eliminating N_1 and N_2, by means of Eqs. (136), (137), and (138), we find

final value of power;

$$F = Q + f\frac{r}{R}Q\frac{A\,B}{R}\left[1 + \tfrac{1}{2}f\frac{r}{R}\,\frac{A\,B}{R}\right] \quad . . (139);$$

and multiplying each member by $R\,s_{,}$,

quantity of work;

$$F R s_{,} = Q R{\cdot}s_{,} + f r\,s_{,} Q\frac{A\,B}{R}\left[1 + \tfrac{1}{2}f\frac{r}{R}\,\frac{A\,B}{R}\right] \quad . . (140).$$

conclusion.

The first will determine the condition of the equilibrium, and the second the quantity of work.

Pulley;

§ 232.—The pulley is a small wheel having a groove in its circumference for the reception of a rope, at one end of which is attached the power F, and at the other

Fig. 248.

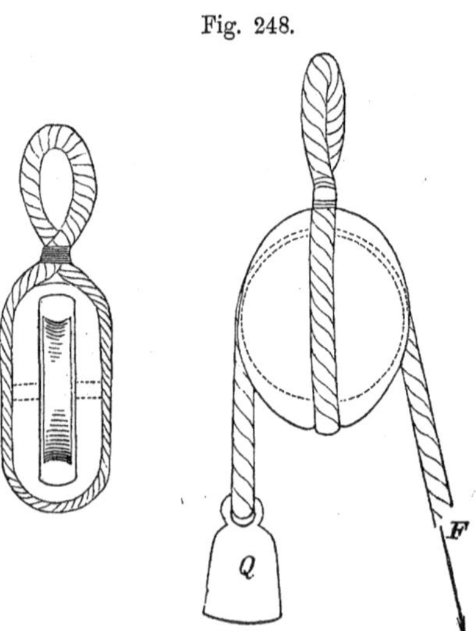

description; and mode of applying the power;

the resistance Q. The pulley may turn either upon trunnions or about an axle, supported in what is called a *block.* This is usually a solid piece of wood, through which is cut an opening large enough to receive the pulley, and allow it to turn freely between its cheeks. Sometimes the block is a simple framework of metal. When the block is stationary, the pulley is said to be *fixed.* The principle of this machine is obviously the same as that of a simple wheel, and to the discussion of § 231 we have but to add the consideration of the stiffness of the rope, to have all the circumstances of its action. The quantity of work due to the stiffness of the rope is given by Eqs. (127) to (130) inclusive.

block; fixed pulley; principle the same as that of the wheel;

Now, when the motion is uniform, or when the pulley is about to turn in the direction of the power F, the quantity of work of the latter must be equal to the work of the resistance Q, increased by that of the friction and stiffness of the rope; and denoting the radius of the pulley by R, that of the trunnion or eye of the pulley, as the case may be, by r, and the arc described at the unit's distance from the axis by $s_{,}$, we must have

$$F R s_{,} = Q R s_{,} + d_{,} \cdot \frac{K + I Q}{2 R} \cdot R s_{,} + f N \,.\, r \,.\, s_{,} ;$$

quantity of work of power;

in which $d_{,}$ denotes either d^2, $d^{\frac{3}{2}}$, d, or n, in Eqs. (127) to (130), according to the kind and condition of the rope; and N, the resultant of all forces except friction.

Dividing by $R s_{,}$, we obtain

$$F = Q + d_{,} \frac{K + I \,.\, Q}{2 R} + f \frac{r}{R} N.$$

value of the power;

Make

$$Q + d_{,} \frac{K + I Q}{2 R} = Q_{,},$$

and the above becomes

different form of same;

$$F = Q_{,} + f\frac{r}{R} \cdot N;$$

and replacing N by its value

resultant of all the forces except friction;

$$\sqrt{F^2 + Q_{,}^2 + 2FQ_{,} \cos \varphi},$$

Fig. 249.

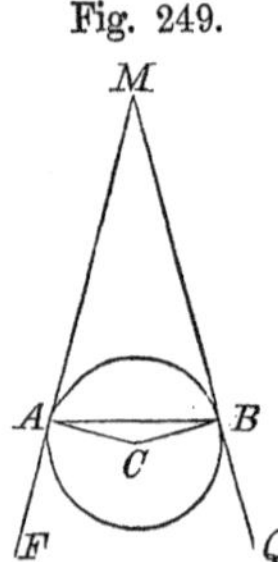

in which φ denotes the angle AMB, made by the branches of the rope not in contact with the pulley, and we get

$$F = Q_{,} + f\frac{r}{R}\sqrt{F^2 + Q_{,}^2 + 2FQ_{,} \cos \varphi}.$$

Transposing $Q_{,}$, squaring and solving the equation with reference to F, and we have

the most general value for the power;

$$F = \frac{Q_{,}}{1-\left(f\frac{r}{R}\right)^2} \cdot \left\{ \begin{array}{l} 1 + \left(f\frac{r}{R}\right)^2 \cos \phi \\ \\ \pm f\frac{r}{R} \cdot \sqrt{(1+\cos\phi)\left[2-\left(f\frac{r}{R}\right)^2(1-\cos\phi)\right]} \end{array} \right. \qquad (141).$$

Taking the upper of the double sign, because the motion takes place in the direction of F; replacing $Q_{,}$ by its value, and calling the angle ACB, enveloped by the rope, θ, in which case,

$$\cos \varphi = - \cos \theta,$$

we finally obtain

the same in known terms;

$$F = \frac{Q + d_{,}\frac{K+IQ}{2R}}{1-\left(f\frac{r}{R}\right)^2} \cdot \left\{ \begin{array}{l} 1 - \left(f\frac{r}{R}\right)^2 \cos \theta \\ \\ + f \cdot \frac{r}{R} \sqrt{(1-\cos\theta)\left[2-\left(f\frac{r}{R}\right)^2(1+\cos\theta)\right]} \end{array} \right. \qquad (142.)$$

Fig. 250.

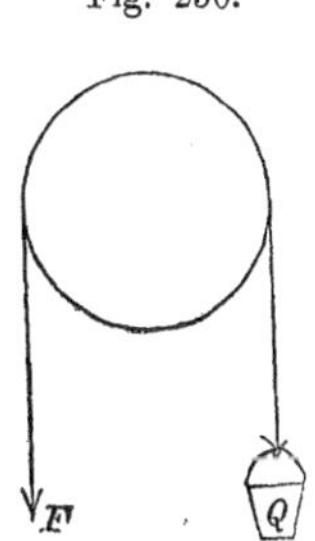

When the two branches of the rope are parallel, then will $\theta = 180°$; $\cos\theta = -1$; and the equation becomes,

$$F = \left(Q + d_, \frac{K + IQ}{2R}\right) \cdot \frac{1 + f\frac{r}{R}}{1 - f\frac{r}{R}} \quad . \quad . \quad (143).$$

value when the branches of the rope are parallel;

If the rope be perfectly flexible, and the friction be zero, then will $K = 0$, $I = 0$, $f = 0$, and

$$F = Q;$$

that is, *the power will always equal the resistance in the fixed pulley, when there is neither friction nor stiffness of cordage.*

To obtain the quantity of work, multiply both members of Eq. (142) by $Rs_{,,}$ and there will result

$$FRs_, = \frac{QRs_, + \frac{1}{2}d_,(K + IQ)s_,}{1 - \left(f\frac{r}{R}\right)^2} \cdot \left\{ \begin{array}{l} 1 - \left(f\frac{r}{R}\right)^2 \cos\theta \\ + f\frac{r}{R} \cdot \sqrt{(1 - \cos\theta)\left[2 - \left(f\frac{r}{R}\right)^2(1 + \cos\theta)\right]} \end{array} \right. \quad (144).$$

general value for the quantity of work of the power;

In finding the value of N, the weight of the pulley was not considered, and for the reason that in practice it is usually small; the friction arising from its action may, therefore, in general, be neglected. Should it be desirable, however, in any case, to take it into account, it is easily done. For this purpose, find, by the parallelogram of forces, the resultant of the weight of the pulley and the force Q, both of which are known, and

weight of the pulley generally neglected;

but it may be taken into the account;

employ this resultant instead of Q in finding the value of F.

Example. Required the quantity of work necessary to raise 500 pounds of coal, through a vertical elevation of 50 feet, by means of a rope passing over a fixed pulley, in such a position that the power F shall be applied in a horizontal direction; the pulley, which is of lignum-vitæ, is 1.25 feet in diameter; the radius of its eye is 0.05 feet; the axle of wrought iron, lubricated with hogs' lard; the rope is white, half worn, and has a diameter of one inch.

example for illustration;

conditions of the proposition;

Fig. 251.

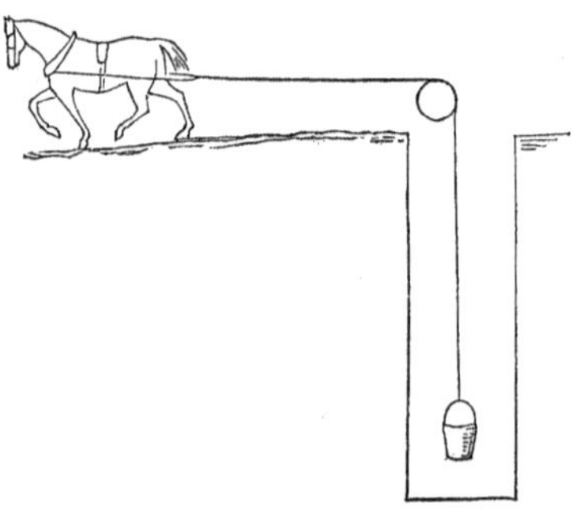

tabular elements;

Here $\theta = 90°$, and $\cos\theta = 0$; in Table IV. § 225, $f = 0.11$; Table No. 3, § 221, $d_{,} = d^{\frac{3}{2}} = \left(\frac{1}{0.79}\right)^{\frac{3}{2}} = (1.2)^{\frac{3}{2}}$ nearly $= 1.315$; $K = 1.13801$; $I = 0.0525889$; $R = 0.625$; $r = 0.05$; $R s_{,} = 0.625 \times s_{,} = 50$; whence

numerical values of the data;

$s_{,} = \dfrac{50}{0.625} = 80$ feet; $Q = 500$ lbs.; and $f\dfrac{r}{R} = 0.0084$. These data in Eq. (144) give

quantity of work;

$$F R s_{,} = (500 \times 50 + 1.315 \frac{1.13801 + 0.05259 \times 500}{2} 80)\ (1 + 0.0084\sqrt{2 - (0.0084)^2};)$$

or

$$F R s_{,} = 26250.17.$$

If there were no friction, or stiffness of cordage, then would

$$F R s_{\prime} = Q R s_{\prime} = 25000.0;$$ value without friction and stiffness;

whence $26250.17 - 25000 = 1250.17$ is the loss due to stiffness of cordage and friction, which would be sufficient to raise 1250.17 pounds through 1 foot of altitude, or $\frac{1250.17}{50} = 25$ pounds through the given height of 50 feet; a result well calculated to impress us with the necessity of including these resistances in all estimates of work. loss due to stiffness and friction;

$$F = \frac{26250.17}{R s_{\prime}} = \frac{26250.17}{50} = 525^{lbs.} \text{ nearly.}$$ numerical value of the power.

§ 233.—Thus far the axis of the pulley is supposed to have remained immoveable. We shall now consider the case in which the pulley is supported upon a rope in its groove, one end of the rope being attached to a fixed hook A, while the other is acted upon by the force F. The pulley is embraced by a kind of iron or other metallic fork whose prongs are perforated near the ends for the reception of the axle, and whose shank terminates in a hook to which the resistance W is attached. The pulley is, in this case, said to be *moveable*. Denote the resistance to be overcome and put in motion, by W; the tension of the rope between the fixed hook and tangential point H by Q; let the other notation be the same as in the case of the fixed pulley. Moveable pulley; description; notation;

Fig. 252.

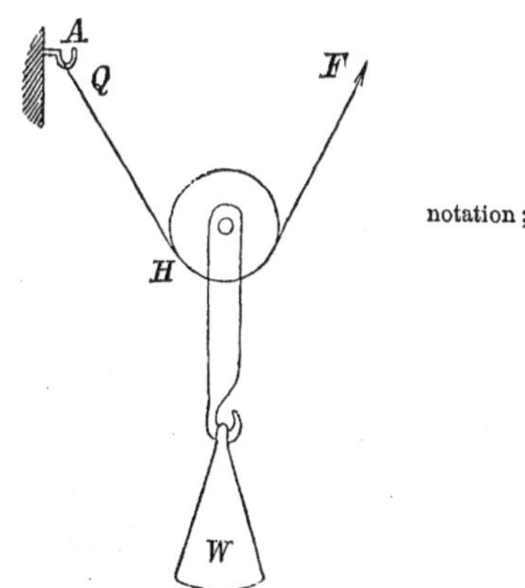

The quantity of work of F must be equal to that of the

tension Q, increased by the work due to the stiffness of the rope and friction; that is,

quantity of work;

$$F R s_{,} = Q R s_{,} + d_{,} \frac{K + I Q}{2 R} R s_{,} + r f W s_{,} \quad . \quad . \quad (145).$$

Dividing both members by $R s_{,}$,

value of the power;

$$F = Q + d_{,} \frac{K + I Q}{2 R} + \frac{r}{R} f W.$$

The pulley being supposed either on the verge of rotary motion in the direction of F, or rotating uniformly, it is obvious that W will be equal and directly opposed to the resultant of F and Q; and that Q will be equal and directly opposed to the resultant of F and W. This latter resultant being found by the parallelogram of forces, Eq. (31), and in its value that of F, in last equation, substituted for F, the force Q will become known in terms of W, the friction, and stiffness of cordage; and this value of Q, being substituted in Eq. (145), will give the work in terms which are known.

to find the tension of rope at the fixed end;

Fig. 252.

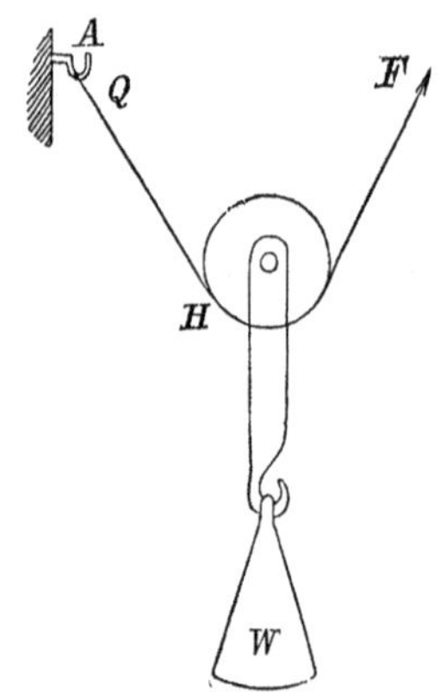

Fig. 253.

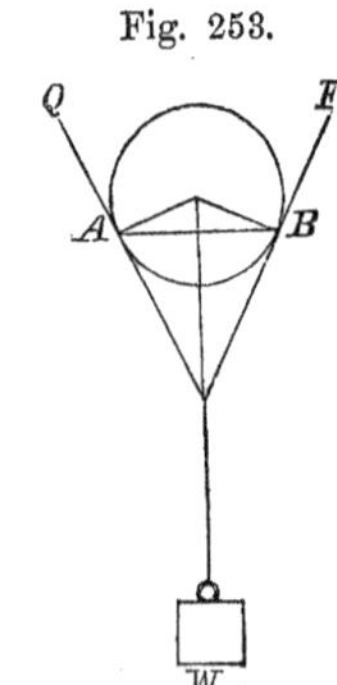

The method here indicated is perfectly rigorous, but is somewhat long, and may be avoided by resort-

the same found by approximation;

ing to an approximation which in practice is sufficiently accurate. If F and Q be supposed for an instant equal, we have seen that

$$Q = \frac{R \,.\, W}{A\,B};$$

approximate value for tension;

which, substituted for Q in Eq. (145), gives

$$F\,R\,s_{,} = \begin{cases} W \cdot \dfrac{R}{A\,B} \cdot R\,s_{,} \\ + \, d_{,} \dfrac{K + I \cdot W \cdot \dfrac{R}{A\,B}}{2} \cdot s_{,} \\ + \, r \,.\, f\,W \,.\, s_{,} \end{cases} \quad . \; . \; (146);$$

quantity of work;

dividing by $R\,s_{,}$,

$$F = W \cdot \frac{R}{A\,B} + d_{,} \frac{K + I \cdot W \cdot \dfrac{R}{A\,B}}{2\,R} + f\,\frac{r}{R} \cdot W \; . \; . \; (147).$$

value of the power;

If we suppose the stiffness of the rope and friction zero, there will result,

$$F = W \cdot \frac{R}{A\,B};$$

power, when stiffness and friction are zero;

or

$$F \; : \; W \; :: \; R \; : \; A\,B;$$

that is to say, *the power is to the resistance as the radius of the pulley is to the chord of the arc enveloped by the rope.*

relation of power and resistance;

Example. Let the pulley be of cast iron and turn upon a wrought-iron axle, greased with tallow; the di-

example;

ameter of the pulley 1.3 feet, and that of its eye 0.045 feet; the diameter of the rope, which is new, white and dry, 1.4 inches; the weight W, 3462 pounds; the height 40 feet, and let the chord AB be equal to the diameter of the pulley.

By reference to the proper tables, we find

data from the tables;

$$f = 0.07; \quad d_{,} = d^2 = \left(\frac{1.4}{0.79}\right)^2 = (1.8)^2 \text{ nearly} = 3.24;$$

$$K = 1.6097; \quad I = 0.0319501;$$

and from the given data,

data of the example;

$$R = 0.65; \quad r = 0.0225; \quad AB = 1.3; \quad Rs_{,} = 40;$$

$$s_{,} = \frac{40}{0.65} = 61.538 \text{ nearly}; \quad \text{and } W = 3462;$$

which, substituted in Eq. (146), give

numerical result;

$$FRs_{,} = \left\{ \begin{array}{l} 3462 \times \frac{(.65)^2}{1.30} \times 61.538 \\ + 3.24 \times \frac{1.6097 + 0.031950 \times 3462 \times \frac{0.65}{1.3}}{2} \times 61.538 \\ + 0.07 \times 0.0225 \times 3462 \times 61.538 \end{array} \right\} = 71279.35;$$

same with neither stiffness nor friction;

with neither friction nor stiffness of cordage, the quantity of work would be simply

$$FRs_{,} = 3462 \cdot \frac{(0.65)^2}{1.30} \times 61.538 = 69239.5;$$

work of stiffness and friction.

the difference $71279.35 - 69239.5 = 2039.85$ is the loss due to the causes just named.

The *muffle*;

§ 234.—The *Muffle* is a collection of pulleys in two separate blocks or frames. One of these blocks is attached to a fixed point A, by which all of its pulleys become *fixed*, while the other block is attached to the resistance Q, and its pulleys thereby made *moveable*. A rope is attached at one end to a hook h at the extremity of the fixed block, and is passed around one of the moveable pulleys, then about one of the fixed pulleys, and so on, in order, till the rope is made to act upon each pulley of the combination. The power F is applied to the other end of the rope, and the pulleys are so proportioned that the parts of the rope between them, when stretched, are parallel. Now suppose the power F to communicate uniform motion to the resistance Q. Denote the tension of the rope between the hook of the fixed block and the point where it comes in contact with the first moveable pulley, by t_1; the radius of this pulley by R_1; that of its eye by r_1; the coefficient of friction on the axle by f; the constant and coefficient of the stiffness of cordage by K and I, as before; then, denoting the tension of the rope between the last point of contact with the first moveable, and first point of contact with the first fixed pulley, by t_2, the quantity of work of the tension t_2 will, Eq. (145), be

Fig. 254.

definition and description;

arrangement of the rope;

application of the power and relative size of the pulleys;

notation;

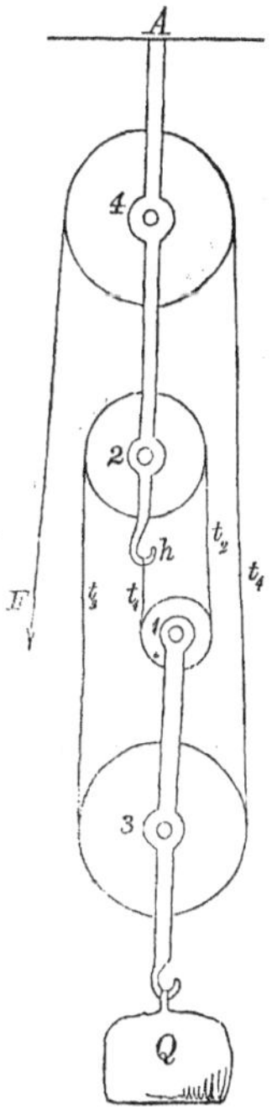

work of the tension on first ascending branch;

$$t_2 R_1 s_, = t_1 R_1 s_, + d_, \frac{K + I t_1}{2 R_1} R_1 s_, + f(t_1 + t_2) r_1 s_, ;$$

dividing by $s_,$

moment of this tension;

$$t_2 R_1 = t_1 R_1 + d_, \cdot \frac{K + I t_1}{2 R_1} \cdot R_1 + f(t_1 + t_2) r_1 \quad . . \ (148).$$

Again, denoting the tension of that part of the rope which passes from the first fixed to the second moveable pulley by t_3; the radius of the first fixed pulley by R_2, and that of its eye by r_2, we shall, in like manner, have

moment of tension on second descending branch;

$$t_3 R_2 = t_2 R_2 + d_, \frac{K + I t_2}{2 R_2} R_2 + f(t_2 + t_3) r_2 \quad . . \ (149).$$

And denoting the tensions, in order, by t_4 and t_5, this last being equal to F, we shall have

moment of tension on second ascending branch;

$$t_4 R_3 = t_3 R_3 + d_, \frac{K + I t_3}{2 R_3} \cdot R_3 + f(t_3 + t_4) r_3 . . \ (150),$$

same on third descending branch

$$F R_4 = t_4 R_4 + d_, \frac{K + I t_4}{2 R_4} R_4 + f(t_4 + F) r_4 . . \ (151);$$

so that we finally arrive at the force F, through the tensions which are as yet unknown. The parts of the rope being parallel, and the resistance Q being supported by their tensions, the latter may obviously be regarded as equal in intensity to the components of Q; hence

components of the resistance;

$$t_1 + t_2 + t_3 + t_4 = Q \quad . . \ (152);$$

which, with the preceding, gives us five equations for the determination of the four tensions and power F. This would involve a tedious process of elimination, which may be avoided by contenting ourselves with an approximation which is found, in practice, to be sufficiently accurate.

If the friction and stiffness be supposed zero, for the moment, Eqs. (148) to (151) become

method of approximation;

$$t_2 R_1 = t_1 R_1,$$

$$t_3 R_2 = t_2 R_2,$$

$$t_4 R_3 = t_3 R_3,$$

$$F R_4 = t_4 R_4;$$

friction and stiffness zero;

from which it is apparent, dividing out the radii R_1, R_2, R_3, &c., that $t_2 = t_1$, $t_3 = t_2$, $t_4 = t_0$, $F = t_4$; and hence, Eq. (152) becomes

the tensions become equal;

$$4 t_1 = Q;$$

whence

$$t_1 = \frac{Q}{4};$$

resistance equal to tension on one branch multiplied by the number of pulleys;

the denominator 4 being the whole number of pulleys, moveable and fixed. Had there been n pulleys, then would

$$t_1 = \frac{Q}{n}.$$

general value for the tension;

With this approximate value of t_1 we resort to Eqs. (148) to (151), and find the values of t_2, t_3, t_4, &c. Adding all these tensions together, we shall find their sum to be greater than Q, and hence we infer each of them to be too large. If we now suppose the true tensions to be proportional to those just found, and whose sum is $Q_1 > Q$, we may find the true tension corresponding to any erroneous tension, as t_1, by the following proportion, viz.:

to find the true from the approximate tension;

$$Q_1 \ : \ Q \ :: \ t_1 \ : \ \frac{Q}{Q_1} t_1;$$

or, which is the same thing, multiply each of the tensions found by the constant ratio $\frac{Q}{Q_1}$, the product will be the true tensions, very nearly. The value of t_4 thus found, substituted in Eq. (151), will give that of F.

example to illustrate;

Example. Let the radii R_1, R_2, R_3, and R_4, be respectively 0.26, 0.39, 0.52, 0.65 feet; the radii $r_1 = r_2 = r_3 = r_4$ of the eyes $= 0.06$ feet; the diameter of the rope, which is white and dry, 0.79 inches, of which the constant and coefficient of rigidity are, respectively, $K = 1.6097$ and $I = 0.0319501$; and suppose the pulley of brass, and its axle of wrought iron, of which the coefficient $f = 0.09$, and the resistance Q a weight of 2400 pounds.

Without friction and stiffness of cordage,

approximate value of first tension;

$$t_1 = \frac{2400}{4} = 600.^{lbs.}$$

Dividing Eq. (148) by R_1, it becomes, since $d_{,} = 1$,

$$t_2 = t_1 + \frac{K + It_1}{2R_1} + \frac{r_1}{R_1} f (t_1 + t_2).$$

Substituting the value of R_1, and the above value of t_1, and regarding in the last term t_2 as equal to t_1, which we may do, because of the small coefficient $\frac{r_1}{R_1} f$, we find

$$t_2 = \left\{ \begin{array}{l} 600 \\ + \dfrac{1.6097 + 0.0319501 \times 600}{2 \times (0.26)} \\ + \dfrac{0.06}{0.26} \times 0.09 \times (600 + 600) \end{array} \right\} = 628.39.$$

approximate value of second tension;

Again, dividing Eq. (149) by R_2, and substituting this value of t_2 and that of R_2, we find

$$t_3 = \overset{lbs.}{673.59}.$$

approximate value of third tension;

Dividing Eq. (150) by R_3, and substituting this value of t_3, as well as that of R_3, there will result

$$t_4 = \overset{lbs.}{709.82};$$

approximate value of fourth tension;

whence

$$Q_1 = t_1 + t_2 + t_3 + t_4 = \left\{ \begin{array}{l} 600 \\ + 628.39 \\ + 673.59 \\ + 709.82 \end{array} \right\} = 2611.80;$$

resultant of these tensions;

and

$$\frac{Q}{Q_1} = \frac{2400}{2611.80} = 0.919;$$

ratio of the approximate to the true resultant;

which will give for the true values of

$$\begin{array}{l} t_1 = 0.919 \times 600 \quad\;\; = 551.400 \\ t_2 = 0.919 \times 628.39 = 577.490 \\ t_3 = 0.919 \times 673.59 = 619.029 \\ t_4 = 0.919 \times 709.82 = 652.324 \\ \hline \qquad\qquad\qquad\qquad\quad 2400.243 \end{array}$$

true tension;

The above value for $t_4 = 652.324$, in Eq. (151), will give, after dividing by R_4 and substituting its numerical value,

$$F = \begin{cases} 652.324 \\ + \dfrac{1.6097 + 0.03195 \times 652.324}{2 \times 0.65} \\ + \dfrac{0.06}{0.65} \times 0.09 \times (652.324 + F); \end{cases}$$

and making in the last factor $F = t_4 = 652.324$, we find

numerical value of the power;

$$F = \overset{lbs.}{652.324} + \overset{lbs.}{17.270} + \overset{lbs.}{10.831} = \overset{lbs.}{680.425}.$$

Thus, without friction or stiffness of cordage, the intensity of F would be 600 lbs.; with both of these causes of resistance, which cannot be avoided in practice, it becomes 680.425 lbs., making a difference of 80.425 lbs., or nearly one seventh; and as the quantity of work of the power is proportional to its intensity, we see that to overcome friction and stiffness of rope, in the example before us, the moter must expend nearly a seventh more work than if these sources of resistance did not exist.

work absorbed by friction and stiffness of cordage.

Wheel and axle;

§ 235.—*Wheel and Axle* is a name given to a machine, which consists of a wheel mounted upon an arbor, supported at either end by a trunnion resting in a box. The plane of the wheel is at right angles to the axis of the arbor; the power F is applied to a rope wound around the wheel; the resistance Q is applied to another rope, wound in the opposite direction about the arbor, and also acts in a plane perpendicular to the axis of motion. The power is generally applied in the plane of the wheel, otherwise, being oblique to the axis, it would be necessary to resolve it into two components, one perpendicular and the other parallel to that line; the latter compo-

description; and application of power and resistance;

nent would press the shoulder of the arbor against the face of the box, and increase the effect of friction by increasing its "*lever arm.*" It may happen, however, that the particular object to be accomplished will sometimes make it inconvenient to satisfy this condition of keeping the action of the power in the plane of the wheel, in which event, it will be easy to find the pressure arising from the parallel component of the power or resistance, and to compute the friction by the rules already given. Supposing the power and resistance to act in planes at right angles to the axis, we remark, that the plane of the wheel in which the power acts, and the plane perpendicular to the axis, through the direction of the resistance, will cut from the arbor equal circles. Through the point E, at which the rope is tangent to the circle in the latter of these planes, and the axis, conceive a plane to be passed; it will cut the circle in the plane of the wheel on the opposite side of the arbor in E', and the line joining E and E' will intersect the axis in I, making $EI = E'I$. At the point E' apply two opposite forces Q_1 and Q_2, parallel and each equal to the resistance Q. These forces will produce no effect upon the system. The resultant of the two equal and parallel forces Q and Q_1 will be equal to their sum, will pass through I, will be resisted by the axis, and produce no work, except what may arise from the friction due to its action on the trunnion. The equi-

effect of an oblique application of the power;

process where power does not act in plane of wheel

when the power acts in the plane of the wheel;

construction;

Fig. 255.

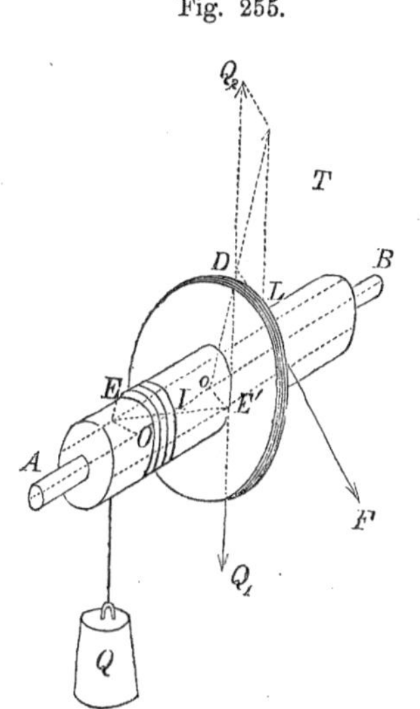

forces which maintain the equilibrium or motion uniform;

librium, if the machine be at rest, or its uniform motion, if at work, must, therefore, be maintained by the power F, the force Q_2, the friction, and the stiffness of cordage. To this end, the resultant of F, Q_2, and stiffness of cordage must intersect the axis. At the point of intersection, conceive this resultant to be replaced by its primitive components, and there will then act upon the axis the forces F, Q_2, $Q + Q_1$, and the resistance due to stiffness of cordage. Each of these forces being resolved into two parallel components acting on the trunnions A and B, there will result two groups of forces, one applied to each trunnion. Denote the resultant of the group acting on the trunnion A by M, that of the group acting on the trunnion B by M', then will the frictions be respectively fM and $f'M'$; and, employing the usual notation, the quantities of work will be $fMrs_{,}$ and $f'M'r's_{,,}$ the radii of the trunnions, and their friction being unequal.

pressure upon the trunnions;

friction on the trunnions and its work;

The quantity of work of the power F, must be equal to that of the resistance Q_2, augmented by the work of the stiffness of cordage and friction, and hence, denoting the radius of the wheel by R, and that of the arbor by R',

$$F \cdot Rs_{,} = Q_2 R' \cdot s_{,} + d_{,} \frac{K + I \, . \, Q}{2R'} R's_{,} + fMrs_{,} + f'M'r's_{,,};$$

but if the trunnions and boxes are supposed of the same size and material,

quantity of work;

$$FRs_{,} = Q_2 R's_{,} + d_{,} \frac{K + IQ}{2R'} R's_{,} + f(M + M')rs_{,}.$$

friction of trunnions, same effect wherever applied;

The quantity $M + M'$, being the sum of the pressures upon the trunnions, the last term shows that the friction is the same as though the resultant of all the forces were applied to a single trunnion in any arbitrary position, and, therefore, at the centre of the wheel. But this would reduce all the forces to the same plane, in which case Q would take the place of Q_2, and Q_1 and Q_2 would disappear from the system. Hence, denoting the resultant of the entire system of forces by N, and writing Q for its equal Q_2, the above equation becomes

$$FRs_{,} = QR's_{,} + d_{,}\frac{K+IQ}{2R'}R's_{,} + fN\cdot r\cdot s_{,}\ .\ .\ (153);$$ quantity of work;

and, dividing by $Rs_{,}$,

$$F = Q\frac{R'}{R} + d_{,}\frac{K+IQ}{2R'}\frac{R'}{R} + fN\cdot\frac{r}{R}\ .\ .\ (154).$$ value of the power;

Now, N being the resultant of all the forces of the system except friction, it is the resultant of F, Q, and $d_{,}\frac{K+IQ}{2R'}$; or, since Q and $d_{,}\frac{K+IQ}{2R'}$ act in the same direction, it is the resultant of F and $Q+d_{,}\frac{K+IQ}{2R'}$. To find N, we will pursue the method explained in § 231.

Make

$$Q + d_{,}\frac{K+IQ}{2R'} = Q_1\ .\ .\ (155);$$ find the resultant of all the forces but friction;

then will

$$F = Q_1\frac{R'}{R} + f\cdot N\cdot\frac{r}{R}\ .\ .\ (156).$$

If we neglect the consideration of friction for a moment, and find the resultant N_1 of F and Q_1, or of by approximation;

$$Q_1\frac{R'}{R}\quad\text{and}\quad Q_1,$$

we shall have, denoting the inclination of the power to the resistance by φ,

$$N_1 = \sqrt{Q_{,}^2 + Q_{,}^2\frac{R'^2}{R^2} + 2Q_{,}^2\frac{R'}{R}\cos\varphi} = Q_{,}\sqrt{1+\frac{R'}{R}\left(\frac{R'}{R}+2\cos\varphi\right)}\ .\ .\ (157);$$ first approximation for resultant;

and this for N, in Eq. (156), gives

first approximation for power;

$$F = Q_1 \frac{R'}{R} + f N_1 \frac{r}{R} = F_1 \quad . \quad . \quad . \quad (158).$$

the first approximation generally sufficient;

Now the value of N_1 was too small for N, because we omitted the term $fN \cdot \frac{r}{R}$, in the value for F; and, hence, F_1 is too small for F; but the deficiency is less and less, in proportion as the fraction $f\frac{r}{R}$ is smaller and smaller. In ordinary practice there will be but little difference between the true value of F and that given by Eq. (158).

when it is not, a second approximation must be made;

In cases wherein r is considerable in comparison with R, a further approximation will be necessary; and to accomplish this, we remark, that F_1 is greater than $Q_1 \frac{R'}{R}$, and Q_1 therefore less than $F_1 \frac{R}{R'}$; and that if this latter be combined with F_1, to obtain a second resultant N_2, this last will be too large, and when substituted in Eq. (156), for N, will give a value F_2 for F, which will also be too large. The mean of the two values of F_1 and F_2 will be the practical value of F.

geometrical indication;

Fig. 256.

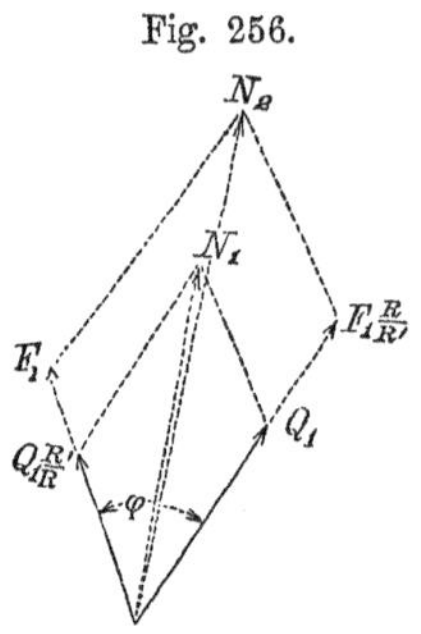

The value of N_2 is given by the equation,

second approximation for resultant;

$$N_2 = F_1 \sqrt{1 + \frac{R}{R'} \left(\frac{R}{R'} + 2 \cos \varphi \right)} \quad . \quad . \quad (159);$$

and

second approximation to value of the power;

$$F_2 = Q_1 \frac{R'}{R} + f N_2 \frac{r}{R};$$

whence

$$F = \frac{F_1 + F_2}{2} = Q_1 \frac{R'}{R} + f \frac{r}{R} \frac{N_1 + N_2}{2} \quad . \quad . \quad (160).$$ mean of the two approximations;

To find the quantity of work, multiply both members by $R s_{,}$ replace Q_1 by its value, and we have

$$F R s_{,} = Q R' s_{,} + d_{,} \frac{K + I Q}{2 R'} R' s_{,} + f r \cdot s_{,} \frac{N_1 + N_2}{2} \quad . \quad . \quad (161).$$ quantity of work;

Example. Required the quantity of work necessary to raise two tons of coal from the bottom to the top of a pit which is 80 feet deep, by means of the wheel and axle. The diameter of the wheel is 4 feet; that of the axle, 1 foot; that of the trunnion, which is of wrought iron, working in cast-iron boxes and lubricated with hogs' lard, 1.5 inches; that of the rope, which is white, half-worn, and dry, 1.5 inches; and the power acts in a horizontal direction.

Fig. 257.

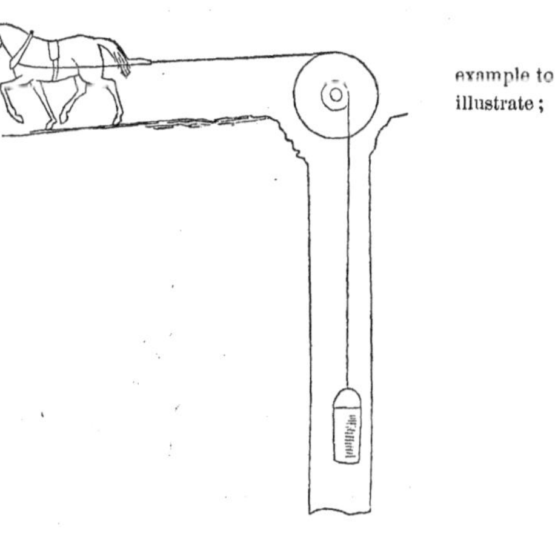

example to illustrate;

Here $R = 2$ feet; $R' = 0.5$ feet; $r = 0.125$ feet; data of the question and tables;
$f = 0.07$; $d_{,} = d^{\frac{3}{2}} = \left(\frac{1.5}{0.79}\right)^{\frac{3}{2}} = (1.9)^{\frac{3}{2}} = 2.619$;
$K = 1.13801$; $I = 0.0525889$; $Q = 4000$ lbs.;
$R' s_{,} = 80$ feet; $s_{,} = \frac{80}{0.5} = 160$ feet; and $\varphi = 90°$, or $\cos \varphi = 0$.

These data, substituted in Eq. (155), give

$$Q_1 = \overset{lbs.}{4000} + 2.619 \, . \, \frac{\overset{lbs.}{1.13801} + 0.0525889 \times \overset{lbs.}{4000}}{1} = 4553.89;$$

and this, in Eq. (157), making $\cos\varphi = 0$, and substituting for $\frac{R'}{R}$, its value $\frac{0.5}{2} = 0.25$ feet, we find

value of first resultant;

$$N_1 = 4553.89 \sqrt{1 + (0.25)^2} = 4694.04.$$

This and the values of $Q_{\prime\prime}$, $\frac{R'}{R}$, f, and $\frac{r}{R}$, in Eq. (158), give

$$F_1 = 4553.89 \times 0.25 + 0.07 \times 4694.04 \times 0.0625 = 1159.008;$$

which, substituted with the values of $\frac{R}{R'}$ and $\cos\varphi = 0$, in Eq. (159), gives

value of second resultant;

$$N_2 = 1159.008 \sqrt{1 + (4)^2} = 4778.68;$$

hence,

$$N = \frac{N_1 + N_2}{2} = \frac{4694.04 + 4778.68}{2} = \overset{lbs.}{4736.36};$$

which, with the values already found for Q_1, in Eq. (160), gives

value of power;

$$F = \overset{lbs.}{4553.89} \times 0.25 + 0.07 \times \frac{0.125}{2} \overset{lbs.}{4736.36} = 1159.19.$$

Here it may be proper to direct the attention to the slight difference between the values of F and F_1, showing that the first approximation, as given by Eq. (158), will generally be sufficient.

Finally, from Eq. (161), we obtain

$$F R s_{,} = \overset{lbs.}{4000} \times \overset{ft.}{80} + 44311.20 + 663.07 = 364974.27.$$ quantity of work;

The first term of the second member = 320000, is the value of the work without any resistance from friction and stiffness of cordage; the sum of the remaining terms = 44974.27, is the work of friction and stiffness of rope; hence it appears, that the loss arising from the latter causes, is nearly one seventh of the work which, without them, would be required to accomplish the object. This loss would be sufficient, without the hinderance from friction and stiffness of cordage, to raise more than a quarter of a ton through the given height. loss of work by friction and stiffness of cordage.

If, in Eq. (154), we make $f = 0$, and disregard the stiffness of cordage, we find

$$F = Q \cdot \frac{R'}{R} \quad . \quad . \quad . \quad . \quad (162);$$

that is to say, in the wheel and axle, *the power is to the resistance as the radius of the axle is to that of the wheel.*

§ 236.—Wheels are often so combined in machinery as to transmit the motion impressed upon some one of them, according to certain conditions, determined by the object to be accomplished. This is usually done by one or other of the following means, viz.: 1st. By *endless ropes, bands,* or *chains,* passing around cylindrical rollers, called *drums,* mounted upon arbors; 2d. By the natural contact of these drums; 3d. By projections called teeth or leaves, according as these projections are upon the surfaces of wheels or arbors. The communication of motion by these means is always accompanied by friction, which it is important in practice to know, since it may not be disregarded. Combination of wheels; motion transmitted by endless bands, ropes, and chains; by natural contact; and by teeth.

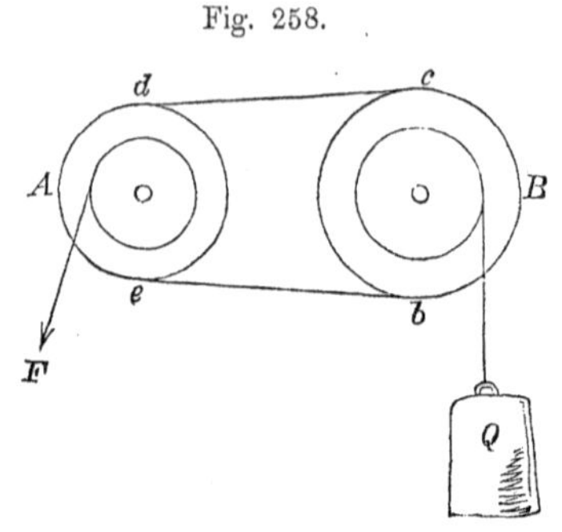

Fig. 258.

Resistance due to stiffness of bands and ropes;

§ 237.—When two wheels are connected with each other by means of an endless band or rope $d\ c\ b\ e$, passing around the drums A and B, mounted upon the arbors of the wheels, a sufficient force F applied to one of them will put it in motion; this motion will be communicated to the other as long as the friction between the band and drums is sufficient to prevent the former from sliding over the latter, and thus a resistance Q, applied to the second wheel, may be overcome. The motion of the drum B is obviously due to the difference of the tensions in the two branches $d\,c$ and $e\,b$; and applying the power as indicated in the figure, the tension of $d\,c$ must be greater than that of $e\,b$. Denoting the first of these by T, and the latter by t, the force which moves the drum B will have an intensity equal to $T - t$; and the quantity of its work must be equal to that of Q, increased by the work of friction on the trunnions of the common arbor. Denote the radius of the drum B by R_2; that of the wheel to which Q is applied by R''; that of its trunnion by r_2; the arc described by the point at the unit's distance from the axis of motion by s_2, &c., then will

friction between the bands and drum;

motion due to difference of tension;

work of difference of tension;

$$(T - t)\,R_2\,s_2 = Q\,R''\,s_2 + f N_2\,r_2\,s_2 \quad . \quad . \quad (163).$$

The action of the force F produces the difference of tension $T - t$, and its work must, therefore, be equal to that of $T - t$ augmented by the work of friction on the trunnions of the arbor of the wheel to which F is applied. Denote the radius of this wheel by R', that of its drum by R_1, that of its trunnion by r_1, the arc described at the unit's distance by s_1, and we have

$$F R' s_1 = (T - t) R_1 s_1 + f N_1 . r_1 s_1 \quad . . (164).$$ work of the power;

Adding these equations together, we get

$$F R' s_1 + (T - t) R_2 s_2 = (T - t) R_1 s_1 + Q R'' s_2 + f N_2 r_2 s_2 + f N_1 r_1 s_1;$$

but because all parts of the band have the same velocity, the circumferences of the drums must move at the same rate; hence

$$R_2 s_2 = R_1 s_1;$$ circumferences of the drum have the same velocity;

which will reduce the above equation to

$$F R' s_1 = Q R'' s_2 + f N_2 r_2 s_2 + f N_1 r_1 s_1 \quad . . (165).$$ work of the power;

Whence we see that the work of F is equal to the work of Q, increased by that of the friction upon the two sets of trunnions; and the same may be shown of any number of wheels thus connected. inferences;

In this equation, N_2 is the resultant of the forces Q, T, and t; and N_1 of F, T, and t. To find these resultants it will be necessary to know T and t.

The difference $T - t$ only exists while the system is in motion; when at rest, and the force does not act, this difference is zero, or T is equal to t. In passing from rest to motion, we may assume that one increases just as much as the other diminishes, and if the common tension at rest be represented by T_1, and the increment of the one and decrement of the other in passing from rest to motion be denoted by H, then will

$$T = T_1 + H, \quad \text{and} \quad t = T_1 - H \quad . . (166);$$ value of the tensions;

from which T and t may be found when T_1 and H are known

tension at rest arbitrary; The tension T_1 is entirely arbitrary. It should be as small as possible, to produce the requisite friction between the band and the drums to avoid sliding during the motion, for if greater than this, it will only increase the pressure and, therefore, the friction on the trunnions, unnecessarily. In general, it will be sufficient if this friction

should be just sufficient to prevent sliding; be great enough to prevent sliding under the effect of Q, at the surface of the drum of the wheel to which Q is applied. But this effect, neglecting friction on the trunnions and stiffness of cordage, is $Q\,\frac{R''}{R_2}$. That is to say, a force whose intensity is given by this expression, when applied to the surface of the drum, will produce the same effect as Q; and the friction between the drum and strap must be at least equal to this force to prevent sliding. The branches dc and eb of the band are solicited respectively by the two forces $T_1 + H$, and $T_1 - H$; and these substituted in Eq. (108), the first for F and the second for W, we find,

relation of the two tensions;

$$T_1 + H = (T_1 - H)\, e^{\frac{fs}{R_2}};$$

subtracting $T_1 - H$ from both members of this equation, and we have

$$T_1 + H - (T_1 - H) = (T_1 - H)\, e^{\frac{fs}{R_2}} - (T_1 - H);$$

the first member reduces to $2H$; that is to say, to the difference of tensions on the two branches of the band, which must be equal to the effect of Q at the surface of the drum; whence

difference of tensions;

$$2H = Q \cdot \frac{R''}{R_2} \quad . \quad . \quad . \quad . \quad (167),$$

$$Q\frac{R''}{R_2} = (T_1 - H)\left(e^{\frac{fs}{R_2}} - 1\right) \quad . \quad (168);$$ same in terms of the friction, &c.; tension at rest;

from which two equations we may compute H and T_1, and therefore, Eq. (166), T and t; and, finally, the resultants N_2 and N_1 by the rules for the composition of forces.

Example. Required the tension of a band necessary to produce friction enough to move a wheel, when subjected to a resistance of 1000 pounds, the radius of the wheel being 0.5 foot, and that of the drum 2 feet, and the arc of the drum enveloped by the band 180°. Let the band be of black leather, and the surface of the drum of oak. example;

Here $R_2 = 2$ feet; $R'' = 0.5$ feet; $Q = 1000$ lbs.; $f = 0.265$, (see Table I, § 212;) $s = \pi R_2 = 3.1416\, R_2$;

$$Q \cdot \frac{R''}{R_2} = 1000 \times \frac{0.5}{2} = 250^{lbs.};$$

$$H = \tfrac{1}{2} Q \cdot \frac{R''}{R_2} = 125^{lbs.};$$ half difference of tensions;

$$T_1 - H = \frac{Q \cdot \frac{R''}{R_2}}{\left(e^{\frac{fs}{R_2}} - 1\right)} = \frac{250}{(2.7182818)^{0.265 \times 3.1416} - 1}.$$ lesser tension;

The first term of the denominator may be easily found by the aid of logarithms, as follows:

$$\begin{aligned} \text{Log}\left[(2.7182818)^{0.83251}\right] &= \text{Log } 2.718281 \times 0.83251 \\ &= 0.4342942 \times 0.83251 \\ &= 0.361554 \text{ nearly}; \end{aligned}$$ value found by the aid of logarithms;

the natural number of which is 2.2991, whence

$$T_1 - H = t = \frac{250}{2.2991 - 1} = \frac{250}{1.2991} = 192.44^{lbs.}$$ value;

Adding $2H = 250$ lbs., we have

greater tension;

$$T_1 + H = T = \overset{lbs.}{442.44}.$$

The arc of the drum enveloped by the band being 180°, the tensions T and t must be parallel, and their resultant $T_2 = T + t = 634.88$ lbs., which being combined with $Q = 1000$, according to the principles of the composition of forces, will give N_2, and with F will give N_1, whence every thing required to determine the quantity of work in Eq. (165) is known.

to find the resultant;

If Eq. (165) be divided by $R's_{,,}$ it becomes

value of power;

$$F = Q \cdot \frac{R''}{R'} \cdot \frac{s_2}{s_1} + f \cdot N_2 \cdot \frac{r_2}{R'} \cdot \frac{s_2}{s_1} + f N_1 \cdot \frac{r_1}{R'};$$

but

velocity of the circumferences equal;

$$R_2 s_2 = R_1 s_1;$$

whence

$$\frac{s_2}{s_1} = \frac{R_1}{R_2};$$

and by substituting above,

final value of power when the motion begins in its direction;

$$F = Q \cdot \frac{R'' \,.\, R_1}{R' \,.\, R_2} + f \cdot N_2 \cdot \frac{r_2}{R'} \cdot \frac{R_1}{R_2} + f \cdot N_1 \cdot \frac{r_1}{R'};$$

which is the relation subsisting between F and Q, in case of an equilibrium bordering on motion in the direction of F, or in the direction of uniform motion.

If we disregard the friction, then will

without friction;

$$F = Q \cdot \frac{R'' \,.\, R_1}{R' \,.\, R_2}.$$

When the motion from one wheel and axle is communicated to a second machine of the same kind, by passing the band about the axle of the wheel to which the power F is applied, and the wheel of that to whose axle Q is applied, then will R_1 be the radius of the first axle, and R_2 that of the second wheel, and the preceding equation gives us this rule, viz.: combination of wheels and axles without friction;

Fig. 259.

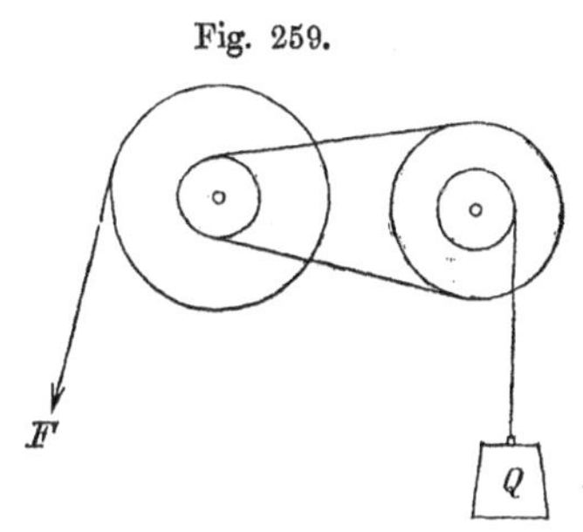

When the friction is so small that it may be disregarded, the power F will be to the resistance Q, as the product of the radii of the axles to that of the radii of the wheels, in the case of an equilibrium or uniform motion. relation of power to the resistance.

§ 238.—In the preceding discussion, no mention is made of the resistance arising from the stiffness of cordage. When the connection or gearing is made by bands, these are so thin as to possess considerable flexibility, and their opposition to bending may, in practice, be safely neglected. If the connection be made by an endless rope, the opposition to motion takes place at the points where the rope bends in passing on to the drums, and not at all at the points where it leaves the latter Rigidity of bands may be neglected; rigidity of ropes;

Fig. 260.

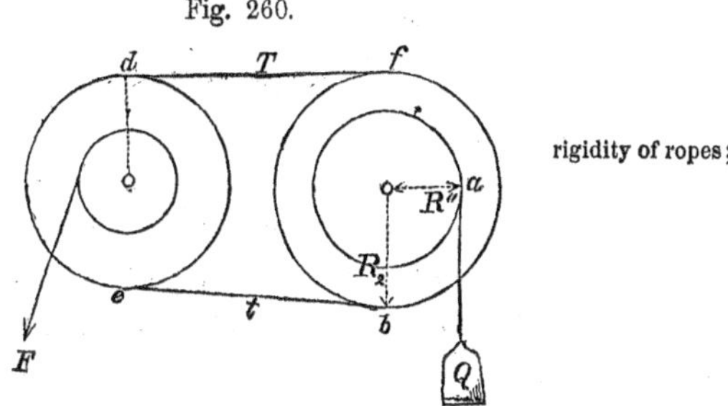

and becomes straight. Thus at the point a, the resistance is

value of the resistance at one point;

$$\frac{K + I \,.\, Q}{2\,R''},$$

at the point b it is

at another;

$$\frac{K' + It}{2\,R_2},$$

and at the point d it is

at another;

$$\frac{K' + IT}{2\,R_1},$$

and, finally, at the points f and e it is nothing. These resistances must be included among those to be overcome by the power F.

rigidity of chains;

If the connection be made by an endless chain, each link, as it turns in the next one in order, may be regarded

Fig. 261.

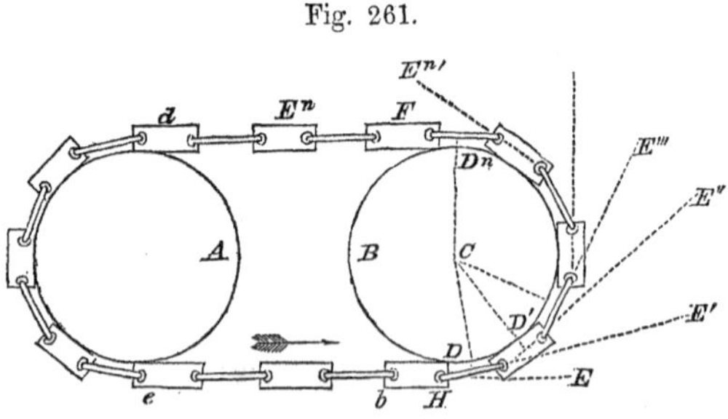

each link a trunnion in its box;

as a trunnion revolving in its box; and each, as it comes to be applied to the drum, revolves about the next one through an angle $E'HE$, equal to DCD', the angle through which the drum revolves to produce the contact; and taking the sum of all these angles, it is obvious that, although each link revolves through a very small angle, yet this sum must be equal to the angle through which the drum has turned to produce it.

Denoting by r the radius of the inner circular arc in which the end of each link is shaped, s_2 the arc described by the point at the distance of unity from the axis of the drum B, f the coefficient of friction, and T and t the tensions on the two branches of the chain, then will the work of friction among the links at the points b and f respectively, (figure before the last,) be notation, &c.;

$$f\,T\,r\,s_2, \quad \text{and} \quad f\,t\,r\,s_2;$$

work of friction among the links, at one set of points;

and denoting by s_1 the arc described by a point at the distance of unity from the axis of the drum A, the work of friction at the points e and d will be, respectively,

$$f\,T\,r\,s_1, \quad \text{and} \quad f\,t\,r\,s_1;$$

the same for another set;

and the whole amount of this kind of work will be

$$f\,r\,(T + t)\,(s_2 + s_1).$$

whole work of this friction;

Recollecting that the points on the surfaces of the drums must have the same velocity, viz.: that of the different links of the chain, we have

$$R_2\,s_2 = R_1\,s_1;$$

velocity of circumference of drums equal;

in which R_2 and R_1 are respectively the radii of the drums B and A. From this relation we find

$$s_1 = \frac{R_2}{R_1}\,s_2;$$

which, substituted above, gives

$$f\,r\,s_2\,(T + t)\left(1 + \frac{R_2}{R_1}\right) \quad . \quad . \quad . \quad (169).$$

value of the work of friction among the links;

xample;

Example. Let T and t have the values of the last example, (that of the strap,) and suppose $r = 0.03$, the chain of wrought iron, for which we find in the table of § 225, (assuming that f is the same for trunnions of wrought iron in boxes of the same material, as for trunnions of wrought iron and boxes of cast iron,) $f = 0.07$; also let the radius of the drum B be four times that of the drum A; then will the expression (169) for a single revolution of the drum B, in which case $s_2 = 2 \times 3.1416 = 6.2832$, become

data;

quantity of work of friction.

$$0.07 \times 0.03 \times 6.2832 \, (\overset{lbs.}{442.44} + \overset{lbs.}{192.44}) \, (1 + 4) = 41.88;$$

that is, the work lost in consequence of the friction among the links of the chain, during one revolution of the drum of the wheel to which the resistance is applied, is sufficient to raise a weight of nearly 42 pounds through one foot of vertical height.

Resistance from friction on the teeth of wheels;

§ 239.—Let us now suppose the circumferences of the wheels to be furnished with teeth, which interlock with each other, so that a force being impressed upon one wheel, it cannot move without communicating motion to the other.

Fig. 262.

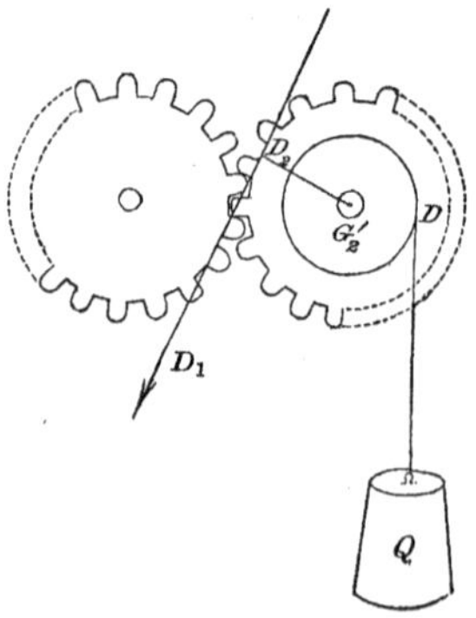

conditions of construction of the teeth;

The teeth are usually curved, and so shaped as to have a common normal $D_1 \, D_2$, at their point of contact m, where the action of one and the reaction of the other take place; and although the point of contact alters its position, as the wheels rotate, yet the place of this normal does not change, but remains stationary, and the point of contact is always on it. We will not stop to explain the constructions by which this is accomplished; it will be sufficient for our present purpose to be assured of its practicability, and that we may proceed on the supposition that it has been executed in the case under consideration.

Fig. 263.

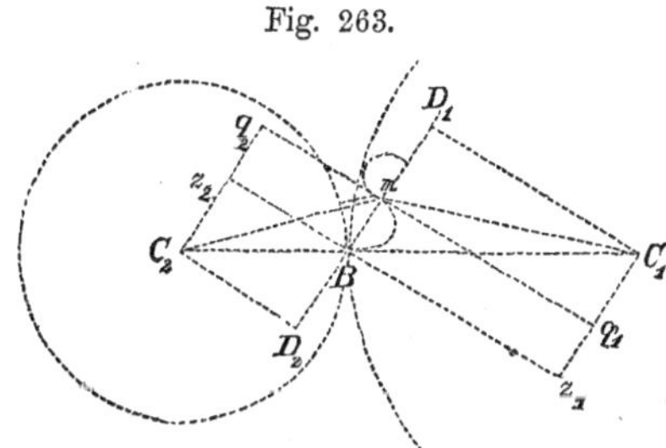

conditions of preserving a constant normal at the point of contact;

From the centres C_1 and C_2 of the wheels, let fall upon the normal $D_1\,D_2$, the perpendiculars $C_1\,D_1$ and $C_2\,D_2$. The points D_1 and D_2 must, during the rotation of the wheels, have the same absolute velocity, and therefore the number of revolutions of the wheel whose centre is C_1, in a given time, must be to that of the wheel whose centre is C_2, in the same time, inversely as the perpendiculars $C_1\,D_1$, and $C_2\,D_2$; or, because of the similar triangles $C_1\,B\,D_1$ and $C_2\,B\,D_2$, inversely as the distances $C_1\,B$ and $C_2\,B$. The circles described about C_1 and C_2 as centres, with radii $C_1\,B$ and $C_2\,B$, respectively, are called the *primitive* circles. These circles and their radii may be easily found from the consideration just named. It will be our object to find a force which, applied tangentially to these circles at B, will produce the same effect as friction on the teeth.

relative velocities of the two wheels;

primitive circles;

Denote by Q the resistance acting at the distance R from the axis of the wheel whose centre is C_2. The effect of this resistance acting at D_2, in the direction of the normal $D_1\,D_2$, will, from the principles of the wheel and axle, be Q_1, given by the relation

$$Q_1 = Q\,\frac{R}{C_2\,D_2} \quad . \quad . \quad . \quad . \quad (169)';$$

effect of the resistance at the distance of common normal;

and this Q_1 will be the pressure at the point m. Its friction will be

value of the friction on the teeth;

$$f\ Q_1,$$

acting in the direction $q_1 q_2$, tangent to both teeth at their point of contact. The elementary quantity of work of this friction will be equal to its intensity, multiplied into the elementary distance by which the rubbing points now at m, separate in the direction of this tangent; which distance is obviously equal to that by which the points q_1 and q_2, the extremities of the perpendiculars let fall from C_1 and C_2 upon the common tangent, approach to or recede from each other. Denoting the elementary path described by a point at the unit's distance from C_1 by s_1, and that described at the same distance from C_2 by s_2, the paths described by q_1 and q_2 will be, respectively, $C_1 q_1 \times s_1$, and $C_2 q_2 \times s_2$; and because the points q_1 and q_2 must move in the same direction when the tangent $q_2 q_1$ passes between the centres C_1 and C_2, the elementary path of friction will be equal to the difference of these paths, and its elementary quantity of work will equal

to obtain the quantity of work of this friction;

Fig. 263.

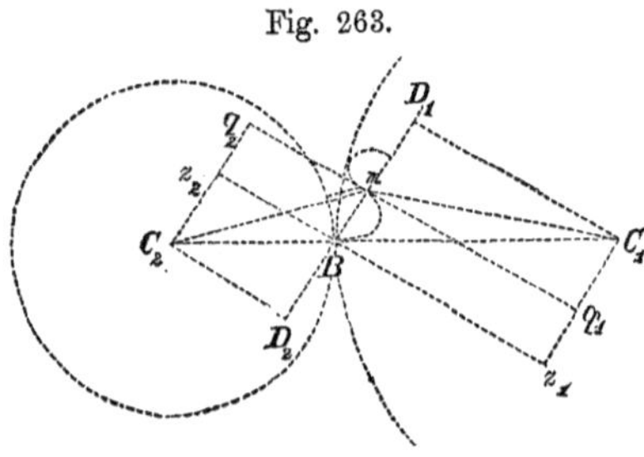

the value of this work;

$$f\ Q_1 \left[C_2 q_2 \times s_2 - C_1 q_1 \times s_1 \right].$$

Designating the radii of the primitive circles whose centres are C_1 and C_2 by R_1 and R_2, respectively, we have, because of the equal velocities of the circumferences of these circles,

$$R_1 s_1 = R_2 s_2;$$

whence

relation of the paths at unit's distance from the two centres;

$$s_2 = \frac{R_1}{R_2} s_1.$$

Moreover, drawing through the point B the line $z_2 z_1$, parallel to the tangent $q_2 q_1$, and denoting the angle $m B C_1$, which is the complement of the angle $C_1 B z_1$, by φ, and the distance $m B$ by h, we find

$$C_2 q_2 = R_2 \cos \varphi + h,$$

lever arms of the friction;

$$C_1 q_1 = R_1 \cos \varphi - h;$$

these values of s_2, $C_2 q_2$, and $C_1 q_1$, substituted in the expression for the elementary work of friction, give

$$f Q_1 h \; s_1 \left(\frac{R_1}{R_2} + 1 \right).$$

work of friction;

Denote by ω the intensity of a force which, applied tangentially to the primitive circles at B, will produce the same effect as the friction. Its elementary work will be $\omega R_1 s_1$, and hence

$$\omega \cdot R_1 s_1 = f Q_1 h \; s_1 \left(\frac{R_1 + R_2}{R_2} \right);$$

or

$$\omega = f \; Q_1 h \; \frac{R_1 + R_2}{R_1 R_2} \quad . \quad . \quad (170).$$

tangential force at the circumference of primitive circle;

Represent the angle $B C_1 m$ by θ. In practice, the angle $m B C_1$ does not differ much from 90°, and we may take

$$h = R_1 \tan \theta;$$

and because θ is generally very small, the tangent may be replaced by the arc, and

$$h = R_1 \theta;$$

which, substituted above, gives

another form for tangential force at primitive circumference;

$$\omega = f Q_1 \frac{R_1 + R_2}{R_1 R_2} \times R_1 \theta \quad . \quad . \quad (171).$$

The value of θ varies from a maximum to zero on one side of the line of the centres $C_1 C_2$, and from zero to a second maximum on the opposite side of this line; the first maximum corresponds to that position of m in which any two teeth come first in contact, and the second to that in which the contact ceases; the intermediate or zero value occurs when m is on the line of the centres. The quantity θ being thus variable, it must be replaced by a constant, and this constant must be a mean of all the values between the two maxima. Designating the first of these by θ_1, and the second by θ_2, lay off the distance $AE = \theta_1$; erect at A the perpendicular $AG = \theta_1$; draw GE: then will the ordinates of this line which are parallel to AG represent the different values of θ, and the area of the triangle EAG will be the sum of all the values of θ between θ_1 and zero. Again, make $EB = \theta_2$; erect at B the perpendicular $BG''' = \theta_2$; draw $G'''E$: the area of the triangle EBG''' will be the sum of all values of θ between zero and θ_2. Make

to find the mean value of the angular distance of point of contact;

Fig. 264.

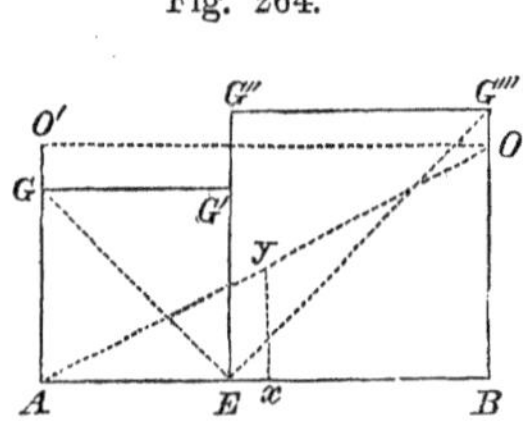

altitude of a mean triangle;

$$BO = \frac{\theta_1^2 + \theta_2^2}{\theta_1 + \theta_2},$$

complete the rectangle $B O'$, and draw $A O$; then will the triangle $A B O$ be equivalent to the sum of the triangles $A E G$ and $E B G'''$, and therefore equivalent to the sum of all values of θ between θ_1 and θ_2, the mean of which is obviously the middle ordinate. construction;

$$xy = \tfrac{1}{2} B O = \frac{\theta_1^2 + \theta_2^2}{2(\theta_1 + \theta_2)} = \frac{(\theta_1 + \theta_2)^2 - 2\,\theta_1\,\theta_2}{2\,(\theta_1 + \theta_2)} = \frac{\theta_1 + \theta_2}{2} - \frac{\theta_1\,\theta_2}{\theta_1 + \theta_2}.$$

mean value of angular distance of point of contact;

Neglecting the last term as insignificant,

$$x\,y = \frac{\theta_1 + \theta_2}{2}$$

Multiplying by R_1, we find that $R_1\,(\theta_1 + \theta_2)$ is the interval between the place of the first and last point of contact of the same pair of teeth, estimated on the circumference of the primitive circle; denoting this interval by a, and substituting in Eq. (171), we find

$$\omega = f\,Q_1\,\frac{R_1 + R_2}{R_1\,R_2} \times \frac{a}{2} = f\,Q_1\left(\frac{a}{2\,R_2} + \frac{a}{2\,R_1}\right).$$

tangential force at primitive circumference;

Denote the number of teeth on the wheel whose centre is C_1 by n_1, and the number on the wheel whose centre is C_2 by n_2; then, because the teeth and intervals between them must be the same on each circumference, in order to work freely,

$$a = \frac{2\,\pi\,R_1}{n_1} = \frac{2\,\pi\,R_2}{n_2};$$

distance from the place of first to that of last point of contact;

which, substituted above, gives

$$\omega = f\,.\,Q_1\left(\frac{1}{n_1} + \frac{1}{n_2}\right)\pi = f\,.\,\pi\,.\,Q_1\,\frac{n_2 + n_1}{n_1\,n_2}.$$

Replacing Q_1 by its value given in Eq. (169)′, and recollecting that, within the limits supposed, $C_2 D_2$ becomes R_2, we finally have

final value of tangential force which is equivalent to friction;

$$\omega = f\pi \cdot Q \cdot \frac{R}{R_2}\left(\frac{n_2 + n_1}{n_1 n_2}\right) \quad . \quad . \quad (172).$$

To find the quantity of work, multiply both members of this equation by $R_2 s_2$, which will give

its quantity of work;

$$\omega R_2 s_2 = f\pi\, s_2\, Q\, R \cdot \frac{n_2 + n_1}{n_2 n_1} \quad . \quad . \quad (173).$$

example;

Example. Required the work consumed in each revolution by friction on the teeth of a wheel whose arbor is subjected to a resistance equivalent to 1000 pounds, the number of teeth on the wheel being 64, and that of the connecting wheel being 192; let the teeth be of *cast iron*, and suppose the radius of the arbor equal to 0.8 foot.

data;

Here, $R = 0.8$; $Q = 1000$ lbs.; $s_2 = 2 \times 3.1416$; $\pi = 3.1416$; $f = 0.152$; $n_2 = 64$; $n_1 = 192$; and, therefore,

work;

$$\omega R_2 s_2 = 0.152 \times 3.1416 \times 6.2832 \times \overset{lbs.}{1000} \times 0.8\, \frac{64 + 192}{64 \times 192} = 50;$$

result.

that is to say, the quantity of work consumed in one revolution by friction on the teeth, in the case supposed, is sufficient to raise 50 pounds through a vertical distance of one foot.

XX.

THE SCREW.

The screw.

The *Screw*, regarded as a mechanical power, is a device by which the principles of the inclined plane are so applied as to produce considerable pressures with great steadiness and regularity of motion.

§ 240.—To form a clear idea of the figure of the screw and its mode of action, conceive a right cylinder $a\,k$, with circular base, and a rectangle $a\,b\,c\,m$ having one of its sides $a\,b$ coincident with a surface element, while its plane passes through the axis of this cylinder. Next, suppose the plane of the rectangle to rotate uniformly about the axis, and the rectangle itself to move also uniformly in the direction of that line; and let this twofold motion of rotation and of translation be so regulated, that in one entire revolution of the plane, the rectangle shall progress in the direction of the axis over a distance greater than the side $a\,b$, which is in the surface of the cylinder. The rectangle will thus generate a projecting and winding solid called a *fillet*, leaving between its turns a similarly shaped groove called the *channel*. Each point as m in the perimeter of the moving rectangle, will generate a curve called a *helix*, and it is obvious, from what has been said, that every helix will enjoy this property, viz.: any one of its points as m, being taken as an origin of reference, as well for the curve itself as for its projection on a plane through this point and at right angles to the axis, the distances $d'\,m'$, $d''\,m''$, &c., of the several points of the helix from this plane, are respectively proportional to the circular arcs $m\,d'$, $m\,d''$, &c., into which the portions $m\,m'$, $m\,m''$, &c. of the helix, between the origin and these points, are projected.

Screw with square fillet;

Fig. 265.

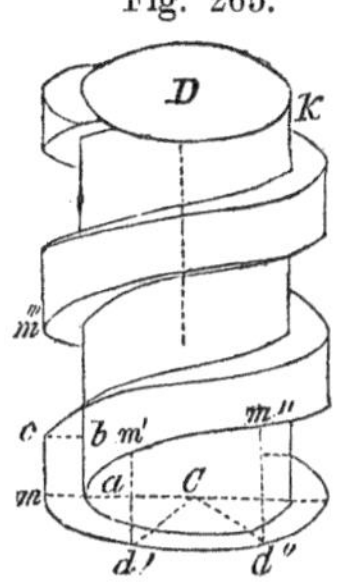

mode of generating;

the fillet, channel, and the helix;

properties of a helix;

The solid cylinder about which the fillet is wound, is called the *newel* of the screw; the distance $m\,m'''$, between the consecutive turns of the same helix, estimated in the direction of the axis, is called the *helical interval*. The surfaces of the fillet which are generated by the sides of the rectangle perpendicular to the axis, are each made up

newel;

helical interval;

of a series of helices, all of which have the same *interval*, though the helices themselves are at different distances from the axis. The inclination of the different helices to the axis of the screw, increases, therefore, from the newel to the exterior surface of the fillet, the same helix preserving its inclination unchanged throughout.

relative position and inclinations of the different helices;

The screw is received into a hole in a solid piece B of metal or wood, called a *nut* or *burr*. The surface of the hole through the nut is furnished with a winding fillet of the same shape and size as the channel of the screw, which it occupies; while the fillet of the latter fills up the channel of the nut, formed by the turns of its fillet, whose inner surface is thus brought in contact with the newel.

the nut;

fillet of the nut;

Fig. 266.

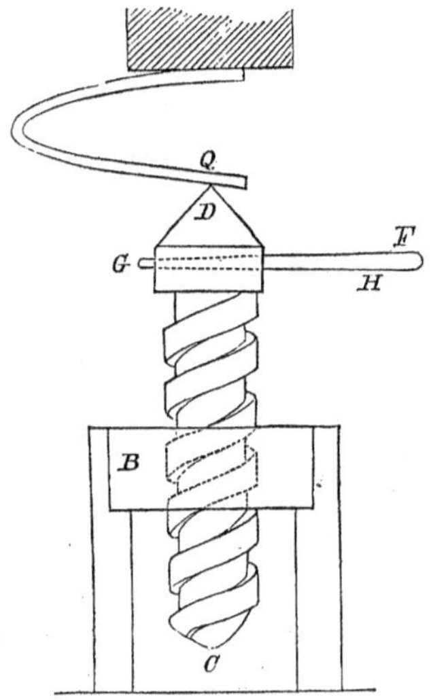

From this arrangement it is obvious that when the nut is stationary, and a rotary motion is communicated to the screw, the latter will move in the direction of its axis; also when the screw is stationary and the nut is turned, the nut must move in the direction of the length of the screw. In the first case, one entire revolution of the screw will carry it longitudinally through a distance equal to the helical interval, and any fractional portion of an entire revolution will carry it through a proportional distance; the same of the nut, when the latter is moveable and the screw stationary.

relative motion of screw and nut;

The resistance Q is applied either to the head of the screw, or to the nut, depending upon which is the moveable element; in either case it acts in the direction $D\,C$ of the axis. The power F is applied at the extremity of a bar $G\,H$ connected with the screw or nut, and acts in a plane at right angles to the axis of the screw. Denote the perpendicular distance of the line of direction of F from the axis of the screw by R, and the helical interval by h; then will the quantity of work of the power F, in one revolution, supposing it to retain the same distance from the axis, be

application of the resistance and power;

$$F \times 2\,\pi\,R;$$

work of the power in one revolution;

and the quantity of work of the resistance will be

$$Q \times h.$$

work of the resistance;

The power F and resistance Q, both act to press the fillet of the screw and that of the nut together, the first acting at right angles to, and the latter in the direction of the axis. To find the work of friction thence arising, it will be necessary to find a force F_1, parallel to F, whose effect at the fillet is the same as that of F, acting at the distance R from the axis, and to resolve both F_1 and Q into two components, one normal and the other parallel to the common surface of the pressing fillets. But the surfaces being warped, the normals at their different points will be oblique to each other, and so inclined to the axis that the normal components of the resistance Q, near the newel, will be less than those towards the outer surface of the fillet, while the reverse will be the case with the power F_1. The resolution must, therefore, be made with reference to a normal at a helix midway between the newel and outer surface. This helix, like all others, is situated upon the surface of a cylinder of which the axis coincides with that of the screw. Denote the radius of this cylinder $C\,m^{\text{IV}}$ by r

intermediate helix;

construction;

Conceive a tangent plane to this cylinder at any point, as m^{IV}, and two cutting-planes normal to the axis, and at a

Fig. 267.

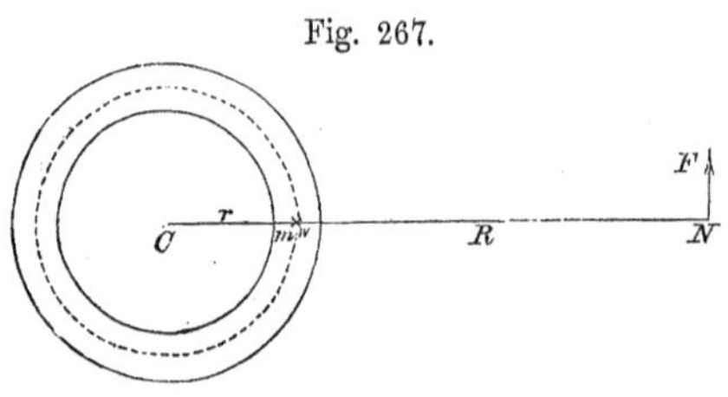

projection of intermediate helix;

distance from each other equal to a helical interval, and equally distant from m^{IV}. If we now develop the portion of the cylindrical surface, included between the cutting-planes, on the tangent plane, the surface of the cylinder will become a rectangle whose base $A\,E$ is equal to $2\,\pi\,r$, and whose altitude $E\,B$ is equal to h; and the helix will become the diagonal $A\,B$. Denote the length of the helix $A\,B$ by l. Then

Fig. 268.

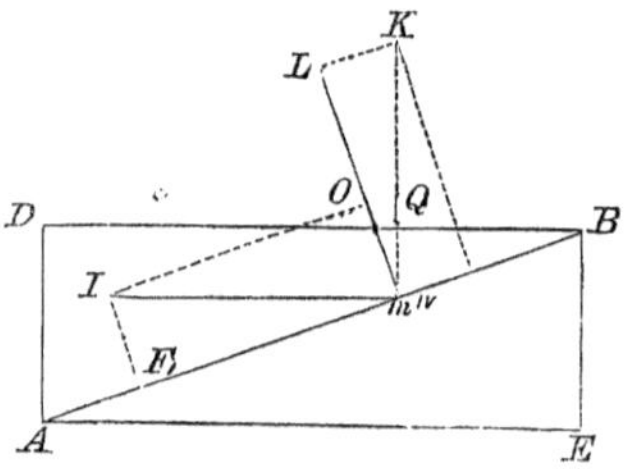

development of the intermediate helix;

resolution of the power and resistance into components;

draw the normal $m^{IV}\,L$, and resolve Q and F_1 as before stated. Since $Q = m^{IV}\,K$ is perpendicular to $A\,E$, and $L\,m^{IV}$ perpendicular to $A\,B$, the angles $L\,m^{IV}\,K$ and $E\,A\,B$ are equal; also, since $F_1 = I\,m^{IV}$ is perpendicular to $B\,E$, the angles $I\,m^{IV}\,L$ and $A\,B\,E$ are equal, and the triangles $A\,B\,E$, $I\,m^{IV}\,O$, and $L\,m^{IV}\,K$, being right angled, are similar, and give the proportions

$$l \;:\; 2\,\pi\,r \;::\; Q \;:\; L\,m^{IV},$$

$$l \;:\; h \;::\; F_1 \;:\; m^{IV}\,O;$$

whence

$$L\,m^{\text{IV}} = \frac{2\,\pi\,r\,Q}{l},$$

normal component of resistance;

$$m^{\text{IV}}\,O = \frac{h\,.\,F_1}{l};$$

normal component of power;

and the total pressure, which is equal to the sum of $m^{\text{IV}}\,O$ and $m^{\text{IV}}\,L$, becomes

$$\frac{2\,\pi\,r\,Q}{l} + \frac{h\,F_1}{l},$$

total normal pressure;

and the friction

$$f\,\frac{(2\,\pi\,r\,Q\, +\, h\,F_1)}{l};$$

friction;

and since in one revolution the path described by this friction is the diagonal $A\,B = l$, its quantity of work will be

$$f(2\,\pi\,r\,Q\, +\, h\,F_1);$$

its quantity of work in one revolution;

and because the work of the power F must equal the work of the resistance Q, increased by that of the friction, we have

$$2\,\pi\,R\,.\,F = Q\,h + f(2\,\pi\,r\,Q\, +\, F_1\,h).$$

work of power equal that of resistance increased by work of friction;

But the effect of F and F_1 being the same, their quantities of work must be equal, and hence

$$2\,\pi\,R\,F = 2\,\pi\,r\,F_1;$$

whence

$$F_1 = F\frac{R}{r};$$

which substituted in the preceding general equation, we get

work of power;

$$2\pi RF = Qh + f(2\pi r Q + F\frac{R}{r}h);$$

and finding the value of F,

value of the power;

$$F = Q\,\frac{hr + 2f\pi r^2}{2\pi Rr - fRh} \quad . \quad . \quad . \quad (174).$$

Multiplying both members by $2\pi R$; then adding and subtracting Qh, in the second member of this equation, we find

work of power;

$$2\pi R . F = Qh + fQ\frac{h^2 + 4\pi^2 r^2}{2\pi r - fh} \quad . \quad . \quad (175),$$

in which the work absorbed by friction is given by the last term; that is to say, by

work absorbed by friction;

$$fQ \cdot \frac{h^2 + 4\pi^2 r^2}{2\pi r - fh}.$$

If we neglect the consideration of friction, or make $f = 0$, we find, from Eq. (174), simply

relation of power and resistance without friction;

$$F = Q \times \frac{h}{2\pi R};$$

that is, the power is to the resistance as the helical interval is to the circumference described by the extremity of the perpendicular, drawn from the axis to the direction of the power. From which it is obvious that the power of the screw may be increased, either by diminishing the distance between the thread or fillet, or by increasing the distance of the power from the axis.

stated in words;

If we examine the expression

$$f\ Q \frac{h^2 + 4\pi^2 r^2}{2\pi r - fh},$$

we shall find that the numerator of the fractional factor increases more rapidly than the denominator for any increment in the value of r, the radius of the mean helix. For this reason, r should be made as small as possible consistently with sufficient strength.

radius of intermediate helix should be small;

Let $b\,O$ be the radius of the interior helix, or that of the newel, and $a\,O$ that of the exterior helix; it is usual to make the projection $a\,b$, of the fillet, equal to the thickness $a\,d$, measured in the direction of the axis; and for facility of execution, the dimensions of the channel are made equal to those of the fillet, that is to say, $a\,b$ is made equal to $a\,d$; in which case, the helical interval $a\,a'$ will be equal to $2\,a\,d = 2\,a\,b$, when there is but a single fillet. Should there be two fillets, which are often employed to increase the helical interval without changing the size of the newel, and therefore of r, then will the helical interval be $4\,a\,b$. Considerations affecting the union of sufficient strength with least friction, have suggested this general rule in regard to the projection of the fillet, viz.: make the projection $a\,b$ equal to one third of the radius $O\,b$ of the newel, or

proportion of the different parts of the screw;

Fig. 269.

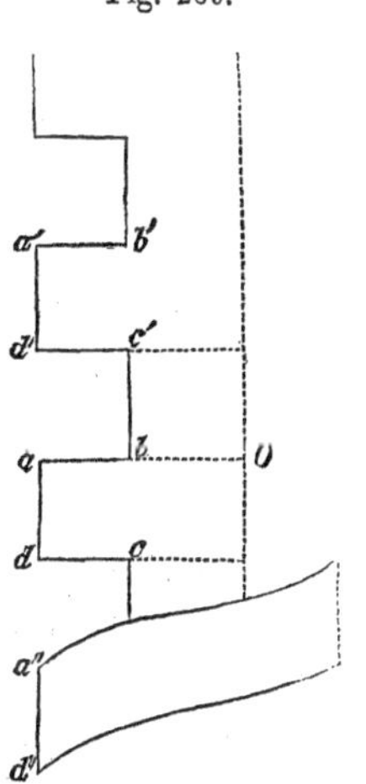

rule;

$$a\,b = \tfrac{1}{3}\,O\,b.$$

This will give

$$O\,b = 3\,a\,b;$$

radius of the newel;

and

$$Ob + \tfrac{1}{2}ab = r = 3ab + \tfrac{1}{2}ab = \tfrac{7}{2}ab;$$

and because $h = 2\,ab$,

radius of intermediate helix;

$$r = \tfrac{7}{4}h;$$

which substituted for r, in the expression for the friction, gives

work of friction;

$$f \cdot Qh \frac{1 + \pi \,.\, \frac{49}{4}}{\pi \,.\, \frac{7}{2} - f};$$

and making $\pi = \dfrac{22}{7}$ to which it is very nearly equal, the expression reduces to

its final value;

$$f \times Qh \frac{122}{11 - f}.$$

example;

To apply this to a particular example, let the screw be made of wrought iron, and the nut of brass, and suppose an unguent of tallow, in which case $f = 0.103$, see Table III, § 212; hence the value of the friction becomes

$$1.152 \times Qh;$$

which, substituted in Eq. (175), gives

result.

$$2\pi R \,.\, F = Qh + 1.152\,Qh = 2.152\,Qh;$$

whence we see, that friction occasions a loss of work greater than the whole work performed by the resistance.

Endless screw;

§ 241.—The *endless screw* is employed to transmit a very slow motion, and, at the same time, to overcome considerable resistance. It is a short screw, with square fillet, and so supported as to revolve freely about its axis, with no motion of translation. It is usually turned by means of a crank. The fillet passes between teeth on the circumference of a wheel of which the axis is perpendicular to that of the screw. The resistance Q is applied to the circumference of the arbor of the wheel. The rubbing faces of the teeth, instead of being parallel to the axis of the wheel, are slightly inclined to that line, so as to make them parallel to the surface of the fillet when the latter is brought in contact with the teeth.

use and description;

Fig. 270.

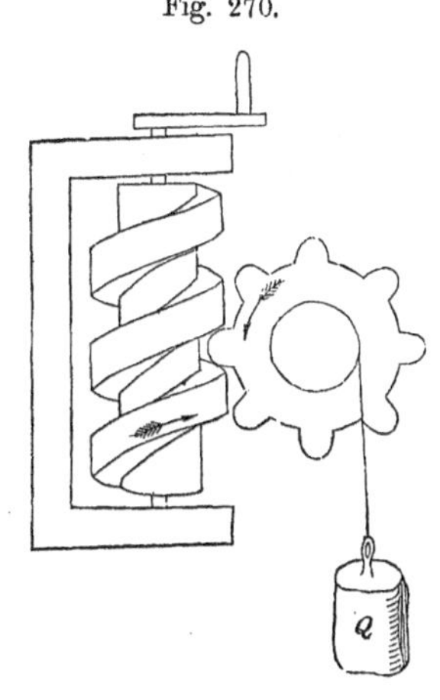

surface of teeth inclined to axis of motion;

A rotary motion being communicated to the screw, its fillet presses against the teeth of the wheel; and as the screw can have no longitudinal motion, the wheel must turn about its axis. As the teeth are withdrawn towards one end of the screw, others are interposed towards the other end, and thus an endless motion may be kept up; hence the name of the machine.

operation and reason for the name;

A plane through the axis of the screw and perpendicular to that of the wheel, will cut from the rubbing surfaces of the fillet and teeth a profile; and if we confine ourselves to what takes place in this plane during the motion, we shall find that the circumstances will be

Fig. 271.

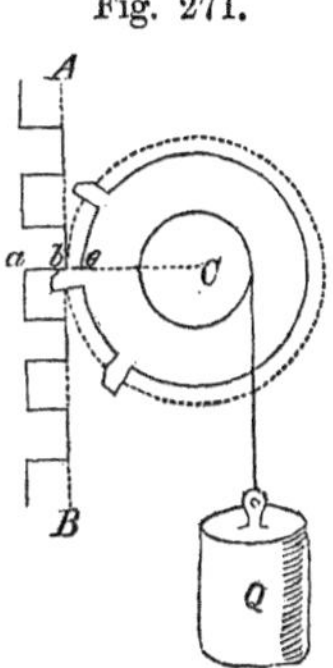

section by a plane through the axis of the screw perpendicular to the axis of the wheel;

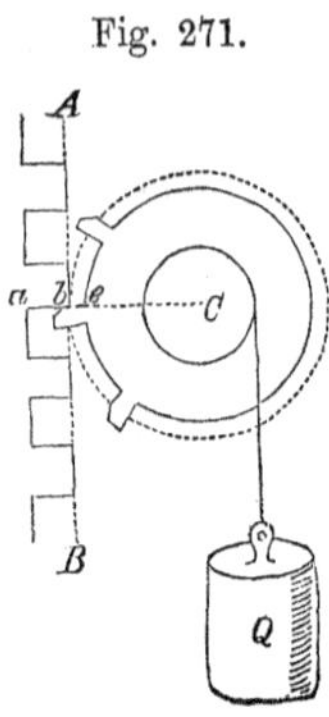

Fig. 271.

circumstances of action same as those of two wheels with teeth;

the same as those of two wheels acting upon one another through the intervention of teeth; for, as the screw turns about its axis to bring different parts of the fillet in this cutting plane, the section ab will move in the direction from A to B, driving the section $b\,e$ of the tooth before it.

Let $Q_{,}$ be the force applied at b in the direction $A\,B$, which is tangent to the circumference whose centre is on the axis of the wheel, and whose radius is $Ce = R_{,}$, and which will sustain the resistance Q in equilibrio: then denoting by N the resultant of $Q_{,}$ and Q, by r the radius of the arbor, and by $r_{,}$ that of the trunnion, will

quantity of work;

$$Q_{,}\,R_{,}\,s_{,} = Q\,r\,s_{,} + f N r_{,}\,s_{,};$$

in which $s_{,}$ is the arc described at the unit's distance from the axis of the wheel.

Dividing by $R_{,}\,s_{,}$,

value of the power;

$$Q_{,} = Q\frac{r}{R_{,}} + f N \frac{r_{,}}{R_{,}} \quad . \quad . \quad . \quad (176).$$

Find, by the process explained in § 234, Eqs. (157) to (160), the value of $Q_{,}$ and N. The pressure upon the tooth at b will thus be known, being equal to $Q_{,}$. This pressure produces a friction upon the teeth of which the value is

friction;

$$f \cdot Q_{,}\frac{n + n'}{n\,n'}\pi = f\,Q_{,}\,\pi\left(\frac{1}{n'} + \frac{1}{n}\right);$$

wherein n denotes the number of teeth on the wheel whose centre is C, and n' the number on the other. But the circumference of this latter wheel being a right line, is infinite as well as the number of its teeth; hence

$$\frac{1}{n'} = 0;$$

reciprocal of the number of teeth on section of screw;

and the foregoing becomes

$$f\pi \cdot Q_{,} \cdot \frac{1}{n},$$

value of the friction;

which must be added to $Q_{,}$ to obtain the force necessary to turn the wheel and to obtain the total pressure on the fillet of the screw. This sum, which is

$$Q_{,} + f\frac{\pi}{n} \cdot Q_{,} = Q_{,}\left(1 + f\frac{\pi}{n}\right),$$

total pressure on the fillet;

being substituted for Q in Eq. (175), will give

$$2\pi RF = Q_{,}\left(1 + f\frac{\pi}{n}\right)h + fQ_{,}\left(1 + f\frac{\pi}{n}\right)\frac{h^2 + 4\pi^2 r^2}{2\pi r - fh};$$

or

$$2\pi RF = Q_{,}\left(1 + f\frac{\pi}{n}\right)\left[h + f \cdot \frac{h^2 + 4\pi^2 r^2}{2\pi r - fh}\right] \quad . \quad . \quad (177).$$

quantity of work of power;

In the discussion of the screw, no reference has been made to the friction on the pivots and collars by which the screw is kept in position. It will always be easy to find this, in any particular case, by the rules for finding the friction upon pivots, sockets, and shoulders or rings, explained in § 223.

friction on pivots and collars neglected.

XXI.

THE LEVER.

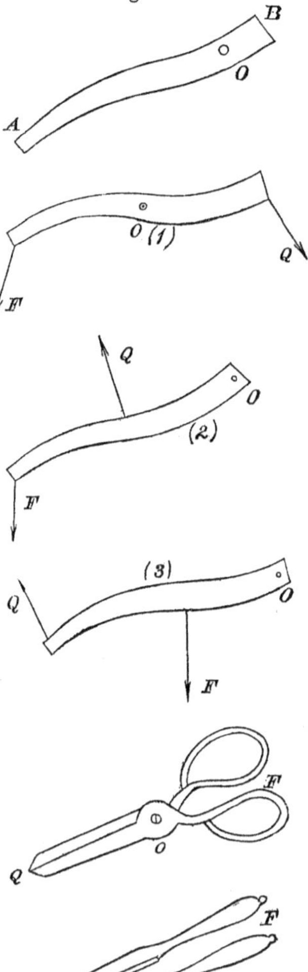

Fig. 272.

The lever; § 242.—The *Lever* is a solid bar $A B$, of any form, supported by a fixed point O, about which it may freely turn, called the *fulcrum*. (fulcrum;) Sometimes it is supported upon trunnions, and frequently upon a knife-edge. Levers have been divided into three different classes, called orders. (levers divided into different orders;)

In levers of the *first order*, the power F and resistance Q are applied on opposite sides of the fulcrum O; in levers of the *second order*, the resistance Q is applied to some point between the fulcrum O and point of application of the power F; and in the *third order* of levers, the power F is applied between the fulcrum O and point of application of the resistance Q. (first order; second; third;)

The common shears furnish an example of a pair of levers of the first order; the nut-crackers of the sec- (examples of different orders of levers.)

ond; and fire-tongs of the third. In all orders, the conditions of equilibrium are the same.

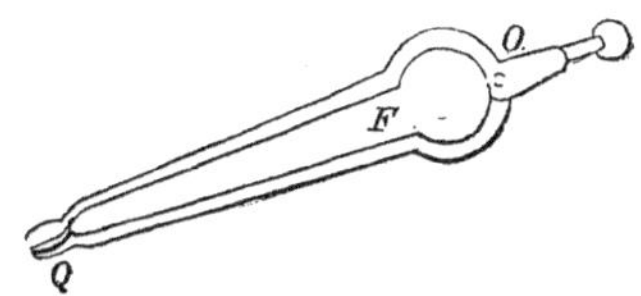

§ 243.—When the lever is supported upon a point, the equilibrium requires that the resultant of the power and resistance shall pass through this point in order to be destroyed by its reaction; to have a resultant, the power and resistance must lie in the same plane, and as the resultant will also be in this plane, the power, resistance, and fulcrum, must be in the same. If the resultant pass through the fulcrum, its moment taken in reference thereto must be zero, which requires that the moment of the power shall be equal to that of the resistance. That is, when a lever AB is in equilibrio and solicited by the power F and resistance Q, O being the fulcrum, if we draw from this latter point Om and On, perpendicular respectively to the direction of the power and resistance, then will

Equilibrium of the lever in which the fulcrum is a point;

Fig. 273.

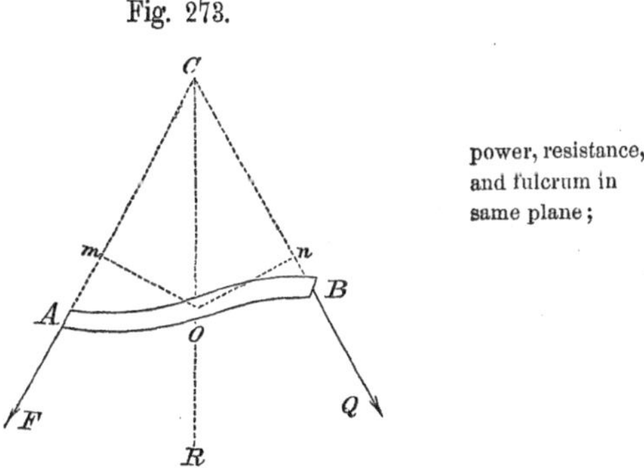

power, resistance, and fulcrum in same plane;

moment of power equal to that of resistance;

$$F \times Om = Q \times On.$$

If the lever turn upon trunnions, then will the moment of the power F, be equal to that of the resistance increased by the moment of the friction on the trunnion. Designating the radius of the latter by r, then will

when lever is supported on trunnions;

moment of power equal to that of resistance, plus that of friction;

$$F \times Om = Q \times On + fN . r;$$

in which N is the resultant of F and Q.

Multiplying both members by $s_{,,}$ we have

work of the power.

$$F \times Om \times s_{,} = Q \times On \times s_{,} + fNr . s_{,};$$

that is to say, *the quantity of work of the power F, must be equal to that of the resistance Q, increased by the quantity of work of the friction.*

Use and advantages of the lever;

§ 244.—The lever is not intended to produce a continuous rotation, but is usually employed to move a heavy burden or great resistance through a short distance during each separate effort of the power.

It is not, therefore, always necessary to make it turn about trunnions which generally operate to disadvantage; since these, to afford sufficient resistance, must be large, which increases the term $fNrs_{,,}$ or the quantity of work absorbed by friction. If the lever be laid upon a simple knife-edge, r becomes zero, and the foregoing equation becomes

Fig. 274.

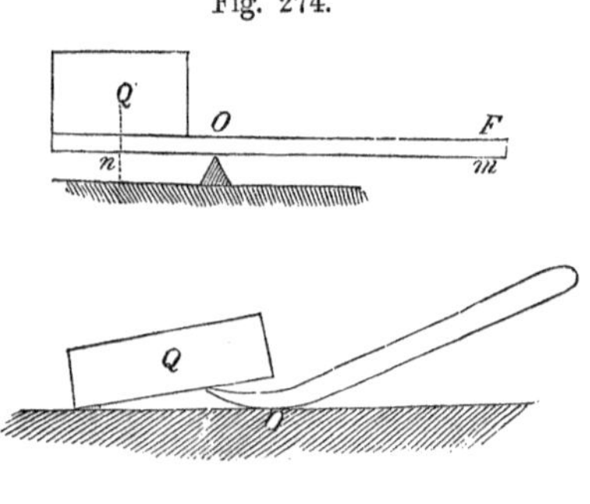

relation of power to resistance on an edge for a fulcrum;

$$F \times Om \times s_{,} = Q \times On \times s_{,,}$$

making the quantity of work of the power equal to that of the resistance. The advantage of this machine, the most simple of all, is, that it transmits without loss, the work of the power to the resistance. But this is not all, a simple change in the point of support or fulcrum, which

usually the lever transmits without loss the power to resistance;

may be made at pleasure, gives the means of establishing any desired relation between the power and resistance. If, for example, the point of support O is placed so that the distance On is one thousandth part of Om, then will

$$F = \frac{Q}{1000};$$

whence we see that with a very small power we may hold in equilibrio an enormous resistance; but as the quantity of work of the resistance must equal that of the power, the path described by the point of application of the latter must increase in the same proportion.

to effect a given purpose, a diminution of power increases its path;

To give an idea of the time necessary to raise a heavy burden through a moderate height with the lever, suppose the weight to be raised is 2000000 pounds, and that it is to be elevated five feet. The quantity of work will be 2000000 lbs. × 5 ft. = 10000000 lbs. Supposing a man to act by his weight = 150 lbs. at the end of a lever, he would have to describe a path equal in length to $\frac{10000000}{150} =$ 66666 feet, nearly. If in each second of time he move the point of the lever at which he applies his weight, through a distance of 0.2 ft., he will require $\frac{66666}{0.2} = 333333$ seconds nearly, = 92.6 hours nearly, = 9.26 days, supposing the man to labor 10 hours a day: in fact a man left to his individual efforts would never accomplish such a task.

example to illustrate this;

This example shows us that the lever is only useful for momentary efforts, and when the burden, being considerable, is to be moved through a very small distance.

practical use of the lever.

XXII.

ATWOOD'S MACHINE.

Atwood's machine;

§ 245.—We shall terminate this branch of our subject with a discussion of an instrument whose object is an experimental verification of the *laws of constant forces*. This instrument is the invention of *Atwood*, an English philosopher, and bears his name. Before proceeding to describe it, let us first find the circumstances of motion under the general case of which the machine in question is but a particular instance. For this purpose, let AB and AD be two inclined planes having a common altitude AE; H and H', two wheels of different diameters mounted upon the same arbor, to which they are firmly attached, and of which the axis is supported upon trunnions parallel to the common intersection of the two planes; W and W' two weights supported upon the inclined planes by means of cords c and c' wound, the first about the wheel H and the second about the wheel H', the cords being parallel to the inclined planes. Now if the weight W be made sufficiently heavy, it will overcome all opposition to motion and slide down the plane AB, while

objects of the machine;

the general case of which this machine is a particular example;

one body ascends while the other descends;

Fig. 275.

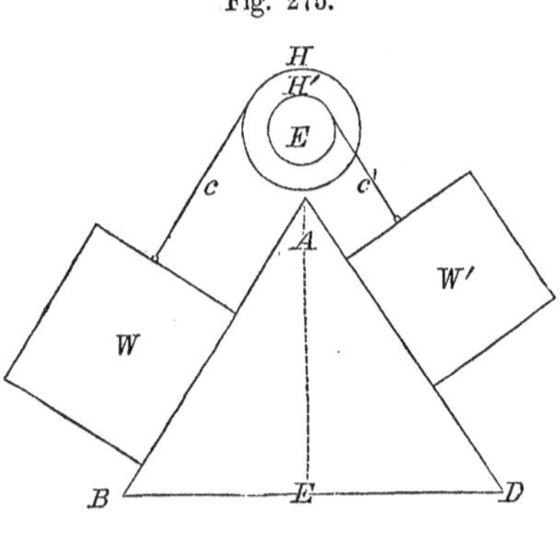

the weight W' must from its connection move up the plane AD. It is required to find the circumstances of motion. Denote the angle which the planes AB and AD make respectively with the vertical AE, by φ and φ'; the radius of the wheel H by R, that of H' by R', and that of the trunnion by r. The pressure of W upon the plane AB we have seen, is

Fig. 276.

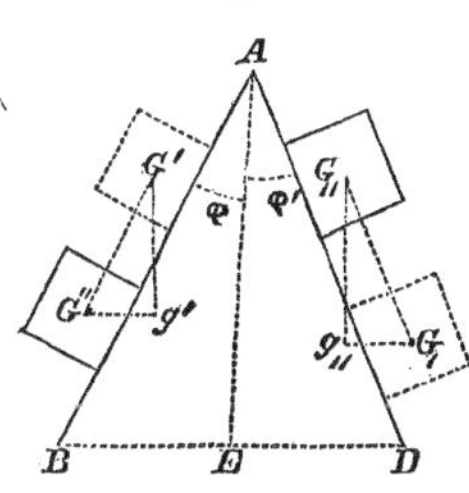

to investigate the circumstances of motion;

$$W \sin \varphi;$$

that of W' on the plane AD is

components of the weights normal to the planes;

$$W' . \sin \varphi';$$

and the friction on the planes AB and AD will be, respectively,

$$f\,W . \sin \varphi, \quad \text{and} \quad f\,W' \sin \varphi'.$$

friction due to these pressures;

The stiffness of the cord c', which alone opposes the motion since the cord c unwinds, is, § 229,

$$d_{\prime} \frac{K + I . (Q)}{2\,R'};$$

stiffness of cord which winds;

in which $d_{\prime}$ represents d^2, $d^{\frac{3}{2}}$, n, or d in Eqs. (127) to (130), inclusive, according to the cord or rope used, and (Q) the tension of the cord c'. This latter is equal to the component of W' parallel to the plane $AD = W' \cos \varphi'$, increased by the friction due to its normal component

$= f\,W' \sin \varphi'$; that is to say,

tension of the cord that winds; $(Q) = W' \cos \varphi' + f\,W' \sin \varphi' = W' (\cos \varphi' + f \sin \varphi');$

which, substituted in the expression above, for the stiffness of the cord c', gives

its stiffness;
$$d_{\prime}\, \frac{K + I \,.\, W' (\cos \varphi' + f \,.\, \sin \varphi')}{2\,R'}.$$

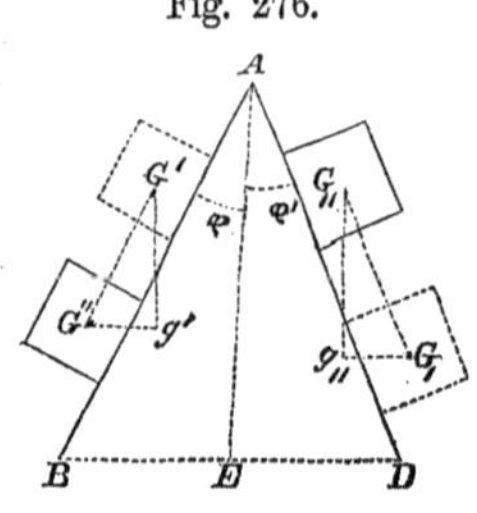

Fig. 276.

At the instant motion begins, let the centres of gravity of W and W' be at G' and $G_{\prime\prime}$ respectively, and in any subsequent instant at G'' and $G_{\prime\prime\prime}$; denote the distance $G'G''$ by x, and $G_{\prime}G_{\prime\prime}$ by x', then will x and x' be the paths described by the centres of gravity parallel to the planes in the interval; and

length of paths in direction of weights;
$$x \cos \varphi, \quad \text{and} \quad x' \cos \varphi',$$

will be the corresponding distances in the direction of the weights.

The quantity of work performed by W will be

quantity of work;
$$W x \cos \varphi,$$

and that performed by W' in the same time,

quantity of work;
$$- \; W' x' \cos \varphi',$$

and the total quantity of work of both will be

total quantity;
$$W x \cos \varphi \; - \; W' x' \cos \varphi'.$$

The quantity of work absorbed by friction on the plane AB is

$$f \,.\, W \,.\, x \sin \varphi,$$

work absorbed by friction on one plane;

and that absorbed by friction on the plane $A\,D$ is

$$f \,.\, W' x' \sin \varphi',$$

that on the other;

and the total quantity absorbed by friction will be, supposing the unit of friction the same on both planes,

$$f\,(W x \sin \varphi \,+\, W' x' \sin \varphi').$$

work absorbed by all the frictions;

The quantity of work absorbed by the stiffness of the cord c' will be

$$d_{,} \frac{K + I \,.\, W' (\cos \varphi' \,+ f \sin \varphi')}{2\,R'}\, x'.$$

work absorbed by stiffness of cord;

The work consumed by friction on the trunnions will be

$$f N \,.\, r \,.\, s_{,}\,;$$

work absorbed by friction on trunnions;

in which N is the resultant of the tensions of the cords c and c'; in other words, is the diagonal of a parallelogram, of which the contiguous sides have

$$W \cos \varphi \,-\, f\, W \sin \varphi, \quad \text{and} \quad W' \cos \varphi' \,+\, f\, W' \sin \varphi',$$

components of the pressure on trunnions;

for their values, and $\varphi + \varphi'$ for their inclination to each other. $s_{,}$ is the arc described at the unit's distance from the axis.

The work absorbed by the inertia of the wheels and arbor, or, which is the same thing, half the living force of the wheels and arbor will, § 159, Eq. (60)″, be

work absorbed by the inertia of wheels and arbor;

$$\frac{V_1^2 I}{2} = \frac{g V_1^2 I}{2g};$$

in which V_1 is the angular velocity, and I the moment of inertia.

Denote by V the velocity of the body whose weight is W, and by V' that of the body whose weight is W'; the living force of the first will be

living force of one body;

$$\frac{W V^2}{g},$$

and that of the second,

that of the other;

$$\frac{W' V'^2}{g};$$

and the quantity of action in the two bodies, will be

quantity of action in the two bodies;

$$\frac{W V^2 + W' V'^2}{2g}.$$

The quantity of work of the weights produces the living force of the bodies, that of the wheels and arbor, as well as the work of friction and that of the stiffness of cordage; hence

work of the weights equal to the living forces of moving parts and the work of friction and stiffness;

$$W x \cos\phi - W' x' \cos\phi' = \begin{cases} \dfrac{W V^2 + W' V'^2}{2g} \\ + f(W x \sin\phi + W' x' \sin\phi') \\ + d_, x' \dfrac{K + I W' (\cos\phi' + f \sin\phi')}{2R'} \\ + f N r s_, + \dfrac{V_1^2 I}{2} \quad . \quad . \quad . \quad . \quad . \quad (178). \end{cases}$$

The variables in this equation, for the same inclination of the planes, are V, V', V_1, x, x', and $s_,$; but these, by the nature of the system, are connected by the following relations, viz.:

$$V : V' :: R : R' \therefore V' = \frac{VR'}{R} \;..\; (179),$$

$$x : x' :: R : R' \therefore x' = \frac{x R'}{R} \;..\; (180),$$

relation between the angular velocities and spaces;

$$1 : R :: s_{,} : x \therefore s_{,} = \frac{x}{R} \;..\; (181),$$

$$R : V :: 1 : V_1 \therefore V_1 = \frac{V}{R} \;..\; (182).$$

These values of V_1, V', x', and $s_{,}$, being substituted in Eq. (178), will give

$$W x \cos\phi - W' x \frac{R'}{R} \cos\phi' = \left\{ \begin{array}{l} \dfrac{W V^2 + W' V^2 \dfrac{R'^2}{R^2}}{2g} \\ + f(W x \sin\phi + W' x \dfrac{R'}{R} \sin\phi') \\ + d_{,} x \dfrac{R'}{R} \cdot \dfrac{K + I W' (\cos\phi' + f \sin\phi')}{2R'} \\ + f . N r \dfrac{x}{R} + \frac{1}{2} g \dfrac{V^2 I}{R^2 g}; \end{array} \right.$$

equation (178) in different terms;

and solving this equation with respect to V^2,

$$V^2 = x \cdot \frac{2g}{W + W' \cdot \dfrac{R'^2}{R^2} + g \cdot \dfrac{I}{R^2}} \cdot \left\{ \begin{array}{l} W . \cos\phi - W' \cdot \dfrac{R'}{R} \cdot \cos\phi' \\ - f(W . \sin\phi - W' \cdot \dfrac{R'}{R} \cdot \sin\phi') \\ - d_{,} \cdot \dfrac{R'}{R} \cdot \dfrac{K + I . W' . (\cos\phi' + f \sin\phi')}{2R'} \\ - f \cdot N \cdot \dfrac{r}{R}. \end{array} \right.$$

value for the square of the velocity;

The coefficient of x containing no variable, we find that the space described by the body on the plane AB, varies as the square of its velocity. Hence the motion is, § 68, Eq. (8), uniformly varied; and the coefficient of x is twice the velocity which the force producing this motion is capable of generating in a unit of time. Making

motion uniformly varied;

value of the velocity generated in a unit of time;

$$A = \frac{g}{W + W'\frac{R'^2}{R^2} + g \cdot \frac{I}{R^2}} \cdot \left\{ \begin{array}{l} W . \cos\phi - W' \cdot \frac{R'}{R} \cdot \cos\phi \\ -f\,(W \sin\phi - W' \frac{R'}{R} \cdot \sin\phi') \\ -d_{,} \cdot \frac{R'}{R} \cdot \frac{K + I . W' . (\cos\phi' + f \sin\phi')}{2\,R'} \\ -f \cdot N \cdot \frac{r}{R}, \end{array} \right.$$

the foregoing equation may be written

square of the velocity;

$$V^2 = x \,.\, 2\,A \quad . \quad . \quad . \quad . \quad (183).$$

Since the motion is uniformly varied, if T denote the time of describing the space x, then will Eq. (7) become

space described;

$$x = \tfrac{1}{2} A\,T^2 \,.\quad . \quad . \quad . \quad . \quad (184);$$

writing A for V_1, and x for s: substituting this for x above, we find

$$V^2 = A^2\,T^2,$$

or

value of velocity;

$$V = A\,T \quad . \quad . \quad . \quad . \quad . \quad (185).$$

laws of uniformly varied motion;

Eqs. (183), (184), and (185), give the laws of uniformly varied motion, or, as it is usually expressed, the *laws of constant forces.* These laws are, 1st. *The velocities are to each other as the times in which the force produces them;* 2d. *The spaces described, are to each other as the squares of the velocities acquired in describing them; or as the squares of the times in which they are described.*

Any device that will make the time in which the motion takes place comparatively great, while the velocity acquired shall be small, will enable us to verify these laws from observation. For this purpose, A must be small. By reference to Eq. (182), we find that A may be diminished at pleasure by increasing the angle φ, or decreasing φ'; this will increase the effect of friction, which opposes, while it will diminish the component of W, which aids the motion. Or A may be diminished, by diminishing the angles φ and φ', the difference between the weights W and W' and that between R and R'. Owing to the uncertainty of friction it is better to accomplish the object by the latter method, and this Atwood has done.

Fig. 277.

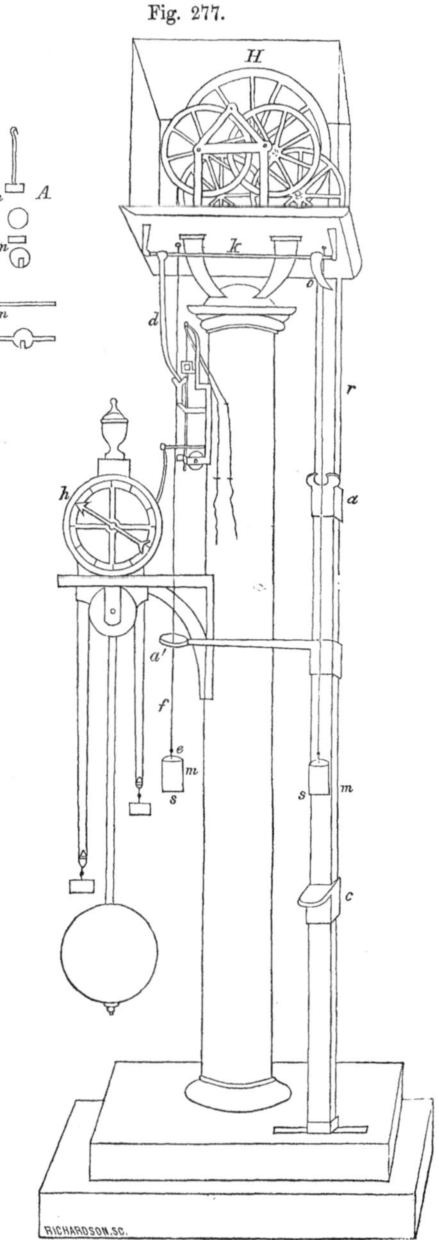

His machine consists essentially of a fixed pulley H, over which passes a cord having attached to each extremity a basin s, for the reception of weights; a vertical graduated scale r of equal parts, say inches, to measure spaces; and a pendulum clock h which beats

Fig. 277.

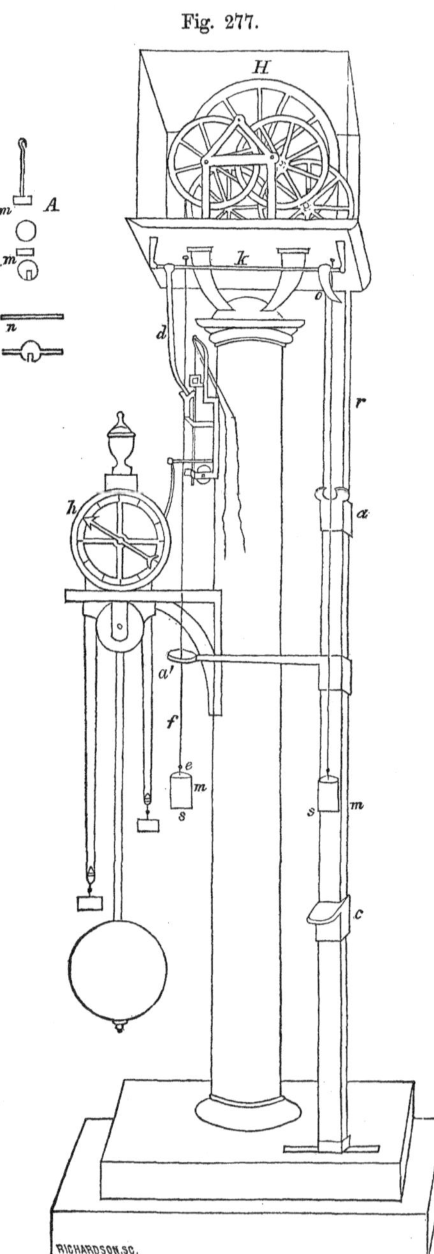

seconds, to mark the time. The basins are short cylinders of brass, having a wire *e* coincident with the axis and projecting some three or four inches beyond the upper bases; the cord is attached to the ends of these wires. The weights are either circular plates *m*, or bars *n*, of greater or less thickness, depending upon the purpose for which they are employed. Both are perforated at the centre, and a channel is cut from the hole to the margin to permit the cord *f* to enter, that the weights may be dropped upon the basins.

The scale piece *r* is provided with three sliding stages, two of which *a* and *a′* are rings, and the third *c* is plane. The rings, whose diameters are less than the length of the bar-weights, serve to take the latter from a descending, or to add them to an ascending basin. The office of the plane stage, is to arrest the motion of a descending basin.

A fourth and revolving stage *o*, connected by an arm *d* with an arbor *k*, in front, is used to support the basin bearing the greater load opposite the zero point of the scale. The arbor is also connected by means of a second arm with the escapement-wheel of the clock. This stage may be thrown from under the basin when the seconds' hand reaches a particular point on the dial plate; thus causing the motion to begin at a particular instant, and from the zero of the scale. device for adjusting the basin to the zero of scale;

If we examine the value of A, we shall find that for Atwood's machine, φ and φ' are both zero, and therefore

$$\sin\varphi = 0; \quad \sin\varphi' = 0; \quad \cos\varphi = 1; \quad \cos\varphi' = 1:$$

reduction of general equation to the case of Atwood's machine;

moreover R is equal to R', and hence

$$\frac{R'}{R} = 1.$$

The cord is very fine, and usually made of raw silk but slightly twisted, so that the term involving the stiffness of cordage has no appreciable value, and may be neglected. The arbor of the pulley or wheel rests upon circumferences of four other wheels of large radii compared with the radii of their trunnions, after the manner explained in § 228, so that the term involving the friction on the trunnions may also be neglected without appreciable error. omissions;

Making the foregoing substitutions and omissions in the value for A, we find

$$A = g \cdot \frac{W - W'}{W + W' + g\frac{I}{R^2}}.$$

corresponding value of the general coefficient;

The circumference of the wheel has the same velocity as the points of the cord, and therefore the same as the basins. Designate by M'', the mass which if concentrated in the circumference of the wheel would have a moment of inertia equal to that of the wheel, then

moment of inertia of the wheel;

$$M'' R^2 = I;$$

whence

$$M'' = \frac{I}{R^2};$$

and this, substituted above, gives

velocity generated in unit of time;

$$A = g \frac{W - W'}{W + W' + g M''} = g \frac{W - W'}{W + W' + W''},$$

in which W'' denotes the weight of the mass M''.

This value of A, substituted in Eqs. (184) and (185), gives

space;

$$x = \frac{W - W}{W + W' + W''} \tfrac{1}{2} g T^2 \quad . \quad . \quad (186),$$

velocity;

$$V = \frac{W - W'}{W + W' + W''} \cdot g T \quad . \quad . \quad (187).$$

experimental determination of the weight W'';

Before proceeding to verify the laws expressed by these equations, it will be necessary to determine the constant weight W''. For this purpose load the machine by placing the same number of circular weights in each basin; then add a bar-weight to the basin, which moves in front of the scale. The basins being of the same weight, the difference $W - W'$ will be the weight of the bar; the sum $W + W'$, will be the sum of the weights of the basins, increased by that of the circular weights added, and that of the bar, all of which are known. Now place the basin which carries the bar at the zero of the scale, by means of the revolving stage; set the clock in motion, and, supposing the bar to commence its descent at a particular beat of the clock, note whether the bar is taken off by the upper ring stage, coincidently with any subsequent beat of the clock; if it is, then the distance of the ring below the zero of the scale being substituted for x, and the number of seconds elapsed from the beginning of motion till the bar is removed, being substituted for T in Eq. (186), will enable us to find W'', since all the other quantities in that equation are known. If the removal of the bar and the beat of the clock be not coincident, the ring stage must be shifted, and the experiment repeated till the coincidence is obtained.

the experiment repeated till coincidence of clock beat with removal of bar is obtained;

Example. Let each basin weigh 11 units, and suppose 14 units of weight to be placed in each basin, and a bar weighing 1 unit to be added to the basin in front of the scale, then will

example;

$$W - W' = 1,$$

data;

$$W + W' = 51;$$

making $g = 32$ feet $= 384$ inches; $\frac{1}{2}g = 192$ inches. Substituting these values in Eq. (186), we find

$$x = \frac{1}{51 + W''} \cdot 192 \times T^2;$$

corresponding value of space;

whence

$$W'' = \frac{192}{x} \cdot T^2 - 51.$$

value of W'';

Now supposing the bar to be removed at the end of the third second, and that we find x, or the space described by the bar to be 27 inches, then will

$$W'' = \frac{192}{27} \times (3)^2 - 51 = 64 - 51 = 13;$$

numerical value of W'';

that is to say, the moment of inertia of the wheel will have conclusion; the same effect to resist motion as the inertia of thirteen units of weight placed in the basins.

Substituting this value for W'' in Eqs. (186) and (187), they become

space for the particular machine;

$$x = \frac{W - W'}{W + W' + 13} \tfrac{1}{2} g T^2 \quad . \quad . \quad (188),$$

velocity in the same;

$$V = \frac{W - W'}{W + W' + 13} g T . \quad . \quad . \quad (189);$$

and, loading the machine as before,

experimental verification;

$$x = \tfrac{1}{64} \times 192 \times T^2 = 3 T^2,$$

$$V = \tfrac{1}{64} \times 384 \times T = 6 T.$$

Making T equal to

times; 1, 2, 3, 4, &c. seconds;

the corresponding spaces will be

spaces; 3, 12, 27, 48, &c. inches;

and the corresponding velocities,

velocities; 6, 12, 18, 24, &c. inches.

verification; Place the basin with the bar-weight at the zero of the scale, and connect with the clock; adjust the ring so as to remove the bar when its basin reaches the 3 inch mark, and place the plane stage at the 9 inch mark $= 3 + 6$. The clock being put in motion, the bar will strike the ring at the first beat of the clock after it begins to descend, and its basin will strike the plane stage at the second beat. The bar being removed, there will be no excess of weight in either basin, and the motion will become uniform, there being no reason why it should be accelerated rather than retarded. To show that the motion will be uniform, repeat the experiment, placing the plane stage first at 1 foot 3 inches, then at 1 foot 9, then at 2 feet 3 inches, and so on, adding 6 inches each time, and it will be found that the basin will be arrested at the third, fourth, fifth, &c., beats of the clock after its motion begins; thus showing that the spaces described are proportional to the times, which is the characteristic of uniform motion. Next adjust the ring so as to remove the bar when its basin reaches the 12 inch or 1 foot mark, and place the plane stage at the 2 feet mark, it will be found that the bar will strike the ring at the second beat after its motion begins, and that the scale will be arrested at the third beat. That the motion is uniform after the removal of the bar may be shown, as before, by repeating the experiment, and adding 12 inches each time to the space to be described after the bar is arrested. In the same way all the other results may be verified.

motion uniform after the bar is removed;

its proof;

repetition of the experiment;

If a bar-weight be placed upon the second ring, and the latter be so adjusted that the ascending basin shall take it up at the moment the bar on the descending basin is removed, the motion will become retarded, and we shall have the case of a body projected vertically upward from rest with a velocity equal to that of the basins. The plane stage being placed at a distance from the ring which takes up the descending bar, equal to that through which the latter has descended, it will be found that the scale will just reach this stage at the instant the motion is destroyed by the action of the ascending bar. All the laws which regulate the fall of heavy bodies may be verified by means of Atwood's instrument.

illustration of retarded motion;

all the laws which regulate the fall of heavy bodies may be verified by this machine.

XXIII.

IMPACT OF BODIES.

Impact of bodies; § 246.—When a body in motion comes into collision with another, either at rest or in motion, an *impact* is said to arise.

We have seen, § 204, that the action and reaction which take place between two bodies, when pressed together, are exerted along the same right line, perpendicular to the surfaces of both, at their common point of contact.

direct impact; When the motions of the centres of gravity of the two bodies are *parallel* to this normal before collision, the impact is said to be *direct*.

direct and central impact; When this normal passes through the centres of gravity of two bodies which come into collision, and the motions of these centres *are along* that line, the impact is said to be *direct* and *central*.

Fig. 278.

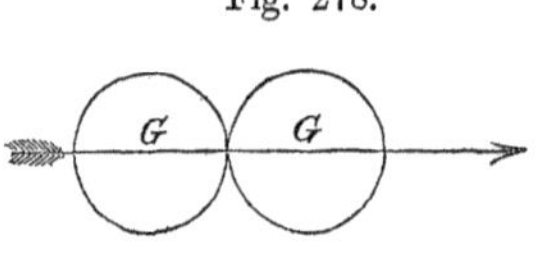

direct and eccentric impact; When the motion of the centre of gravity of one of the bodies is along the common normal, and the normal does not pass through the centre of gravity of the other, the impact is said to be *direct* and *eccentric*.

Fig. 279.

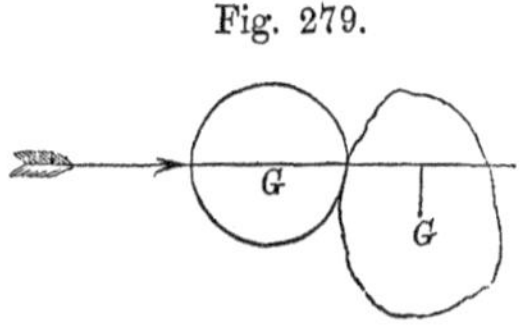

oblique impact; When the path described by the centre of gravity of one of the bodies, makes an angle with this normal, the impact is said to be *oblique*.

Fig. 280.

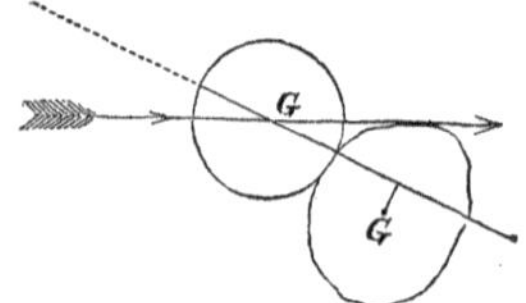

Bodies resist, by their inertia, all effort to change either the quantity or the direction of their motion. When, therefore, two bodies come into collision, each will experience a pressure from the reaction of the other; and as all bodies are more or less compressible, this pressure will produce a change in the figure of both; the change of figure will increase till the instant the bodies cease to approach each other, when it will have attained its maximum, and the bodies will have the same velocity. The molecular spring of each will now act to restore the former figures, the bodies will repel each other, and finally separate. circumstances of figure during the collision;

In the impact of bodies, three periods must therefore be distinguished, viz.: 1st., that occupied by the process of compression; 2d., that during which the greatest compression exists, and in which it is obvious the bodies have the same velocity; 3d., that occupied by the process, as far as it extends, of restoring the figures. We are also carefully to distinguish the *force of restitution* from *the force of distortion;* the latter denoting the reciprocal action exerted between the bodies in the first, and the former that exerted in the third period. three periods of the impact; force of restitution and of distortion;

The greater or less capacity of the molecular springs of a body to restore to it the figure of which it has been deprived by the application of some extraneous force when the latter ceases to act, is called its *elasticity.* elasticity defined;

The ratio of the force of distortion to the force of restitution, is the measure of a body's elasticity. This ratio is sometimes called the *coefficient of elasticity.* When these two forces are equal, the ratio is unity, and the body is said to be *perfectly elastic;* when the ratio is zero, the body is said to be *non-elastic.* There are no bodies that satisfy these extreme conditions, all being more or less elastic, but none perfectly so. coefficient of elasticity; perfect elasticity; non-elastic.

§ 247.—Suppose two bodies A and B to move in the same direction, the body A to overtake B, and the impact

Direct impact of two bodies;

to be direct. The forces of distortion and of restitution, arising as they do from the reciprocal action of the bodies upon each other, are real pressures, measurable in pounds, and are capable of generating in each body, in a given time, a certain quantity of motion. Denote the intensity of this force, at any instant of the impact, by F; the small velocity lost by the body A, in the short time during which F may be regarded as constant, by v; and the small velocity gained by B, in the same time, by the action of the same force, by v'; also denote the mass of A by M, and that of B by M'; then will F, which may be called indifferently the action of one body or the reaction of the other, be measured by Mv, or $M'v'$; and, because of the equality of action and reaction,

notation;

Fig. 281.

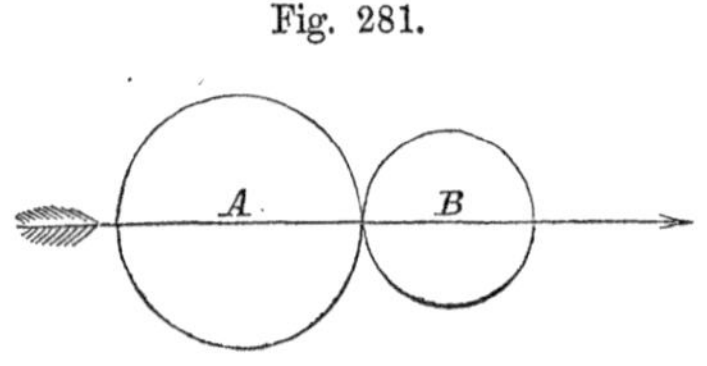

equality of action and reaction;

$$MV = M'V'.$$

That is to say, *the quantity of motion lost or gained by each of the bodies, during each indefinitely small portion of time, will be equal to each other*; and if we take the sum of all the quantities of motion lost or gained by each of the bodies, we shall have the whole quantity of motion gained or lost by the one, equal to that gained or lost by the other. Denoting the entire gain or loss of velocity of the body A by $V_{,}$, that of the body B by $V_{,,}$, we shall have

gain and loss of motion equal;

$$MV_{,} = M'V_{,,}.$$

But the force F acts in opposite directions upon the two bodies, and hence, if we give the positive sign to the velocity generated in one body, that of the other must be negative; that is, if $V_{,}$ be counted positive, $V_{,,}$ must be negative, which will make

forces producing these act in opposite directions;

$$MV_{,} = -M'V_{,,},$$

or

$$MV_{,} + M'V_{,,} = 0.$$

quantity of motion of the system constant.

That is to say, the *algebraic sum, or the whole quantity of motion lost and gained will be zero; and in every stage of the impact the quantity of motion in the entire system will, therefore, be the same as before the impact began.*

To find the common velocity at moment of greatest compression;

§ 248.—At the instant the bodies have ceased to approach each other, they will have attained their greatest compression, and, considering their condition before the retrocession begins under the action of the molecular springs, it is obvious that they may be regarded as a single body, having a common velocity. Denote this velocity by U; also denote the velocity of the body A, before the impact, by V; that of the body B, before the impact, by V', the masses being M and M' as before. The whole quantity of motion before the impact will be

$$MV + M'V',$$

and that at the instant of greatest compression will be $(M + M')U$. But these, by the last article, must be equal, or

$$(M + M')U = MV + M'V';$$

whence

$$U = \frac{MV + M'V'}{M + M'} \quad . \quad . \quad . \quad (190).$$

the value of this velocity;

That is to say, when two bodies moving in the same

expressed in words;

direction have a direct impact, *the common velocity, at the instant of greatest compression, is equal to the sum of the quantities of motion before the impact, divided by the sum of the masses.*

If the bodies moved in opposite directions, either V or V' would be negative, say V', and

value when bodies move in opposite directions.

$$U = \frac{MV - M'V'}{M + M'} \quad . \quad . \quad . \quad (191).$$

§ 249.—The velocity lost by the body A, up to the instant of greatest compression, is obviously equal to

$$V - U,$$

Velocity gained up to greatest compression;

and that gained by the body B is equal to

$$U - V';$$

the force of distortion will, therefore, be measured by

$$M(V - U),$$

or by

force of distortion;

$$M'(U - V').$$

Denote by $V_{,}$ the velocity which A loses by the force of restitution; and by $V_{,,}$, that which B gains by the action of the same force; the force of restitution will be measured by

force of restitution;

$$MV_{,} \quad \text{or} \quad M'V_{,,};$$

and if e denote the coefficient of elasticity, then, from the definition

coefficient of elasticity;

$$\frac{MV_{,}}{M(V - U)} = e,$$

$$\frac{M' V_{\prime\prime}}{M'(U - V')} = e;$$

coefficient of elasticity;

whence

$$V_{\prime} = e(V - U) \quad . \quad . \quad . \quad (192),$$

$$V_{\prime\prime} = e(U - V') \quad . \quad . \quad . \quad (193).$$

velocities lost and gained;

That is to say, *the velocity which A loses by the force of restitution, is equal to the coefficient of elasticity, into the velocity which it lost by the force of distortion; and the velocity gained by B by the same force, is equal to that which it gained by the force of distortion, into the coefficient of elasticity.*

the same expressed in words;

The total loss of velocity which A will experience by the impact will be

$$V - U + e(V - U);$$

loss of velocity of the impinging body;

and the entire gain of B will be

$$U - V' + e(U - V').$$

gain of the other;

Denote by v the velocity retained by A, and by v' that which B has after the impact; then, since the velocity retained by A, must be equal to that which it had before the impact, diminished by its loss,

$$v = V - V + U - e(V - U) = (1 + e)\,U - eV;$$

and as B must, after the impact, have its primitive velocity increased by its gain,

$$v' = V' + U - V' + e(U - V') = (1 + e)\,U - eV';$$

and substituting for U its value in Eq. (190), we have

$$v = (1 + e)\frac{MV + M'V'}{M + M'} - eV. \quad . \quad (194),$$

velocity of the impinging body after the impact;

velocity of the other after the impact;

$$v' = (1 + e)\frac{MV + M'V'}{M + M'} - eV' \quad . \quad . \quad (195).$$

in words;

Thus, *the velocity of either body after impact, is equal to the coefficient of elasticity increased by unity, multiplied into the common velocity at the instant of greatest compression, and this product diminished by the product of the coefficient of elasticity into the velocity of the body before impact.*

If the body B move to meet the body A, its velocity will be negative, and the above reduce to

when the bodies meet.

$$v = (1 + e)\,\frac{MV - M'V'}{M + M'} - eV \quad . \quad . \quad (196),$$

$$v' = (1 + e)\cdot\frac{MV - M'V'}{M + M'} + eV' \quad . \quad . \quad (197).$$

§ 250.—If the body B be at rest when the body A impinges against it, then will V' be zero, and

When one of the bodies is at rest;

$$v = (1 + e)\,\frac{MV}{M + M'} - eV \quad . \quad . \quad (198),$$

$$v' = (1 + e)\,\frac{MV}{M + M'} \quad . \quad . \quad . \quad . \quad (199).$$

From the last equation we find

coefficient of elasticity;

$$e = \frac{v'(M + M')}{MV} - 1 \quad . \quad . \quad . \quad (200);$$

and when the masses of the bodies are equal, or $M = M'$,

its value when the masses are equal;

$$e = \frac{2v'}{V} - 1 \quad . \quad . \quad . \quad . \quad (201);$$

which suggests a very easy method of finding the coefficient of elasticity of any solid body. For this purpose, turn a pair of spherical balls of the same weight from the body whose coefficient of elasticity is to be found; suspend them by silken strings, so that when the latter are vertical the balls shall just touch each other, be upon the same level, and have their centres opposite the zeros of two circular graduated arcs whose centres of curvature are at the points of suspension. The body A being drawn back to any given degree upon its scale and abandoned, will descend and impinge against the body B with a velocity due to a height equal to the versed sine of the arc which it describes before the impact; the body B will ascend on the opposite arc to a height due to the velocity with which it leaves A; this height will be the versed sine of the arc described by B before it begins to descend again. The arcs being known, their versed sines are easily computed from the properties of the circles. Denoting these versed sines by h and h', then will

experimental determination of the coefficient of elasticity;

Fig. 282.

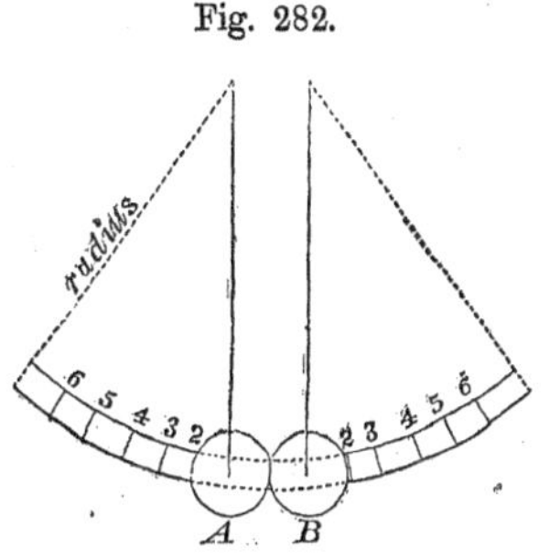

description of instrument, and mode of using it;

$$V = \sqrt{2gh},$$

$$v' = \sqrt{2gh'};$$

velocity of impinging body and that of the body struck;

which, substituted in the value of e, gives

$$e = 2\sqrt{\frac{h'}{h}} - 1 \quad . \quad . \quad . \quad (202).$$

coefficient of elasticity;

Example. Two *ivory* balls of equal weights, and therefore of equal masses, were made to collide in the manner above described. One descended through an arc of 20 degrees, and the other ascended through an arc of 18 degrees and 30 minutes; required the value of e.

example of two ivory balls;

By tables of natural sines and cosines, we find

$$\text{nat. cos } 20^\circ = 0.9396926;$$

$$\text{versed sin } 20^\circ = 1 - 0.9396926 = 0.0603074;$$

and denoting the radius of the circular scale by R, we have

height of fall of the colliding body;

$$h = 0.0603074\, R.$$

Again,

$$\text{nat. cos } 18^\circ\ 30' = 0.9483236;$$

$$\text{versed sin } 18^\circ\ 30' = 1 - 0.9483236 = 0.0516764;$$

height due to the velocity of the body struck;

$$h' = 0.0516764\, R;$$

and

numerical value of the coefficient;

$$e = 2\sqrt{\frac{0.0516764.R}{0.0603074.R}} - 1 = 2\sqrt{\frac{0.0516764}{0.0603074}} - 1 = 0.85138;$$

whence we conclude that the coefficient of elasticity of the specimen of ivory employed, is about 0.85; that of glass will be found to be about 0.93, and that of steel about 0.56.

example of the collision of ivory balls;

Example. Two ivory balls, whose masses are represented by 6 and 4, move in the same direction with velocities of 10 and 7 feet a second respectively. What is the velocity of each after impact? The conditions of the question require that the larger mass 6 shall overtake the smaller mass 4, because the former has the greater velocity. Hence

$$M = 6; \quad V = 10;$$
$$e = 0.85. \qquad \text{given data;}$$
$$M' = 4; \quad V' = 7;$$

These data, in Eqs. (194) and (195), give

$$v = 1.85 \frac{60 + 28}{10} - 0.85 \times 10 = \overset{ft.}{7.78},$$

velocities after impact;

$$v' = 1.85 \frac{60 + 28}{10} - 0.85 \times 7 = \overset{ft.}{10.33}.$$

Example. Let the same balls move in opposite directions so as to meet, each with the same velocity as before. The same data, substituted in Eqs. (196) and (197), give another example.

$$v = 1.85 \frac{60 - 28}{10} - 0.85 \times 10 = - \overset{ft.}{2.58},$$

$$v' = 1.85 \frac{60 - 28}{10} + 0.85 \times 7 = \overset{ft.}{11.87}.$$

§ 251.—Now suppose the bodies A and B to move, the first with a velocity V in the direction from E towards F, and the second with a velocity V' in the direction from C towards D; and let the collision take place at H. Through the point H, draw the common normal HN, and resolve each of the velocities V and V' into two components, one in the direction of the normal and the other in the direction of the tangent plane at H. For this purpose designate the angle FON by φ, and $DO_{\prime}N$

Oblique impact;

Fig. 283.

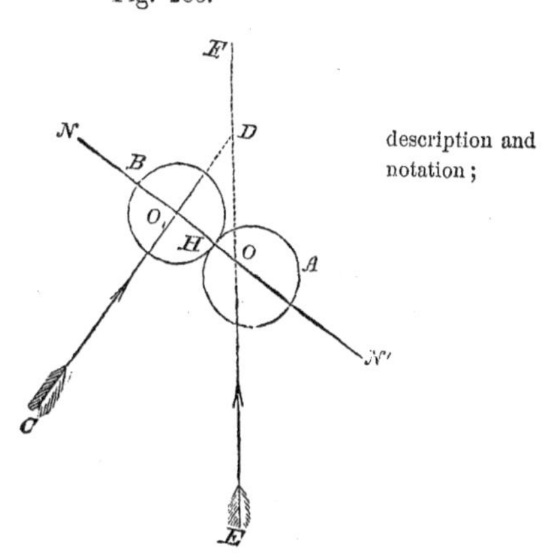

description and notation;

22 by φ'; the components in the direction of the normal, will be

normal velocities;

$$V \cos \varphi, \quad \text{and} \quad V' \cos \varphi';$$

and those parallel to the tangent plane, will be

tangential velocities;

$$V \sin \varphi, \quad \text{and} \quad V' \sin \varphi'.$$

If the bodies were animated by these last velocities alone, they would not collide, but would in general move by one another without exerting any pressure; and hence the impact will be wholly due to the components in the direction of the normal; but these acting along the same line perpendicular to the surfaces at their common point of contact, will give rise to a direct impact, and denoting the velocities of the bodies A and B after impact by v and v', and the angles which their directions make with the normal by θ and θ', respectively, we shall have, from Eqs. (194) and (195),

components of velocity in direction of the normal after impact;

$$v \cos \theta = (1 + e) \cdot \frac{M V \cos \phi + M' V' \cos \phi'}{M + M'} - e V \cos \phi \; . \; . \; . \; (203),$$

$$v' \cos \theta' = (1 + e) \; \frac{M V \cos \phi + M' V' \cos \phi'}{M + M'} - e V' \cos \phi' \; . \; . \; (204).$$

Moreover, because the effects of the impact arising from the components of the velocities in the direction of the normal will be wholly in that direction, the components of the velocities of each body before and after the impact at right angles to the normal, will be the same, and hence

tangential components of velocity after the impact;

$$v \sin \theta = V \sin \varphi \quad . \; . \; . \quad (205),$$

$$v' \sin \theta' = V' \sin \varphi' \quad . \; . \; . \quad (206).$$

Squaring Eqs. (203) and (205), adding, extracting the

square root, and reducing by the relation,

$$\cos^2 \theta + \sin^2 \theta = 1,$$

we find

$$v = \sqrt{\left[(1+e)\frac{M V \cos \phi + M' V' . \cos \phi'}{M + M'} - e V \cos \phi\right]^2 + V^2 \sin^2 \phi} \quad .. (207);$$

velocity of the impinging body after the impact;

and treating Eqs. (204) and (206) in the same way,

$$v' = \sqrt{\left[(1+e)\frac{M V \cos \phi + M' V' \cos \phi'}{M + M'} - e V' \cos \phi'\right]^2 + V'^2 \sin^2 \phi'} \quad .. (208).$$

velocity of the body struck after the impact;

Again, dividing Eq. (205) by Eq. (203), we have

$$\tan \theta = \frac{V \sin \varphi}{(1+e)\dfrac{M V \cos \varphi + M' V' \cos \varphi'}{M + M'} - e V \cos \varphi} \quad .. (209);$$

direction of the first body's motion;

and, dividing Eq. (206) by (204),

$$\tan \theta' = \frac{V' \sin \varphi'}{(1+e)\dfrac{M V \cos \varphi + M' V' \cos \varphi'}{M + M'} - e V' \cos \varphi'} \quad .. (210).$$

that of the second;

The Eqs. (207) and (208) will make known the velocities, and (209) and (210) will give the directions in which the bodies will move, after the impact.

Now suppose the body B at rest, and its mass so great that the mass of A is insignificant in comparison, then will V' be zero, M' may be written for $M + M'$, and $\frac{M}{M'}$ will be a fraction so small that all the terms into which it enters as a factor may be neglected. Applying these considerations to Eq. (207), we find

suppose one body very large and at rest;

reductions;

$$v = V \sqrt{e^2 \cos^2 \varphi + \sin^2 \varphi};$$

velocity of the impinging body after impact;

and to Eq. (209),

direction of the impinging body's motion after impact;

$$\tan\theta = -\frac{\tan\varphi}{e} \quad . \quad . \quad . \quad . \quad (211).$$

The tangent of θ being negative, shows that the angle NHK, which the direction of A's motion makes with the normal NN' after the impact, is greater than 90 degrees; in other words, that the body A is driven back or reflected from B. This explains why it is that a cannon-ball, stone, or other body thrown obliquely against the surface of the earth, will rebound several times before it comes to rest.

graphical illustration of this result;

Fig. 284.

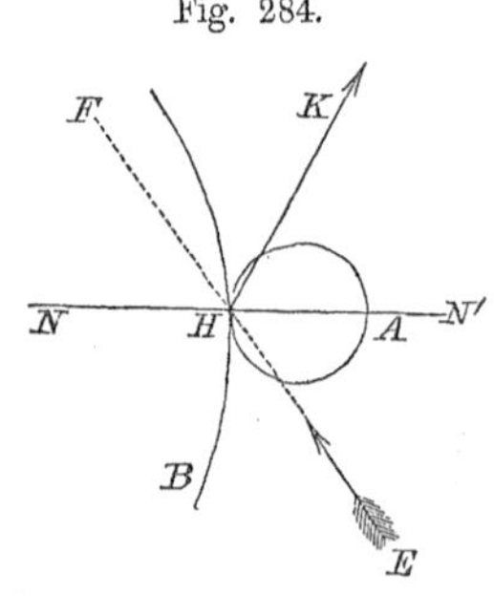

If the bodies be non-elastic, or, which is the same thing, if e be zero, the tangent of θ becomes infinite; that is to say, the body A will move along the tangent plane, or if the body B were reduced at the place of impact to a smooth plane, the body A would move along this plane.

body will not rebound when non-elastic;

If the body were perfectly elastic, or if e were equal to unity, which expresses this condition, then would Eq. (211) become

$$\tan\theta = -\tan\varphi \quad . \quad . \quad . \quad . \quad (212);$$

which means that the angle $NHF = EHN'$ becomes equal to KHN'. The angle EHN' is called the angle of incidence, the angle KHN', commonly, the angle of reflection. Whence we see, that when a perfectly elastic body is thrown against a smooth, hard, and fixed plane, the angle of incidence will be equal to the angle of reflection.

in perfectly elastic bodies the angle of incidence equal to angle of reflection;

If the angles φ and φ' be zero, then will $\cos\varphi = 1$, $\cos\varphi' = 1$, $\sin\varphi = 0$, and $\sin\varphi' = 0$, and Eqs. (207) and (208) become

$$v = (1 + e)\frac{MV + M'V'}{M + M'} - eV,$$

case of direct impact;

$$v' = (1 + e)\frac{MV + M'V'}{M + M'} - eV';$$

the same as Eqs. (194) and (195); and passing to the limits, non-elasticity on the one hand and perfect elasticity on the other, we have, in the first case, $e = 0$, and

$$v = \frac{MV + M'V'}{M + M'} \quad . \quad . \quad . \quad (213),$$

bodies non-elastic;

$$v' = \frac{MV + M'V'}{M + M'} \quad . \quad . \quad . \quad (214);$$

and in the second, $e = 1$, consequently

$$v = 2\frac{MV + M'V'}{M + M'} - V \quad . \quad . \quad (215),$$

bodies perfectly elastic.

$$v' = 2\frac{MV + M'V'}{M + M'} - V' \quad . \quad . \quad (216).$$

§ 252.—The equations which have just been deduced, are sufficient to make known the circumstances of motion of the centres of gravity of the colliding bodies, for we have seen, § 146, that whenever a body is acted upon in a direction normal to its surface, its centre of gravity will move as though the force were applied directly to that point. But we have also seen, in the same article,

Oblique and eccentric impact;

in the eccentric impact the bodies will rotate; that when the direction of the force does not pass through the centre of gravity, which is the case in the eccentric impact, the body will also have a rotary motion.

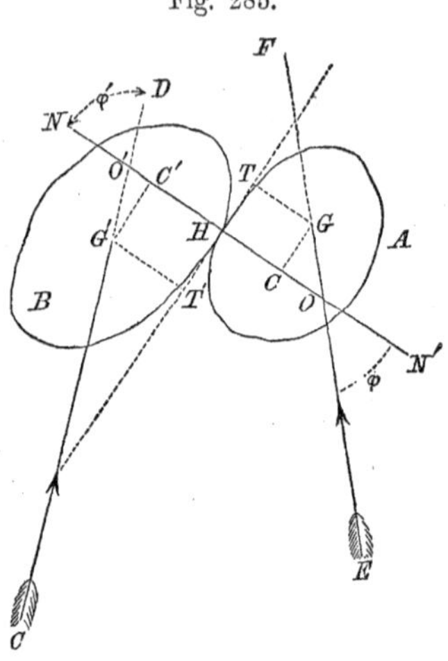

Fig. 285.

Employing the same notation as before, and subtracting Eq. (203) from the identical equation,

$$V \cos \varphi = V \cos \varphi,$$

we find

loss of velocity of one body in direction of normal;

$$V \cos \varphi - v \cos \theta = (1 + e) \frac{M' (V \cos \varphi - V' \cos \varphi')}{M + M'};$$

the first member is the loss of velocity of the body A in the direction of the normal, during the impact; and multiplying both members by the mass of $A = M$, we have, for the quantity of motion lost in the direction of the normal,

motion lost in that direction;

$$M(V \cos \varphi - v \cos \theta) = (1 + e) \frac{MM' (V \cos \varphi - V' \cos \varphi')}{M + M'}.$$

If the force of which either member of this equation measures the intensity, and of which the direction coincides with the normal, does not pass through the centre of gravity, it will give rise to rotary motion. From the centre of gravity G', of the body B, let fall the perpendicular $G'C'$ upon the normal, and denote its length by b; also denote the angular velocity of the body B by $s_{,,}$ and its moment of inertia with reference to an axis through the centre of gravity, and perpendicular to the plane of the normal and centre of gravity, by I_1; then, because the angular velocity

construction;

i. equal to the moment of the impressed force divided by the moment of inertia,

$$s_{,} = (1 + e)\,b\,\frac{MM'}{M + M'} \times \frac{V\cos\varphi - V'\cos\varphi'}{I_1} \quad . \; . \; (217).$$

angular velocity of one of the bodies;

Also let fall from the centre of gravity G of the body A, the perpendicular $G\,C$ upon the normal, and call its length a. Since the reaction of the body B, which is equal to the action of A, does not pass through the centre of gravity of the latter, it will communicate a rotary motion; and, denoting the angular velocity of A by $s_{,,}$, we shall have,

$$s_{,,} = (1 + e)\,a\,\frac{MM'}{M + M'} \times \frac{V\cos\varphi - V'\cos\varphi'}{I_1'} \quad . \; . \; (218);$$

angular velocity of the other;

in which I_1' is the moment of inertia of the body A, in reference to an axis through its centre of gravity and perpendicular to the plane containing this point and the normal.

In what precedes, no reference is made to friction, but it is obvious that this principle cannot be wholly disregarded; for the bodies acting upon each other in the direction of the normal with a pressure of which the measure is

thus far no account has been taken of friction;

$$(1 + e) \cdot \frac{MM'}{M + M'}\,(V\cos\varphi - V'\cos\varphi');$$

this pressure will give rise to friction, whose intensity is measured by

$$f(1 + e) \cdot \frac{MM'}{M + M'}\,(V\cos\varphi - V'\cos\varphi');$$

measure of the friction;

and this acting in the direction of the tangential components of the velocities will accelerate the one and retard the other. Let $U_{,}$ denote the tangential velocity lost by the body A; then, the force exerted to overcome the friction will be measured by

tangential force to overcome friction;

$$M U_{,}.$$

Now if the tangential velocities be equal, it is obvious that the bodies will move together in the direction of the tangent, $M U_{,}$ will be zero, the friction will not be called into action, and the bodies will not rotate from friction. If the tangential velocities differ by a quantity that will make $M U_{,}$ equal to the friction, then will the whole of the latter be exerted to produce rotation. If the tangential velocities be such as to give to $M U_{,}$ any value between these limits, a part only of friction will be exerted, and this part alone will determine the rotation. If the difference of the tangential velocities be such as to make $M U_{,}$ greater than the friction, the bodies will slide along each other and rotate at the same time; the latter motion being due to the entire friction, and the former to the excess of $M U_{,}$ over the value of this force.

limits within which friction may act to produce rotation;

Fig. 285.

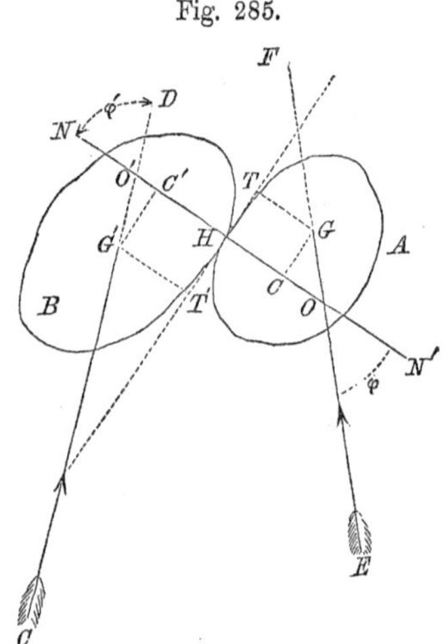

Denote by n, the ratio of the friction to $M U_{,}$, then will

quantity of tangential motion lost;

$$M U_{,} = n f (1 + e) \cdot \frac{M M'}{M + M'} (V \cos \varphi - V' \cos \varphi').$$

Let fall from the centres of gravity of the two bodies the perpendiculars $G T$ and $G' T'$, upon the tangent $T T'$;

denote the length of the first by $b_{,}$ and that of the second by $a_{,}$. Then will the angular velocity of the body B, produced by friction, be

$$nf(1 + e)\, b_{,} \frac{M M'}{M + M'} \cdot \frac{V \cos \varphi - V' \cos \varphi'}{I_1};$$

angular velocity of one body due to friction;

and that of the body A,

$$nf(1 + e)\, a_{,} \frac{M M'}{M + M'} \cdot \frac{V \cos \varphi - V' \cos \varphi'}{I_1'};$$

angular velocity of the other, due to friction;

whence the whole angular velocities of the two bodies will become

$$s_{,} = (1 + e) \cdot \frac{M M'}{M + M'} \cdot \frac{V \cos \phi - V' \cos \phi'}{I_1} \cdot (b + nf b_{,}),$$

$$s_{,,} = (1 + e) \cdot \frac{M M'}{M + M'} \cdot \frac{V \cos \phi - V' \cos \phi'}{I_1'} \cdot (a + nf a_{,}).$$

whole angular velocity of the bodies;

If the balls be spherical and homogeneous, the normal will always pass through the centre of gravity, b and a will reduce to zero, and the rotation will be due to friction alone. If the impact be direct, then φ and φ' will be zero, there will be no tangential components of the velocities $M U_{,,}$ and consequently n will reduce to zero, and the rotation will be due to the eccentricity of the impact.

particular cases of figure.

PART SECOND.

MECHANICS OF FLUIDS.

I.

INTRODUCTORY REMARKS.

Condition of all bodies depends upon the molecular forces; a solid; a liquid; a gas or vapor;

§ 253.—We have seen, § 12, that the physical condition of every body depends upon the relation subsisting among its molecular forces. When the attractions prevail greatly over the repulsions, the particles are held firmly together, and the body is called a *solid*. In proportion as the difference between these two sets of forces becomes less, the body is softer, and its figure yields more readily to external pressure. When these forces are equal, the particles will yield to the slightest force, the body will, under the action of its own weight, and the resistance of the sides of a vessel into which it is placed, readily take the figure of the latter, and is called a *liquid*. Finally, when the repulsive exceed the attractive forces, the elements of the body tend to separate from each other, and require either the application of some extraneous force or to be confined in a closed vessel to keep them together; the body is then called a *gas* or *vapor*, according to the greater or less pertinacity with which the repulsive retain their ascendency over the attractive forces. In the vast range of relation among the molecular forces, from that which distinguishes a solid to that which determines a gas or vapor, bodies are found in all possible conditions—solids run imperceptibly into liquids, and liquids into gases. Hence all classification of bodies founded on their physical properties alone, must, of necessity, be arbitrary. solids, liquids, and vapors run into each other.

§ 254.—Any body whose elementary particles admit of motion among each other, is called a *fluid*—such as water, wine, mercury, the air, and, in general, liquids, gases, and vapors; all of which are distinguished from solids by the great mobility of their particles among themselves. This distinguishing property exists in different degrees in different liquids—it is greatest in the ethers and alcohol; it is less in water and wine; it is still less in the oils, the sirups, greases, and melted metals, that flow with difficulty, and rope when poured into the air. Such fluids are said to be *viscous*, or to possess *viscosity*. Finally, a body may approach so closely both a solid and liquid, as to make it difficult to assign it a place among either class of these bodies, as *paste*, *putty*, and the like. Definitions, &c.; a fluid; viscous fluids; paste; putty.

§ 255.—Fluids are divided in mechanics into two classes, viz.: *compressible* and *incompressible*. The term incompressible cannot, in strictness of propriety, be applied to any body in nature, all being more or less compressible; but the enormous power required to change, in any sensible degree, the volumes of liquids, seems to justify the term, when applied to them in a restricted sense. The *gases* and *vapors* are highly compressible. All liquids will, therefore, be regarded as incompressible; the *gases* and *vapors* as compressible. Classification of fluids; compressible and incompressible: liquids incompressible; gases and vapors compressible

§ 256.—There are many fluids that readily pass from the compressible to the incompressible class, when subjected to moderate increase of pressure, and reduction of temperature. These are called *vapors*, and are such as arise from the application of heat to liquids, particularly when vapors;

confined in closed vessels, as in the instance of steam in boilers. Vapors are generally invisible, and must not be confounded with the mists and clouds which are often seen suspended above the surface of the earth, and which are nothing more than water, in the form of small vesicles filled with air, and supported by the buoyant action of the atmosphere. Others of the compressible fluids are more permanent, requiring very great pressure and reduction of temperature to bring them to a liquid form. All such fluids are called *gases*. The most familiar instance of this class of bodies is the atmosphere which surrounds us on every side and in which we live. It envelops the entire earth, reaches far beyond the tops of our highest mountains, and pervades every depth from which it is not excluded by the presence of solids or liquids. It is even found in the pores of these bodies. It plays a most important part in all natural phenomena, and is ever at work to influence the motions and to modify the results of machinery. It is essentially composed of *oxygen* and *nitrogen*, in a state of mechanical mixture. The former is a supporter of combustion, and, with the various forms of carbon, is one of the principal agents employed in the development of mechanical power.

vapors distinguished from mists and clouds; gases distinguished from vapors; atmosphere; its composition; mechanical use of oxygen;

The existence of air, gases, and vapors, is proved by a multitude of facts. Contained in a flexible and impermeable envelope, they resist pressure like solid bodies. The gas in an inverted glass vessel plunged into water, will not yield its place to the liquid, unless some avenue of escape is provided for it. Those winds, hurricanes, and tornadoes which uproot trees, overturn houses, and devastate entire districts, are but air in motion. Air opposes, by its inertia, the motion of solid bodies through it, and this opposition is called its resistance. Finally, we know that wind is employed as a moter to turn windmills and to give motion to ships of the largest kind.

proof of the existence of vapors and gases; atmospheric resistance; used as a moter.

§ 257.—Many bodies take, successively, the solid, liquid,

or vaporous state, according to the heat to which they are subjected. Water, for instance, is solid in the state of ice and snow, liquid in its ordinary condition, and vapor when heated in a closed vessel. The process by which a body passes from a solid to a liquid state, is called *liquefaction* or *fusion;* from a liquid to a state of vapor, *vaporization* or *volatilization;* that by which a vapor returns to a liquid, *condensation;* and a liquid to a solid, *solidification* or *congelation.* Some bodies appear to take but two of these states, while others constantly present themselves only under one, which is the case with the infusible solids and permanent gases, including among the latter, the atmospheric air; but the number of these bodies is constantly diminishing in the progress of physical science.

Change of state; liquefaction; vaporization; condensation; solidification.

§ 258.—The subject of the mechanics of fluids, is usually divided, as before remarked, into *hydrostatics* and *hydrodynamics*, the former treating of the equilibrium of fluids, and the latter of their motions; and not unfrequently the compressible fluids are discussed under a separate head called *pneumatics.* In the present instance, these divisions will not be adhered to, as it is believed the whole subject may be presented in a manner more connected and perspicuous by disregarding them. And in the discussions which are to follow, the fluid will be considered as without viscosity; that is to say, the particles will be supposed to have the utmost freedom of motion among each other. Such a fluid is said to be *perfect.* The results deduced upon the hypothesis of perfect fluidity will, of course, require modification when applied to fluids possessing sensible viscosity. The nature and extent of these modifications can be known only from experiments.

Mode of considering the subject; hydrostatics; hydrodynamics; pneumatics; perfect fluid.

II.

MECHANICAL PRINCIPLES OF FLUIDS.

Level surface;

§ 258.—From the nature of a fluid, it is obvious that when a force is applied to any one of its particles, the latter must move in the direction of the force, unless prevented by the reaction of the surrounding particles; but these being equally free, can only react to prevent motion, by being supported or acted upon by opposing forces. From this arises a general law, viz.: that when a fluid is in

when at rest;

equilibrio, its free surface is always normal to the resultants of the forces which solicit each of its surface particles. For if the resultant OF of the forces which act upon any one of these particles O were oblique to

normal to the resultant of the forces which act upon the surface particles;

the surface AB, this resultant might be resolved into two components, one OF' normal, and the other OF'' tangent to the surface; the former would be destroyed by the reaction of the fluid mass supposed in equilibrio, while the latter would move the particle along the surface, and with the greater facility in proportion as similar components tend to move the particles to which they are applied in the same direction. Hence the supposition of an oblique resultant is inconsistent with the equilibrium. This free surface which every

Fig. 286.

level surface defined;

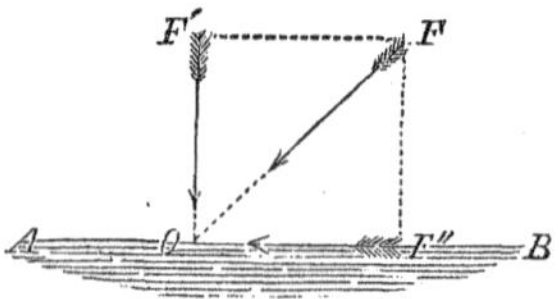

fluid in equilibrio presents in a direction normal to the resultant of the forces which act upon each of its surface particles, is called a *level surface*. Hence every heavy fluid upon the earth's surface in a state of repose, presents its upper or free surface normal to the direction of the force of gravity. If the earth did not rotate about an axis PP', thus giving rise to a centrifugal force, every such surface would be a portion of the surface of a sphere, having its centre at the centre of the earth; but its centrifugal force MC, combined with the weight MG of each element, giving rise to a resultant MN slightly oblique to the direction of the weight, every free surface is in strictness a portion of the surface of a spheroid of revolution, flattened at the poles and protuberant at the equator. The great size of the earth, and the limited field that may be brought under observation at the same instant, will scarcely permit us however to distinguish any visible portion of fluid surface from a plane. Instance, the ponds, lakes, ocean. The same is true of the atmosphere. This fluid being elastic, its elements tend to recede from each other and from the earth's surface; in proportion as it expands, the repulsive action becomes less; the weight of the elements tends to draw them towards the earth; at the upper surface of the atmosphere these opposing forces, which act towards and from the centre of the earth, become equal, and the further retrocession of the particles is impossible. The atmosphere would, under the operation of these causes alone, come to a state of rest, and present an exterior boundary similar to that of the earth.

level surface of heavy fluids;

Fig. 287.

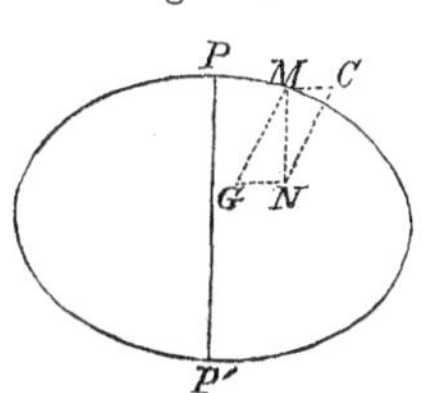

figure of the level surface of heavy fluids;

visible portions sensibly plane;

case of the atmosphere.

§ 260.—Let the vessel $ABDC$ contain a heavy fluid, or a fluid acted on only by its own weight; the upper surface RS will, from what we have seen, be horizontal when at rest; and it is obvious that this position of the surface will not be disturbed, or in the least altered, if the portion of the fluid indicated by the shaded parts of the second figure were to become solid, leaving the fluid portions *E T*, *F H*, *H G*, communicating freely with each other; that is to say, the surfaces at *E*, *F*, and *G*, of the communicating fluid would be upon the same level. Whence we conclude, that a *heavy fluid*, as water or mercury, *poured into several vessels which communicate freely with each other, will, when in equilibrio, have its upper surface in all the vessels on the same level.* This important fact is easily illustrated by experiment. *A* is a vessel at the bottom of which is a horizontal tube connecting freely with the vessels *B* and *C*, and having a stop-cock *D* interposed, so that the connection may be interrupted or established at pleasure. Fill *A* with water, the stop-cock being closed. When the water in *A* is at rest, open the cock *D*; the water will descend in *A* and ascend in *B* and *C* till it comes to the same level in all.

A homogeneous heavy fluid in vessels communicating freely will stand in all at the same level;

Fig. 288.

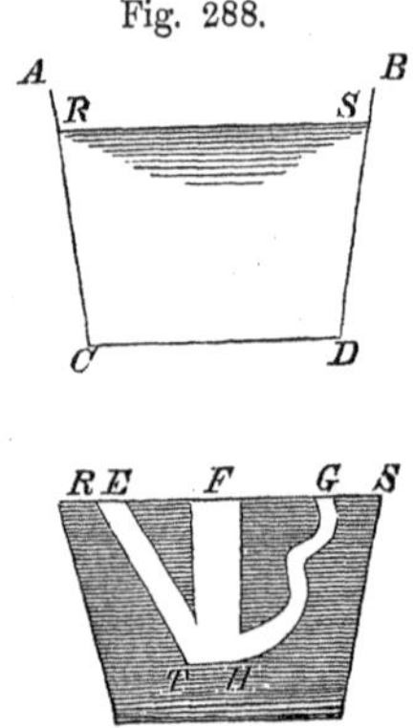

experimental illustration;

Fig. 289.

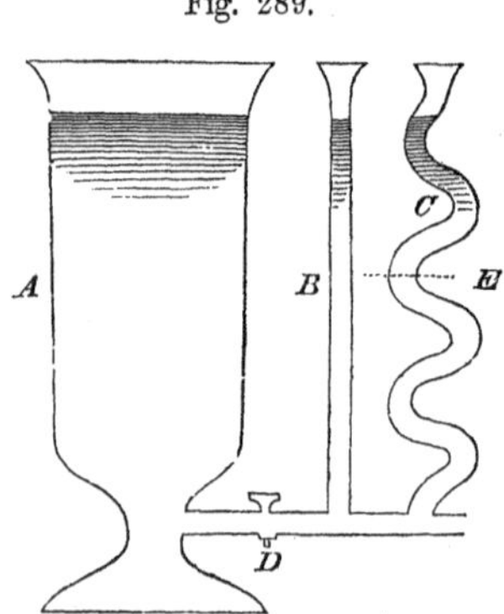

If the vessel *C* be broken off at *E*, the water will overflow at this point till it sinks in the vessels A and B to the level of E.

To the operation of this principle we are indebted for the transfer of water from remote locations to artificial reservoirs for the supply of cities and towns. Springs also owe their existence to it. The greater part of the solid crust of the earth consists of various strata ranged one above another; many of these are of a loose and porous nature and are penetrated with clefts, whilst others are more dense and free from flaws. Through the former of these, rains and melted snows find their way to the latter, where their further progress is for a time checked, till the water accumulates in sufficient quantity to force its way through the sides of hills and mountains, and often at points of considerable elevation. When the harder and impervious strata form the outer crust of mountain ranges, they often force the water to take an oblique underground course through porous strata, that extend to considerable depth and reach to remote districts. Here, if a channel be provided for the water by boring through the hard crust which confines it, it will spout forth or overflow, in its effort to gain the level of its source in the distant mountain. This constitutes an *Artesian well*, a name derived

this principle determines the transfer of water to artificial reservoirs from remote points;

Fig. 290.

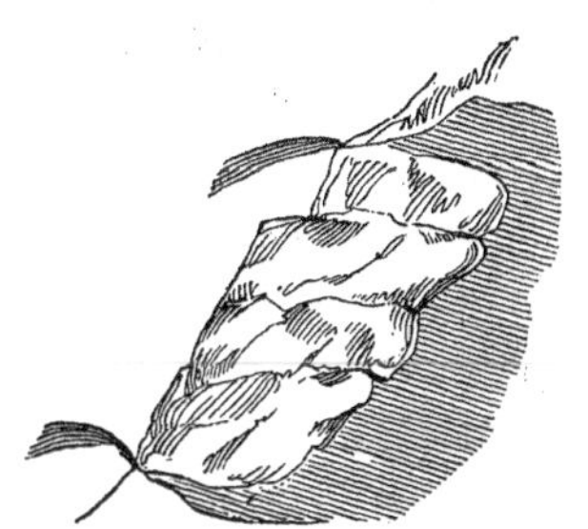

it is also the cause of springs;

Fig. 291.

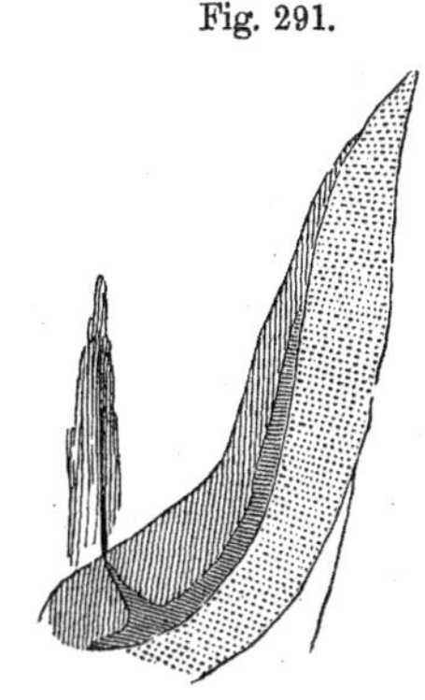

and is the cause of the discharge from Artesian wells.

from the French province Artois, where, according to account, they were first employed.

Principle of equal pressures;

§ 261.—From the principle of fluid level, it is easy to pass to that of equal pressure. Suppose a vessel, $ABDCFE$, in which the branches EF and BDC have a free communication with the part AB; then if water, mercury, wine, or any other fluid, be poured in either at E, A, or C, and the whole be suffered to come to rest, the surface at IK of the fluid in the part AB, at L in the branch EF, and at M in the branch BDC, will be upon the same level.

a heavy fluid; several vessels communicating;

Fig. 292.

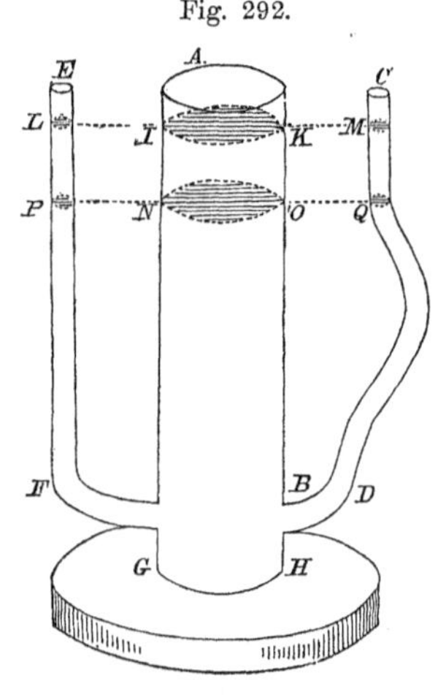

Through the point N, taken at pleasure below the surface of the fluid, conceive a horizontal plane to be passed. It is obvious that the weight of the fluid contained in the vessel below PNQ can contribute nothing to the support of the columns LP, IO, and MQ, since this weight acts downward; and the equilibrium would obtain if the fluid contained in the part of the vessel below PNQ were without weight. This fluid may therefore be regarded as solely a means of communication between the columns LP, IO, and MQ, in such manner that it will transmit the pressure resulting from the weight of the columns LP and MQ to support the weight of IO, and reciprocally. If now, instead of the columns LP, IO, and MQ of the fluid, pistons were applied to the surfaces at P, NO, and Q, and were separately urged downward by pressures respectively equal to the weights of these columns, the

several columns of unequal weights supporting each other;

equilibrium would manifestly obtain in like manner. Or if a pressure equal to that arising from the column MQ be applied to the surface Q, while the columns LP and IO remain, the equilibrio will still subsist, and this, whatever be the directions and sinuosities at D, F, &c. The weight W of the column QM is measured by $b.h.d.g$; in which b is the area of the base at Q, h the height QM, d the density of the fluid, and g the force of gravity. The weight W' of the column IO is measured by $b'.h.d.g$, in which b' is the area of the base NO, the other quantities being the same as before. Dividing the latter by the former, we find

weight of columns of fluid replaced by pressures upon pistons;

$$\frac{W'}{W} = \frac{b'.h.d.g}{b.h.d.g} = \frac{b'}{b} \quad . \quad . \quad (219);$$

ratio of the weights of columns of equal altitudes;

hence, the weights are to each other as the bases b' and b. Now these weights act in the same direction, and are unequal; they cannot, therefore, maintain each other in equilibrio, unless the pressure arising from the column IO were transmitted by the fluid down the vessel NB, up the sinuous vessel BDQ to Q, and there diminished in the ratio of the base NO to that at Q. In like manner, the pressure from the column MQ must be transmitted by the fluid down the tube QDH, up the vessel BN to the base NO, and there increased in the proportion of the base at Q to that at N.

That is, *the forces applied to two pistons in a vessel filled with fluid, will be in equilibrio when their intensities are directly proportional to the areas of the pistons to which they are respectively applied.* If the areas b and b' of the pistons become equal, the forces will be equal, and this, whatever be the actual dimensions of the pistons. Whence we conclude, *that the force impressed upon a fluid, is transmitted by it equally in all directions; and that every surface exposed to the fluid will receive a pressure which is directly proportional to its extent.* Moreover, this pressure will be perpendicular to the surface, for if it were oblique, it might be replaced

forces on two pistons are in equilibrio when proportional to the areas of the pistons;

pressure transmitted equally in all directions;

pressure always normal to the surface;

by its two components, one normal, the other parallel to the surface; the former would be destroyed by the resistance of the surface, while the latter would give motion to the fluid, which is contrary to the supposition that the fluid is in equilibrio.

From Eq. (219) we find

value of the pressure transmitted;

$$W' = W \cdot \frac{b'}{b} \quad . \quad . \quad . \quad . \quad (220);$$

1st. rule;

whence we have this rule for finding the amount of pressure transmitted to any surface, viz.: *Multiply the intensity of the pressing force into the ratio obtained by dividing the area to which the pressure is transmitted, by that to which the force is directly applied.* Making $b = 1$, W will be the pressure upon the unit of surface, and Eq. (220) becomes

pressure transmitted when the pressure is applied to a unit of surface;

$$W' = W \, . \, b' \quad . \quad . \quad . \quad . \quad (221);$$

2d. rule;

whence we have this second rule for finding the pressure transmitted to any given surface, viz.: *Multiply the intensity of the force applied to the unit of surface by the area of the surface to which the pressure is transmitted.*

illustration by the anatomical siphon;

The truth of these deductions is finely illustrated by the *Anatomical Siphon.* A short cylindrical vessel A, made of metal, and open at one end, is connected with an upright glass tube fh, say half an inch in diameter, open at the top. The vessel is filled with water, and closed by tying over it a bladder, on which a plate of wood or metal is laid to receive weights W'. Water is now poured down the glass tube fh; the water in A, with its superincumbent weights W', will be raised by the pressure

Fig. 293.

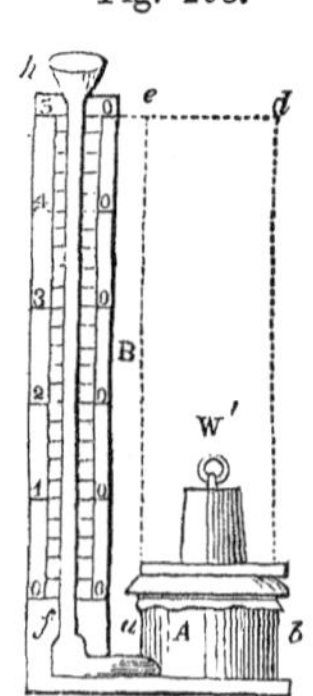

arising from the weight of that portion of the fluid in the glass tube above the level of the bladder. Let this difference of level be 50 inches, then will the volume, in cubic feet, of the pressing water, be

illustrated by a numerical example;

$$\frac{\pi R^2 \times \overset{in.}{50}}{1728} = \frac{3.1416 \times (0.25)^2 \times \overset{in.}{50}}{1728} = \overset{c.\,ft.}{0.00568}.$$

Now one cubic foot of water weighs sixty-two and a half pounds, whence the weight of the pressing column or W becomes, in pounds,

$$W = \overset{lbs.}{62.5} \times 0.00568 = \overset{lb.}{0.355}.$$

weight of the pressing column;

The area of a section of the glass tube is

$$b = \pi R^2 = 3.1416 \times (0.25)^2 = \overset{in.}{0.196};$$

area of a section of the tube;

or, in square feet,

$$b = \frac{0.196}{144} = 0.00136, \text{ nearly}.$$

Let the diameter of the vessel A be one foot then will

diameter of the larger vessel;

$$b' = 3.1416 \times (0.50)^2 = \overset{ft.}{0.7854};$$

and these values of W, b, and b', substituted in Eq. (220), give

$$W' = 0.355\,\frac{0.7854}{0.00136} = \overset{lbs.}{204.8}, \text{ nearly};$$

weight sustained;

that is to say, the trifling weight of three tenths of a pound sustains in equilibrio a weight of more than two hundred and four pounds; a result usually denominated the *hydrostatic paradox.*

hydrostatic paradox;

if the bladder were removed the water would rise in the larger vessel;

If the bladder were removed, and the vessel extended upward to the line ed, on a level with the fluid in the tube, the water would rise in it to that height, when it would come to rest. The volume of the added water, in cubic feet would be

$$\frac{\overset{in.}{50}}{12} \times 0.7854 = 3.272;$$

and allowing $62\frac{1}{2}$ pounds to each cubic foot, the weight of distilled water at 60° Fah. gives

verification.

$$3.272 \times 62\tfrac{1}{2} = \overset{lbs.}{204.5}, \text{ nearly,}$$

as before.

III.

WORK OF THE POWER AND OF THE RESISTANCE.

Multiplication of power by the principle of equal transmission of pressure;

§ 262.—It follows from Eq. (220), that a given power may be multiplied at pleasure by this principle of equal transmission of pressure. It will be sufficient for this purpose, to provide a strong vessel for the reception of a fluid, and to connect with it a pair of pistons whose surfaces bear to each other any desired ratio; the power F being applied to the smaller piston b will be transmitted to the larger b' and made to hold in equilibrio or overcome almost any given resistance R applied to the latter. But we are not, there-

Fig. 294.

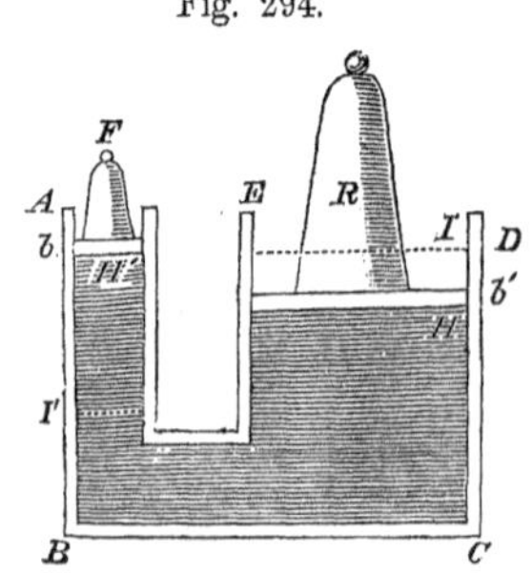

fore, to infer that there is any gain in the quantity of work performed, for if we multiply Eq. (220) by the distance $HI = s'$ through which the larger piston may have been moved by the pressure transmitted to it, we have, by writing R for W', and F for W, no work gained, however;

$$Rs' = F \cdot \frac{s'b'}{b} \quad . \quad . \quad . \quad . \quad . \quad (222).$$

work of the resistance;

The product $s'b'$, being the area of the larger piston into the distance HI, is the measure of the volume of fluid which has passed into the chamber CE, by the action of the power F upon the smaller piston; and if we regard the water as incompressible, this must be equal to the volume of fluid which has been pressed out of the chamber AB. Supposing the smaller piston to have been depressed to I', and denoting the distance $H'I'$ by s, this latter volume will be measured by sb, and, therefore, from what has just been remarked,

$$s'b' = sb;$$

volumes of the fluid equal;

whence

$$s = \frac{s'b'}{b};$$

which, substituted above, gives

$$Rs' = Fs \quad . \quad . \quad . \quad . \quad . \quad . \quad (223).$$

work of power equal to that of resistance;

The first member of this equation is the work performed by the resistance, the second that performed by the power, whence we conclude, that *in hydraulic machines depending upon the transmission of pressure, as in other machines, the work of the power is equal to that of the resistance.* conclusion;

If the friction of the pistons against the sides of their respective chambers and the viscosity of the fluid be taken into the account, the work of these must be added to the friction and viscosity;

term Rs', which would make the effective quantity of work, measured by Rs', actually less than the work of the power. What then is gained? The answer is the same as before, viz.: the machine gives to a feeble power the ability to perform, by a succession of efforts, an amount of work which it could not accomplish by a single one. It would be quite within the physical capabilities of an individual to raise to the summit of a wall a ton of bricks, by taking a few bricks at a time, whereas an effort to elevate the whole at once by his unassisted strength would prove an utter failure. And this is true of all kinds of machinery; whenever a given amount of work is accomplished by the application of a diminished power, the space through which the latter is exerted must be proportionally increased.

the hydraulic machine enables a feeble power to perform what it could not without it;

general principle of all machines;

Had this principle, together with the incompressibility of the fluid, been assumed at the outset, it would have been an easy matter to deduce Eq. (220), and therefore the principle of the equal transmission of pressure; for, the volume of the fluid remaining the same, we should have

this principle employed to prove that of equal pressure;

$$s' b' = s b,$$

and the quantity of work of the power and resistance being equal, gives

$$R s' = F s;$$

dividing the first of these equations by the second, we find

$$\frac{b'}{R} = \frac{b}{F},$$

whence

$$F : R :: b : b';$$

pressures are proportional to the surfaces.

that is to say, the pressures are directly proportional to the areas of the pistons to which they are applied, when there is an equilibrium, or when the pistons have a uniform motion.

§ 263.—One of the most interesting and important applications of the principle of equal transmission of pressure is exhibited by the *Hydraulic* or *Bramah's Press.* Hydraulic press; The main features of this machine are the following: A large and small metallic cylinder A and a, are made to communicate freely with each other by a duct-pipe r. Water stands in both of the cylinders, and each is provided with a strong piston. The piston S of the larger cylinder carries a strong head-plate P, which works in a frame, so as to move directly towards or from a plate R which is stationary. The substance to be pressed is placed between these two plates. The piston in the small tube a is worked by a lever $c\,d$, of the second order, having its fulcrum at c, the piston-rod being attached at b, while power is applied at d. The pressure exerted by the smaller piston on the water is transmitted by the latter to the piston S.

Fig. 295.

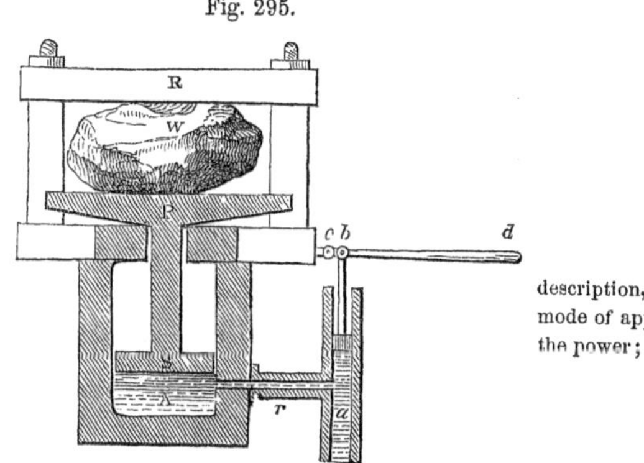

description, and mode of applying the power;

Let the diameter of the cylinder a be half an inch, that of the larger 200 inches, then will

its power illustrated by an example;

$$\frac{b'}{b} = \frac{(200)^2}{(\frac{1}{2})^2} = 160000;$$

and suppose the distance $c\,d$ to be equal to 50 inches, and

cb to be one inch, and let a man throw his weight, say 150 pounds, on the point d; then from the property of the lever will the force F, applied to the smaller piston, be given by the proportion

data;

$$\overset{in.}{1} : \overset{in.}{50} :: \overset{lbs.}{150} : F;$$

whence

power applied at smaller piston;

$$F = 150 \times 50 = \overset{lbs.}{7500}.$$

Substituting these values for F and $\frac{b'}{b}$ in Eq. (222), and omitting the common factor s', we find

value of resistance;

$$R = \overset{lbs.}{7500} \times 160000 = \overset{lbs.}{1200000000};$$

thus an effort equal in intensity to a weight of one hundred and fifty pounds applied at d, is capable of holding in equilibrio a power, or of maintaining in uniform motion a body subjected to a constant resistance, equal to one billion two hundred million pounds.

Dividing both members of Eq. (223) by F, we find

path of the power at smaller piston;

$$s = \frac{R \cdot s'}{F};$$

substituting the above values for R and F, and suppose the piston-head to have been raised through the distance of one foot, we have

its numerical value for one foot of path of the resistance;

$$s = \frac{1200000000}{7500} = \overset{ft.}{160000};$$

and because the power applied at d must pass over 50 times this distance, we find

$$160000 \times 50 = \overset{ft.}{8000000},$$

or

$$\frac{8000000}{5280} = \overset{miles.}{1515},$$ path of the power;

for the distance described by the power to compress the resistance one foot, or to raise a weight equivalent to the resistance through that height. The hydraulic press is used in the arts to press paper, cloth, hay, to uproot trees, to test the strength of ropes, chains, building materials, and guns; and two were recently employed with success to raise, through a vertical height of more than one hundred feet, the great iron viaduct-tube, weighing upward of eighteen hundred tons, over the Menai Straits. uses of the hydraulic press.

IV.

PRESSURE OF HEAVY FLUIDS.

§ 264.—Let us now examine the pressure which a heavy fluid exerts on the base of a vessel in which it is contained. For this purpose, let $ABDC$ be a vessel containing a heavy fluid, as water, in equilibrio. The upper surface AB of the fluid will be horizontal. Conceive a horizontal plane GH to be passed, and suppose the fluid below this plane, or that contained in the portion $GCDH$, to be devoid of weight; then it is obvious, from our previous principles, that the weight of any slender vertical column, as EI, will exert a pressure Pressure of heavy fluids; the fluid below the horizontal stratum devoid of weight;

Fig. 296.

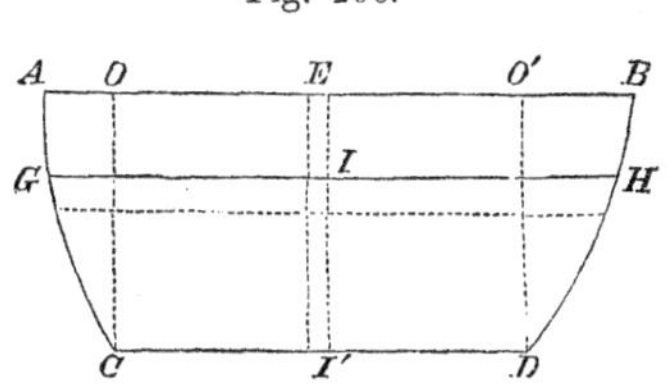

at I, which is distributed equally in all directions through the fluid $G\,C\,D\,H$, and that this pressure acts equally upward to oppose the descent of the other columns which stand vertically over the plane GH; the column EI alone keeps, therefore, in equilibrio all the other columns of the mass $A\ G\ H\ B$; consequently, the mass $G\,C\,D\,H$, being still supposed without weight, there will result no pressure upon the base CD, except that which arises from the weight of a single filament EI, which being transmitted equally to all the points of the base CD, the pressure on the latter will be given by Eq. (220); that is, by

each elementary column sustaining all the others;

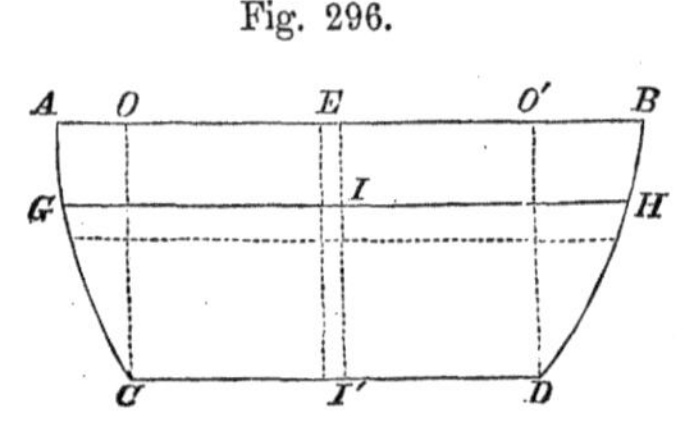

Fig. 296.

pressure upon the base;

$$W' = W \cdot \frac{b'}{b};$$

in which W is the weight of the column EI, b the area of its base, b' the area of the base CD, and W' the pressure which it sustains.

Denoting the height of the column EI by h, its weight W will be given by

weight of the pressing filament;

$$W = h \,.\, b \,.\, D \,.\, g;$$

in which D denotes the density of the fluid, and g the force of gravity.

Substituting this above for W, we find

pressure upon the base;

$$W' = h \,.\, b' \,.\, D \,.\, g \quad . \quad . \quad . \quad (224).$$

If now the plane GH be depressed so as to leave all the heavy fluid above it, this plane will coincide with the bottom, I will come to I', and h will become the vertical height EI' of the surface of the fluid above the base.

But the product $b'h$ is obviously the volume $COO'D$ of the fluid contained in a right cylinder or prism having for its base, the base of the vessel; $D.b'.h$ is the mass of this cylinder or prism, and $D.b'.h.g$ is its weight. Whence we conclude, that the pressure exerted by a heavy fluid upon the horizontal base of a vessel containing it, *is equal to the weight of a column of this fluid, whose base is the base of the vessel, and whose altitude is equal to the depth of this base below the surface of the fluid.* weight of the column which measures this pressure;

In this measure for the pressure on the base of a vessel containing a heavy fluid, there is nothing at all relating to the figure or actual volume of the vessel, and we are, hence, to infer that this pressure is wholly independent of both, and will always be the same whenever the area of the base and altitude of the fluid are the same. The right cylinder, inverted and erect truncated cones, having equal inferior bases B, B, B, and the same altitude h, will, when filled, contain very different volumes of fluid, yet the bases will all experience the same amount of pressure from the weight of the fluid, if it be the same in kind, or of the same density. pressure independent of figure of vessel and quantity of the pressing fluid; illustration; right cylinder, truncated cone, both erect and inverted;

Fig. 297.

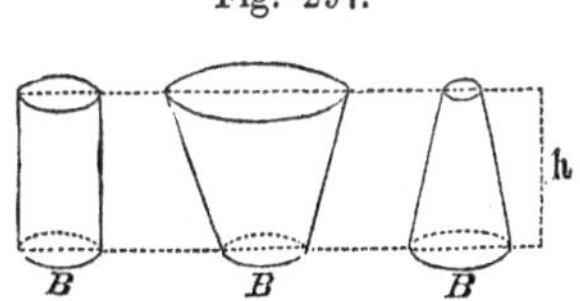

Fig. 298.

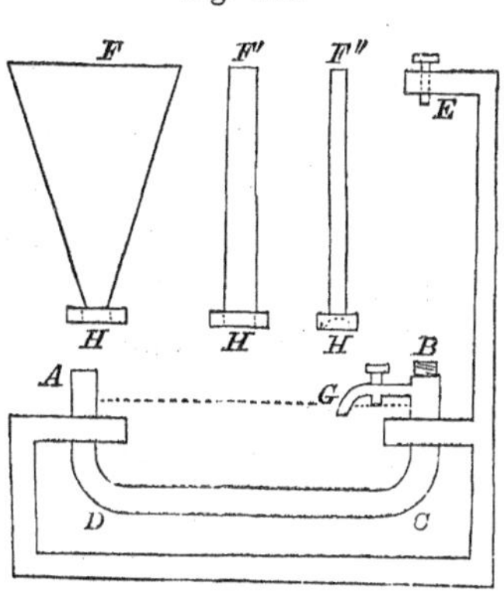

The experimental verification of this

apparent paradox is easy. $A\,D\,C\,B$ is a glass tube, of which the ends are open and bent upward; the end B is furnished with a brass ferrule upon which a screw is cut for the reception of a mate-screw H around the bottom of the vessels F, F', and F'', also open at both ends. On the end A is a sliding ring of metal or wood. At E is a short wire that may be moved up and down, and is held in any desired position by friction.

experimental verification of this fact;

description of the apparatus for the purpose;

Fig. 298.

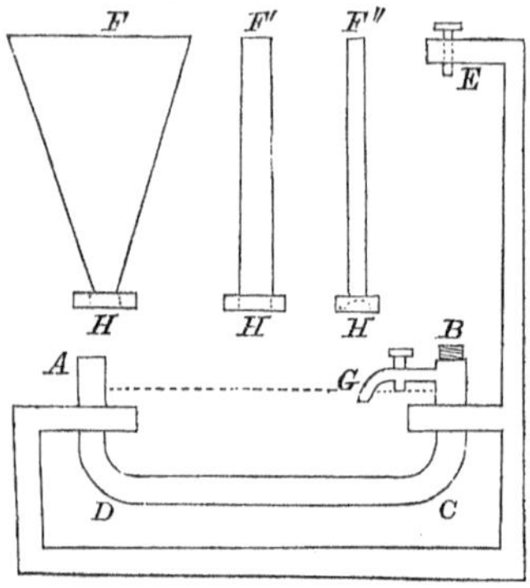

Pour mercury in either end of the bent tube till it rises to any desired level, say that of the dotted line; next, screw either of the vessels, say F, on its place at B, and fill it with water. The water passing freely through to the surface of the mercury will press upon the latter by its weight and force it up the end A. When both fluids come to rest, move the ring on the end A to a level with the mercury to mark its place, and press the wire E down to the surface of the water to determine its height. Now draw off the water by the stop-cock G, remove the vessel F and replace it by F', and fill with water as before; when the level of the water reaches the end of the wire E, the mercury will be found to have reached the ring on the end A. The experiment being repeated with the slender vessel F'', not even half as thick as the tube $A\,D\,C\,B$, the mercury will again be found at the ring. In all these experiments, the base pressed is the same, being a section of the bent tube at the level of the mercury; and the altitude is the same, being the difference of level of the mercury in the end B and lower extremity of the wire E,

details of the experiment;

deductions;

when the mercury in the end A stands at the level of the ring. The quantities of water employed in the three cases are very different, and yet the pressures exerted by their weights are the same. conclusion.

§ 265.—The pressure of a heavy fluid upon a horizontal plane, enables us to pass to that on a plane inclined under any angle whatever to the horizon, and thence to the pressure on a curved surface. Pressure of a heavy fluid against inclined surfaces;

Let $A\,B\,D\,C$ be a vessel with plane or curved sides, and filled with a heavy fluid; suppose $G\,H$ and $G'\,H'$ to be two horizontal planes indefinitely near each other. The layer of fluid between these planes may be considered as without weight, and as transmitting the pressure of the superincumbent fluid to the surface of the vessel with which this layer is in contact; and the pressure upon this surface will be the same as though it were in either of the two planes in question. Designating the extent of this elementary surface by b', and the depth $E\,I$ by h', the measure of this pressure will be

Fig. 299.

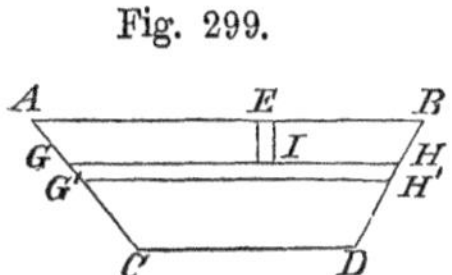

$$D \,.\, g \,.\, b' \,.\, h';$$

pressure upon an elementary inclined surface;

in which D and g denote respectively the density of the fluid and force of gravity. In like manner, the pressure upon any other elementary portions b'', b''', b'''', &c., of the surface at distances h'', h''', and h'''', &c., respectively, below the upper surface of the fluid, will be

$$D \,.\, g \,.\, b'' \,.\, h'', \qquad D \,.\, g \,.\, b''' \,.\, h''', \text{ \&c.};$$

similar pressures;

and the pressure upon the entire surface will obviously be the sum of these; or, if the total pressure be denoted by

32

P, then will

total pressure upon the entire surface;

$$P = Dg\,(b'h' + b''h'' + b'''h''' + \&c.).$$

But if we take the upper surface of the fluid as a plane of reference, and denote by b the entire area of which b', b'', &c., are the elements, and of which the distance of the centre of gravity from this plane of reference is h, then, from the principle of the centre of gravity, will

$$bh = b'h' + b''h'' + b'''h''' + \&c;$$

which, substituted above, gives

value of this pressure in weight;

$$P = D.g.b.h. \quad . \quad . \quad . \quad (225);$$

expressed in words;

that is to say, *the pressure exerted by a heavy fluid against the surface of any vessel in which it is contained, is measured by the weight of a column of the fluid having for its base the surface pressed, and for its altitude the depth of the centre of gravity of this surface below the upper level of the fluid.*

example first;

Example 1*st.* Required the pressure against the inner surface of a cubical vessel filled with water, one of its faces being horizontal. Call the edge of the cube a, the area of each face will be a^2, the distance of the centre of gravity of each vertical face below the upper surface will be $\frac{1}{2}a$, and that of the lower face a; whence, the principle of the centre of gravity gives,

Fig. 300.

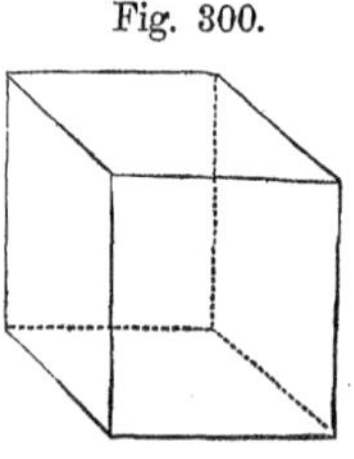

distance of centre of gravity below the surface;

$$h = \frac{4a^2 \times \frac{1}{2}a + a^2 \times a}{5a^2} = \tfrac{3}{5}a.$$

Again,

surface pressed;

$$b = 5a^2;$$

and these, substituted in Eq. (225), give

$$P = D \,.\, g \,.\, b \,.\, h = D \,.\, g \,.\, 3\, a^3.$$ value of the pressure;

Now $Dg \times 1^3 = Dg$, is the weight of a cubic foot of water $= 62.5$ lbs. whence

$$P = \overset{lbs.}{62.5} \times 3\, a^3.$$ in pounds;

Make $a = 7$ feet, then will

$$P = 62.5 \times 3 \times (7)^3 = \overset{lbs.}{27562.5}.$$ its numerical value

The weight of the water in the vessel is $62.5\, a^3$, yet the pressure is $62.5 \times 3\, a^3$, whence we see that the outward pressure to break the vessel, is three times the weight of the fluid. conclusion;

Example 2*d*. Let the vessel be a sphere filled with mercury, and let its radius be R. Its centre of gravity is at the centre, and therefore below the upper surface at the distance R. The surface of the sphere being equal to that of four of its great circles, we have example second;

Fig. 301.

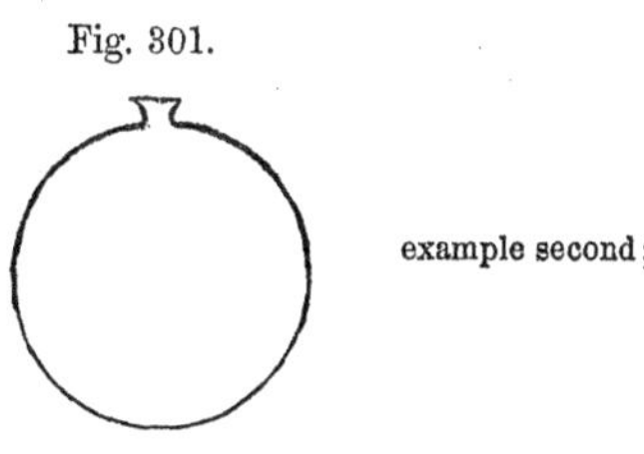

$$b = 4\, \pi\, R^2;$$ surface pressed;

whence

$$b \,.\, h = 4\, \pi\, R^3;$$ volume whose weight is equal to the pressure;

and, Eq. (225),

$$P = 4\, \pi \,.\, D \,.\, g \,.\, R^3.$$ whole pressure;

The quantity $Dg \times 1^3 = Dg$, is the weight of a cubic foot of mercury = 843.75 lbs., and therefore, substituting the value of $\pi = 3.1416$,

pressure in pounds;

$$P = 4 \times 3.1416 \times \overset{lbs.}{843.75} \,.\, R^3.$$

Now suppose the radius of the sphere to be two feet, then will $R^3 = 8$, and

its numerical value;

$$P = 4 \times 3.1416 \times \overset{lbs.}{843.75} \times 8 = \overset{lbs.}{84822.4}.$$

The volume of the sphere is $\frac{4}{3}\pi R^3$; and the weight of the contained mercury will therefore be $\frac{4}{3}\pi R^3 g D = W$. Dividing the whole pressure by this, we find

ratio of weight of pressing fluid to pressure;

$$\frac{P}{W} = 3;$$

whence the outward pressure is three times the weight of the fluid.

example third; *Example 3d.* Let the vessel be a cylinder, of which the radius r of the base is 2, and altitude l, 6 feet. Then will

$$b \,.\, h = \pi r l (r + l) = 3.1416 \times 2 \times 6 \times 8;$$

which, substituted in Eq. (225),

value of pressure;

$$P = 301.5936 \times Dg,$$

and

weight of pressing fluid;

$$W = 3.1416 \times 2^2 \times 6 \times Dg = 75.398 \times Dg;$$

whence,

ratio of weight to pressure.

$$\frac{P}{W} = \frac{301.5936 \times Dg}{75.3984 \,.\, Dg} = 4;$$

that is, the pressure against the vessel is four times the weight of the fluid.

§ 266.—Although the pressure of a heavy fluid depends upon the position of the centre of gravity of the surface pressed, yet the resultant of all the elementary pressures passes through a different point, the position of which for a plane surface may be thus found. Let EIF be any plane, and MN the intersection of this plane produced with the upper surface of the fluid which presses against it. Denote the area of any elementary portion n of the plane EIF by b'; and let m be the projection of its place upon the upper surface of the fluid; draw mM perpendicular to MN, and join n with M by the right line nM, the latter will also be perpendicular to MN, and the angle nMm will measure the inclination of the plane EIF to the surface of the fluid. Denote this angle by φ, the distance mn by h', and Mn by r'; then will

Centre of pressure;

Fig. 302.

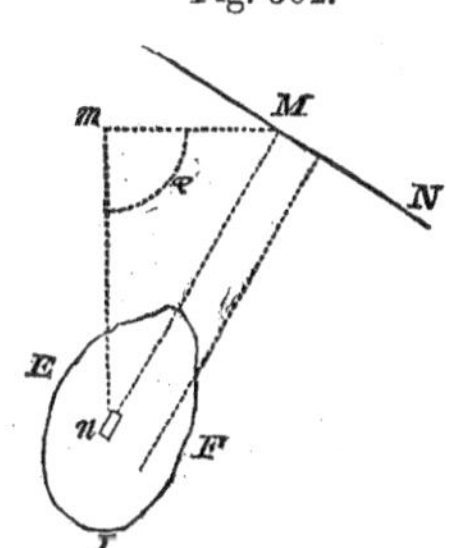

geometrical representation and notation;

$$h' = r' \sin \varphi.$$

distance of an elementary pressed surface below the fluid surface;

The pressure of the fluid upon the element n will, Eq. (225), be

$$D \,.\, g \,.\, b' \,.\, h' = D g b' r' \sin \varphi;$$

pressure upon this element;

and its moment, in reference to the line MN as an axis,

$$D g b' r'^2 \sin \varphi;$$

its moment;

and for any other elements of which b'', b''', &c., denote the areas, we have, in like manner,

moments of the elementary pressures;

$$D\,g\,b''\,r''^2 \sin\varphi,$$

$$D\,g\,b'''\,r'''^2 \sin\varphi,$$

$$\&c., \qquad \&c.$$

Denoting by h the depth of the centre of gravity of the area EIF below the surface of the fluid, and by r the distance of that point from the line MN, we shall have

depth of centre of gravity of the whole area pressed;

$$h = r \sin\varphi;$$

and, for the total pressure upon EIF,

whole pressure;

$$P = D\,.\,g\,.\,b\,.\,h = D\,g\,b\,r \sin\varphi,$$

in which b denotes the area of EIF; and if x denote the distance of the point of application of this pressure from the line MN, its moment will be

moment of the entire pressure;

$$D\,g\,b\,r \sin\varphi\,.\,x.$$

But the moment of the entire pressure must be equal to the sum of the moments of the partial pressures, and hence

$$D\,g\,b\,r\,x \sin\varphi = D\,g \sin\varphi\,(b'\,r'^2 + b''\,r''^2 + b'''\,r'''^2 + \&c.);$$

whence

distance of the point of total pressure from the axis;

$$x = \frac{b'\,r'^2 + b''\,r''^2 + b'''\,r'''^2 + \&c.}{b\,r} \quad . \; . \; (226).$$

The numerator of the second member, is the moment of inertia of the plane EIF; the denominator is the product of the area of the plane itself by the distance of its centre of gravity from the axis, and as a similar expression would result if the pressures were referred to any other line in the plane EIF as an axis, it follows from § 184, Eq. (86), that the resultant pressure passes through the centre of percussion of the surface pressed. This point is called the *centre of pressure. It is that point in the surface to which, if a single force be applied in a direction contrary and equal to the total pressure exerted upon it, the surface will remain in equilibrio.*

Interpretation of the last equation;

Fig. 302.

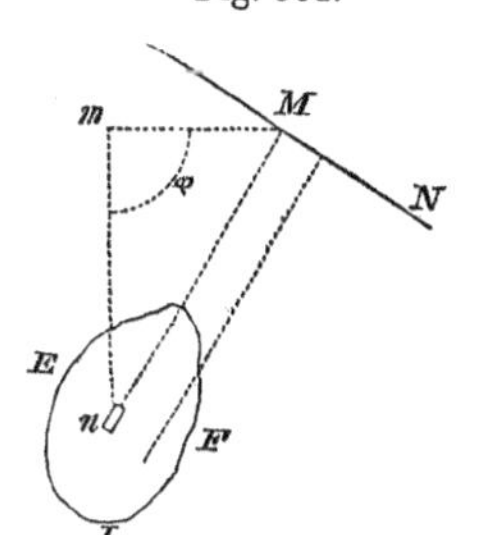

centre of pressure;

defined;

coincident with centre of percussion.

§ 267.—The principles which have now been explained, are of high practical importance. It is not only interesting, but necessary, often to know the precise amount of pressure exerted by fluids against the sides of vessels and obstacles exposed to their action, to enable us so to adjust the dimensions of the latter as to give them sufficient strength to resist. Reservoirs in which considerable quantities of water are collected and retained till needed for purposes of irrigation, the supply of cities and towns, or to drive machinery; dykes to keep the sea and lakes from inundating low districts; artificial embankments constructed along the shores of rivers to protect the adjacent country in times of freshets; boilers in which are pent up elastic vapors in a high state of tension, to be worked off at pleasure to propel boats and cars, and to give motion to machinery generally, are examples.

Application of the preceding principles;

objects to which they are applicable;

thickness of the sustaining wall of a reservoir;

Let $ABCD$ be a section or profile of the wall of a reservoir, MN the upper surface of the water, and EE' the bottom. Denote the length of the wall by l, the depth NE of the water against its face, supposed vertical, by d; then will the surface pressed be measured by $l\,d$; the distance of the centre of gravity of this surface from the upper level of the water will be $\frac{1}{2}d$, whence the whole pressure will be

Fig. 303.

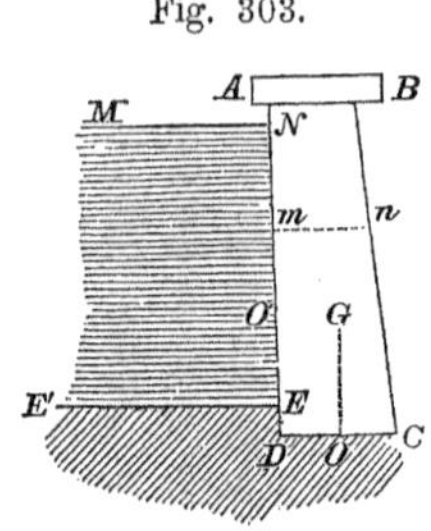

pressure against the face;

$$\frac{D\,.\,g\,.\,l\,.\,d^2}{2},$$

in which D is the density of the water, and g the force of gravity. The inner surface of the wall being vertical, this pressure is exerted in a horizontal direction, and must be resisted by the wall. Now the wall, if it move at all, may either slide along its base DC, or turn about the horizontal edge passing through C. First, let us suppose it slides. Denote the depth of the face AD by d', the mean thickness mn by t; then will the weight of the wall be

suppose the wall may slide;

weight of the wall;

$$D'\,.\,g\,.\,l\,.\,d'\,.\,t;$$

and, denoting the coefficient of friction between the wall and earth by f, the whole friction will be

friction on the ground;

$$f\,.\,D'\,.\,g\,.\,l\,.\,d'\,.\,t,$$

in which D' is the mean density of the wall; and the condition of stability will be satisfied as long as we have

condition of stability;

$$\frac{D\,g\,l\,d^2}{2} = f D' g\,l\,d'\,t;$$

from which we find

$$t = \frac{D}{D'} \times \frac{d^2}{2fd'}.$$

value of mean thickness;

The density of water is usually taken as unity, and on ordinary earth, the value of f, for masonry, does not vary much from $\frac{1}{3}$, whence

$$t = \frac{3\,d^2}{2\,D'\,d'}.$$

value of the thickness in ordinary cases;

The thickness is the only unknown quantity, since d and d' must result from the capacity of the reservoir.

If the wall tend to turn about the edge C, then must the moment of its weight be equal to the moment of the pressure when both are taken in reference to that line. Let G be the centre of gravity of the profile $ABCD$, and denote the distance CO of its projection upon the base of the wall from C, by r. Then, from the assumed figure of the profile, we shall have

suppose the wall may rotate about the front line of its base;

$$\frac{r}{t} = n, \qquad \text{or} \qquad r = n\,t,$$

ratio of lever arm of the wall to its thickness;

in which n is known; and the moment of the weight of the wall will be

$$D' . g . l . d' . t^2 . n.$$

moment of the weight of wall;

The centre of pressure O', being that of a rectangle of which the side through N is horizontal, is at a distance below N equal to $\frac{2}{3}$ of NE, or from the bottom point E equal to $\frac{1}{3}d$; and adding the distance ED denoted by a, the moment of the pressure, in reference to C, will be

$$\frac{D\,g\,l\,d^2}{2}\left(\tfrac{1}{3}d + a\right);$$

moment of the fluid pressure;

and, to insure stability, we must have

condition of stability;

$$D' g l d' t^2 n = \frac{D g l d^2}{2} (\tfrac{1}{3} d + a);$$

whence

thickness of the wall;

$$t = \sqrt{\frac{1}{6n} \cdot \frac{D}{D'} \cdot \frac{d^2 (d + 3a)}{d'}}.$$

If the water come to the bottom of the wall, and the reservoir be full, then will

$$a = 0, \qquad d = d',$$

and

$$t = d \cdot \sqrt{\frac{1}{6n} \cdot \frac{D}{D'}}.$$

thickness of water-pipes, boilers, &c.;

Next, let $A B C$ be a section of a cylindrical water-pipe or boiler perpendicular to the axis, the inner surface of which is subjected to a pressure of p pounds on each superficial unit. Denote by R the radius of the interior circle, and by l the length of the pipe or boiler parallel to the axis; then will the surface pressed be measured by

Fig. 304.

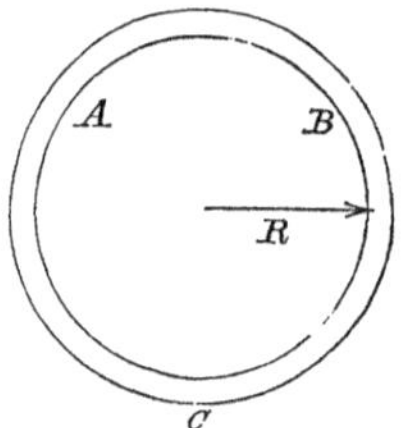

surface pressed;

$$2 \pi R l,$$

and the whole pressure, by

whole pressure;

$$2 \pi R l p.$$

If, in virtue of this pressure, the pipe stretches so that its interior radius becomes $R + r$, it is obvious that the small distance r will denote the path described by the whole pressure, and its quantity of work will be suppose the pipe to stretch;

$$2 \pi R l p r.$$

quantity of work;

The interior circumference before the application of the pressure was $2 \pi R$, and afterward, $2 \pi (R + r)$; the difference of which, or

$$2 \pi (R + r) - 2 \pi R = 2 \pi r,$$

path of the resisting molecular action;

is obviously the distance through which the resisting molecular forces of the material of which the pipe or boiler is made, have acted during the stretching process. Denote the resistance which the material of the pipe or boiler is capable of opposing, without losing its elasticity, to a stretching force on a section of one superficial unit, by B; the length of the pipe or boiler by l; and its thickness by t. The intensity of the force which a section parallel to the axis is capable of resisting will be $B l t$, and its quantity of work

$$B l t \times 2 \pi r.$$

the quantity of work of this force;

But by virtue of the principle of the transmission of work, this must be equal to the work of the pressure, and we have

$$2 \pi B l t r = 2 \pi R l p r;$$

condition of stability;

whence

$$t = \frac{R p}{B}.$$

thickness.

The value of p is easily estimated in the case of water in a pipe, by the rules just given. In the case of steam in a boiler, it may with equal ease be found by rules to be given presently. The value of B is readily obtained from the following table giving the results of experiments on the strength of materials:—

TABLE.

THE TENACITIES OF DIFFERENT SUBSTANCES, AND THE RESISTANCES WHICH THEY OPPOSE TO DIRECT COMPRESSION.

SUBSTANCES EXPERIMENTED ON.	Tenacity in Tons per Square Inch.	Name of Experimenter.	Crushing Force in Tons per Sq. Inch.	Name of Experimenter.
Wrought iron, in wire from 1-20th to 1-30th of an inch in diameter - -	60 to 91	Lamé		
in wire, 1-10th of an inch	36 to 43	Telford		
in bars, Russian (mean)	27	Lamé		
English (mean)	25½	—		
hammered - -	30	Brunel		
rolled in sheets, and cut lengthwise - - - -	14	Mitis		
ditto, cut crosswise	18	—		
in chains, oval links 6 in. clear, iron 1½ in. dia.	21½	Brown		
ditto, Brunton's, with stay across link - -	25	Barlow		
Cast iron, quality No. 1 - -	6 to 7¾	Hodgkinson	38 to 41	Hodgkinson
2 - -	6 to 8	—	37 to 48	—
3* -	6 to 9¾	—	51 to 65	—
Steel, cast - - - - - - -	44	Mitis		
cast and tilted - - -	60	Rennie		
blistered and hammered	59½	—		
shear - - - - - - -	57	—		
raw - - - - - - - -	50	Mitis		
Damascus - - - - -	31	—		
ditto, once refined -	36	—		
ditto, twice refined -	44	—		
Copper, cast - - - - -	8½	Rennie	52	Rennie
hammered - - - -	15	—	46	—
sheet - - - - - - -	21	Kingston		
wire - - - - - - -	27⅓			
Platinum wire - - - - -	17	Guyton		
Silver, cast - - - - -	18	—		
wire - - - - - - -	17	—		
Gold, cast - - - - - - -	9	—		
wire - - - - - - -	14	—		
Brass, yellow (fine) - - -	8	Rennie	73	—
Gun metal (hard) - - -	16	—		
Tin, cast - - - - - - -	2	—	7	—

* The strongest quality of cast iron, is a Scotch iron known as the Devon Hot Blast, No. 3: its tenacity is 9¾ tons per square inch, and its resistance to compression 65 tons. The experiments of Major Wade on the gun iron at West Point Foundry, and at Boston, give results as high as 10 to 16 tons, and on small cast bars, as high as 17 tons.—See Ordnance Manual, 1850, p. 402.

TABLE—*continued.*

SUBSTANCES EXPERIMENTED ON.	Tenacity in Tons per Square Inch.	Name of Experimenter.	Crushing Force in Tons per Sq. Inch.	Name of Experimenter.
Tin wire - - - - - - -	3	Rennie		
Lead, cast - - - - - - -	4-5ths	—	3½	Rennie
milled sheet - - - -	1½	Tredgold		
wire - - - - - - -	1.1	Guyton		
Stone, slate (Welsh) - - -	5.7			
Marble (white) - - -	4	- -	1.4	—
Givry - - - - - - -	1			
Portland - - - - -	⅛	- -	1.6	—
Craigleith freestone -	- -	- -	2.4	—
Bramley Fall sandstone	- -	- -	2.7	—
Cornish granite - - -	- -	- -	2.8	—
Peterhead ditto - - -	- -	- -	3.7	—
Limestone (compact blk)	- -	- -	4	—
Purbeck - - - - -	- -	- -	4	—
Aberdeen granite - -	- -	- -	5	—
Brick, pale red - - - -	.13	- -	.56	—
red - - - - - - -	- -	- -	.8	—
Hammersmith (pavior's)	- -	- -	1	—
ditto (burnt) -	- -	- -	1.4	—
Chalk - - - - -	- -	- -	.22	—
Plaster of Paris - - - -	.03			
Glass, plate - - - - -	4			
Bone (ox) - - - - - -	2.2			
Hemp fibres glued together	41			
Strips of paper glued together	13			
Wood, Box, spec. gravity .862	9	Barlow		
Ash - - - - - - .6	8	—		
Teak - - - - - .9	7	—		
Beech - - - - - .7	5	—		
Oak - - - - - .92	5	—	1.7	—
Ditto - - - - - .77	4	—		
Fir - - - - - .6	5	—		
Pear - - - - - .646	4¼	—		
Mahogany - - - - .637	3½	—		
Elm - - - - - - -	6	- -	.57	—
Pine, American - - -	6	- -	.73	—
Deal, white - - - -	6	- -	.86	—

In the result just obtained for the value of t, no attention has been paid to the pressure upon the ends of the boiler or pipe, but these are usually made thick enough to throw the chances of breaking altogether upon the cylindrical portion of the surface.

V.

EQUILIBRIUM OF FLOATING BODIES.

§ 268.—The rules for finding the pressure against the sides of vessels are equally applicable to the determination of the pressure on the surfaces of bodies, however sub-
Equilibrium of floating bodies; jected to the action of a homogeneous heavy fluid. But when it is the question to ascertain the circumstances that determine a heavy body to be in equilibrio or in motion, when immersed in a heavy fluid, it is usual to employ the results deduced from the following considerations.

Fig. 305.

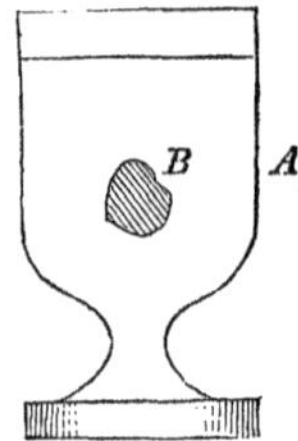

Suppose a vessel A to contain any heavy fluid in a state of rest. All parts of the fluid being in equilibrio, it is obvious that this state
a body wholly immersed in a fluid; will in no respect be altered by supposing any portion B to become solid without changing its density. This solid is entirely immersed in the fluid, with which it has the same density, and is in equilibrio. Now this solid is urged downward by its weight, which passes through its centre of gravity. This weight can only be in equilibrio with a single force when the latter is directed vertically upward through the centre of gravity of the body, which centre coincides with that of the fluid converted into a solid, or that of the displaced fluid. But the only forces that act upon the solid besides its weight, are the pressures of the surrounding fluid; whence we conclude that

first result; 1st. *The pressures upon the surface of a body entirely immersed in a fluid, have a single resultant, and that this resultant is directed vertically upward.*

2d. *The resultant of all the pressures is equal, in intensity, to the weight of the displaced fluid.* second result;

3d. *The line of direction of the resultant, passes through the centre of gravity of the displaced fluid.* third result;

4th. *The horizontal pressures destroy each other.* fourth result;

Again, if without altering the volume of this solid, we give it an additional quantity of matter, it is obvious that the weight of this latter will cause it to descend, that is, sink to the bottom of the vessel. Or if, without altering its volume, we conceive a portion of matter taken from its interior, the equilibrium will again be destroyed, the weight of the solid will be diminished by that of the subducted matter, the resultant of the pressures will prevail, and the body will rise to the surface, through which it will continue to ascend, till the weight of the fluid displaced by the part immersed, is equal to that of the entire body.

In the first case, the density of the body will be increased, containing a greater quantity of matter under the same volume, and in the second the density will be diminished; and as the density of the original body was the same as that of the fluid, we see that *when the density of an immersed body is greater than that of the fluid, it will sink to the bottom; when less, it will rise to the surface, and float.* an immersed body will sink or float, according as its density is greater or less than that of the fluid;

It follows, also, from what has been said above, that *when a body is immersed in a fluid, it will lose a portion of its weight equal to that of the displaced fluid.* This is beautifully illustrated by what is usually called the "*cylinder and bucket*" experiment. Place a hollow cylinder *a*, in one of the scales of a balance; suspend to this scale a second cylinder *b*, of solid metal, exactly fitting the former, and in the opposite scale put a weight *c*, that the body will lose a portion of its weight equal to that of the displaced fluid;

Fig. 306.

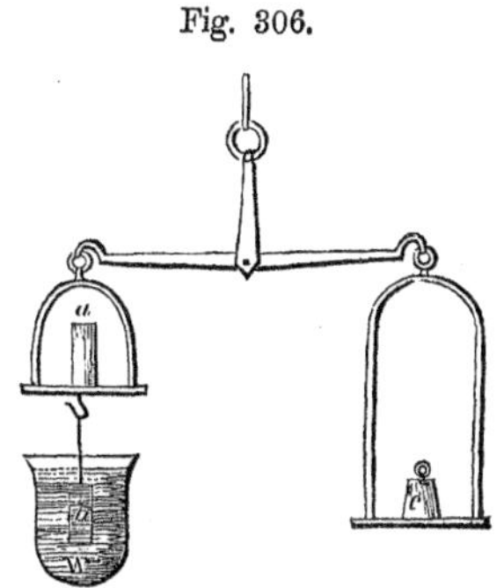

Fig. 306.

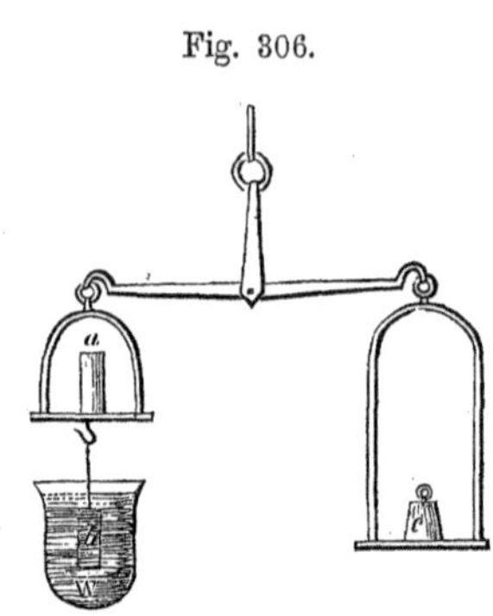

cylinder and bucket experiment;

shall restore the equilibrium of the balance. Now immerse the cylinder b in a vessel W of water, the scale of the weight c will descend; fill the cylinder a with water taken from the vessel W, the beam of the balance will return to its horizontal position.

weight of the immersed solid transmitted to the vessel;

The weight lost by the solid is transmitted through the fluid to the vessel, in the same way that the weight of a person in bed is transmitted through the latter to the bedstead, and thence to the floor. This is proved, experimentally, thus: Place a tumbler of water in one of the scales A of a balance, bring the beam to a horizontal position by means of the empty hollow cylinder a of the last experiment and a weight c; suspend the solid cylinder b by means of a thread from a detached ring R, and depress it till it is wholly immersed into the water of the tumbler; the scale A will fall; fill the cylinder a with water of the same temperature and density as that in the tumbler; the equilibrium will be restored.

experimental proof;

Fig. 263.

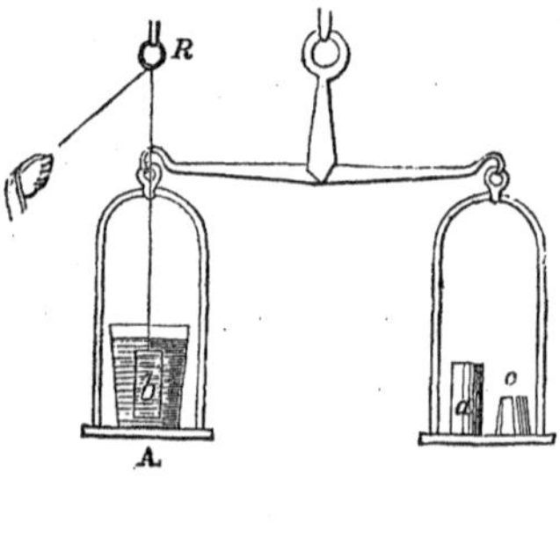

this principle used to find weight of ships, &c.;

This important principle, which determines the circumstances under which a body will rest upon a fluid, is frequently employed to ascertain the weights of large floating masses, such as ships, boats, and the like, which are entirely beyond the capacity of our ordinary weighing machines. For this purpose the volume, in cubic feet, of the immersed part is computed from the known figure and dimensions of the body, and this is multiplied by the known weight of a cubic foot of water, which is 62.5 pounds avoirdupois; the product is the weight of the floating body, in pounds. By taking in this way the difference of weights of a ship, with and without her cargo, the weight of the latter may be ascertained. **weight of a ship's cargo;**

The upward action by which an immersed body apparently loses a portion of its weight, is called the *buoyant effort of the fluid;* and as the line of direction of this effort passes through the centre of gravity of the displaced fluid, this point is called the *centre of buoyancy.* The vertical line through the centre of buoyancy, is called the *line of support.* The weight of a body acting at its centre of gravity downward, and the buoyant effort at the centre of buoyancy upward, the body can only be in equilibrio when the line joining these centres is vertical, for it is only then that the forces are directly opposed. When the line joining the centre of buoyancy and the centre of gravity of the floating body is vertical, it is called the *line of rest.* **buoyant effort of a fluid;** **centre of buoyancy;** **line of support;** **line of rest;**

When the equilibrium exists, it may be *stable, unstable,* or *indifferent.* If stable, the body will not overturn when careened; if unstable, it will; if indifferent, the body will retain any position in which it may be placed. **stable, unstable, and indifferent equilibrium;**

Let MQN represent a section of any body, as a boat at rest upon the water, of which the upper surface is AB, called the *plane of floatation.* When this plane is produced through the boat, it will divide her into two partial volumes, the lower of which being supposed for an instant to consist of water, would weigh as much as the entire boat and her load, and **plane of floatation;**

Fig. 308.

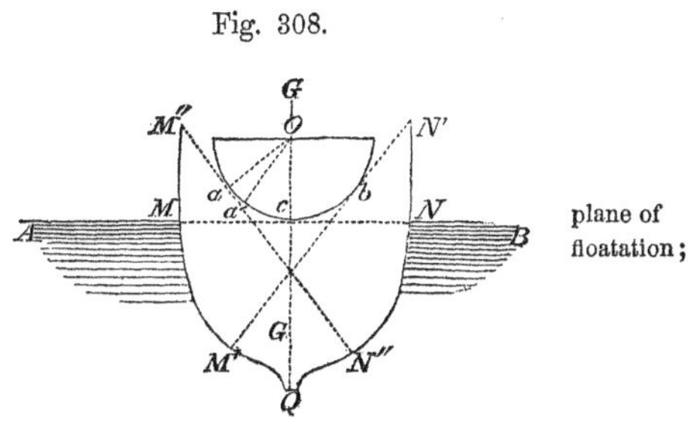

this whatever be her position, whether careened or erect. Whence it follows, that if a series of planes $M'N'$, $M''N''$, &c., be passed, making the volumes $M'QN'$, $M''QN''$, &c., respectively equal to MQN, these planes will, each in its turn, come to coincide with the plane of floatation, whenever the boat, in the process of careening, takes a suitable position. But these planes may be regarded as so many tangent planes to a curved surface abc, which may be conceived as invariably connected with the boat. Now the effect, as regards the careening motion, will be the same as though this surface were the boundary of a physical axis which is made to roll back and forth on the plane of floatation, regarded as a physical surface, after the manner of the pendulum axis on its supporting plane, during an oscillation. When the boat has a position of equilibrium, the line of support and of rest coincide, and are normal to this surface at its lowest point c. As the boat careens, the line of support, being always vertical, will still be normal to this axis surface at its lowest point, being that in which it is tangent to the plane of floatation; hence each of these normal lines must in turn become a line of support. If two normals aO and $a'O$, which lie in the same plane, be drawn at tangential points answering to two consecutive positions of the boat, these normals will intersect at some point O, which point will, obviously, be the momentary centre of rotation, when the plane of floatation coincides with $M''N''$. When one of these normals coincides with the line of rest, the point O is called the *metacentre,* being *the point of intersection of the line of rest, with an adjacent*

series of cutting planes;

oscillations of the boat;

Fig. 308.

position of the line of rest during the equilibrium;

metacentre;

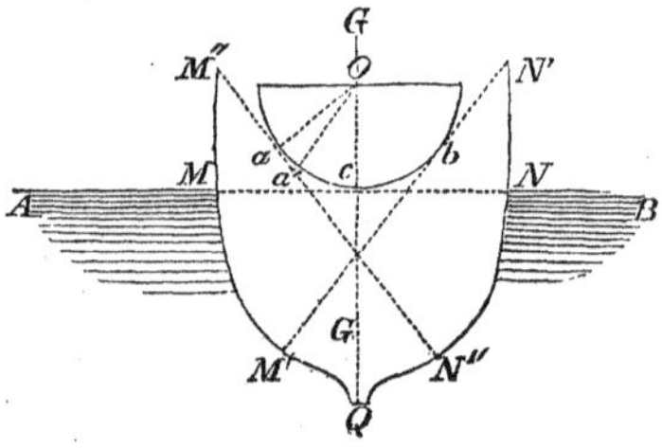

line of support. But we have seen that the equilibrium of a heavy body which may turn about a fixed point, will be stable or unstable, according as the centre of gravity during a slight departure from a position of equilibrium is compelled by the connection to ascend or descend; and it is obvious that, in the present case, the centre of gravity will ascend or descend on making a slight derangement of the line joining the centres of buoyancy and of gravity from the line of rest, according as the centre of gravity is below or above the metacentre. Whence we see, that *the equilibrium will be stable when the centre of gravity is below the metacentre, unstable when the relative positions of these points are reversed, and indifferent when these centres coincide, for then a slight derangement will cause no motion in the centre of gravity.* It is also obvious that the stability of the equilibrium will be the greater, in proportion as the centre of gravity of the floating body be at a greater distance below the centre of buoyancy. It is for this reason that ships sent to sea without cargoes are provided with ballast of stone, sand, or other heavy matter, to diminish the chances of upsetting. The buoyant effort of water is used to great advantage in raising heavy sunken masses. For this purpose it is usual to connect two or more boats A and B, by means of a substantial cross-beam; to fill them nearly full of water, that they may sink as low as possible, and while in this condition to attach the body to be raised to the cross-beam by means of a taught chain or rope, and then to pump the water from the boats; the tension upon the chain will be equal to the weight of the water pumped from the boats. If it is the question to raise a sunken boat, one of the most effective means is to

defined; the nature of the equilibrium determined by the relative positions of the centres; object of ship-ballast; buoyant effort used to raise sunken masses; a common mode of employing this principle.

Fig. 309.

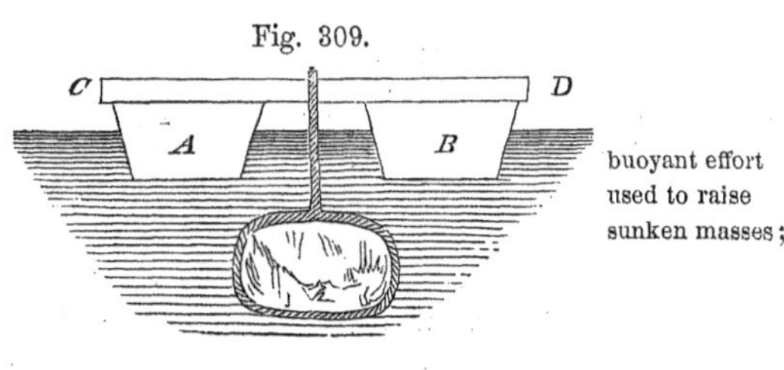

force empty and water-tight barrels between her deck and hull.

Level strata in heterogeneous fluids;

§ 269.—We have just seen that when a body is immersed in a fluid, it loses a portion of its weight equal to that of the displaced fluid, and that it will sink or rise to the surface, depending upon its relative density. This is universally true whatever be the size and number of the bodies immersed. If, therefore, one fluid be poured into another for which it has no affinity, as oil into water, it will sink to the bottom or rise to the surface and float, according as its density is greater or less than that of the fluid into which it is poured. The elements of the lighter fluid will act as so many immersed bodies till they reach the surface of the heavier fluid, where, being freed from the buoyant action of the latter, they will arrange themselves, under the efforts of their own weight, into a stratum of which the upper surface will, like that of the fluid below it, be perpendicular to the direction of the force of gravity. What is here said of two, is equally applicable to three, four, or any number of fluids of different densities mixed together; whence we conclude, that *such fluids will come to rest only after arranging themselves into* LEVEL STRATA *in the order of their densities; the most dense being at the bottom and the least dense at the top.* This is confirmed by daily observation, and may be easily illustrated by pouring mercury, water, and oil, into a common tumbler. The mercury will come to rest at the bottom, the oil at the top, the upper surfaces of all being level.

mixture of different fluids having no affinity for each other;

will form level strata; the most dense lowest;

The same conclusion follows from the consideration, that these fluids when mixed constitute a heavy system, which, we have seen, can only come to a state of stable equilibrium when its centre of gravity is at the lowest point, a condition only fulfilled by the arrangement, in respect to density, just described.

the same results from the properties of the centre of gravity;

If the elements of one fluid have an affinity for those of another, this affinity will, when the fluids come into contact, counteract the buoyant action of the heavier fluid, and the lighter will be held in a state of mixture. Instance wine and water, water and alcohol, brandy and water, and the like.

they will not obtain when the fluids have an affinity for each other.

VI.

SPECIFIC GRAVITY.

§ 270.—The specific gravity of a body, is the weight of so much of the body, as would be contained under a unit of volume.

Specific gravity defined;

It is measured by the quotient arising from dividing the weight of the body by the weight of an equal volume of some other substance, assumed as a standard; for the ratio of the weights of equal volumes of two bodies being always the same, if the unit of volume of each be taken, and one of the bodies become the standard, its weight will become the unit of weight.

its measure;

The term *density* denotes the degree of proximity among the particles of a body. Thus, of two bodies, that will have the greater density which contains, under an equal volume, the greater number of particles. The force of gravity acts, within moderate limits, equally upon all elements of matter. The weight of a substance is, therefore, directly proportional to its density, and the ratio of the weights of equal volumes of two bodies is equal to the ratio of their densities. Denote the weight of the first by W, its density by D, its volume by V, and the force of gravity by g, then will

density;

illustration;

$$W = g \cdot D \cdot V;$$

measure for the weight of a body;

and denoting the like elements of the other body by $W_{\prime}$,

$D_{,,}$ and $V_{,,}$ we have

weight of a second body;

$$W_{,} = g \, . \, D_{,} \, . \, V_{,}.$$

Dividing the first by the second,

ratio of the weights;

$$\frac{W}{W_{,}} = \frac{g\, D\, V}{g\, D_{,}\, V_{,}} = \frac{D\, V}{D_{,}\, V_{,}};$$

and making the volumes equal,

same when the volumes are equal;

$$\frac{W}{W_{,}} = \frac{D}{D_{,}} \quad . \quad . \quad . \quad . \quad . \quad (227).$$

Now suppose the body whose weight is $W_{,}$ to be assumed as the standard both for specific gravity and density, then will $D_{,}$ be unity, and

specific gravity;

$$S = \frac{W}{W_{,}} = D \quad . \quad . \quad . \quad . \quad (228);$$

specific gravity and density expressed by same numbers for same standard.

in which S denotes the specific gravity of the body whose density is D; and from which we see, that when specific gravities and densities are referred to the same substance as a standard, the numbers which express the one will also express the other.

Choice of a standard;

§ 271.—Bodies present themselves under every variety of condition—gaseous, liquid, and solid; and in every kind of shape and of all sizes. The determination of their specific gravity, in every instance, depends upon our ability to find the weight of an equal volume of the standard. When a solid is immersed in a fluid, it loses a portion of its weight equal to that of the displaced fluid. The volume of the body and that of the displaced fluid are equal. Hence the weight of the body in vacuo, divided by its loss of weight when immersed, will give the ratio of the weights of equal volumes of the body and fluid; and if the latter be taken as the standard, and the loss of weight occupies the denominator, this ratio becomes the measure of the specific gravity of the body immersed. For this reason, and in view of the consideration that it may be obtained pure at all times and places, *water* is assumed as the general standard of specific gravities and densities for all bodies. Sometimes the gases and vapors are referred to atmospheric air, but the specific gravity of the latter being known as referred to water, it is very easy, as we shall presently see, to pass from the numbers which relate to one standard to those that refer to the other.

water assumed as the standard for specific gravities and density;

gases sometimes referred to atmospheric air.

§ 272.—But water, like all other substances, changes its density with its temperature, and, in consequence, is not an invariable standard. It is hence necessary either to employ it at a constant temperature, or to have the means of reducing the specific gravities, as determined by it at different temperatures, to what they would have been if taken at a fixed or standard temperature. The former is generally impracticable; the latter is easy.

Varying density of water;

Let D denote the density of any solid, and S its specific gravity, as determined at a standard temperature corresponding to which the density of the water is $D_{,}$. Then, Eq. (227),

reduction to a standard temperature;

$$S = \frac{D}{D_{,}}.$$

specific gravity at one temperature;

Again, if S' denote the specific gravity of the same body, as indicated by the water when at a temperature different from the standard, and corresponding to which it has a density $D_{,,}$, then will

$$S' = \frac{D}{D_{,,}}.$$

same at another temperature;

Dividing the first of these equations by the second, we

have

ratio of these specific gravities;

$$\frac{S}{S'} = \frac{D_{\prime\prime}}{D_{\prime}};$$

whence

$$S = S' \cdot \frac{D_{\prime\prime}}{D_{\prime}} \quad . \quad . \quad . \quad . \quad (229);$$

and if the density $D_{\prime}$ be taken as unity,

specific gravity reduced to a standard;

$$S = S' \cdot D_{\prime\prime} \quad . \quad . \quad . \quad . \quad (230).$$

expressed in words;

That is to say, *the specific gravity of a body as determined at the standard temperature of the water, is equal to its specific gravity determined at any other temperature, multiplied by the density of the water corresponding to this temperature, the density at the standard temperature being regarded as unity.*

density of water at different temperatures;

To make this rule practicable, it becomes necessary to find the relative densities of water at different temperatures. For this purpose, take any metal, say silver, that easily resists the chemical action of water, and whose rate of expansion for each degree of Fahr. thermometer is accurately known from experiment; give it the form of a slender cylinder, that it may readily conform to the temperature of the water when immersed. Let the length of the cylinder at the temperature of 32° Fah. be denoted by l, and the radius of its base by $m\,l$; its volume at this temperature will be,

volume of a slender cylinder;

$$\pi\, m^2\, l^2 \times l = \pi\, m^2\, l^3.$$

Let $n\,l$ be the amount of expansion in length for each degree of the thermometer above 32°. Then, for a temperature denoted by t, will the whole expansion in length be

its expansion;

$$n\,l \times (t - 32°),$$

and the entire length of the cylinder will become

$$l + n\,l\,(t - 32^\circ) = l\,[1 + n\,(t - 32^\circ)]\,;$$ its increased length;

which, substituted for l in the first expression, will give the volume for the temperature t equal to

$$\pi\, m^2\, l^3\, [1 + n\,(t - 32^\circ)]^3.$$ its increased volume;

The cylinder is now weighed in vacuo and in the water, at different temperatures, varying from 32° upward, through any desirable range, say to one hundred degrees. The temperature at each process being substituted above, gives the volume of the displaced fluid; the weight of the displaced fluid is known from the loss of weight of the cylinder. Dividing this weight by the volume, gives the weight of the unit of volume of the water at the temperature t. It was found by *Stampfer*, that the weight of the unit of volume is greatest when the temperature is $38\overset{\circ}{.}75$ Fahrenheit's scale. Taking the density of water at this temperature as unity, and dividing the weight of the unit of volume at each of the other temperatures by the weight of the unit of volume at this, $38\overset{\circ}{.}75$, the following table will result:—

Fig. 310.

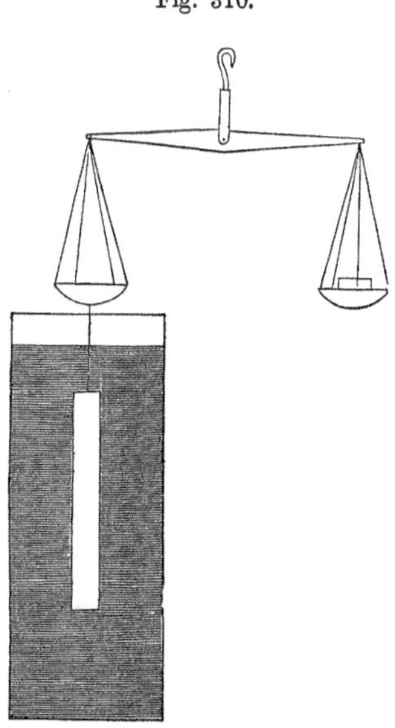

experimental determination of the density of water at different temperatures;

greatest density at $38\overset{\circ}{.}75$;

TABLE

OF THE DENSITIES AND VOLUMES OF WATER AT DIFFERENT DEGREES OF HEAT, (ACCORDING TO STAMPFER,) FOR EVERY $2\frac{1}{4}$ DEGREES OF FAHRENHEIT'S SCALE.

(*Jahrbuch des Polytechnischen Institutes in Wein*, *Bd.* 16. *S.* 70.)

t Temperature.	$D_{\prime\prime}$ Density.	Diff.	V Volume.	Diff.
32°.00	0.999887		1.000113	
34.25	0.999950	63	1.000050	63
36.50	0.999988	38	1.000012	38
38.75	1.000000	12	1.000000	12
41.00	0.999988	12	1.000012	12
43.25	0.999952	35	1.000047	35
45.50	0.999894	58	1.000106	59
47.75	0.999813	81	1.000187	81
50.00	0.999711	102	1.000289	102
52.25	0.999587	124	1.000413	124
54.50	0.999442	145	1.000558	145
56.75	0.999278	164	1.000723	165
59.00	0.999095	183	1.000906	183
61.25	0.998893	202	1.001108	202
63.50	0.998673	220	1.001329	221
65.75	0.998435	238	1.001567	238
68.00	0.998180	255	1.001822	255
70.25	0.997909	271	1.002095	273
72.50	0.997622	287	1.002384	289
74.75	0.997320	302	1.002687	303
77.00	0.997003	317	1.003005	318
79.25	0.996673	330	1.003338	333
81.50	0.996329	344	1.003685	347
83.75	0.995971	358	1.004045	360
86.00	0.995601	370	1.004418	373
88.25	0.995219	382	1.004804	386
90.50	0.994825	394	1.005202	398
92.75	0.994420	405	1.005612	410
95.00	0.994004	416	1.006032	420
97.25	0.993579	425	1.006462	430
99.50	0.993145	434	1.006902	440

With this table it is easy to find the specific gravity by means of water at any temperature. Suppose, for example, the specific gravity S' in Eq. (230), had been found at the temperature of 59°, then would $D_{\prime\prime}$ in that equation, be 0.999095, and the specific gravity of the body referred to water at its greatest density, would be given by

$$S = S' \times 0.999095.$$

The column under the head V, will enable us to determine how much the volume of any mass of water, at a temperature t, exceeds that of the same mass at its maximum density. For this purpose, we have but to multiply the volume at the maximum density by the tabular number corresponding to the given temperature. relation of volumes of the same amount of fluid at different temperatures.

§ 273.—Before proceeding to the practical methods of finding the specific gravity of bodies, and to the variations in the processes rendered necessary by the peculiarities of the different substances, it will be necessary to give some idea of the best instruments employed for this purpose. These are the *Hydrostatic Balance* and *Nicholson's Hydrometer*. Instruments used to find the specific gravity of a body;

The first is similar in principle and form to the common balance. It is provided with numerous weights, extending through a wide range, from a small fraction of a grain to several ounces. Attached to the under surface of one of the basins is a small hook, from which may be suspended any body by means of a thin platinum wire, horse-hair, or any other delicate thread that will neither absorb nor yield to the chemical action of the fluid in which it may be desirable to immerse it. hydrostatic balance; mode of attaching the body;

Fig. 311.

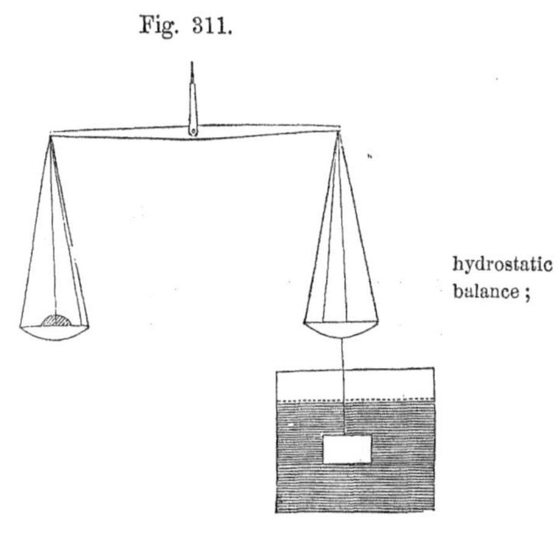

Nicholson's Hydrometer consists of a hollow metallic ball A, through the centre of which passes a metallic wire, prolonged in both directions beyond the surface, and supporting at either end a basin B and B'. The concavities Nicholson's hydrometer;

of these basins are turned in the same direction, and the basin B' is made so heavy that when the instrument is placed in water the stem CC' shall be vertical, and a weight of 500 grains being placed in the basin B, the whole instrument will sink till the upper surface of distilled water, at the standard temperature, comes to a point C marked on the upper stem near its middle. This instrument is provided with weights similar to those of the Hydrostatic Balance.

description, and conditions the instrument must satisfy.

Fig. 312.

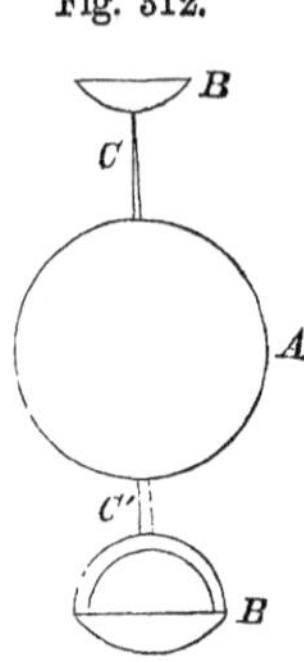

Process for finding specific gravity of a solid heavier than waterby the balance;

§ 274.—(1). *If the body be solid, insoluble in water, and will sink in that fluid*, attach it, by means of a hair, to the hook of the basin of the hydrostatic balance; counterpoise it by placing weights in the opposite scale; now immerse the body in water, and restore the equilibrium by placing weights in the basin above the body, and note the temperature of the water. Divide the weights in the basin to which the body is not attached by those in the basin to which it is, and multiply the quotient by the density corresponding to the temperature of the water, as given by the table; the result will be the specific gravity.

Thus denote the specific gravity by S, the density of the water by $D_{,,}$, the weight in the first case by W, and that in the scale above the solid by w, then will

specific gravity;

$$S = D_{,,} \times \frac{W}{w}.$$

when the body is lighter than water;

(2). *If the body be insoluble, but will not sink in water*, as would be the case with most varieties of wood, wax, and the like, attach to it some body, as a metal, whose weight in the air and loss of weight in the water are previously found. Then proceed exactly as in the case before, to find the weights which will counterpoise the compound in air and restore the equilibrium of the balance when it is immersed in the water. From the weight of the compound in air, subtract that of the heavier body in air; from the loss of weight of the compound in water, subtract that of the heavier body; divide the first difference by the second, and multiply by the density of the water answering to its temperature, and the result will be the specific gravity of the lighter body. process described;

Example.

A piece of wax and copper in air $= \overset{grs.}{438} = W + W'$, example; the case of wax;
Lost on immersion in water - - $= 95.8 = w + w'$,
Copper in air - - - - - - - $= 388 = W'$,
Loss of copper in water - - - $= 44.2 = w'$.

Then

$$W + W' - W' = 438 - 388 = 50 = W,$$

$$w + w' - w' = 95.8 - 44.2 = 51.6 = w.$$

Temperature of water $43^{\circ}.25$,

$$D_{\prime\prime} = 0.999952,$$

$$S = D_{\prime\prime} \times \frac{W}{w} = 0.999952 \times \frac{50}{51.6} = 0.968.$$ specific gravity of wax;

(3). *If the body readily dissolve in water*, as many of the salts, sugar, &c., find its apparent specific gravity in some liquid in which it is insoluble, and multiply this apparent specific gravity by the density or specific gravity of the liquid referred to water at its maximum density as a standard; the product will be the true specific gravity. when the body is soluble in the standard fluid;

If it be inconvenient to provide a liquid in which the solid is insoluble, saturate the water with the substance and find the apparent specific gravity with the water thus saturated. Multiply this apparent specific gravity by the density of the saturated fluid, and the product will be the specific gravity referred to the standard. This is a common method of finding the specific gravity of gunpowder, the water being saturated with nitre.

saturate the liquid with the body and proceed as before;

(4). *If the body be a liquid*, select some solid that will resist its chemical action, as a massive piece of glass suspended from fine platinum wire; weigh it in air, then in water, and finally in the liquid; the differences between the first weight and each of the latter, will give the weights of equal volumes of water and the liquid. Divide the weight of the liquid by that of the water, and the quotient will be the specific gravity of the liquid, provided the temperature of water be at the standard. If the water have not the standard temperature, multiply this apparent specific gravity by the tabular density of the water corresponding to the actual temperature.

when the body is a liquid;

rule;

Example.

example;

Loss of glass in water at 41°, $\overset{grs.}{150} = w'$,

" " sulphuric acid, $277.5 = w$,

specific gravity of sulphuric acid;

$$S = \frac{277.5}{150} \times 0.999988 = 1.85.$$

(5.) *If the body be a gas or vapor*, provide a large glass flask-shaped vessel, weigh it when filled with the gas; withdraw the gas, which may be done by means to be explained presently, fill with water, and weigh again; finally, withdraw the water and exclude the air, and weigh again. This last weight subtracted from the first will give the weight of the gas that filled the vessel, and subtracted from the second will give the weight of an equal volume of water; divide the weight of the gas by that of the water, and multiply by the tabular density of the water

when the body is a gas or vapor;

process;

answering to the actual temperature of the latter; the result will be the specific gravity of the gas.

The atmosphere in which all these operations must be performed, varies at different times, even during the same day, in respect to temperature, the weight of its column which presses upon the earth, and the quantity of moisture or aqueous vapor it contains. That is to say, its density depends upon the state of the thermometer, barometer, and hygrometer. On all these accounts corrections must be made, before the specific gravity of atmospheric air, or that of any gas exposed to its pressure, can be accurately determined. The principles according to which these corrections are made, will be discussed when we come to treat of the properties of elastic fluids.

Influence of the atmosphere; temperature; pressure; moisture;

To find the specific gravity of a solid by means of Nicholson's Hydrometer, place the instrument in water, and add weights to the upper basin till it sinks to the mark on the upper stem; remove the weights and place the solid in the upper basin, and add weights till the hydrometer sinks to the same point; the difference between the first weights and those added with the body, will give the weight of the latter in air. Take the body from the upper basin, leaving the weights behind, and place it in the lower basin; add weights to the upper basin till the instrument sinks to the same point as before, the last added weights will be the weight of the water displaced by the body; divide the weight in air by the weight of the displaced water, and multiply the quotient by the tabular density of the water answering to its actual temperature; the result will be the specific gravity of the solid.

mode of using Nicholson's hydrometer for solids;

To find the specific gravity of a fluid by this instrument, immerse it in water as before, and by weights in the upper basin sink it to the mark on the upper stem; add the weights in the basin to the weight of the instrument, the sum will be the weight of the displaced water. Place the instrument in the fluid whose specific gravity is to be found, and add weights in the upper basin till it sinks to

also for fluids;

the mark as before; add these weights to the weight of the instrument, the sum will be the weight of an equal volume of the fluid; divide this weight by the weight of the water, and multiply by the tabular density corresponding to the temperature of the water, the result will be the specific gravity.

The scale areometer;

§ 275.—Besides the hydrometer of Nicholson, which requires the use of weights, there is another form of this instrument which is employed solely in the determination of the specific gravities of liquids, and its indications are given by means of a scale of equal parts. It is called the *Scale-Areometer*. It consists, generally, of a glass vial-shaped vessel A, terminating at one end in a long slender neck C, to receive the scale, and at the other in a small globe B, filled with some heavy substance, as lead or mercury, to keep it upright when immersed in a fluid. The application and use of the scale depend upon this, that a body floating on the surface of different liquids, will sink deeper and deeper, in proportion as the density of the fluid approaches that of the body; for when the body is at rest its weight and that of the displaced fluid must be equal. Denoting the volume of the instrument by V, that of the displaced fluid by V', the density of the instrument by D, and that of the fluid by D', we must always have

description;

the principle of this instrument;

Fig. 313.

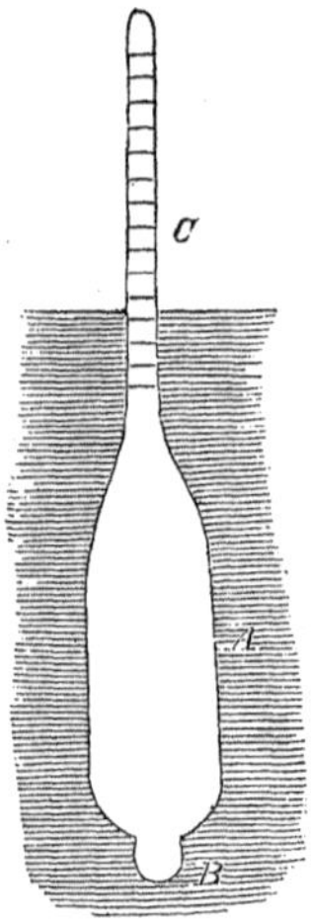

conditions of equilibrium;

$$g\,V D = g\,V'\,D';$$

in which g denotes the force of gravity, the first member the weight of the instrument, and the second that of the displaced fluid. Dividing both members by $D' V$, and omitting the common factor g, we have

$$\frac{D}{D'} = \frac{V'}{V}.$$

ratio of densities equal to that of the volumes;

In which, if the densities be equal, the volumes must be equal; if the density D' of the fluid be greater than D, or that of the solid, the volume V of the solid must be greater than V', or that of the displaced fluid; and in proportion as D' increases in respect to D, will V' diminish in respect to V, that is, the solid will rise higher and higher out of the fluid in proportion as the density of the latter is increased, and the reverse. The neck C of the vessel should be of the same diameter throughout. To establish the scale, the instrument is placed in distilled water at the standard temperature, and when at rest the place of the surface of the water on the neck is marked and numbered 1; the instrument is then placed in some heavy solution of salt, whose specific gravity is accurately known by means of the Hydrostatic Balance, and when at rest the place on the neck of the fluid surface is again marked and characterized by its appropriate number. The same process being repeated for rectified alcohol, will give another point towards the opposite extreme of the scale, which may be completed by graduation.

construction of the scale;

To use this instrument, it will be sufficient to immerse it in a fluid and take the number on the scale which coincides with the surface.

To bring into view the circumstances which determine the sensibility both of the Scale-Areometer and Nicholson's Hydrometer, let s denote the specific gravity of the fluid, c the volume of the vial, l the length of the immersed portion of the narrow neck, r its semi-diameter, and w the total weight of the instrument. Then will πr^2, denote the area of a section of the neck, and $\pi r^2 l$, the volume of fluid displaced by the immersed part of the neck. The weight,

use;

sensibility of the instrument;

therefore, of the whole fluid displaced by the vial and neck will be

weight of fluid displaced

$$s c + s \pi r^2 l;$$

but this must be equal to the weight of the instrument, whence

condition of the equilibrium;

$$w = s (c + \pi r^2 l),$$

from which we deduce

specific gravity;

$$s = \frac{w}{c + \pi r^2 l},$$

length of neck immersed;

$$l = \frac{w - s c}{\pi r^2 s} \quad . \quad . \quad . \quad . \quad (231).$$

Now, immersing the instrument in a second fluid whose specific gravity is s', the neck will sink through a distance l', and from the last equation we have

length immersed for second fluid;

$$l' = \frac{w - s' c}{\pi r^2 s'};$$

subtracting this equation from that above and reducing, we find

difference of specific gravity;

$$l - l' = \frac{w}{\pi r^2} \left(\frac{s' - s}{s s'} \right).$$

The difference $l - l'$ is the distance between two points on the scale which indicates the difference $s' - s$ of specific gravities, and this we see becomes longer, and the instrument more sensible, therefore, in proportion as w is made greater and r less. Whence we conclude that the Areometer is the more valuable in proportion as the vial portion is made larger and the neck smaller.

inference;

sensibility of Nicholson's hydrometer;

If the specific gravity of the fluid remain the same, which is the case with Nicholson's Hydrometer, and it becomes a question to know the effect of a small weight added to the instrument, denote this weight by w', then will Eq. (231) become

$$l' = \frac{w + w' - s\,c}{\pi\, r^2\, s};$$

subtracting from this Eq. (231), we find

$$l' - l = \frac{w'}{\pi\, r^2\, s}.$$

From which we see that the narrower the upper stem of Nicholson's instrument, the greater its sensibility.

TABLE

OF THE SPECIFIC GRAVITIES OF SOME OF THE MOST IMPORTANT BODIES.

[The density of distilled water is reckoned in this Table at its maximum 38¾° F.=1.000.]

Name of the Body.	Specific Gravity.
I. SOLID BODIES.	
(1) METALS.	
Antimony (of the laboratory) - - -	4.2 — 4.7
Brass - - - - - - - -	7.6 — 8.8
Bronze for cannon, according to Lieut. Matzka	8.414 — 8.974
Ditto, mean - - - - - - -	8.758
Copper, molten - - - - - - -	7.788 — 8.726
Ditto, hammered - - - - - -	8.878 — 8.9
Ditto, wire-drawn - - - - - -	8.78
Gold, molten - - - - - - -	19.238 — 19.253
Ditto, hammered - - - - - -	19.361 — 19.6
Iron, wrought - - - - - - -	7.207 — 7.788
Ditto, cast, a mean - - - - - -	7.251
Ditto, gray - - - - - - -	7.2
Ditto, white - - - - - - -	7.5
Ditto for cannon, a mean - - - - -	7.21 — 7.30
Lead, pure molten - - - - - -	11.3303
Ditto, flattened - - - - - - -	11.388
Platinum, native - - - - - -	16.0 — 18.94
Ditto, molten - - - - - - -	20.855
Ditto, hammered and wire-drawn - -	21.25
Quicksilver, at 32° Fahr. - - - - -	13.568 — 13.598
Silver, pure molten - - - - - -	10.474
Ditto, hammered - - - - - -	10.51 — 10.622
Steel, cast - - - - - - - -	7.919
Ditto, wrought - - - - - - - -	7.840
Ditto, much hardened - - - - -	7.818
Ditto, slightly - - - - - - - -	7.833
Tin, chemically pure - - - - - -	7.291
Ditto, hammered - - - - - -	7.299 — 7.475
Ditto, Bohemian and Saxon - - -	7.312

TABLE—*continued.*

Name of the Body.	Specific Gravity.	
Tin, English	7.291	
Zinc, molten	6.861	— 7.215
Ditto, rolled	7.191	
(2) Building Stones.		
Alabaster	2.7	— 3.0
Basalt	2.8	— 3.1
Dolerite	2.72	— 2.93
Gneiss	2.5	— 2.9
Granite	2.5	— 2.66
Hornblende	2.9	— 3.1
Limestone, various kinds	2.64	— 2.72
Phonolite	2.51	— 2.69
Porphyry	2.4	— 2.6
Quartz	2.56	— 2.75
Sandstone, various kinds, a mean	2.2	— 2.5
Stones for building	1.66	— 2.62
Syenite	2.5	— 3.
Trachyte	2.4	— 2·6
Brick	1.41	— 1.86
(3) Woods.	Fresh-felled.	Dry.
Alder	0.8571	0.5001
Ash	0.9036	0.6440
Aspen	0.7654	0.4302
Birch	0.9012	0.6274
Box	0.9822	0.5907
Elm	0.9476	0.5474
Fir	0.8941	0.5550
Hornbeam	0.9452	0.7695
Horse-chestnut	0.8614	0.5749
Larch	0.9206	0.4735
Lime	0.8170	0.4390
Maple	0.9036	0.6592
Oak	1.0494	0.6777
Ditto, another specimen	1.0754	0.7075
Pine, *Pinus Abies Picea*	0.8699	0.4716
Ditto, *Pinus Sylvestris*	0.9121	0.5502
Poplar (Italian)	0.7634	0.3931
Willow	0.7155	0.5289
Ditto, white	0.9859	0.4873
(4) Various Solid Bodies.		
Charcoal, of cork	0.1	
Ditto, soft wood	0.28	— 0.44
Ditto, oak	1.573	
Coal	1.232	— 1.510
Coke	1.865	
Earth, common	1.48	
rough sand	1.92	
rough earth, with gravel	2.02	
moist sand	2.05	
gravelly soil	2.07	
clay	2.15	
clay or loam, with gravel	2.48	

TABLE—*continued.*

Name of the Body.	Specific Gravity.	
Flint, dark	2.542	
Ditto, white	2.741	
Gunpowder, loosely filled in		
coarse powder	0.886	
musket ditto	0.992	
Ditto, slightly shaken down		
musket-powder	1.069	
Ditto, solid	2.248	— 2.563
Ice	0.916	— 0.9268
Lime, unslacked	1.842	
Resin, common	1.089	
Rock-salt	2.257	
Saltpetre, melted	2.745	
Ditto, crystallized	1.900	
Slate-pencil	1.8	— 2.24
Sulphur	1.92	— 1.99
Tallow	0.942	
Turpentine	0.991	
Wax, white	0.969	
Ditto, yellow	0.965	
Ditto, shoemaker's	0.897	
II. LIQUIDS.		
Acid, acetic	1.063	
Ditto, muriatic	1.211	
Ditto, nitric, concentrated	1.521	— 1.522
Ditto, sulphuric, English	1.845	
Ditto, concentrated (Nordh.)	1.860	
Alcohol, free from water	0.792	
Ditto, common	0.824	— 0.79
Ammoniac, liquid	0.875	
Aquafortis, double	1.300	
Ditto, single	1.200	
Beer	1.023	— 1.034
Ether, acetic	0.866	
Ditto, muriatic	0.845	— 0.874
Ditto, nitric	0.886	
Ditto, sulphuric	0.715	
Oil, linseed	0.928	— 0.953
Ditto, olive	0.915	
Ditto, turpentine	0.792	— 0.891
Ditto, whale	0.923	
Quicksilver	13.568	— 13.598
Water, distilled	1.000	
Ditto, rain	1.0013	
Ditto, sea	1.0265	— 1.028
Wine	0.992	— 1.038
III. GASES.	Water = 1. Temp. 38¾° F.	Barometer 30 In. Temp. = 34°.
Atmospheric air $= \frac{1}{770} =$	0.00130	1.0000
Carbonic acid gas	0.00198	1.5240
Carbonic oxide gas	0.00126	0.9569
Carbureted hydrogen, a maximum	0.00127	0.9784
Ditto, from coals	0.00039 0.00085	0.3000 0.5596

TABLE—*Continued.*

Name of the Body.	Specific Gravity.	
	Water = 1. Temp. 38¾° F.	Barometer 30 In. Temp. = 34°.
Chlorine - - - - - - - -	0.00321	2.4700
Hydriodic gas - - - - - - -	0.00577	4.4430
Hydrogen - - - - - - -	0.0000895	0.0688
Hydrosulphuric acid gas - - - - -	0.00155	1.1912
Muriatic acid gas - - - - - -	0.00162	1.2474
Nitrogen - - - - - - - -	0.00127	0.9760
Oxygen - - - - - - - -	0.00143	1.1026
Phosphureted hydrogen gas - - - -	0.00113	0.8700
Steam at 212° Fahr. - - - - - -	0.00082	0.6235
Sulphurous acid gas - - - - - -	0.00292	2.2470

use of a table of specific gravities;

The knowledge of the specific gravities or densities of different substances is of great importance, not only for scientific purposes, but also for its application to many of the useful arts. This knowledge enables us to solve such problems as the following, viz.:—

1st. The weight of any substance may be calculated, if its volume and specific gravity be known.

2d. The volume of any body may be deduced from its specific gravity and weight. Thus we have always

weight of any body;

$$W = g\,D\,V;$$

in which g is the force of gravity, D the density, V the volume, and W the weight, of which the unit of measure is the weight of a unit of volume of water at its maximum density.

Making D and V equal to unity, this equation becomes

$$W_{\prime} = g;$$

but if the density be one, the substance must be water at $38\overset{\circ}{.}75$ Fahr. The weight of a cubic foot of water at 60° is 62.5 lbs., and, therefore, at $38\overset{\circ}{.}75$, it is

weight of a cubic foot of distilled water at maximum density;

$$\frac{\overset{lbs.}{62.5}}{0.99914} = \overset{lbs.}{62.556};$$

whence, if the volume be expressed in cubic feet, volume in cubic feet;

$$W = \overset{lbs.}{62.556} \times D\,V \;.\;.\;.\; (232),$$ weight of a body in pounds, volume being in cubic feet;

in which W is expressed in pounds; and if the unit of volume be a cubic inch,

$$W = \frac{62.556}{1728} D\,V = 0.036201\, D\,V_{,} \;.\;.\; (233).$$ weight in pounds, volume in cubic inches;

Also

$$V = \frac{W.}{\overset{lbs.}{62.556} \,.\, D} \;.\;.\;.\; (234),$$ volume in cubic feet;

$$V_{,} = \frac{W.}{\overset{lbs.}{0.036201} \,.\, D} \;.\;.\;.\; (235).$$ volume in cubic inches;

Example 1*st.* Required the weight of a block of dry fir, containing 50 cubic inches. The specific gravity or density of dry fir is 0.555, and $V = 50$; substituting these values in Eq. (233), example first;

$$W = 0.036201 \times 0.555 \times 50 = \overset{lbs.}{1.00457}.$$ weight of 50 cubic inches of fir;

Example 2*d.* How many cubic inches are there in a 12-pound cannon-ball? Here W is 12 pounds, the mean specific gravity of cast iron is 7.251, which, in Eq. (235), give example second;

$$V_{,} = \frac{12}{0.036201 \times 7.251} = \overset{in.}{45.6}.$$ volume of a 12-pound cannon-ball.

VII.

COMPRESSIBLE FLUIDS.

Peculiarities of gases and vapors;

§ 276.—The properties of liquids which have now been considered, are common to all fluids. But gases and vapors have, in addition, properties peculiar to themselves which we now proceed to consider.

contract and expand according to pressure;

Gases and vapors differ mostly from liquids, in the readiness with which they yield a portion of their volume and contract into smaller spaces when subjected to an augmentation of external pressure, and diffuse themselves in all directions when this pressure is withdrawn. These distinguishing properties are due to the repulsive forces or molecular springs by which the particles are urged to separate from each other, and which make it impossible for compressible fluids, that are also highly elastic, ever to be at rest, unless these forces are opposed by the reaction of inclosing surfaces, as the sides of vessels, or the application of some other antagonistic forces acting inwardly, as in the case of the earth's attraction upon our atmosphere.

conditions of rest;

Besides these essential peculiarities, there are other characteristics that distinguish compressible fluids, usually denominated *aeriform bodies*, from the other forms of aggregation. Between solids and liquids, a gradation is observable, and in the degree of fluidity of the latter, a strongly marked variety obtains—as in tar, oil, water, ether, and the like; but between compressible and incompressible fluids, no similar connecting links are found. Again, as a general rule, gases are highly transparent, for most part colorless, and therefore invisible, and are distinguished from all other bodies by their small degree of density and consequent low specific gravity.

no marked variety of fluidity;

usually transparent;

small density;

The atmosphere, as being the most important of the aeriform bodies, may be taken as the representative of the whole class, as regards their mechanical properties. It is to this class of bodies, what water is to liquids. It exists all over the earth, and its ever-active agency in the production of phenomena, makes it not less interesting than important to determine the laws of its equilibrium and motion.

atmosphere the type of the class;

(1) The compressibility and elasticity are easily shown by inclosing the air in a bag of some impervious substance, as india-rubber, and pressing it with the hand; the hand will experience a resistance, while the volume of the confined air will diminish: on removing the hand, the bag will be distended by the elasticity of the air, and restored to its former dimensions. Air-pillows and cushions, in common use, are familiar illustrations.

compressibility and elasticity shown;

india-rubber bag;

Fig. 314.

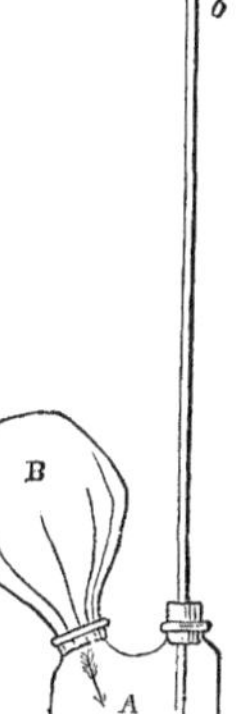

(2) A is a two-necked bottle containing some liquid, as water, B an inflated bladder, or india-rubber bag, attached by the neck to one of the mouths. A glass tube $a\,b$, open at both ends, is fitted air-tight to the other mouth, its lower end a reaching nearly to the bottom of the bottle. On compressing with the hand, the air in the bladder or bag, the liquid will be seen to mount up the tube.

india-rubber bag connected with a two-necked bottle;

Fig. 315.

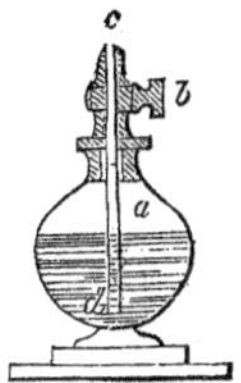

(3) *Hero's Ball.*—A hollow globe a, from which the external air can be excluded by turning a cock b, contains a tube that reaches nearly to the bottom, and fits in the neck by a screw. Fill the vessel about half full of water, screw in the tube $c\,d$,

Hero's ball;

breathe through c, and close the stop-cock b; the breath will ascend through the water, mingle with the air in the space a, take from it a portion of its volume and thus increase its elasticity, which, reacting upon the surface of the water, will force the latter up the tube $c\,d$ on turning the cock b. On this principle depend the operations of the air-chamber in fire-engines and similar machines.

Fig. 315.

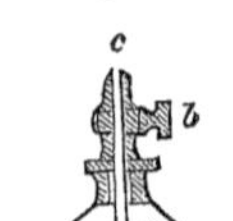

principle of the fire-engine;

(4) *Hero's Fountain.*—In this apparatus, also, the compression of air and consequent increase of elasticity, are manifested in producing a water-jet. Two vessels a and g are united by a tube t, open at both ends, extending from the upper surface of the lower vessel to near the top of the other. A pipe $c\,d$, provided with a stop-cock b, screws into the top of the vessel a, and extends nearly to its bottom, as in Hero's Ball. Upon the top of this vessel is a basin $n\,o$, from the bottom of which a pipe $e\,f$, open at both ends, passes clear through, nearly to the bottom of the vessel g. The tube $c\,d$, being unscrewed, is removed, and after pouring water into the vessel a till its surface comes nearly to the upper end of the tube t, the pipe $c\,d$ is replaced, and the stop-cock b closed. Water is now poured into the basin $n\,o$; this will descend through the tube $e\,f$ into the vessel g, and expel a portion of its air by forcing it up the tube t into the vessel a; there, finding no means of escape, it will be compressed, and its increased elasticity made to act upon the water

Hero's fountain;

description;

mode of action;

Fig. 316.

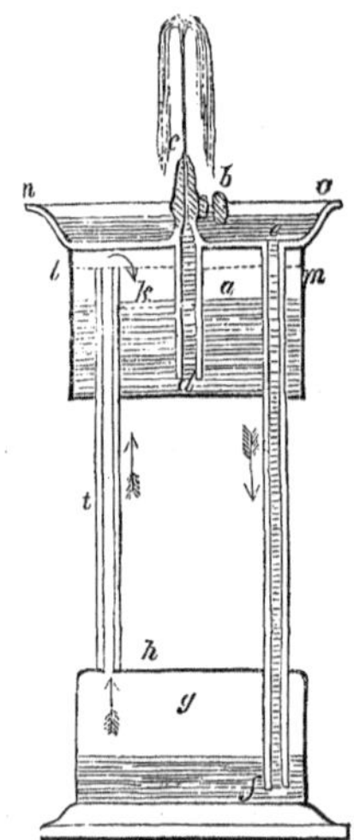

precisely as in the case of the Ball. The water will continue to descend through the tube *e f* from the basin, till the increasing elasticity of the air becomes equal to the pressure arising from a head of water equal to the difference between the level of the water in the basin and that in the lower vessel, when the flow will cease, and every thing will come to rest. In this condition of things turn the cock *b*, and the water will spout through the tube *c d*. The fluid in the upper vessel being thus ejected, there will be room for more air; this will pass from the lower vessel through the tube *t*, and the water will again descend from the basin to the vessel *g*. The water discharged by the jet falls into the basin *n o*, and is ready, in its turn, to pass down the tube *e f*. A constant flow is thus maintained as long as the fluid in the vessel *a* remains above the bottom of the tube *c d*.

conditions of rest;

conditions to cause the flow;

(5) *The Cartesian Devil.*—This is a well-known figure, constructed so as to float in a glass vessel of water, above the surface of which a portion of air is confined in such manner, that if this air be compressed, the figure will descend, and rise again when the compression ceases. It is thus contrived: In the middle of the figure *a* is a small capillary tube *b*, through which so much water is admitted into the interior of the body as to make its mean density a little less than that of the water in which it is to float. Being thus adjusted, the figure is immersed in a wide-mouthed glass vessel, over which a piece of bladder or sheet of india-rubber is then stretched to confine the air over the fluid. The finger being now pressed upon the bladder or india-rubber, the air will be compressed, the increased elasticity thus pro-

Cartesian devil;

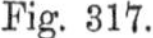
Fig. 317.

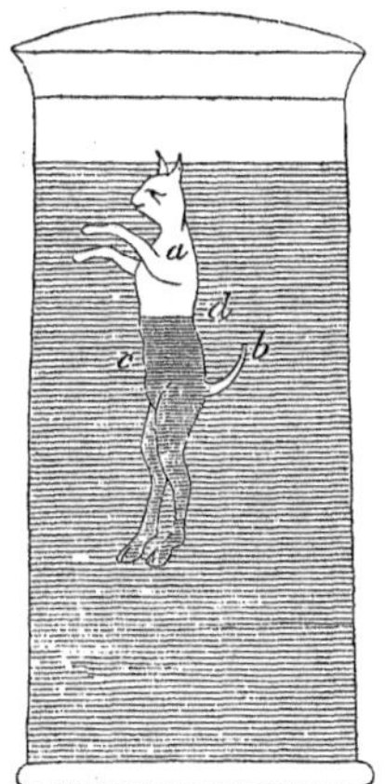

description;

motion of the figure;

duced will be exerted upon the water, which will be forced by it through the tube *b*, the mean density of the figure will be increased, and it will sink to the bottom; on removing the finger, the air above the water as well as that in the figure, being relieved from the pressure, expands, the water is forced back through the tube *b* into the vessel again, and the figure will rise to the surface in consequence of diminished mean density.

explanation of the motion.

VIII.

THE AIR-PUMP.

Air-pump, or air-syringe;

§ 277.—Seeing that the air expands and tends to diffuse itself in all directions when the surrounding pressure is lessened, it may be rarefied and brought to almost any degree of tenuity. This is accomplished by an instrument called the *Air-Pump* or *Exhausting Syringe*, one of the most important pieces of apparatus used by the natural philosopher. It will be best understood by describing one of the simplest kind. It consists, essentially, of

1st. A *Receiver R*, or chamber from which the exterior air is excluded, that the air within may be rarefied. This is commonly a bell-shaped glass vessel, with ground edge, over which a small quantity of grease is smeared, that no air may pass through any remaining inequalities on its surface, and a ground glass plate *m n* imbedded in a metallic table, on which it stands.

the receiver;

2d. A *Barrel B*, or chamber into which the air in the receiver is to expand itself. It is a hollow cylinder of metal or glass, connected with the receiver *R* by the communication *o f g*. An air-tight piston *P* is made to move back and forth in the barrel by means of the handle *a*.

the barrel and piston;

Fig. 318.

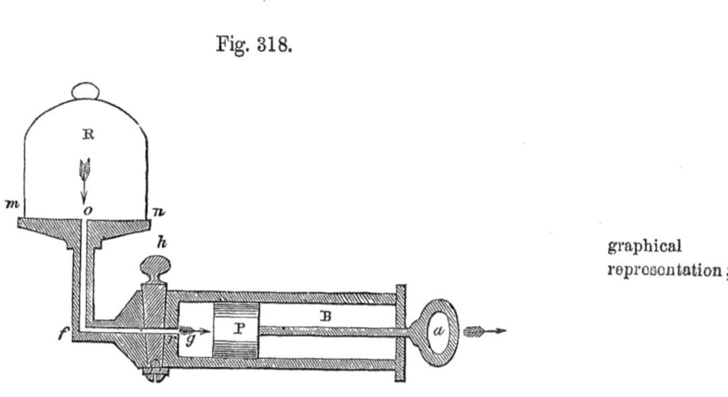

graphical representation;

3d. A *Stop-cock h*, by means of which the communication between the barrel and receiver is established or cut off at pleasure. This cock is a conical piece of metal fitting air-tight into an aperture just at the lower end of the barrel, and is pierced in two directions; one of the perforations runs transversely through, as shown in the first figure, and when in this position the communication between the barrel and receiver is established; the second perforation passes in the direction of the axis from the smaller end, and as it approaches the first, inclines sideways, and runs out at right angles to it, as indicated in the second figure. In this position of the cock, the communication between the receiver and barrel is cut off, whilst that with the external air is opened.

stop-cock, or valve;

Fig. 319.

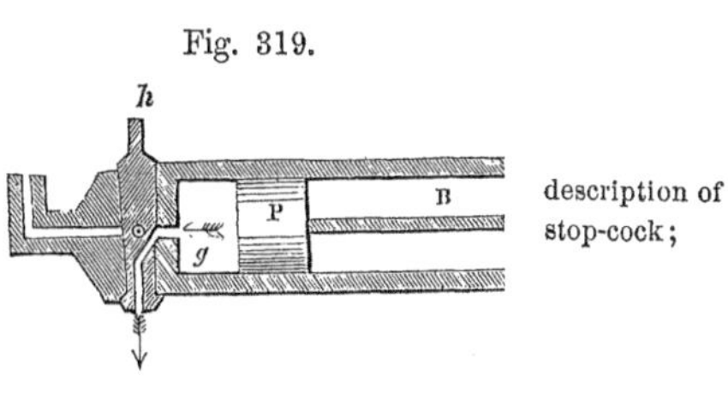

description of stop-cock;

Now, suppose the piston at the bottom of the barrel, and the communication between the barrel and the receiver established; draw the piston back, the air in the receiver will rush out, in the direction indicated by the arrow-head, through the communication ofg, into the vacant space within the barrel. The air which now occupies both the

mode of action;

barrel and receiver is less dense than when it occupied the receiver alone. Turn the cock a quarter round, the communication between the receiver and barrel is cut off, and that between the latter and the open air is established; push the piston to the bottom of the barrel again, the air within the barrel will be delivered into the external air. Turn the cock a quarter back, the communication between the barrel and receiver is restored; and the same operation as before being repeated, a certain quantity of air will be transferred from the receiver to the exterior space at each double stroke of the piston.

mode of operation;

To find the degree of exhaustion after any number of double strokes of the piston, denote by D the density of the air in the receiver before the operation begins, being the same as that of the external air; by r the capacity of the receiver, by b that of the barrel, and by p that of the pipe. At the beginning of the operation, the piston is at the bottom of the barrel, and the internal air occupies the receiver and pipe; when the piston is withdrawn to the opposite end of the barrel, this same air expands and occupies the receiver, pipe, and barrel; and as the density of the same body is inversely proportional to the space it occupies, we shall have

to find the degree of exhaustion;

ratio of the densities;

$$r + p + b \ : \ r + p \ :: \ D \ : \ x;$$

in which x denotes the density of the air after the piston is drawn back the first time. From this proportion, we find

first diminished density;

$$x = D \cdot \frac{r + p}{r + p + b}.$$

The cock being turned a quarter round, the piston pushed back to the bottom of the barrel, and the cock again turned to open the communication with the receiver, the operation is repeated upon the air whose density is x, and

we have

$$r + p + b \ : \ r + p \ :: \ D \cdot \frac{r + p}{r + p + b} \ : \ x';$$ ratio of densities;

in which x' is the density after the second backward motion of the piston, or after the second double stroke; and we find

$$x' = D \cdot \left(\frac{r + p}{r + p + b}\right)^2;$$ second diminished;

and if n denote the number of double strokes of the piston, and x_n the corresponding density of the remaining air, then will

$$x_n = D \cdot \left(\frac{r + p}{r + p + b}\right)^n.$$ the nth diminished density;

From which it is obvious, that although the density of the air will become less and less at every double stroke, yet it can never be reduced to nothing, however great n may be; in other words, the air cannot be wholly removed from the receiver by the air-pump. The exhaustion will go on rapidly in proportion as the barrel is large as compared with the receiver and pipe, and after a few double strokes, the rarefaction will be sufficient for all practical purposes. Suppose, for example, the receiver to contain 19 units of volume, the pipe 1, and the barrel 10; then will

the air can never be wholly exhausted from the receiver;

$$\frac{r + p}{r + p + b} = \frac{20}{30} = \tfrac{2}{3}:$$

and suppose 4 double strokes of the piston; then will $n = 4$, and

illustration;

$$\left(\frac{r + p}{r + p + b}\right)^n = (\tfrac{2}{3})^4 = \frac{16}{81} = 0.197, \text{ nearly};$$ density after 4th double stroke;

that is, after 4 double strokes, the density of the remaining air will be but about two tenths of the original density. With the best machines, the air may be rarefied from four to six hundred times.

rarefaction by best pumps;

The degree of rarefaction is indicated in a very simple manner by what are called *gauges*. These not only indicate the condition of the air in the receiver, but also warn the operator of any leakage that may take place either at the edge of the receiver or in the joints of the instrument. The mode in which the gauge acts, will be better understood when we come to discuss the barometer; it will be sufficient here simply to indicate its construction. In its more perfect form, it consists of a glass tube, about 60 inches long, bent in the middle till the straight portions are parallel to each other; one end is closed and the branch terminating in this end is filled with mercury. A scale of equal parts is placed between the branches, having its zero at a point midway from the top to the bottom, the numbers of the scale increasing in both directions. It is placed so that the branches of the tube shall be vertical, with its ends upward, and inclosed in an inverted glass vessel, which communicates with the receiver of the air-pump.

gauges;

objects, and construction;

scale of the gauge, and position;

Fig. 320.

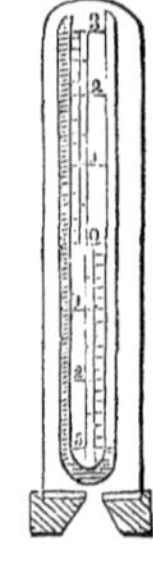

Repeated attempts have been made to bring the air-pump to still higher degrees of perfection since the time of Otto von Guericke, burgomaster of Magdeburg, who first invented this machine in 1560. Self-acting valves, opening and shutting by the elastic force of the air, have been used instead of cocks. Two barrels have been given to the air-pump instead of one, so that an uninterrupted and more rapid rarefaction of the air is brought about, the piston in one barrel being made to ascend as that of the other descends. The most serious defect in the air-

first inventor;

improvements;

pump was, that the atmospheric air could not be entirely ejected from the barrel, but remained between the piston and the bottom of the barrel. This intervening space is filled with air of the ordinary density at each descent of the piston; when the cock is turned, and the communication re-established with the receiver, this portion of air forces its way in and diminishes the degree of rarefaction

the most serious defect of the older pumps;

Fig. 321.

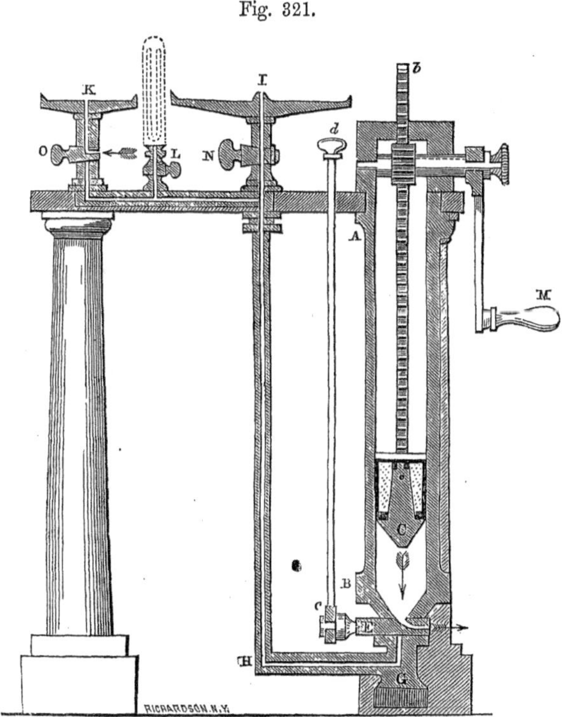

section of one of the most approved pumps;

already attained. If the air in the receiver is so far rarefied, that one stroke of the piston will raise only such a quantity as equals the air contained in this space, it is plain that no further exhaustion can be effected by continuing to pump. This limit to rarefaction will be arrived at the sooner, in proportion as the space below the piston is larger; whence one chief point in the improvements has

limits to rarefaction due to the defect above;

been to diminish this space as much as possible. AB is a highly polished cylinder of glass, which serves as the barrel of the pump; within it the piston works perfectly air-tight. The piston consists of washers of leather soaked in oil, or of cork covered with a leather cap, and tied together about the lower end C of the piston rod by means of two parallel metal plates. The piston-rod Cb, which is

description of the improved pump;

Fig. 321.

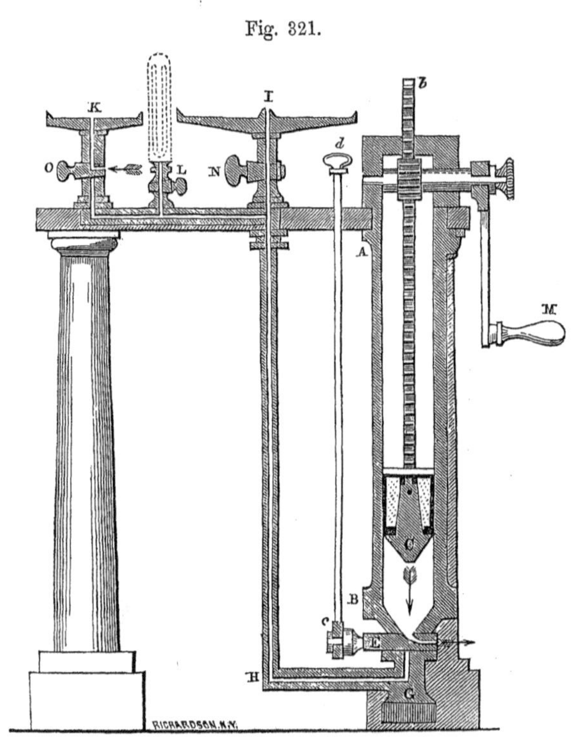

section of the pump;

toothed, is elevated and depressed by means of a cog-wheel that is turned by the handle M. If a thin film of oil be poured on the upper surface of the piston the friction will be lessened, and the whole will be rendered more perfectly air-tight. To diminish to the utmost the space between the bottom of the barrel and the piston-rod, the form of a truncated cone is given to the latter, so that its extremity

use of oil;

shape of the lower end of piston-rod;

may be brought as nearly as possible into absolute contact with the cock *E;* this space is therefore rendered indefinitely small, the oozing of the oil down the barrel contributing still further to lessen it. The exchange-cock *E* has the double bore already described, and is turned by a short lever, to which motion is communicated by the rod *c d*. The communication *G H* is carried to the two plates *I* and *K*, on one or both of which receivers may be placed; the two cocks *N* and *O* below these plates, serve to cut off the rarefied air within the receivers when it is desired to leave them for any length of time. The cock *O* is also an exchange-cock, so as to admit the external air into the receivers.

exchange-cock; communication; cut-off cock cock to readmit the air;

Pumps thus constructed have advantages over such as work with valves, in that they last longer, exhaust better, and may be employed as condensers when suitable receivers are provided, by merely reversing the operations of the exchange valve during the motion of the piston.

advantages of this kind of pump.

§ 278.—The following are some of the most interesting experiments performed with the aid of an air-pump, showing the expansive force of the atmosphere, and also the relations between air of ordinary density and that which is highly rarefied:—

1st. Under a receiver place a bladder tied tightly about the neck and partly filled with air; exhaust the air in the receiver, and that confined within the bladder will gradually distend, proving experimentally the expansive force of atmospheric air. When the air is readmitted into the receiver, the bladder will resume its former dimensions.

Experiments with air-pump; first experiment;

An analogous appearance will be exhibited if a jar, over which some india-rubber has been tied, be placed beneath a receiver, and the air be then exhausted.

second;

2d. The expansive force of our atmosphere is further shown if a long-necked flask, or retort, be inverted so that its mouth shall be below the surface of some water contained in a vessel, and the whole be placed under the receiver of an air-pump; when the air within the receiver is rarefied, that which was contained in the bulb, expanding, escapes through the water; and on readmitting the atmosphere the water will rise and occupy the space vacated by the air.

showing also the expansion of air;

Fig. 322.

3d. The transfer of a fluid from one flask to another. Let there be a fluid in the flask *A*. The neck of this flask contains a glass tube fitted air-tight into it, and reaching almost to the bottom; the tube being bent twice at right angles, the other end passes freely through the neck of a second bottle *B*. Place this apparatus under the receiver of an air-pump, and exhaust; the fluid will mount up from the bottle *A* and pass through the tube over into the bottle *B*. Readmit the air, the fluid will pass back again.

third, illustrating the same principle;

Fig. 323.

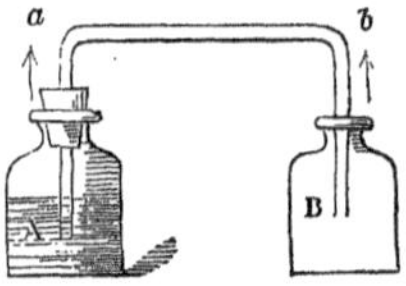

4th. Place Hero's ball under the receiver when half filled with water, and exhaust; the expansion of the air within will send the water up through the tube in a jet.

fourth.

§ 279.—When a piece of metal and a feather are abandoned to their own weight in the air, they fall with very different velocities. The cause is the great disparity in the extent of surfaces exposed to the resistance of the air as compared with the weights.

Let *a* and *b* be two wheels resembling the arms of a windmill, with this difference only, that the vanes of *b* shall strike the air with their broad faces, whilst those of *a* shall cut it edgewise; each has a separate axis on which it revolves. By means of a mechanical contrivance a rapid rotary motion is com-

Atmospheric resistance illustrated;

Fig. 324.

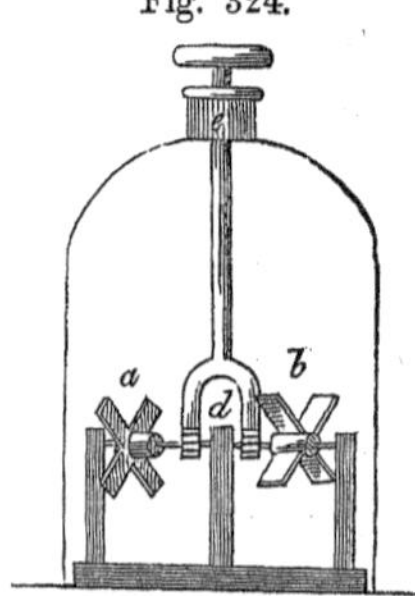

municated to them. In order that this may act under a receiver, a rod must be made to pass through an air-tight leather stuffing-box *e*; at the end of the rod is a curved arm *d*, which drives the wheels. If the rotation take place in vacuo, the two wheels *a* and *b* will cease to revolve simultaneously; whereas, if the motion take place in the ordinary atmosphere, the resistance of the latter will bring *b* to a stand long before *a* ceases to turn.

description and use of the instrument.

§ 280.—The atmosphere is the ordinary medium through which sound is transmitted to the ear. In proportion as the air becomes more rarefied, the transmission of sound through it becomes more feeble.

Effects of rarefaction on sound;

Under a receiver furnished with a leather stuffing-box, place a bell whose clapper may be struck by a rod passing through the box, taking care to place the bell on some soft unelastic substance, to prevent its communicating sound to the plate of the pump and thus to the external air. The annexed figure represents such an apparatus, which may, however, be considerably varied: *a* is the bell, *b* the clapper attached by a spring to a thin plate of wood *c*, into which the support of the bell is screwed; *g* is a leather drum stuffed with horse-hair, fitting into the upper wooden plate *c*, and into a lower plate *d*, by which the whole apparatus is fastened down to the plate of the pump; lastly, *h* is the lever by which the clapper is agitated. After about 10 strokes of the piston, the sound becomes sensibly more feeble, and if the exhaustion be continued long enough it will cease altogether.

instrument by which this may be illustrated;

Fig. 325.

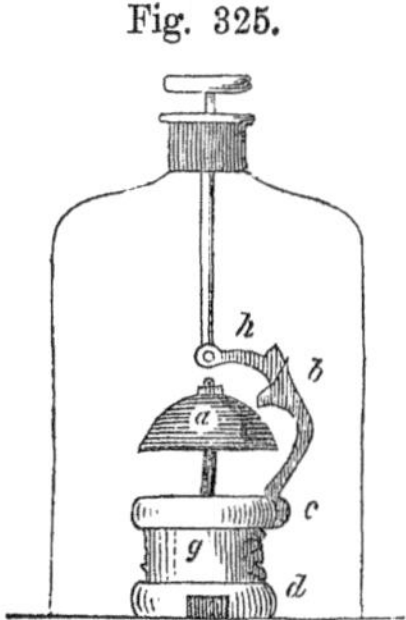

Air is necessary to respiration. Place a bird beneath the receiver of an air-pump; a few strokes of the piston will cause it to make convulsive struggles, and death will soon

air is necessary to respiration;

ensue unless air be admitted. Warm-blooded animals, as birds, die if rarefaction be carried to a small degree; cold-blooded animals, on the contrary, endure a high degree of rarefaction. Many birds ascend to considerable heights in the atmosphere, and it may be hence inferred that the density of the air at these altitudes is greater than that in the exhausted receiver of an air-pump.

place a bird in the receiver of a pump and exhaust;

Fig. 326.

Air is necessary to combustion. Introduce a taper into a bell-shaped receiver full of atmospheric air, and observe the time it will continue to burn. Light the taper again, place it beneath the receiver and exhaust quickly, after it has been replenished with fresh air; the flame will expire much sooner than before.

air is necessary to combustion.

To the same cause it is owing that in vacuo no light is produced by striking a flint and steel together.

IX.

WEIGHT AND PRESSURE OF THE ATMOSPHERE.

§ 281.—From the resistance which the atmosphere opposes to the motion of bodies through it, we might infer that it has *weight* as well as inertia. That it has weight is obvious from the fact that the atmosphere incases, as it were, the whole earth: if it were destitute of weight and subjected only to the repulsive action among its own particles, it would recede further and further and extend itself throughout space. But the existence of weight in the atmosphere may be shown experimentally, thus:—

The atmosphere has weight;

Take a flask of some two or three inches in diameter, having an air-tight stop-cock. Suspend it from one end of the balance-beam and ascertain its weight when filled with air. Exhaust the air, by means of the air-pump, and the flask will be found lighter than before; readmit the air, it will regain its former weight. Force into the flask an additional quantity of air, by means of the air-pump, used as a condenser, and the weight will be found to be increased. experiment to show this;

Fig. 327.

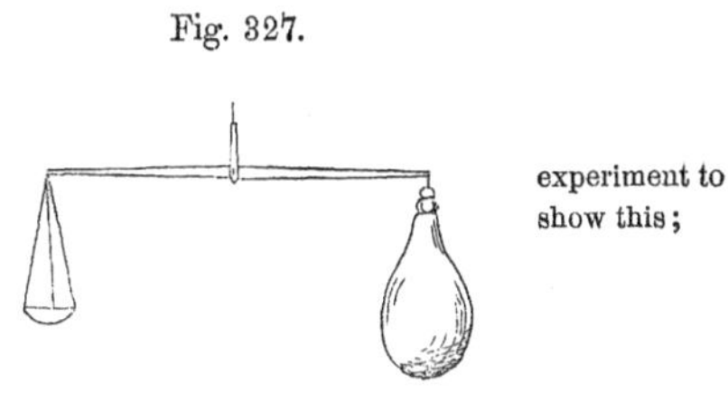

Since the atmosphere has weight, it must exert a pressure upon all bodies in it. To illustrate the truth of this, fill with mercury a glass tube, about 32 or 33 inches long, and closed at one end by an iron stop-cock. Close the open end by pressing the finger against it, and invert the tube in a basin of mercury; remove the finger, the mercury will not escape, but remain apparently suspended nearly 30 inches above the level of the mercury in the basin. the air exerts a pressure upon all bodies within it; experiment to illustrate this;

Fig. 328.

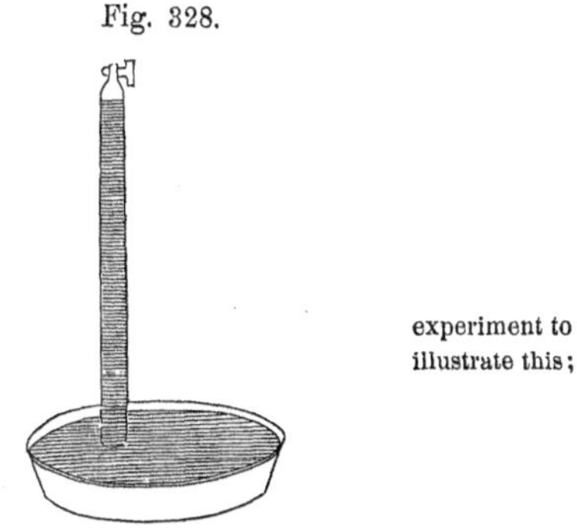

If we consider the circumstances attending this experiment, it will be seen that the tube containing the mercury forms with the basin a system of communicating tubes, as in § 259. Now the atmosphere rests on the mercury in the basin, and is excluded by the glass from that in the tube, above which there is therefore a vacuum. Withdraw the atmosphere from the surface of the mercury in the basin, and, by the effect of withdrawing the external atmosphere;

law of equilibrium of fluids, the mercury will descend in the tube till it comes to a level with that without; restore the pressure of the atmosphere, and the mercury in the tube will again rise to its former height. This is well illustrated by the following device. *R* is a receiver closed air-tight at the top by means of a metallic plate; *a* is a tube filled with mercury after the manner just described, and terminating at the open end in an inverted vial-shaped vessel —this tube passes air-tight through the plate on the receiver; *b* is a second tube bent in the manner indicated in the figure, and, like the tube *a*, it terminates at one end in a vial-shaped vessel, but is open at both ends; this tube communicates with the receiver by passing through the metallic plate at top, and thus a connection is established between the open air and the interior of the receiver. Mercury being poured into the vial of the tube *b*, it will rise to the same level on either side of the bend *m*, and the communication between the interior of the receiver and exterior air will be interrupted. The receiver being placed upon the plate of the air-pump and the air exhausted, the mercury will descend in the tube *a*, and ascend in the tube *b* towards the bend at the top; readmit the air into the receiver, the mercury will rise in the tube *a* and fall in the tube *b*.

an instrument well suited to exhibit the facts of this experiment;

description and use;

Fig. 329.

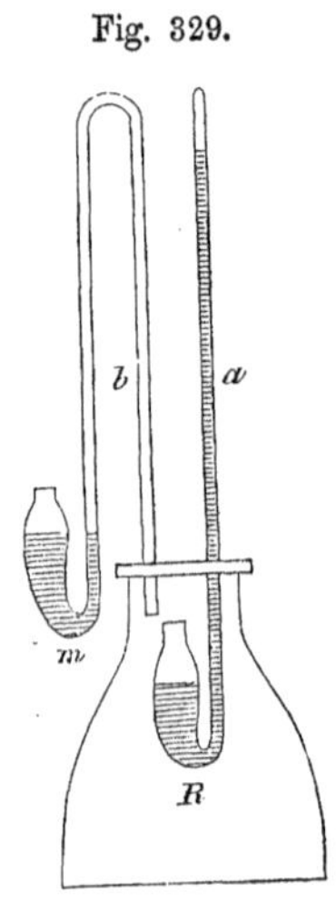

From this we see, that the atmospheric air presses on the mercury, and indeed upon the surfaces of all bodies exposed to it, with a force sufficient to maintain the quick-

inference from this experiment;

silver in the tube at a height of nearly 30 inches; whence, *the intensity of its pressure must be equal to the weight of a column of mercury whose base is equal to that of the surface pressed and whose altitude is about 30 inches. The force thus exerted is called the atmospheric pressure.* atmospheric pressure;

The absolute amount of atmospheric pressure was first discovered by Torricelli, a pupil of Galileo; the tubes employed in the experiments are called, on this account, *Torricellian tubes*, and the vacant space above the mercury in the tube is called, the *Torricellian vacuum*, to distinguish it from that of a receiver, which is frequently called the Guerickian vacuum, from Otto von Guericke, who first invented the air-pump. Torricellian tubes; Torricellian vacuum;

The pressure of the atmosphere at the level of the sea will support a column of mercury 30 inches high. Now, if we suppose the bore of the tube to have a cross-section of one square inch, the atmospheric pressure up the tube will be exerted upon this extent of surface, and will support 30 cubic inches of mercury. Each cubic inch of mercury weighs 0.49 of a pound—say half a pound—from which it is apparent that *the surfaces of all bodies, at the level of the sea, are subjected to an atmospheric pressure of fifteen pounds to each square inch.* atmospheric pressure at the level of the sea;

The body of a man of ordinary stature has a surface of about 2000 square inches; whence, the whole pressure to which he would be exposed, at the level of the sea, is 15 pounds $\times$ 2000 $=$ 30000 pounds. pressure upon the surface of a man;

The pressure of the atmosphere, resulting as it does from its weight, it is an easy matter to estimate the weight of the entire atmosphere of the earth. It will be sufficient to compute, from the known diameter of the earth, the extent of its surface in square inches, and to multiply this by fifteen; the product will be the weight in pounds. weight of the entire atmosphere;

When the height of the mercury in the Torricellian tube is 30 inches, the atmospheric pressure will support in vacuo a column of water 34 feet, the specific gravity of mercury being 13.6 referred to water as a standard. This column of water supported by the atmospheric pressure;

has been verified by Hanson and Sturm, who actually performed the experiment at Leipzig.

Magdeburg hemispheres;

The atmospheric pressure is exhibited in a most striking way by means of the *Magdeburg hemispheres.* These are two hollow hemispheres, of brass or copper, whose edges fit air-tight, each hemisphere being furnished with a strong ring or handle, one of them also having a tube with stop-cock. Place the two hemispheres together, connect them with the communication-pipe of the air-pump, exhaust the air, and turn the stop-cock, and disconnect from the pump. It will be found that great force will be necessary to pull the hemispheres asunder. If the diameter of the hemispheres, as in the case of those employed by Guericke, in one of his experiments, were 2 feet, the number of square inches in a great circle would be

description and mode of using;

Fig. 330.

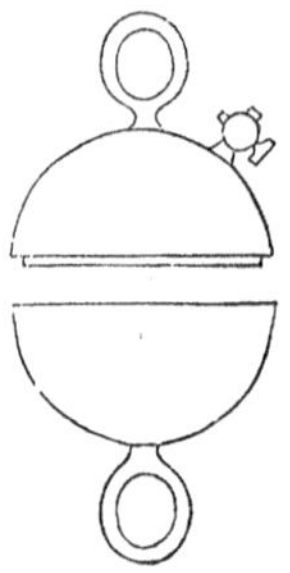

examples of Guericke's hemispheres;

$$3.1416 \times \left(\frac{24}{2}\right)^2 = 452.39,$$

and the force, estimated in pounds to overcome the pressure, would be

$$15 \times 452.39 = 6785.85.$$

Fig. 331.

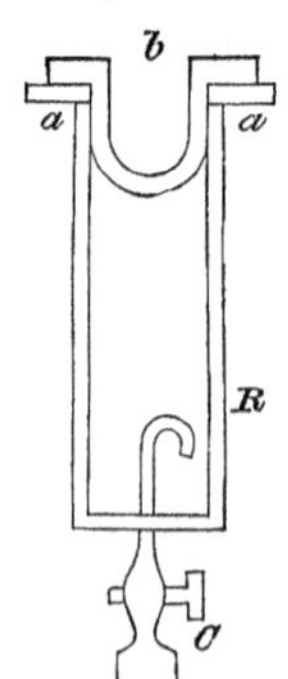

In the experiment referred to above, there were successively from 14 to 30 horses harnessed to the hemispheres, without effecting the separation.

the forcing of fluid through pores of solids;

The pressure of the atmosphere will force fluids through such solid bodies as are porous. Let R be a long receiver, provided with a tube and stop-cock *C* at one end, for the purpose of connecting with the air-pump, and at the other a perforated metallic plate *a a*, into which fits, air-tight, a wooden cup *b*, whose pores are in the direction of the axis of the tube. This cup being filled with mercury, and the air exhausted by the air-pump, the mercury will fall in a fine shower down the receiver. The tube below is made to enter the receiver, and to curve over at the top to prevent the mercury from falling into the communication-pipe of the pump.

instrument to exhibit this;

The atmosphere presses not only downward, but upward, and laterally in all directions. This is shown by the following experiment: The two hemispheres *A* and *B*, are connected by a tube in such manner that one of them may turn about a joint *C*, while the other is stationary. Place the hemisphere *A* upon the plate of the air-pump, and upon *B* lay a plane plate of glass or metal fitting it air-tight. Exhaust the air, and the hemisphere *B* may be turned in any direction without its plate falling off. This equal pressure of the atmosphere in all directions, is of great practical utility, as we shall presently see when we come to speak of *siphons* and *water-pumps*. To this pressure it is owing that flies, and other insects, are enabled to support themselves upon smooth vertical walls, and in inverted positions upon the ceilings of rooms, &c. The feet being flat and flexible, are brought close against the wall or ceiling so as to exclude the air, the centre of the foot is then drawn away, leaving the margin in contact; a partial vacuum is thus formed, and the external pressure of the air is sufficient to support the weight of the insect.

atmospheric pressure is exerted in every direction;

Fig. 332.

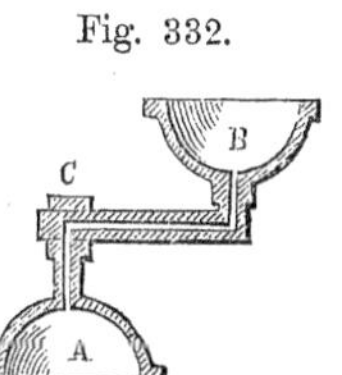

exemplification of this in the adhesion of insects to walls and ceilings.

Fig. 333.

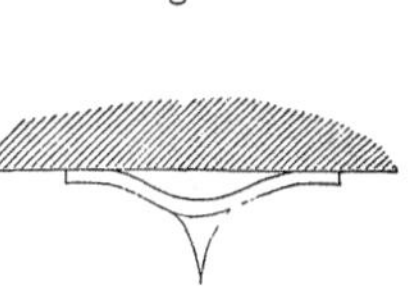

X.

MARIOTTE'S LAW.

Mariotte's law;

§ 282.—We have seen that the atmosphere readily contracts into a smaller volume when pressed externally, that it as readily regains its former dimensions when the pressure is removed, and that it is, therefore, both compressible and elastic. Let us now consider the law which connects the *pressure*, *density*, and *elasticity*. For this purpose, procure a siphon-shaped tube $A\,B\,D$, open at A, the end of the longer branch, and hermetically sealed at the end D of the shorter branch. Place between the branches, and parallel to them, a scale of equal parts, say inches, having its zero on the line $o\ o$.

connecting the pressure, density, and elasticity;

the instrument for compressing the air;

Fig. 334.

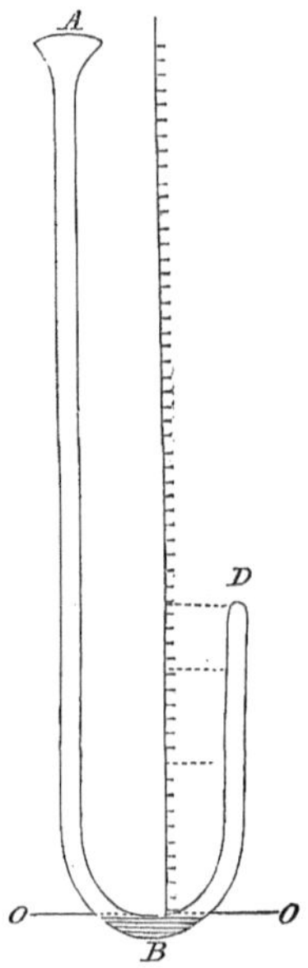

Pour in, at the open end A, as much quicksilver as will fill the horizontal part of the tube, and bring its upper surface to the zero line in both branches; a quantity of atmospheric air of ordinary density will then be confined in the shorter branch. The expansive action of this air, resisting, as it does, the pressure of the external air, is measured by the weight of a column of mercury, whose base is a section of the tube and height 30 inches. Pour into the longer branch an additional quantity of mercury; it will rise in

mode of using it;

the shorter branch, and cause the air above it to be compressed into a smaller space, but the heights at which it will stand in the two branches will be different. The difference between these two heights, added to 30 inches, will be the altitude of the column of mercury, whose weight is just sufficient to resist the expansive action of the confined air. Now it is found by trial, that when the air in the shorter branch is compressed into half its primitive volume, the difference of level of the mercury in the two branches is just 30 inches, thus making the compressing force double what it was before; that when it is compressed into one third of its original volume, the difference of level is 60 inches, thus trebling the pressure; when compressed into one fourth, the difference of level is 90 inches, thus quadrupling the pressure, and so on. Hence we see, that in compressing the same quantity of air into smaller spaces, *the volumes occupied by it are inversely proportional to the pressures.*

details and rationale of the experiment;

volumes are inversely proportional to the pressures;

This law holds equally when the air, instead of being compressed, is permitted to expand. Let $a\,b$ be a glass tube, about 33 inches long, one end a, being fitted with an air-tight cock, and the entire length of the tube being graduated in inches. Open the cock a, immerse the tube with its open end downward into the vessel A, previously half filled with mercury, which will, of necessity, stand at an equal height within and without the tube. Now close the cock a, and so confine a portion of air at its ordinary density within the tube above the surface of the mercury.

instrument for expanding the air;

Fig. 335.

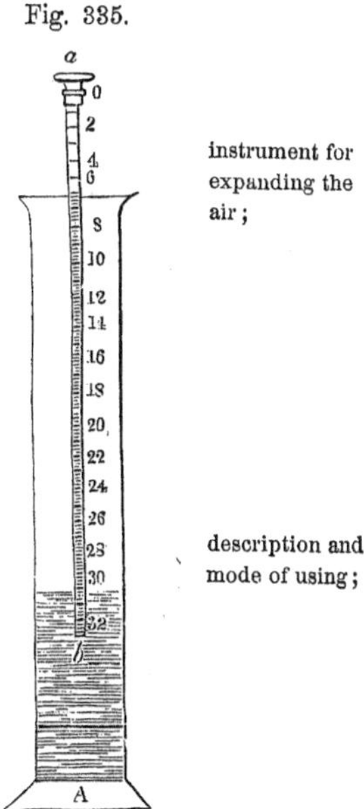

description and mode of using;

Elevate the tube any distance whatever, taking care that its open end shall be below the surface; the air will expand, and fill a larger portion of the tube, though a column of mercury will still stand at an elevation above the outer level, so that the weight of this column, with the elastic force of the inclosed air, counterbalances the natural pressure of the atmosphere. The pressure therefore which the included air sustains, is equal to the weight of a column of mercury 30 inches high, minus that of the column supported in the tube. Let the space full of air above the mercury in the closed tube be 3 inches; lift up the tube so that this space shall be 6 inches, the mercury will be found to stand in the tube 15 inches above that in the outer vessel. Here the volume of the air is doubled, and the pressure upon it is $30 - 15 = 15 =$ one half of 30, what it was before. Again raise the tube till the volume of air becomes 9 inches long, the mercury in the tube will be found to stand 20 inches higher than in the outer vessel; here the volume is three times its primitive volume, and its pressure $30 - 20 = 10 =$ one third of 30, its original pressure; whence the law is manifest.

weight of the suspended column of mercury plus elastic force of confined air, equal to atmospheric pressure;

By experiments made at Paris, it has been found that this law obtains when air is condensed 27 times, and rarefied 112 times. Other gases obey it also, till the pressure becomes a few atmospheres less than that at which they assume a liquid form.

experiments made at Paris;

The density of the same quantity of matter is inversely proportional to the volume it occupies. If, therefore, P be the pressure upon a unit of surface necessary to produce a density unity, p the pressure corresponding to a density D, then, according to this law, will,

expression of Mariotte's law.

$$p = PD \quad . \quad . \quad . \quad . \quad . \quad (236).$$

This law was investigated by Boyle and Mariotte, the former in 1660, and the latter in 1668, and is now known as *Mariotte's law.*

XI.

LAW OF THE PRESSURE, DENSITY, AND TEMPERATURE.

§ 283.—It is a universal law of nature that heat expands all bodies, and is ever active in producing changes of density. We have now to consider the law of this change in air.

Law connecting the pressure, density, and temperature;

It has been ascertained, experimentally, that air, subjected to any constant pressure, will expand 0.00208th of its volume at 32° Fahr., for each degree of the same scale above this temperature; so that if V_1 be the volume of the air at 32°, and V its volume at any other temperature t, then will

rate of the air's expansion;

$$V = V_1 [1 + (t - 32°)\, 0.00208)] \;.\;.\; (237).$$

If $D_{,}$ be the density at 32°, under a pressure p, and D that at the temperature t, under the same pressure, then, because the densities are inversely as the volumes, will

$$V_1 \;:\; V_1 [1 + (t - 32°)\, 0.00208] \;::\; D \;:\; D_{,};$$

value for any temperature under a constant pressure;

whence

$$D = \frac{D_{,}}{1 + (t - 32°)\,.\,0.00208} \;.\;.\; (238).$$

density at any temperature under a constant pressure;

If $p_{,}$ denote the pressure necessary to restore this air to the density $D_{,}$, we shall have from Mariotte's law

$$\frac{D_{,}}{1 + (t - 32°)\, 0.00208} \;:\; D_{,} \;::\; p \;:\; p_{,};$$

whence

pressure to produce at a given temperature a density at 32° under a given pressure;

$$p_{\prime} = p\,[1 + (t - 32°)\,0.00208] \quad . \quad . \quad (239).$$

Again, let the pressure p be produced by the weight of a column of mercury, having a base unity, and an altitude $h_{\prime\prime}$, taken at a given latitude, say that of 45°, in order that the force of gravity may be constant. Denoting the density of the mercury by $D_{\prime\prime}$, its weight will be

weight of a column of mercury;

$$D_{\prime\prime}\,h_{\prime\prime}\,g'\,;$$

in which g' denotes the force of gravity at the latitude of 45°.

Substituting this for p, in Eq. (236), we have

$$D_{\prime\prime}\,h_{\prime\prime}\,g' = PD;$$

whence

pressure to produce a unit of density;

$$P = \frac{D_{\prime\prime}\,h_{\prime\prime}\,g'}{D};$$

and substituting the value of D, given in Eq. (238), this becomes

same in different form;

$$P = \frac{D_{\prime\prime}\,h_{\prime\prime}\,g'}{D_{\prime}}\,[1 + (t - 32°)\,0.00208] \quad . \quad . \quad (240).$$

From Eq. (236), we have

$$D = \frac{p}{P};$$

and substituting the value for P above, we get

density at 32° under a constant pressure;

$$D = \frac{p\,D_{\prime}}{D_{\prime\prime}\,h_{\prime\prime}\,g'\,[1 + (t - 32°)\,0.00208]}.$$

Denote by h the height of the column of mercury at 32°, necessary to produce upon a unit of surface the pressure p, then will

$$p = D_{\prime\prime} h g' ;$$

weight of a column of mercury equal to the constant pressure;

which, substituted for p above, gives, after striking out the common factors,

$$D = \frac{D_{\prime} h}{h_{\prime\prime} [1 + (t - 32°) 0.00208]}.$$

density at 32° under a constant pressure;

Now, when $h_{\prime\prime}$ becomes 30 inches, then will $D_{\prime}$ take the value given in the table of § 275 opposite the name of the gas or vapor under consideration, and we have, for the practical application of that table,

$$D = \frac{D_{\prime}}{30} \times \frac{h}{1 + (t - 32°) 0.00208} \quad . \quad . \quad (240)';$$

density answering to a given temperature and barometric column;

in which $D_{\prime}$ is the tabular specific gravity or density, h the height of the column of mercury expressed in inches, and D the density of the gas pressing upon the mercury.

Example. What is the density of atmospheric air, when the barometer stands at 26 inches and thermometer at 42°? In this case, $D_{\prime}$ will be found in the table to be 0.0013, whence

example to illustrate the use of this formula;

$$D = \frac{0.0013}{30} \times \frac{26}{1 + (42° - 32°) 0.00208} = 0.0011.$$

We are now prepared to understand how the values of $D_{\prime\prime}$ in the table just referred to, were obtained, and of which no explanation has, thus far, been made.

It will be recollected that, when referred to the same standard, the numbers which express the specific gravities of bodies also express their densities, and that the specific

to obtain the tabular specific gravity of gases, &c.;

gravity of a body is the ratio obtained by dividing the weight of the body by that of an equal volume of the standard substance. The gases and vapors are incessantly changing their densities, on account of the varying pressures and temperatures to which they are subjected. Tabulated densities must, therefore, correspond to a standard of temperature and of pressure. Thirty-two degrees Fahrenheit's scale is adopted for the former; and the weight of a column of mercury, at the same temperature, having an altitude equal to thirty inches, and resting upon a base whose area is a superficial unit, is taken for the latter.

specific gravity of any body;

standard temperature and pressure for density of gases, &c.;

By a very simple transformation of Eq. (240)′, we find

tabular value for density;

$$D_{,} = \frac{30 \times [1 + (t - 32°)\, 0.00208]}{h} \times D.$$

To make this formula applicable to any gas, it will only be necessary to observe h, by means of a barometer in the atmosphere; t, by a thermometer in contact with the gas; and to find $D_{,}$ corresponding to these quantities, by the following process: Provide a glass vessel A, whose mouth may be closed by a stopcock B, air-tight, and of which the bottom terminates in a long narrow tube C, closed at the end. Let the capacity of this vessel be carefully ascertained by filling it with water, and pouring this water afterward into a graduated vessel; also let the tubular portion C be graduated and numbered by tenths, hundredths, &c., so that the numbers shall increase towards the smaller end, and express that portion of the entire capacity

vessel for finding the weights and volumes of gases;

Fig. 336.

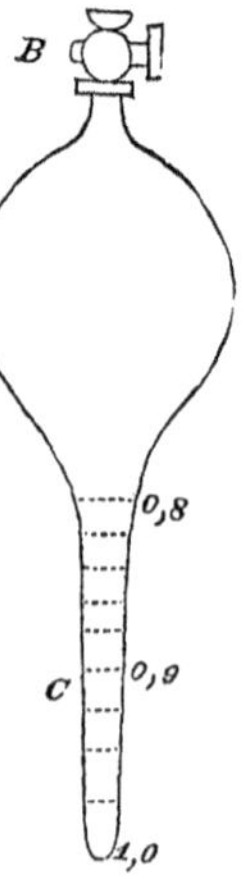

B
A
0,8
C
0,9
1,0

of the vessel, regarded as unity, which is comprised between its mouth B and these numbers.

This being understood, denote the weight of this vessel by W_v; that of a volume of air, or of the gas under consideration, equal to the contents of the vessel, and under the pressure h and temperature t, by W_a; the buoyant effort of the atmosphere, under the same pressure and temperature, by e; and the weight required to counterpoise the vessel filled with air by W_1, then will

notation;

$$W_1 = W_v + W_a - e \quad . \quad . \quad . \quad . \quad (a).$$

weight of vessel filled with air;

Connect with the air-pump, and exhaust as far as convenient; close the stop-cock, disconnect and weigh again, and denote the weight necessary to counterpoise the vessel with its rarefied air by W_2, and we shall have

$$W_2 = W_v + W_{a_,} - e;$$

weight of vessel filled with rarefied air;

in which $W_{a_,}$ denotes the weight of the rarefied air remaining in the vessel.

Subtracting this from the equation above, we find

$$W_1 - W_2 = W_a - W_{a_,};$$

weight of the extracted air;

which is obviously the weight of the extracted air.

Now immerse the vessel in water, mouth downward, and open the stop-cock; the liquid will enter, and taking care to keep its level on the inside and outside the same, the water will come to rest at or near some one of the graduated points on the tube. The air or gas within will then have the same elasticity as the external atmosphere, and the reading h of the barometer becomes applicable to the gas. This graduated point will make known the volume V of air or gas extracted; and, knowing its weight, that of a volume equal to the contents of the whole vessel, which we have denoted by W_a, may be

volume of the air extracted under the barometric pressure;

found from the proportion

$$V : W_1 - W_2 :: 1 : W_a;$$

whence

weight of the vessel filled with air under the barometric pressure;

$$\frac{W_1 - W_2}{V} = W_a \quad . \quad . \quad . \quad . \quad . \quad (b).$$

Next fill the vessel with water, and weigh again; denote the counterpoising weight by W_3, and the weight of the contained water by W_w, and we shall have

$$W_3 = W_v + W_w - e;$$

and subtracting Eq. (a), we find

$$W_3 - W_1 = W_w - W_a;$$

adding Eq. (b), we find

weight of the vessel full of water;

$$W_3 - W_1 + \frac{W_1 - W_2}{V} = W_w;$$

and dividing Eq. (b) by this one, we get

ratio of the weights of equal volumes of water and gas;

$$\frac{W_1 - W_2}{(W_3 - W_1)V + W_1 - W_2} = \frac{W_a}{W_w}.$$

Multiplying both members by the tabular density d of water corresponding to the temperature of that employed, and dividing both numerator and denominator of the first member by $W_1 - W_2$, we finally get

density of the air;

$$\frac{d}{\frac{W_3 - W_1}{W_1 - W_2}V + 1} = \frac{W_a}{W_w} \times d.$$

But the second member is the specific gravity or density D of air or gas, under the pressure h and temperature t.

Whence, to find the value of D, we have this rule, viz.: Weigh the vessel full of the gas under consideration; exhaust, and weigh a second time; find, by admitting water, the volume of gas exhausted by the pump; fill with water, and weigh a third time; then divide the difference between the last and first weights by the difference between the first and second; multiply this quotient by the volume exhausted; increase this product by unity, and divide the tabular density of water, corresponding to its observed temperature, by this sum. The value of D, thus found, and the observed values of h and t, being substituted in the value for $D_{\prime\prime}$, this latter may be found and tabulated.

Process for finding the density or specific gravity of a gas.

XII.

BAROMETER.

§ 284.—The atmosphere being a heavy and elastic fluid, is compressed by its own weight. Its density cannot be the same throughout, but diminishes as we approach its upper limits where it is least, being greatest at the surface of the earth. If a vessel filled with air be closed at the base of a high mountain and afterward opened on its summit, the air will rush out; and the vessel being closed again on the summit and opened at the base of the mountain, the air will rush in.

The barometer; density and pressure of the atmosphere at different places;

The evaporation which takes place from large bodies of water, the activity of vegetable and animal life, as well as vegetable decompositions, throw considerable quantities of aqueous vapor, carbonic acid, and other foreign ingredients temporarily into the permanent portions of the atmosphere. These, together with its ever-varying temperature, keep the density and elastic force of the air in a state of almost incessant change. These changes are indi-

foreign ingredients in the air, and its change of density;

barometer; weather-glass;

cated by the *Barometer*, an instrument employed to measure the intensity of atmospheric pressure, and frequently called a *weather-glass*, because of certain agreements found to exist between its indications and the state of the weather.

description of the barometer;

The barometer consists of a glass tube about thirty-four or thirty-five inches long, open at one end, partly filled with distilled mercury, and inverted in a small cistern also containing mercury. A scale of equal parts is cut upon a slip of metal, and placed against the tube to measure the height of the mercurial column, the zero being on a level with the surface of the mercury in the cistern. The elastic force of the air acting freely upon the mercury in the cistern, its pressure is transmitted to the interior of the tube, and sustains a column of mercury whose weight it is just sufficient to counterbalance. If the density and consequent elastic force of the air be increased, the column of mercury will rise till it attain a corresponding increase of weight; if, on the contrary, the density of the air diminish, the column will fall till its diminished weight is sufficient to restore the equilibrium.

column of mercury in equilibrio with atmospheric pressure;

Fig. 337.

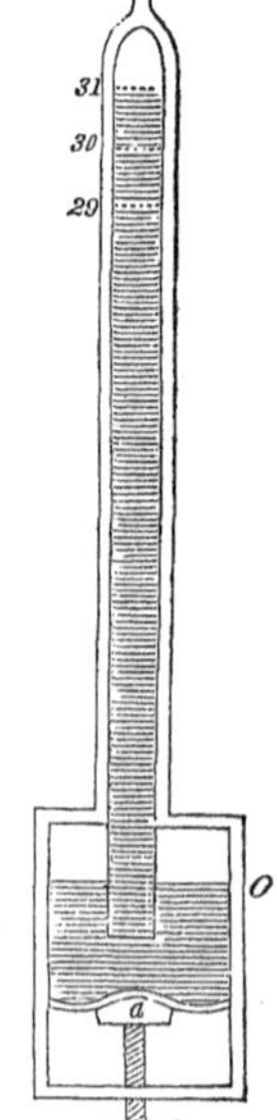

common mountain barometer;

In the *Common Barometer*, the tube and its cistern are partly inclosed in a metallic case, upon which the scale is cut, the cistern, in this case, having a flexible bottom of leather, against which a plate a at the end of a screw b is made to press, in order to elevate or depress the mercury in the cistern to the zero of the scale.

De Luc's siphon barometer;

De Luc's Siphon Barometer consists of a glass tube bent upward so as to form two unequal parallel legs: the longer is hermetically sealed, and constitutes the Torricellian tube; the shorter is open, and on the surface of the quicksilver the pressure of the atmosphere is exerted. The difference between the levels in the longer and shorter legs is the barometric height. The most convenient and practicable way of measuring this difference, is to adjust a moveable scale between the two legs, so that its zero may be made to coincide with the level of the mercury in the shorter leg.

Fig. 338.

moveable or sliding scale;

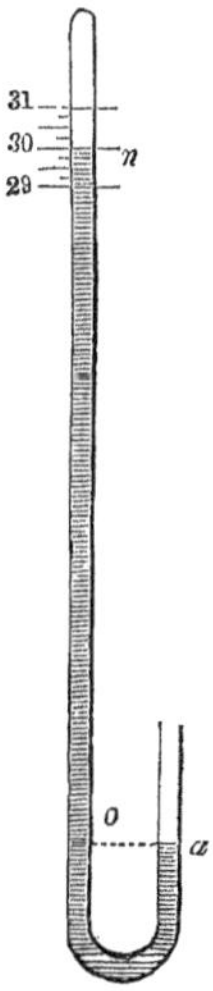

Different contrivances have been adopted to render the minute variations in the atmospheric pressure, and consequently in the height of the barometer, more readily perceptible by enlarging the divisions on the scale, all of which devices tend to hinder the exact measurement of the length of the column. Of these we may name Morland's Diagonal, and Hook's Wheel-Barometer, but especially Huygen's Double-Barometer.

different devices for appreciating slight changes of barometric column;

The essential properties of a good barometer are: width of tube; purity of the mercury; accurate graduation of the scale; and a good *vernier*.

Heat affects the density of mercury as well as that of all other bodies. When its temperature is increased, it expands; when diminished it contracts. The same atmospheric pressure will sustain the same weight—in other words, the same quantity of mercury; but the same quantity of mercury will occupy different volumes, according to its temperature, and the same atmospheric pressure will, hence, sustain a longer column when the temperature is high than it will when the temperature is low. The indications of the barometer must, therefore, be reduced to what they would have been, if taken at a standard or fixed temperature, without which reduction they would be utterly worthless.

effects of temperature;

From the experiments of Dulong and Petit, it is found

expansion of mercury;

that mercury expands $\frac{1}{9990}$th part of its volume for each degree of Fahrenheit's scale by which its temperature is increased, and that it contracts according to the same law as its temperature is diminished. If, therefore, T denote the standard temperature, and T' the temperature of observation; b the altitude which the barometer would have at the standard temperature, and b' the observed altitude, then will,

barometric column reduced to standard temperature;

$$b = b'\left[1 + \frac{T - T'}{9990}\right] = b'\,[1 + (T - T')\,0.0001001] \,..\, (241);$$

when T' becomes T, b' will be equal to b.

attached thermometer;

A thermometer is usually attached to the barometer tube for the purpose of observing the temperature of the mercury.

example for illustration;

Example. Observed the barometric column to stand at 29.81 inches, while its thermometer gave a temperature of 93°. What would have been the column under the same pressure, had the temperature of the mercury been 32°? Here we have

data;

$$b' = \overset{in.}{29.81},$$

$$T' = 93\overset{\circ}{.}00,$$

$$T = 32\overset{\circ}{.}00,$$

$$T - T' = -\ 61.00;$$

and

reduced column.

$$b = \overset{in.}{29.81}\,[1 - 61 \times 0.0001001] = \overset{in.}{29.63}.$$

Barometer used to measure the elasticity of confined gases, &c.;

§ 285.—The barometer may be used not only to measure the pressure of the external air, but also to determine the density and elasticity of pent-up gases and vapors, and furnishes the most direct means of ascertaining the degree of rarefaction in the receiver of an air-pump. When thus employed, it is called the *barometer-gauge.* In every case it will only be necessary to establish a free connection between the cistern of the barometer and the vessel containing the fluid whose elasticity is to be indicated; the height of the mercury in the tube, expressed in inches, reduced to a standard temperature, and multiplied by the known weight of a cubic inch of mercury at that temperature, will give the pressure in pounds on each square inch. In the case of the steam in the boiler of an engine, the upper end of the tube is sometimes left open. The cistern A is a steam-tight vessel, partly filled with mercury, a is a tube communicating with the boiler, and through which the steam flows and presses upon the mercury; the barometer tube bc, open at top, reaches nearly to the bottom of the vessel A, having attached to it a scale whose zero coincides with the level of the quicksilver. On the right is marked a scale of inches, and on the left a scale of atmospheres.

barometer gauge;

its use and application;

Fig. 339.

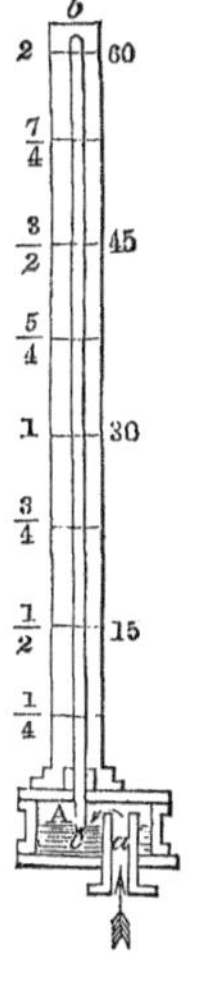

scale of inches and another of atmospheres;

If a very high pressure were exerted, one of several atmospheres, for example, an apparatus thus constructed would require a tube of great length, in which case *Mariotte's manometer* is considered preferable. The tube being filled with air and the upper end closed, the surface of the mercury in both branches will stand at the same level as long as no steam is admitted. The steam being admitted through d, presses on the surface of the mercury a and forces

Fig. 340.

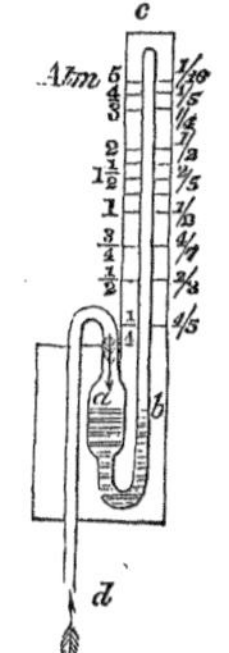

Mariotte's manometer;

it up the branch $b\,c$, and the scale from b to c marks the force of compression in atmospheres. The greater width of tube is given at a, in order that the level of the mercury at this point may not be materially affected by its ascent up the branch $b\,c$, the point a being the zero of the scale.

its mode of action.

Fig. 340.

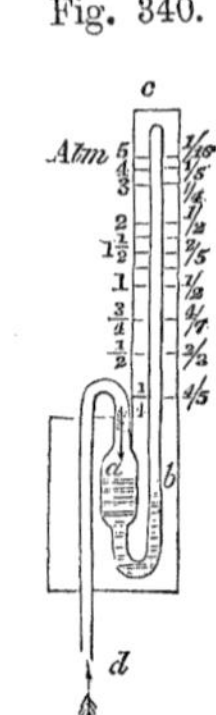

§ 286.—Another very important use of the barometer, is to find the difference of level between two places on the earth's surface, as the foot and top of a hill or mountain.

Levelling by means of the barometer;

Since the altitude of the barometer depends on the pressure of the atmosphere, and as this force depends upon the height of the pressing column, a shorter column will exert a less pressure than a longer one. The quicksilver in the barometer falls when the instrument is carried from the foot to the top of a mountain, and rises again when restored to its first position: if taken down the shaft of a mine, the barometric column rises to a still greater height. At the foot of the mountain the whole column of the atmosphere, from its utmost limits, presses with its entire weight on the mercury; at the top of the mountain this weight is diminished by that of the intervening stratum between the two stations, and a shorter column of mercury will be sustained by it.

effect of change of place upon the height of the barometer;

It is well known that the surface of the earth is not uniform, and does not, in consequence, sustain an equal atmospheric pressure at its different points; whence the mean altitude of the barometric column will vary at different places. This furnishes one of the best and most expeditious means of getting a profile of an extended section of the earth's surface, and makes the barometer an instrument of great value in the hands of the traveller in search of geographical information.

effects of irregularity of the earth's surface;

To find the relation which subsists between the altitudes of two barometric columns, and the difference of level of the places where they exist, conceive the atmosphere to be divided into an indefinite number of elementary horizontal strata of equal thickness, and so thin that the density from the top to the bottom of the same stratum may without error be regarded as uniform, the density varying from one stratum to another.

relation between the barometric columns and difference of level of the places;

Then, commencing at any elevated position O, above the level of the sea, denote by p the pressure exerted upon the unit of surface by the whole column of atmosphere above this point. The density of the stratum of air A, immediately below this point, will be due to this pressure; call this density D, then, from Mariotte's law, Eq. (236), will

Fig. 341.

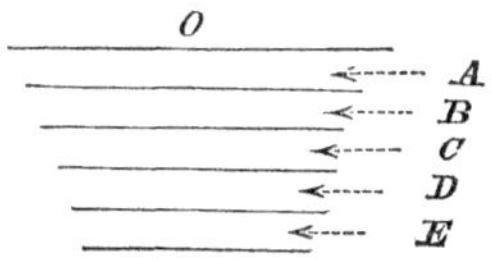

O A B C D E

$$p = PD;$$

elastic pressure;

in which P is the pressure necessary to produce, on a unit of surface, a unit of density. From this equation, we have

$$D = \frac{p}{P}.$$

density corresponding;

The weight of so much of this stratum as stands upon a unit of surface will be

$$g\,D\,h = p \cdot \frac{g\,h}{P};$$

weight of a small column on unit of surface;

in which h denotes the indefinitely small height common to all the strata, and g the force of gravity.

The pressure upon the unit of surface of the second stratum B, will be p, transmitted through the first stratum, augmented by the weight of this stratum found above; and, denoting this pressure by p', we shall have

pressure on unit of surface of second stratum;

$$p' = p + p \cdot \frac{g\,h}{P} = p\left(1 + \frac{g\,h}{P}\right).$$

Denoting by D' the density of the second stratum B, we have again by Mariotte's law

$$p' = P\,D',$$

or

$$D' = \frac{p'}{P};$$

and for the weight of this stratum upon a unit of surface,

weight of second stratum;

$$g\,h\,D' = p' \cdot \frac{g\,h}{P},$$

and substituting the value of p', found above,

same under another form;

$$g\,h\,D' = p\left(1 + \frac{g\,h}{P}\right) \cdot \frac{g\,h}{P}.$$

The pressure upon the unit of surface of the third stratum C, will be the pressure p', transmitted through the second stratum, increased by the weight found above for this same stratum; hence, denoting this pressure by p'', will

pressure upon unit of surface of third stratum;

$$p'' = p\left(1+\frac{g\,h}{P}\right) + p\left(1+\frac{g\,h}{P}\right)\frac{g\,h}{P} = p\left(1+\frac{g\,h}{P}\right)\left(1+\frac{g\,h}{P}\right) = p\left(1+\frac{g\,h}{P}\right)^2;$$

and in the same way will the pressure p''', upon the fourth stratum, be given by the equation

same for fourth stratum;

$$p''' = p\left(1 + \frac{g\,h}{P}\right)^3,$$

and so on to the surface of the earth: and supposing n to denote the number of strata between these limits, then will

$$p_n = p\left(1 + \frac{g\,h}{P}\right)^n;$$ pressure upon unit of surface of nth stratum;

in which p_n denotes the pressure at the lowest station.

Developing the second member of this equation by the binomial formula, and dividing by p, we have

$$\frac{p_n}{p} = 1 + n\frac{g\,h}{P} + \frac{n(n-1)}{1\,.\,2}\cdot\frac{g^2 h^2}{P^2} + \frac{n(n-1)(n-2)}{1\,.\,2\,.\,3}\cdot\frac{g^3 h^3}{P^3} + \&c.$$ ratio of the upper and lower pressure;

The strata being indefinitely thin, the number in any definite altitude will be indefinitely great, and this being the case in the above series, it is obvious that the numbers 1, 2, 3, &c., connected with n by the minus sign, may be disregarded without sensibly impairing the result, which will give

$$\frac{p_n}{p} = 1 + \frac{n\,g\,h}{P} + \frac{n^2\,g^2\,h^2}{1.2\,P^2} + \frac{n^3\,g^3\,h^3}{1\,.\,2\,.\,3\,P^3} + \&c.$$ same reduced;

But the second member is equal to

$$e^{\frac{n\,g\,h}{P}},$$

in which $e = 2.7182818$, the base of the Naperian system of logarithms. Whence,

$$\frac{p_n}{p} = e^{\frac{n\,g\,h}{P}}.$$ same under different form;

But n being the number of strata, and h the common height of each, $n\,h$ will be equal to the difference of level between the first and last points. Calling this z, and taking the Naperian logarithm of both members, we find, after substituting z,

$$\log\,\frac{p_n}{p} = \frac{g\,z}{P};$$ Naperian logarithm;

and passing to the common logarithms

common logarithm;

$$M \,.\, \text{Log}\, \frac{p_n}{p} = \frac{g\,z}{P},$$

in which M denotes the reciprocal of the modulus of the common system; whence we have

difference of level;

$$z = \frac{M\,P}{g} \cdot \text{Log}\, \frac{p_n}{p}.$$

Denote by b_n the height of the barometric column at the lower station, where the pressure is p_n, and by b that at the upper station where the pressure is p, then will

ratio of pressures in terms of barometric columns:

$$\frac{p_n}{p} = \frac{b_n}{b};$$

and reducing the barometric column b to the temperature of b' taken as the standard, we have, Eq. (241),

same reduced to a standard temperature T;

$$\frac{p_n}{p} = \frac{b_n}{b\,[1 + (T - T')\,0.0001001]},$$

in which T becomes the temperature of the mercury at the lower, and T' that at the upper station. Moreover, we have, Eq. (81)′,

value of force of gravity;

$$g = \overset{ft.}{32.1803}\,[1 - 0.002551 \cos 2\,\psi],$$

or

$$g = g'\,(1 - 0.002551 \cos 2\,\psi);$$

in which $g' = \overset{ft.}{32.1803}$, the force of gravity in the latitude of 45°.

Substituting these values of $\frac{p_n}{p}$, g, and the value of P given by Eq. (240), in the value for D above, and we find

value for difference of level;

$$z = \frac{M\,D_{,,}\,h_{,,}}{D_{,}} \cdot \frac{1 + (t - 32)\,0.00208}{1 - 0.002551 \cos 2\,\psi} \times \text{Log}\left[\frac{b_n}{b} \times \frac{1}{1 + (T - T')\,0.0001001}\right].$$

In this it will be remembered that t denotes the temperature of the air; but this may not be, indeed scarcely ever is, the same at both stations, and thence arises a difficulty in applying the formula. But if we represent, for a moment, the entire factor of the second member, into which the factor involving t is multiplied, by X, then we may write

difficulty arising from difference of temperature at the two stations;

$$z = [1 + (t - 32°)\ 0.00208]\ X.$$

difference of level for constant temperature;

If the temperature of the *lower station* be denoted by $t_{,}$, and this temperature be the same throughout to the upper station, then will

$$z_{,} = [1 + (t_{,} - 32°)\ 0.00208]\ X.$$

temperature throughout the same as at lower station;

And if the actual temperature of the *upper station* be denoted by t', and this be supposed to extend to the lower station, then would

$$z' = [1 + (t' - 32°)\ 0.00208] \,.\, X.$$

temperature same as upper station;

Now if $t_{,}$ be greater than t', which is usually the case, then will the barometric column, or b, at the upper station be greater than would result from the temperature t', since the air being more expanded, a portion which is actually below would pass above the upper station and press upon the mercury in the cistern; and because b enters the denominator of the value X, $z_{,}$ would be too small. Again, by supposing the temperature the same as that at the upper station throughout, then would the air be more condensed at the lower station, a portion of the air would sink below the upper station that before was above it, and would cease to act upon the mercurial column b, which would, in consequence, become too small; and this would make z' too great. Taking a mean between $z_{,}$ and z' as the true value, we find

mean value of difference of level, the true one;

true value for difference of level;

$$z = \frac{z_{,} + z'}{2} = [1 + \tfrac{1}{2}(t_{,} + t' - 64°)\,0.00208]\,X.$$

Replacing X by its value,

value for difference of level;

$$z = \frac{M D_{,,} h_{,,}}{D_{,}} \cdot \frac{1 + \frac{1}{2}(t_{,} + t' - 64°)\,0.00208}{1 - 0.002551 \cos 2\Psi} \times \text{Log.}\left[\frac{b_n}{b} \times \frac{1}{1 + (T - T')\,0.0001001}\right].$$

The factor $\frac{M D_{,,} h_{,,}}{D_{,}}$, we have seen, is constant, and it only remains to determine its value. For this purpose, measure with accuracy the difference of level between two stations, one at the base and the other on the summit of some lofty mountain, by means of a Theodolite, or levelling instrument—this will give the value of z; observe the barometric column at both stations—this will give b and b_n; take also the temperature of the mercury at the two stations—this will give T and T'; and by a detached thermometer in the shade, at both stations, find the values of $t_{,}$ and t'. These, and the latitude of the place, being substituted in the formula, every thing is known except the coefficient in question, which may, therefore, be found by the solution of a simple equation. In this way, it is found that

find the value of the coefficient;

its value;

$$\frac{M D_{,,} h_{,,}}{D_{,}} = 60345.51 \text{ English feet};$$

which will finally give for z,

final value for difference of level;

$$z = 60345.51^{ft.} \cdot \frac{1 + \frac{1}{2}(t_{,} + t' - 64°)\,0.00208}{1 - 0.002551 \cos 2\Psi} \times \text{Log.}\left[\frac{b_n}{b} \times \frac{1}{1 + (T - T')\,0.0001001}\right].$$

To find the difference of level between any two stations, the latitude of the locality must be known; it will then only be necessary to note the barometric columns, the temperature of the mercury, and that of the air at the two stations, and to substitute these observed elements in this formula.

data for its use;

Much labor is, however, saved by the use of a table for the computation of these results, and we now proceed to explain how it may be formed and used. labor saved by a table;

Make

$$60345.51\ [1 + (t_{,} + t' - 64)\ 0.00104] = A,$$

$$\frac{1}{1 - 0.002551 \cos 2 \psi} = B,$$

$$\frac{1}{1 + (T - T')\ 0.0001} = C.$$

mode of computing one;

Then will

$$z = A\,B \cdot \text{Log.}\ \frac{C \cdot b_n}{b},$$

$$z = A\,B \cdot [\text{Log.}\ C + \text{Log.}\ b_n - \text{Log.}\ b]\ ;$$

abbreviated formula;

and taking the logarithms of both members,

$$\text{Log.}\ z = \text{Log.}\ A + \text{Log.}\ B + \text{Log.}\ [\text{Log.}\ C + \text{Log.}\ b_n - \text{Log.}\ b] \quad . . (242).$$

its logarithm;

Making $t_{,} + t'$ to vary from 40° to 162°, which will be sufficient for all practical purposes, the logarithms of the corresponding values of A, are entered in a column, under the head A, opposite the values $t_{,} + t'$, as an argument. variations of the temperature of air;

Causing the latitude ψ to vary from 0° to 90°, the logarithms of the corresponding values of B are entered in a column headed B, opposite the values of ψ. variations in latitude;

The value of $T - T'$ being made, in like manner, to vary from − 30° to + 30°, the logarithms of the corresponding values of C are entered under the head of C, and opposite the values of $T - T'$. In this way a table is easily constructed. That subjoined, was computed by Samuel Howlet, Esq., from the formula of Mr. Francis Baily, which is very nearly the same as that just described, there being but a trifling difference in the coefficients. variations in temperature of mercury;

Table for finding Altitudes

Detached Thermometer.							
$t_{,}+t'$	A	$t_{,}+t'$	A	$t_{,}+t'$	A	$t_{,}+t'$	A
40	4.7689067	75	4.7859208	110	4.8022936	145	4.8180714
41	.7694021	76	.7863973	111	.8027525	146	.8185140
42	.7698971	77	.7868733	112	.8032109	147	.8189559
43	.7703911	78	.7873487	113	.8036687	148	.8193975
44	.7708851	79	.7878236	114	.8041261	149	.8198387
45	.7713785	80	.7882979	115	.8045830	150	.8202794
46	.7718711	81	.7887719	116	.8050395	151	.8207196
47	.7723633	82	.7892451	117	.8054953	152	.8211594
48	.7728548	83	.7897180	118	.8059509	153	.8215988
49	.7733457	84	.7901903	119	.8064058	154	.8220377
50	.7738363	85	.7906621	120	.8068604	155	.8224761
51	.7743261	86	.7911335	121	.8073144	156	.8229141
52	.7748153	87	.7916042	122	.8077680	157	.8233517
53	.7753042	88	.7920745	123	.8082211	158	.8237888
54	.7757925	89	.7925441	124	.8086737	159	.8242256
55	.7762802	90	.7930135	125	.8091258	160	.8246618
56	.7767674	91	.7934822	126	.8095776	161	.8250976
57	.7772540	92	.7939504	127	.8100287	162	.8255331
58	.7777400	93	.7944182	128	.8104795	163	8259680
59	.7782256	94	.7948854	129	.8109298	164	.8264024
60	.7787105	95	.7953521	130	.8113796	165	.8268365
61	.7791949	96	.7958184	131	.8118290	166	.8272701
62	.7796788	97	.7962841	132	.8122778	167	.8277034
63	.7801622	98	.7967493	133	.8127263	168	.8281362
64	.7806450	99	.7972141	134	.8131742	169	.8285685
65	.7811272	100	.7976784	135	.8136216	170	.8290005
66	.7816090	101	.7981421	136	.8140688	171	.8294319
67	.7820902	102	.7986054	137	.8145153	172	.8298629
68	.7825709	103	.7990681	138	.8149614	173	.8302937
69	.7830511	104	.7995303	139	.8154070	174	.8307238
70	.7835306	105	.7999921	140	.8158523	175	.8311536
71	.7840098	106	.8004533	141	.8162970	176	.8315830
72	.7844883	107	.8009142	142	.8167413	177	.8320119
73	.7849664	108	.8013744	143	.8171852	178	.8324404
74	4.7854438	109	4.8018343	144	4.8176285	179	4.8328686

WITH THE BAROMETER.

Latitude.		Attached Thermometer.		
Ψ	B	T—T'	C	C
0°	0.0011689		+	—
3	.0011624	0°	0.0000000	0.0000000
6	.0011433	1	.0000434	9.9999566
9	.0011117	2	.0000869	.9999131
12	.0010679	3	.0001303	.9998697
15	.0010124	4	.0001737	.9998262
18	.0009459	5	.0002171	.9997828
21	.0008689	6	.0002605	.9997393
24	.0007825	7	.0003039	.9996959
27	.0006874	8	.0003473	.9996524
30	.0005848	9	.0003907	.9996090
33	.0004758	10	.0004341	.9995655
36	.0003615	11	.0004775	.9995220
39	.0002433	12	.0005208	.9994785
42	.0001223	13	.0005642	.9994350
45	.0000000	14	.0006076	.9993916
48	9.9998775	15	.0006510	.9993481
49	.9998372	16	.0006943	.9993046
50	.9997967	17	.0007377	.9992611
51	.9997566	18	.0007810	.9992176
52	.9997167	19	.0008244	.9991741
53	.9996772	20	.0008677	.9991305
54	.9996381	21	.0009111	.9990870
55	.9995995	22	.0009544	.9990435
56	.9995613	23	.0009977	.9990000
57	.9995237	24	.0010411	.9989564
58	.9994866	25	.0010844	.9989129
59	.9994502	26	.0011277	.9988694
60	.9994144	27	.0011710	.9988258
63	.9993115	28	.0012143	.9987823
66	.9992161	29	.0012576	.9987387
69	.9991293	30	.0013009	.9986952
75	.9989852	31	0.0013442	9.9986516
81	.9988854			
90	9.9988300			

Taking Eq. (242) in connection with this table, we have this rule for finding the altitude of one station above another, viz. :—

rule for computing difference of level with a barometer;

Take the logarithm of the barometric reading at the lower station, to which add the number in the column headed C opposite the observed value of $T - T'$, and subtract from this sum the logarithm of the barometric reading at the upper station; take the logarithm of this difference, to which add the numbers in the columns headed A and B, corresponding to the observed values of $t_{,} + t'$ and ψ; the sum will be the logarithm of the height in English feet.

Example. At the mountain of Guanaxuato, in Mexico, M. Humboldt observed at the

example first;

	Upper Station.	Lower Station.
Detached thermometer,	$t' = 70^\circ.4$;	$t_{,} = 77^\circ.6$.
Attached "	$T' = 70.4$;	$T = 77.6$.
Barometric column,	$b = 23.66$;	$b_n = 30.05$.

What was the difference of level?

Here

observed data;

$$t_{,} + t' = 148^\circ; \quad T - T' = 7^\circ.2; \quad \text{Latitude } 21^\circ.$$

To Log. $30^{in.}.05$	$=$	1.4778445		
Add C for $7^\circ.2$	$=$	0.0003165		
		1.4781610		
Sub. Log. $23^{in.}.66$	$=$	1.3740147		
Log. of - - - -		0.1041463	$=$	$-$ 1.0176439
Add A for 148° - - - - -			$=$	4.8193975
Add B for 21° - - - - - -			$=$	0.0008689
$6885^{ft.}.1$ - - - - - - -				3.8379103;

height of Guanaxuato;

whence the mountain is 6885.1 feet high.

It will be remembered that the final Eq. (242) was deduced on the supposition that each stratum of air pressed with its entire weight on that below it, a condition which can only be fulfilled when the air is in equilibrio—that is to say, when there is no wind. The barometer can, therefore, only be used for levelling purposes in calm weather. Moreover, to insure accuracy, the observations at the two stations whose difference of level is to be found, should be made simultaneously, else the temperature of the air may change during the interval between them; but with a single instrument this is impracticable, and we proceed thus, viz.: Take the barometric column, the reading of the attached and detached thermometers, and time of day at one of the stations, say the lower; then proceed to the upper station, and take the same elements there; and at an equal interval of time afterward, observe these elements at the lower station again; reduce the mercurial columns at the lower station to the same temperature by Eq. (241), take a mean of these columns, and a mean of the temperatures of the air at this station, and use these means as a single set of observations made simultaneously with those at the higher station.

barometric formula true only when there is no wind;

observations at both stations should be made simultaneously

or at equal intervals apart;

Example. The following observations were made to determine the height of a hill near West Point, N. Y.

example second;

	Upper Station.	Lower Station. (1)	(2)
Detached thermometer,	$t' = 57^\circ$;	$t_{,} = 56^\circ$	and 61°.
Attached "	$T' = 57.5$;	$T = 56.5$	and 63.
Barometric column,	$b = 28.94$ in.;	$b_n = 29.62$ in.	and 29.63 in.

observed data;

First, to reduce 29.63 inches at 63°, to what it would have been at $56^\circ.5$. For this purpose, Eq. (241) gives

$$b(1 + \overline{T - T'} \times 0.0001) = 29.63\,(1 - 6.5 \times 0.0001) = 29.611 \text{ in.}$$

reduction;

Then

reduced column;

$$b_n = \frac{29.62 + 29.611}{2} = \overset{in.}{29.6105},$$

temperature at lower station;

$$t_{,} = \frac{56^{\circ} + 61^{\circ}}{2} \;\cdot\;\cdot\; = 58^{\circ}.5,$$

$$t_{,} + t' = 58^{\circ}.5 + 57^{\circ} \;\cdot\;\cdot\; = 115^{\circ}.5,$$

$$T - T' = 56^{\circ}.5 - 57^{\circ}.5 \;\cdot\; = -1^{\circ}.$$

computation;

To Log. $\overset{in.}{29.6105}$ =	1.4714458		
Add C for -1° =	9.9999566		
	1.4714024		
Sub. Log. of $\overset{in.}{28.94}$ =	1.4614985		
Log. of - - - -	0.0099039	=	$-$ 3.9958062
Add A for $115^{\circ}.5$ - - - - -		=	4.8048112
Add B for $41^{\circ}.4$ - - - - -		=	0.0001465
$\overset{ft.}{632.07}$ - - - - - - - -			2.8007639;

height of the hill.

whence the height of the hill is 632.07 English feet.

XIII.

PUMPS.

Pumps;

§ 287.—Any machine employed for raising water from one level to a higher, in which the agency of atmospheric pressure is employed, is called a *Pump*. There are various kinds of pumps; the more common are the *sucking*, *forcing*, and *lifting* pumps. different kinds.

§ 288.—The *Sucking-Pump* consists of a cylindrical body or barrel B, from the lower end of which a tube D, called the sucking-pipe, descends into the water contained in a reservoir or well. In the interior of the barrel is a moveable piston C, surrounded with leather to make it water-tight, yet capable of moving up and down freely. The piston is perforated in the direction of the bore of the barrel, and the orifice is covered by a valve F called the *piston-valve*, which opens upward; a similar valve E, called the *sleeping-valve*, at the bottom of the barrel, covers the upper end of the sucking-pipe. Above the highest point ever occupied by the piston, a discharge pipe P is inserted into the barrel; the piston is worked by means of a lever H, or other contrivance, attached to the piston-rod G. The distance AA', between the highest and lowest points of the piston, is called the *play*. To explain the action of this pump, let the piston be at its lowest point A, the valves E and F closed by their own weight, and the air within the pump of the same density or elastic force as that on the exterior. The water of the reservoir will stand at the same level LL both within and without the sucking-pipe. Now suppose the piston raised

Sucking-pump; piston; piston-valve; sleeping-valve; discharge-pipe; play; operation of the pump;

Fig. 342.

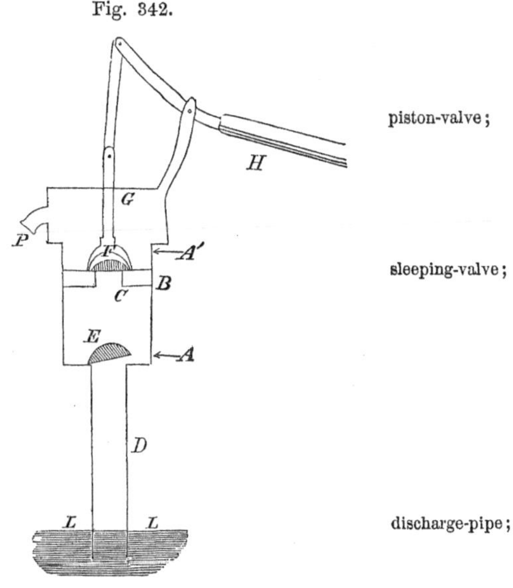

to its highest point A', the air contained in the barrel and sucking-pipe will tend by its elastic force to occupy the space which the piston leaves void, the valve E will, therefore, be forced open, and air will pass from the pipe to the barrel, its elasticity diminishing in proportion as it fills a larger space. It will, therefore, exert a less pressure on the water below it in the sucking-pipe than the exterior air does on that in the reservoir, and the excess of pressure on the part of the exterior air, will force the water up the pipe till the weight of the suspended column, increased by the elastic force of the internal air, becomes equal to the pressure of the exterior air. When this takes place, the valve E will close of its own weight; and if the piston be depressed, the air contained between it and this valve, having its density augmented as the piston is lowered, will at length have its elasticity greater than that of the exterior air; this excess of elasticity will force open the valve F, and air enough will escape to reduce what is left to the same density as that of the exterior air. The valve F will then fall of its own weight; and if the piston be again elevated, the water will rise still higher, for the same reason as before. This operation of raising and depressing the piston being repeated a few times, the water will at length enter the barrel, through the valve F, and be delivered from the discharge-pipe P. The valves E and F closing after the water has passed them, the latter is prevented from returning, and a cylinder of water equal to that through which the piston is raised, will, at each upward motion, be forced out, provided the discharge-pipe is large enough. As the ascent of the water to the piston is produced by the difference of pressure of the internal and external air, it is plain that the lowest point to which the piston may reach, should never have a greater altitude above the water in the reservoir than that of the column of this fluid which the atmospheric pressure may support, in vacuo, at the place.

action during the ascent of the piston;

equilibrium;

action during the descent of the piston;

the result of a few strokes of the piston;

greatest altitude of lower limit of the play;

From a little reflection upon what has been said of the

operations of this pump, it will appear that the rise of water, during each ascent of the piston after the first, depends upon the expulsion of air through the piston-valve during its previous descent. But air can only issue through this valve when the air below it has a greater density, and, therefore, greater elasticity, than the external air; and if the piston may not descend low enough, for want of sufficient play, to produce this degree of compression, the water must cease to rise, and the working of the piston can have no other effect than alternately to compress and dilate the same air between it and the surface of the water. To ascertain, therefore, the relation which the play of the piston should bear to the other dimensions, in order to make the pump effective, suppose the water to have reached a stationary level X, at some one ascent of the piston to its highest point A', and that, in its subsequent descent, the piston-valve will not open, but the air below it will be compressed only to the same density with the external air, when the piston reaches its lowest point A. The piston may be worked up and down indefinitely, within these limits for the play, without moving the water. Denote the play of the piston by a; the greatest height to which the piston may be raised above the level of the water in the reservoir, by b, which may also be regarded as the altitude of the discharge-pipe; the elevation of the point X, at which the water stops, above the water in the reservoir, by x; the cross-section of the interior of the barrel by B. The volume of the air between the level X and A will be

fact upon which depends the rise of the water;

to find the relation of the play to the other dimensions;

Fig. 343.

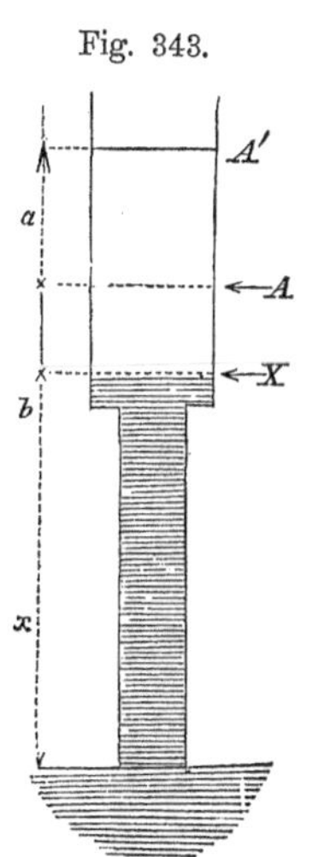

hypothesis in respect to rise of water;

notation;

volume of the confined air, when the piston is at its lowest point;

$$B \times (b - x - a);$$

the volume of this same air, when the piston is raised to A', provided the water does not move, will be

volume of same air expanded when piston is at highest point;

$$B(b - x).$$

Represent by h the greatest height to which water may be supported in vacuo at that place. The weight of the column of water which the elastic force of the air, when occupying the space between the limits X and A, will support in a tube, with a bore equal to that of the barrel, is measured by

weight of the column of water which the first will support;

$$B h \,.\, g \,.\, D;$$

in which D is the density of the water, and g the force of gravity. The weight of the column which the elastic force of this same air will support, when expanded between the limits X and A', will be

weight supported by the second;

$$B h' \,.\, g \,.\, D;$$

in which h' denotes the height of this new column. But from Mariotte's law we have

$$B(b - x - a) \;:\; B(b - x) \;::\; B h' g D \;:\; B h g D;$$

whence

ratio of the heights;

$$h' = h \cdot \frac{b - x - a}{b - x}.$$

But there is an equilibrium between the pressure of the external air and that of the rarefied air between the limits X and A', when the latter is increased by the weight of the column of water whose altitude is x. Whence, omitting the common factors, B, D, and g,

condition of equilibrium;

$$x + h' = x + h \cdot \frac{b - x - a}{b - x} = h;$$

or, clearing the fraction and solving the equation in reference to x, we find

altitude of point of stopping;

$$x = \tfrac{1}{2} b \pm \tfrac{1}{2} \sqrt{b^2 - 4 a h} \quad . \quad . \quad (243).$$

When x has a real value, the water will cease to rise, but x will be real as long as b^2 is greater than $4\,a\,h$. If, on the contrary, $4\,a\,h$ is greater than b^2, the value of x will be imaginary, and the water cannot cease to rise, and the pump will always be effective when its dimensions satisfy this condition, viz.:— condition of stoppage;

$$4\,a\,h > b^2,$$

or

$$a > \frac{b^2}{4\,h};$$

condition of incessant flow;

that is to say, *the play of the piston must be greater than the square of the altitude of the upper limit of the play of the piston above the surface of the water in the reservoir, divided by four times the height to which the atmospheric pressure at the place, where the pump is used, will support water in vacuo.* This last height is easily found by means of the barometer. We have but to notice the altitude of the barometer at the place, and multiply its column, reduced to feet, by $13\frac{1}{2}$, this being the specific gravity of mercury referred to water as a standard, and the product will give the value of h in feet. rule for play of the piston; the value of h found by the barometer;

Example. Required the least play of the piston in a sucking-pump intended to raise water through a height of 13 feet, at a place where the barometer stands at 28 inches. example;

Here $b = 13,$ and $b^2 = 169.$

Barometer, $\frac{\overset{in.}{28}}{12} = 2.333$ feet. data;

$$h = \overset{ft.}{2.333} \times 13.5 = 31.5 \text{ feet.}$$

$$\text{Play} = a > \frac{b^2}{4\,h} = \frac{169}{4 \times 31.5} = \overset{ft.}{1.341} + ;$$

resulting limit for play;

that is, the play of the piston must be greater than one and one third of a foot.

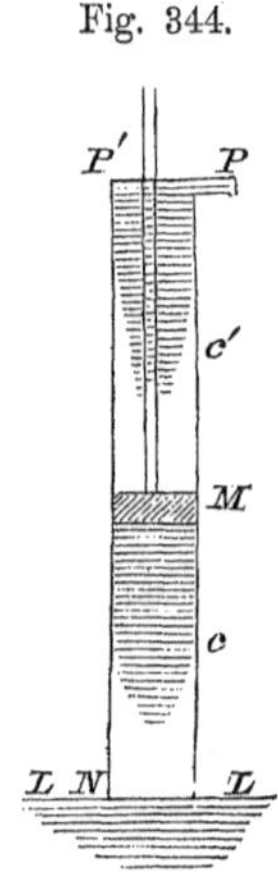

Fig. 344.

quantity of work of the moter in the sucking-pump;

The quantity of work performed by the moter during the delivery of water through the discharge-pipe P, is easily computed. Suppose the piston to have any position, as M, and to be moving upward, the water being at the level LL in the reservoir, and at P in the pump. The pressure upon the upper surface of the piston will be equal to the entire atmospheric pressure denoted by A, increased by the weight of the column of water MP', whose height is c', and whose base is the area B of the piston; that is, the pressure upon the top of the piston will be

pressure on top of piston;

$$A + Bc'gD,$$

in which g and D are the force of gravity and density of the water, respectively. Again, the pressure upon the under surface of the piston is equal to the atmospheric pressure A, transmitted through the water in the reservoir and up the suspended column, diminished by the weight of the column of water NM below the piston, and whose base is B and altitude c; that is, the pressure from below will be

pressure on the under surface of piston;

$$A - BcgD,$$

and the difference of these pressures will be

$$A + Bc'gD - (A - BcgD) = BgD(c + c');$$

but, employing the notation of the sucking-pump just described,

pressure to be overcome by the power;

$$c + c' = b;$$

whence the foregoing expression becomes

weight to be overcome;

$$Bb \,.\, g \,.\, D;$$

which is obviously the weight of a column of the fluid whose base is the area of the piston and altitude the height of the discharge-pipe above the level of the water in the reservoir. And adding to this the effort necessary to overcome the friction of the parts of the pump when in motion, denoted by φ, we shall have the resistance which the force F, applied to the piston-rod, must overcome to produce any useful effect; that is,

to which friction must be added;

$$F = B b g D + \varphi.$$

value of the motive force;

Denote the play of the piston by p, and the number of its double strokes, from the beginning of the flow through the discharge-pipe till any quantity Q is delivered, by n; the quantity of work will, by omitting the effort necessary to depress the piston, be

$$F n p = n p \left[B b \, . \, g D + \varphi \right];$$

quantity of work;

or estimating the volume in cubic feet, in which case p and b must be expressed in linear feet and B in square feet, and substituting for $g D$ its value 62.5 pounds, we finally have for the quantity of work necessary to deliver a number of cubic feet of water $Q = B n p$,

$$F n p = n p \left[62.5 \, . \, B b + \varphi \right] \quad . \quad . \quad (244);$$

quantity requisite to deliver a given number of cubic feet;

in which φ must be expressed in pounds, and may be determined either by experiment in each particular pump, or computed by the rules already given.

It is apparent that the action of the sucking-pump must be very irregular, and that it is only during the ascent of the piston that it produces any useful effect; during the descent of the piston, the force is scarcely exerted at all, not more than is necessary to overcome the friction.

sucking-pump irregular in its action.

§ 289.—What is usually called the *lifting-pump*, does not differ much from the sucking-pump just described,

Lifting-pump;

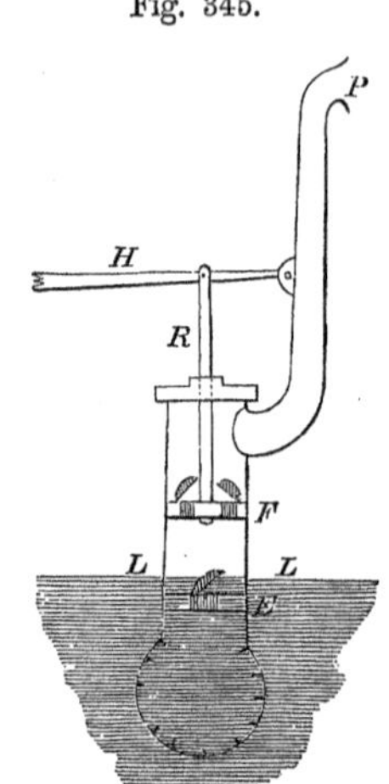

Fig. 345.

positions of the barrel and pipe reversed in this pump; except that the barrel and sleeping-valve E are placed at the bottom of the pipe, and some distance below the surface of the water $L\,L$ in the reservoir; the piston may or may not be below this same surface when at the lowest point of its play. The piston and sleeping valves open upward. Supposing the piston at its lowest point, it mode of action; will, when raised, lift the column of water above it, and the pressure of the external air, together with the head of fluid in the reservoir above the level of the sleeping-valve, will force the latter open, the water will flow into the barrel and follow the piston. When the piston reaches the upper limit of its play, the sleeping-valve will close and prevent the return of the water above it. The piston being depressed, its valves F will open and the water will flow through them till the piston reaches its lowest point. The same operation being repeated a few the result of several strokes of the piston; times, a column of water will be lifted to the mouth of the discharge-pipe P, after which every elevation of the piston will deliver a volume of the fluid equal to that of a cylinder whose base is the area of the piston and whose altitude is equal to its play.

As the water on the same level within and without the pump will be in equilibrio, it is plain that the resistance to be overcome by the power, will be the friction of the rubbing surfaces of the pump, augmented by the weight of a column of fluid whose base is the area of the piston, and work estimated by the same rule as for sucking-pump; altitude, the difference of level between the surface of the water in the reservoir and the discharge-pipe. Hence the quantity of work is estimated by the same rule, Eq. (244).

If we omit for a moment the consideration of friction, and take but a single elevation of the piston after the water has reached the discharge-pipe, n will equal one, φ will be zero, and that equation reduces to

work for one elevation of piston;

$$Fp = 62.5\ Bp \times b;$$

but $62.5 \times Bp$ is the quantity of fluid discharged at each double stroke of the piston, and b being the elevation of the discharge-pipe above the water in the reservoir, we see that, the work will be the same as though that amount of fluid had actually been lifted through this vertical height, which, indeed, is the useful effect of the pump for every double stroke.

measure of the useful effect.

Forcing-pump;

§ 290.—The forcing-pump is a further modification of the simple sucking-pump. The barrel B and sleeping-valve E are placed upon the top of the sucking-pipe M. The piston F is without perforation and valve, and the water, after being forced into the barrel by the atmospheric pressure without, as in the sucking-pump, is driven by the depression of the piston through a lateral pipe H into an air-vessel N, at the bottom of which is

description;

action of the piston and sleeping-valve;

air-vessel;

Fig. 346.

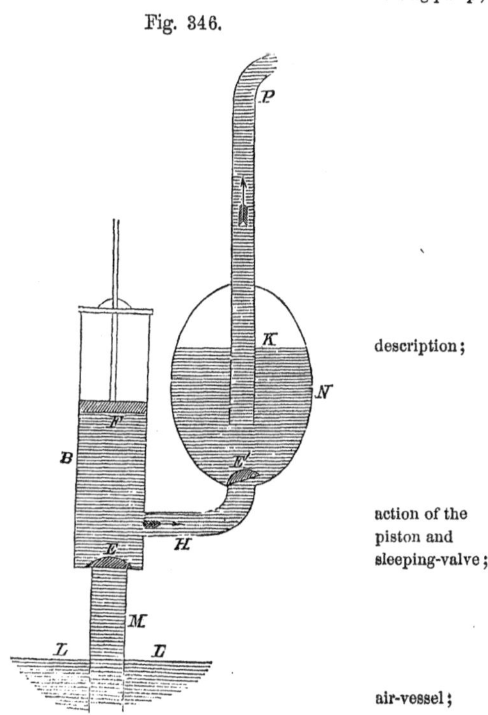

second sleeping-valve; a second sleeping-valve E', opening, like the first, upward. Through the top of the air-vessel a discharge-pipe; discharge-pipe K passes, air-tight, nearly to the bottom. The water when forced into the air-vessel by the descent of the piston, rises above the lower end of this pipe, confines and compresses the air, and this, reacting by its elasticity, action of the air-vessel, second valve; forces the water up the pipe, while the valve E' is closed by its own weight and the pressure from above, as soon as the piston reaches the lower limit of its play. A few strokes of the piston will, in general, be sufficient to raise water in the pipe K to any desired height, the only limit being that determined by the power at command and the strength of the pump.

Fig. 346.

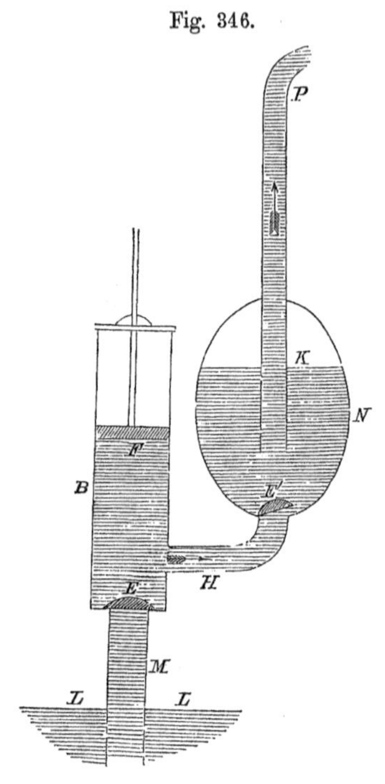

During the ascent of the piston, the valve E' is closed and E is open; the pressure upon the upper surface of quantity of work in forcing-pump; the piston is that exerted by the entire atmosphere; the pressure upon the lower surface is that of the entire atmosphere transmitted from the surface of the reservoir through the fluid up the pump, diminished by the weight of the column of water whose base is the area of the piston and altitude the height of the piston above the surface of the water in the reservoir; hence the resistance to be overcome by the power will be the difference of these pressures, which is obviously the weight of this column of water. Denote the area of the piston by B, its height above the water of the reservoir at one instant by y, and the weight of a unit of volume of the fluid by w, then will the resistance to be overcome at this point of the ascent be

resistance to be overcome by the power;

$$w \, . \, B \, . \, y;$$

its measure;

and denoting the indefinitely small space described by the piston from this position by s, the elementary quantity of work will be

$$w \, B \, y \, . \, s.$$

elementary quantity of work;

In like manner, denoting by y', y'', y''', &c., the different heights of the piston, and by s', s'', s''', &c., the corresponding elementary spaces described by it, the elementary quantities of work of the power will be

$$w \, B \, y' \, s', \qquad w \, B \, y'' s'', \qquad w \, B \, y''' s''', \text{ \&c.};$$

same for different positions of piston;

and the whole quantity of work during the entire ascent, will be

$$w \, [B \, y \, s + B \, y' \, s' + B \, y'' s'' + B \, y''' s''' + \text{\&c.}];$$

work during one entire ascent;

but $B s$ is the volume of a horizontal stratum of the fluid in the barrel, and $B s \times y$ is the product of this volume into the distance of its centre of gravity from the surface of the fluid in the reservoir; and the same of the others. Hence, if $y_{,}$ denote the height of the centre of gravity of the play p of the piston, in other words, of its middle point, then will

$$B p \, y_{,} = B \, y \, s + B \, y' s' + B \, y'' s'' + \text{\&c.};$$

equivalent expression for the same;

and

$$w \, . \, B p \, . \, y_{,}$$

will measure the quantity of work of the moter during one ascent of the piston. During the descent of the piston, the valve E is closed and E' open, and as the columns of the fluid in the barrel and discharge-pipe, below the horizontal plane of the lower surface of the piston, will maintain each other in equilibrio, the resistance to be overcome by the power will, obviously, be the weight of a column of fluid whose base is the area of the piston, and altitude, the difference of level between the piston and point of delivery P; and denoting by $z_{,}$ the distance of the central point of the play below the point P, we shall find, by exactly the same process,

work during one descent;

its measure;

$$w\,B\,p\,z_{,}$$

for the quantity of work of the moter during the descent of the piston; and hence the quantity of work during an entire double stroke will be the sum of these, or

work during one double stroke;

$$w\,B p\,(y_{,} + z_{,}).$$

But $y_{,} + z_{,}$ is the height of the point of delivery P above the surface of the water in the reservoir, and denoting this, as before, by b, we have

same;

$$w\,B\,p\,b;$$

and calling the number of double strokes n, and the whole quantity of work Q, we finally have

work for any number of double strokes;

$$Q = n\,w\,.\,B\,p\,b \quad . \quad . \quad . \quad . \quad (245).$$

If we make $z_{,} = y_{,}$, or $b = 2\,y_{,}$, which will give $y_{,} = \frac{b}{2}$, the quantity of work during the ascent will be equal to that during the descent, and thus, in the forcing-pump, the work may be equalized and the motion made in some

motion made regular in forcing-pump;

degree regular. In the lifting and sucking pumps, the moter has, during the ascent of the piston, to overcome the weight of the entire column whose base is equal to the area of the piston, and altitude the difference of level between the water in the reservoir and point of delivery, and being wholly relieved from this load during the descent, when the load is thrown upon the sleeping-valve and its box, the work becomes exceedingly variable, and the motion irregular.

It is very irregular in the lifting and sucking-pumps.

XIV.

THE SIPHON.

§ 291.—The *siphon* is a bent tube of unequal branches, open at both ends, and is used to convey a liquid from a higher to a lower level, over an intermediate point higher than either; and although its discussion more naturally appertains to the motion of fluids, its analogy with the pumps, renders a description of it here proper. The siphon having its parallel branches vertical and plunged into two liquids whose upper surfaces are at $L\,M$ and $L'\,M'$, the fluid will stand at the same level both within and without each branch of the tube when a vent or small opening is made at O. If the air be withdrawn from the siphon through this vent, the water will rise in the branches by the atmospheric pressure without, and when the two columns

Siphon;

description;

mode of using;

Fig. 347.

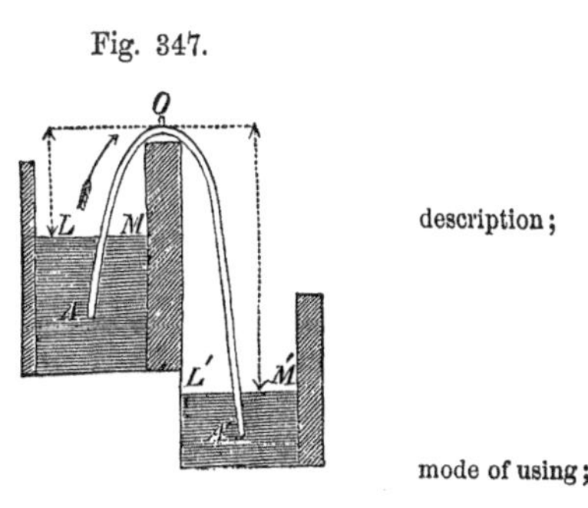

conditions of the flow; unite and the vent is stopped, the liquid will flow from the reservoir A to A', as long as the level $L' M'$ is below $L M$, and the end of the shorter branch of the siphon is below the surface of the liquid in the reservoir A.

The cause of this apparent paradox will be manifest from the following consideration, viz.: The atmospheric pressures upon the surfaces $L M$ and $L' M'$, tend to force the liquid up the two branches of the tube. When the siphon is filled with the liquid, each of these pressures is counteracted in part by the weight of the fluid column in the branch of the siphon that dips into the fluid upon which the pressure is exerted. The atmospheric pressures are very nearly the same for a difference of level of several feet, by reason of the slight density of air. The weights of the suspended columns of water will, for the same difference of level, differ considerably, in consequence of the greater density of the liquid. The atmospheric pressure opposed to the longer column will therefore be more diminished than that opposed to the shorter, thus leaving an excess of pressure at the end of the shorter branch, which will produce the motion. Thus, denote by A the intensity of the atmospheric pressure upon a surface a equal to that of a cross-section of the bore of the siphon; by h the difference of level between the surface $L M$ and the bend O of the siphon; by h' the difference of level between the same point O and the level $L' M'$; by D the density of the liquid; and by g the force of gravity: then will the pressure, which tends to force the fluid up the branch which dips below $L M$, be

explanation;

motion due to the excess of pressure up the shorter branch;

Fig. 347.

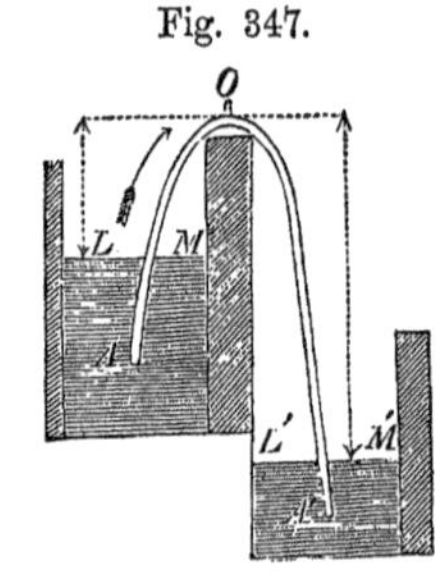

pressure up the shorter branch;

$$A - a h D g;$$

and that which tends to force the fluid up the branch immersed in the other reservoir, be

$$A - a h' D g;$$

pressure up the longer branch;

and subtracting the second from the first, we find

$$a D g (h' - h),$$

pressure which determines the flow;

for the actual intensity of the force which urges the fluid within the siphon, in a direction from the upper to the lower reservoir.

Denote by l the entire length of the siphon. It is obvious that this will be the distance over which any one stratum will move, while subjected to the action of the above force, and that the quantity of action will be measured by

quantity of action in passing a siphon full from the upper to lower reservoir;

$$a D g (h' - h) l.$$

The mass moved will be all the fluid in the siphon which is measured by $a l D$; and if we denote the velocity by V, we shall have, for the living force of the moving mass,

$$a l D . V^2;$$

living force;

and because the quantity of action is equal to half the living force, we find

$$a D g (h' - h) l = \frac{a D l V^2}{2};$$

whence

$$V = \sqrt{2 g (h' - h)};$$

velocity of the flow;

from which it appears, that *the velocity with which the liquid will flow through the siphon, is equal to the square root of twice the force of gravity, into the difference of level of the fluid in the two reservoirs.* When the fluid in the reservoirs comes to the same level, the flow will cease, since, in that case, $h' - h = 0$.

flow will cease when the water in the reservoirs comes to same level;

The siphon may be employed to great advantage to drain canals, ponds, marshes, and the like. For this purpose, it may be made flexible by constructing it of leather, well saturated with grease, like the common *hose*, and furnished with internal hoops to prevent its collapsing by the pressure of the external air. It is thrown into the water to be drained, and filled; when, the ends being plugged up, it is placed across the ridge or bank over which the water is to be conveyed; the plugs are then removed, the flow will take place, and thus the atmosphere will be made literally to press the water from one basin to another, over an intermediate ridge.

practical application of the siphon;

mode of using it for draining purposes;

Fig. 348.

It is obvious that the difference of level between the bottom of the basin to be drained and the highest point O, over which the water is to be conveyed, should never exceed the height to which water may be supported in vacuo by the atmospheric pressure at the place.

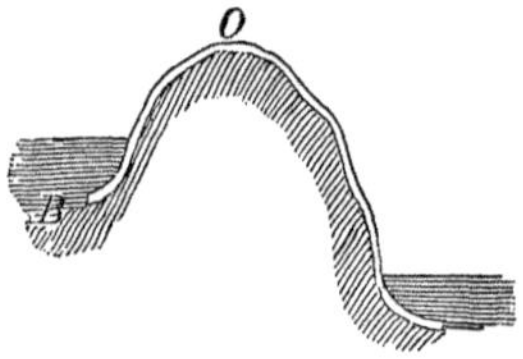

greatest elevation over which the water may be raised.

XV.

MOTION OF FLUIDS.

Motion of fluids;

§ 292.—The purpose now is to discuss the laws which govern the motion of fluids; and we shall begin with those that relate to liquids. Suppose $A\,B\,D\,C$ to be any vessel containing a heavy fluid whose upper level is

A B. If a small opening *a b* be made in the vertical side of the vessel, the pressure from within will urge the fluid out, and this pressure being greater as we descend to a greater distance from the upper surface *A B*, the fluid will flow with a greater velocity and in greater quantity during a given time, in proportion as the opening is made nearer the bottom. The quantity of fluid discharged in a unit of time, as a *second*, is called the *expense*. The liquid on leaving the vessel forms a continuous stream called the *vein* or *jet*, which takes the form of the curve described by a body thrown perpendicularly from the side of the vessel with the velocity which the fluid has at its exit, and afterward acted upon by its own weight. This, we have seen, is a parabola. At every point of this parabola, the weight of the fluid tends to alter its velocity, but at the orifice, the velocity is determined solely by what takes place within the vessel.

Fig. 349.

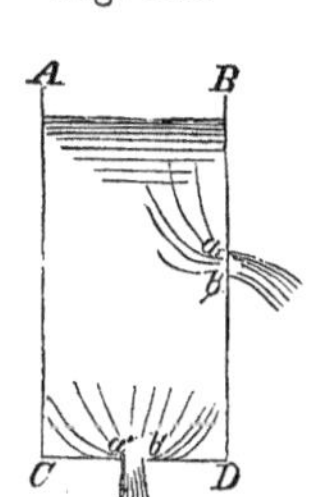

flow of liquids from vessels, through apertures;

expense;

vein or jet;

in shape, a parabola;

If the orifice be in the horizontal bottom, as at *a' b'*, the *jet* will be vertical, and the liquid will flow downward; if, as at *d*, the orifice be in a horizontal face pressed vertically upward, the jet will also be vertical, and the liquid will ascend on leaving the vessel. In general, when the sides of the vessel are thin, the direction of the vein will be perpendicular to the surface through which the orifice is made.

Fig. 350.

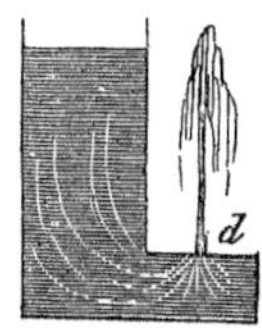

direction of the vein determined by the face of the vessel.

§ 293.—The interior surface of every vessel containing a heavy fluid is subjected, as we have seen, to a pressure therefrom, which depends upon the extent of surface and

Motion through orifices;

the distance of its centre of gravity below the upper level of the fluid. At the moment an orifice $a\,b$ is made, the fluid at its mouth is urged by this pressure to leave the vessel, the neighboring particles crowd towards the opening, describing paths which converge towards and lead through it. This movement is soon propagated in some modified degree to all parts of the fluid, and speedily each point of space within the vessel becomes distinguished by the constant velocity which every particle of the fluid mass that passes through it will there possess. It is from this instant, when the motion of the fluid becomes permanent, that we are to consider the flow.

to find the velocity of a fluid flowing freely through an orifice in a thin plate;

Fig. 351.

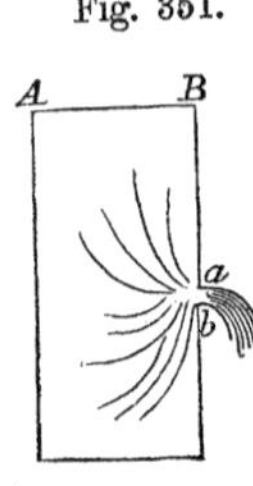

permanent flow;

If the fluid be incompressible, it is obvious that the same volume will flow through each horizontal section of the vessel above the orifice in the same time, and that this volume must be equal to that discharged through the orifice. Denote by A the area of the section NB of the interior of the vessel, at the upper surface of the fluid; by a the area of the orifice $M\,O$; by s the distance through which the upper stratum NB descends in any indefinitely small portion of time; and by S the distance $O\,O'$ through which the stratum at the mouth of orifice passes in the same time. The volume of the fluid which flows through the section NB in this time will be measured by $A\,s$; and that through the orifice, by $a\,S$; and because these must be equal, we have

equal volumes flow through the different sections in same time;

data;

Fig. 352.

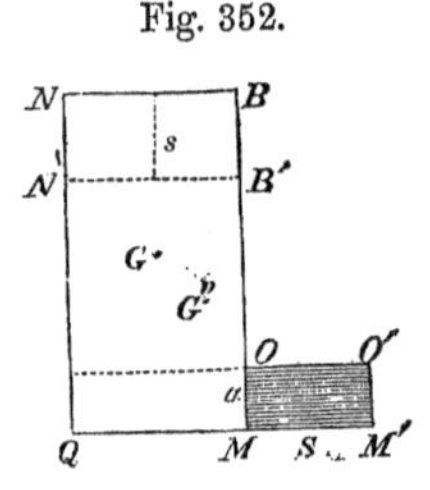

equal volumes;

$$A\,s = a\,S;$$

whence

$$\frac{s}{S} = \frac{a}{A}.$$ ratio of spaces and areas of sections;

But because the distances s and S are described in the same time, they will be proportional respectively to the velocities of the strata which describe them; and denoting the velocity of the stratum at the upper surface by v, and that of the stratum at the orifice by V, we have

$$\frac{s}{S} = \frac{v}{V};$$ ratio of spaces and velocities;

which, substituted above, gives

$$\frac{v}{V} = \frac{a}{A}.$$ ratio of velocities and areas;

That is to say, the velocities of the strata are inversely proportional to the areas of the sections through which they flow, and from which we obtain

$$v = V \cdot \frac{a}{A} \quad . \quad . \quad . \quad . \quad (246).$$ velocity through any section;

Again, since the flow is permanent, it is obvious that the living force of the fluid mass $N'B'MQ$ must always be the same. Denote this by L, and let w represent the weight of the fluid mass in $NBB'N'$, equal to that in $MM'O'O$; then will the living force of the mass $NBMQ$ be

$$L + \frac{w}{g}v^2,$$ living force of the interior fluid;

and that of the mass $B'N'QM'O'O$ be

$$L + \frac{w}{g}V^2;$$ that of a portion within and that at the jet;

and subtracting the first from the second, we find for the difference of living force of the same mass $NBMQ$, and

$B' N' Q M M' O' O$, moving with the velocities v and V respectively, the expression

difference of living force of the same mass;

$$\frac{w}{g}(V^2 - v^2).$$

The quantity of work performed by the weight of this same mass in the interval between its occupying the space $N B M Q$, and $B' N' Q M' O' O$, is, as we have seen, equal to this weight multiplied by the vertical distance through which its centre of gravity may have descended in the interval. Let G' be the centre of gravity of the whole mass when in the position $N B M Q$, and G'' when it occupies the space $B' N' Q M' O' O$. Denote the vertical distance of G' below the upper surface $N B$ by h', that of G'' below the same surface by h'', and the weight of the entire fluid by W, then will the quantity of work of this weight be

Fig. 352.

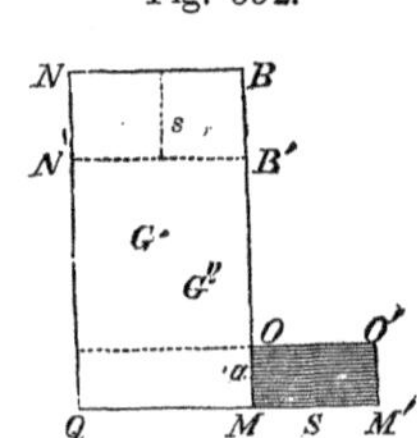

work of the weight of the entire fluid;

$$W(h'' - h') = W h'' - W h';$$

and calling the distance of the centre of gravity of the mass $M M' O' O$ below the upper surface, h'''; that of the centre of gravity of the mass $N' B' M Q$ below the same surface, l; and the weight of each of these equal masses, W'; we have, from the principles of the centre of gravity,

work at the beginning of any short interval;

$$W h'' = W' l + w h''',$$

that at the end;

$$W h' = W' l + w \tfrac{1}{2} s;$$

in which $\frac{1}{2} s$ denotes the distance of the centre of gravity of the mass $N B B' N'$ below the surface $N B$; whence

work during the interval;

$$W h'' - W h' = w(h''' - \tfrac{1}{2} s);$$

but $h''' - \frac{1}{2} s$ is the vertical distance between the centres of gravity of the masses $N B B' N'$ and $M M' O' O$, and when these masses are considered as elementary, this distance becomes the depth of the centre of gravity of the orifice below the upper level of the fluid. Denote this distance by h, and the quantity of work of the weight of the fluid while the stratum NB is passing to $N' B'$, and the stratum $M O$ to $M' O'$, becomes

$$w\,h.$$

the same;

If the upper surface be subjected to any pressure, as that of a piston or the atmosphere, then will the quantity of work due to this pressure be

$$p\,A\,s;$$

elementary work from external pressure above;

in which p denotes the pressure exerted upon the unit of surface. If, moreover, the fluid at the orifice be also subjected to a like pressure inward, this pressure would be transmitted to the lower face of the stratum whose area is A, and its work would be measured by

$$p'\,A\,s;$$

elementary work from external pressure below;

and taking the difference, we have, for the effective work of these pressures,

$$(p' - p)\,A\,s.$$

effective work of external pressures;

Now $A\,s\,D\,g = w$, from which

$$A\,s = \frac{w}{D\,g};$$

volume of the stratum;

and, substituting this above, we have

$$(p' - p)\,A\,s = (p' - p)\frac{w}{D\,g};$$

effective work;

whence the whole quantity of work due to the weight of

the fluid and the pressures at the upper surface and the orifice, becomes

total effective work;

$$w\,h + (p' - p)\,\frac{w}{D\,g};$$

and because the difference of the living force at the beginning and end of any interval, is equal to twice the quantity of action in this interval, we have

quantity of work equal half gain of living force;

$$\frac{w}{g}\,(V^2 - v^2) = 2\,w\left(h + \frac{p' - p}{D\,g}\right);$$

or, dividing out the common factor, multiplying by g, and substituting for v its value, given in Eq. (246), we have

$$V^2 - V^2 \cdot \frac{a^2}{A^2} = 2\left(g\,h + \frac{p' - p}{D}\right);$$

whence

velocity of egress through the orifice;

$$V = \sqrt{\frac{2\left(g\,h + \frac{p' - p}{D}\right)}{1 - \frac{a^2}{A^2}}} \quad . \quad . \quad (247).$$

If p and p' denote the atmospheric pressures upon the unit of surface, they become equal when the altitude of the fluid above the orifice is not very great, in which case

same when the pressures at top and orifice are the same;

$$V = \sqrt{\frac{2\,g\,h}{1 - \frac{a^2}{A^2}}} \quad . \quad . \quad . \quad . \quad (248);$$

and if the area of the orifice be very small as compared with that of the upper surface of the fluid, the fraction $\frac{a^2}{A^2}$ will be so small, that it may, without sensible error, be omitted; in which case, the fluid at the surface will be at comparative rest while it flows through the orifice, and

$$V = \sqrt{2gh};$$

velocity of egress through a very small orifice;

that is to say, when a liquid is flowing through a small orifice in the side or bottom of a large vessel, *its velocity is equal to the square root of twice the force of gravity multiplied by the depth of the centre of gravity of the orifice below the upper surface of the fluid.*

rule;

It is apparent from the form of the above expression, that this velocity is the same as that acquired by a heavy body while falling, in vacuo, from a state of rest, through the distance of the orifice below the fluid level. The distance h is called, in the case of discharging fluids, the *generating load.*

velocity same as that acquired by a heavy body in falling through the depth of orifice;

If a be equal to A, that is, if the bottom of the vessel be removed, then will, Eq. (246),

$$v = V.$$

The space described uniformly by the stratum of fluid at the orifice in a unit of time being V, the expense, estimated in volume, will be

$$a\,V;$$

expense in volume;

and in weight,

$$A\,v\,D\,g.$$

in weight;

So that, if t denote the time of flow, expressed in seconds; Q the quantity in volume, and Q' the quantity in weight discharged, then will

$$Q = a\,V\,t \quad . \quad . \quad . \quad . \quad . \quad . \quad (249),$$

quantity in volume in a given time;

$$Q' = a\,V\,D\,g\,t \quad . \quad . \quad . \quad . \quad (250);$$

quantity in weight in a given time;

in which $D\,g$ is the weight of the unit of volume.

example;

Example. The upper surface of the water, which is 15 feet above the centre of gravity of the orifice, is pressed with an intensity equal to 20 pounds upon the square foot; the area of the orifice being 0.02 of a foot. What is the velocity of egress, and what the expense? Here, the atmospheric pressure upon the piston and at the orifice being the same,

data;

$$p' - p = 20 \text{ pounds},$$

$$D = 1,$$

$$h = 15,$$

$$g = 32 \text{ nearly};$$

and neglecting the small fraction $\frac{a^2}{A^2}$, we find, from Eq. (247),

velocity of egress;

$$V = \sqrt{30 \times 32 + 40} = 31.6 \text{ feet};$$

and for one second,

quantity in volume in one second;

$$Q \doteq 0.02 \times 31.6 = 0.632 \text{ cubic feet},$$

quantity in pounds in one second.

$$Q' = \overset{lbs.}{62.5} \times 0.632 = 39.5 \text{ pounds}.$$

XVI.

MOTION OF GASES AND VAPORS.

Motion of gases and vapors;

§ 294.—In the preceding case, we have supposed, 1st, that the volume of the fluid which escapes through the orifice, is equal to that which passes, during the same time, through any interior horizontal section of the vessel; 2d, that the density in all parts of the vessel remains the same: both of which suppositions are sensibly true for liquids, but are not so in the cases of gases and vapors.

both the volume through different sections and density vary;

When fluids of this latter class are confined and subjected to any compressing action, as that of a piston, and are permitted to escape through an orifice at which the resistance of external pressure is too feeble to retain them, the density, tending as it always does to conform to Mariotte's law, will be greater at the piston where the pressure is greatest, than at the place of egress where it is least. Again, the motion being permanent, the same amount, in weight, of gas will flow through any section $A'B'$ of the vessel as through the orifice ab; but the densities at these places being different, the volumes of these equal weights will also be different. In these particulars, the circumstances attending the motion of gases and vapors differ from those of liquids.

density greatest at piston and least at orifice;

equal quantities, by weight, will flow through the different sections in the same time;

the volumes of these will be different;

Fig. 353.

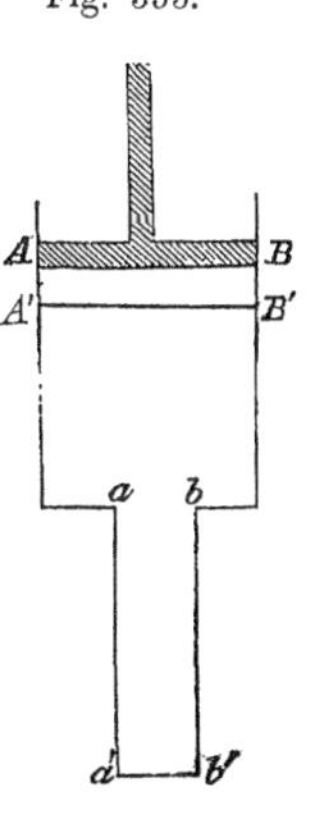

To find the velocity of egress at the orifice, we remark, that the fluid is subjected, as in the case of liquids, to the action, 1st, of its own weight; 2d, to that of the opposing pressures at the piston and orifice; and 3d, to the additional action arising from the repulsions of the particles for each other, this latter producing expansion whenever the pressure from without will permit it. The quantity of work upon the stratum issuing through the orifice, due to the weight of the fluid mass, is, as we have seen, measured by wh; in which w denotes the weight of the stratum, and h the height of the fluid above the orifice. To find the work due to the pressures, denote the pressure upon a unit of surface at the piston by p; that on the same

the forces which act: weight, pressure from the piston, and molecular repulsion;

work of the weight;

to find the work due to the pressures;

extent of surface at the orifice by p'; the area of the piston by A; that of the orifice by a; the distance between any two consecutive positions, as $A\,B$ and $A'\,B'$, of the piston by s; the distance between the two corresponding positions $a\,b$ and $a'\,b'$ of the stratum at the orifice by S. Then, because the weights of the volumes $A\,B\,B'\,A'$ and $a\,b\,b'\,a'$ of the fluid are equal, we have

notation;

Fig. 353.

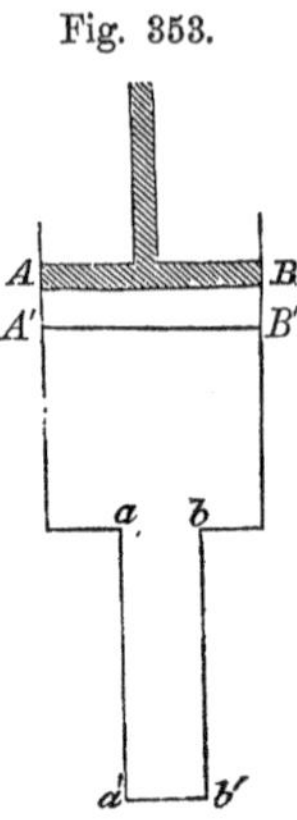

the equal weights;

$$A\,s\,D\,g = a\,S\,D'\,g \quad . \quad . \quad . \quad (251);$$

in which D and D' denote the densities of the gas at the piston and orifice, respectively, and g the force of gravity. Whence

volumes inversely as densities;

$$\frac{A\,s}{a\,S} = \frac{D'}{D}.$$

But by Mariotte's law the densities are directly proportional to the pressures, hence

densities directly as pressures;

$$\frac{D'}{D} = \frac{p'}{p};$$

which substituted above, gives

relation of volumes and pressures;

$$\frac{A\,s}{a\,S} = \frac{p'}{p} \quad . \quad . \quad . \quad . \quad . \quad (252).$$

Clearing the fraction and transposing, we find

loss or gain of work due to pressure;

$$p\,.\,A\,s - p'\,a\,S = 0.$$

But $p\,A$ is the pressure on the whole extent of the piston, and $p\,A\,s$ is, therefore, the whole work of this pressure; also $p'\,a$ is the pressure on the surface of the stratum of fluid in the orifice, and $p' a S$ is the quantity of work of this pressure; and as these quantities of work are produced in the same time, we see that the loss or gain of work, due to these pressures, is zero. The quantity of work due to the molecular actions, arises in consequence of the expansion which takes place when the gas passes from the pressure p, within the vessel and near the piston, to the pressure p', at the mouth of the orifice. The amount of this work is directly proportional to the primitive volume expanded during the change of pressure; if the primitive volume to be expanded be doubled, tripled, or quadrupled, &c., the quantity of the work will be doubled, tripled, quadrupled, &c. Hence, taking a cubic foot of the gas under the pressure p, and denoting the quantity of work due to the expansion, corresponding to a change from the pressure p to the pressure p', by E, then will the work due to the expansion of the volume $A\,B\,B'A'$ to $a\,b\,b'a'$, be measured by

this is zero;

to find quantity of work due to molecular action;

it is directly proportional to the primitive volume to be expanded;

$$A \,.\, s \,.\, E.$$

work of the expansion from the volume at the piston to that at orifice;

But since w denotes the weight of the gas in the volume $A\,B\,B'A'$, we have

$$w = A\,s\,g\,D;$$

weight of the stratum;

whence

$$A\,s = \frac{w}{g\,D},$$

its volume;

and

$$A \,.\, s \,.\, E = \frac{w \,.\, E}{g\,D};$$

work due to expansion;

whence the whole quantity of action or work due to the weight and expansion of the fluid will be

$$w\,h + w \cdot \frac{E}{g\,D} = w\left(h + \frac{E}{g\,D}\right).$$

work due to weight and expansion of the stratum;

Denoting, as before, the velocity at the piston by v, and that at the orifice by V, we have, from the principle of living forces,

living force equal to twice quantity of action;

$$\frac{w}{g}(V^2 - v^2) = 2w\left(h + \frac{E}{gD}\right);$$

or

$$V^2 - v^2 = 2gh + \frac{2E}{D} \quad . \quad . \quad (253).$$

From Eq. (252) we have

relation of elementary paths;

$$\frac{s}{S} = \frac{p'a}{pA};$$

and the spaces s and S, being described in the same time, they are to each other as the velocities v and V, hence

same as ratio of velocities;

$$\frac{v}{V} = \frac{p'a}{pA},$$

or

velocity at piston;

$$v = V \cdot \frac{p'a}{pA};$$

which substituted in Eq. (253) for v, we find

$$V^2\left(1 - \frac{p'^2 a^2}{p^2 A^2}\right) = 2gh + \frac{2E}{D}.$$

Making

$$1 - \frac{p'^2 a^2}{p^2 A^2} = K^2,$$

the above gives

value for the velocity of egress;

$$V = \frac{1}{K}\sqrt{2gh + \frac{2E}{D}} \quad . \quad . \quad (254).$$

It remains to find the value of E. For greater simplicity, let us take for the primitive volume of gas a unit or cubic foot; and suppose this unit of volume to be contained in a tube, of which the area of the internal cross-section is a unit of surface, or square foot, so that in its primitive condition, under the pressure p, the length of the tube it occupies will be the linear unit, one foot. When the pressure is reduced to p', the volume becomes dilated, and because the volume, and therefore the length, since the base is supposed constant, is inversely as the pressure, we have, calling the new length l,

to find the work due to the expansion of a unit of volume, from one pressure to another;

$$p : p' :: l : 1;$$

whence

$$l = \frac{p}{p'}.$$

new length of the volume of gas;

The path described by the moveable face of the cubic foot of the gas, during the expansion, will be

$$l - 1 = \frac{p}{p'} - 1 = \frac{p - p'}{p'}.$$

expansion during the change;

Dividing this path into two equal parts, and adding one of them to unity, the original length, we have

$$1 + \frac{p - p'}{2p'} = \frac{p + p'}{2p'},$$

length when the expansion is half completed;

for the length of the fluid when its expansion is half completed; and denoting the corresponding pressure by $p_{,}$, we have, by Mariotte's law,

$$1 : \frac{p + p'}{2p'} :: p_{,} : p;$$

whence

$$p_{,} = \frac{2pp'}{p + p'}.$$

corresponding pressure;

If we now observe that the consecutive pressures are

the three consecutive pressures;

$$p, \quad \frac{2\,p\,p'}{p + p'}, \quad \text{and } p';$$

and that the constant space passed over, during the interval which separates the instants in which these pressures are exerted, is

spaces described by the pressure while its value is changing from the first to last;

$$\frac{p - p'}{2\,p'};$$

Fig. 354.

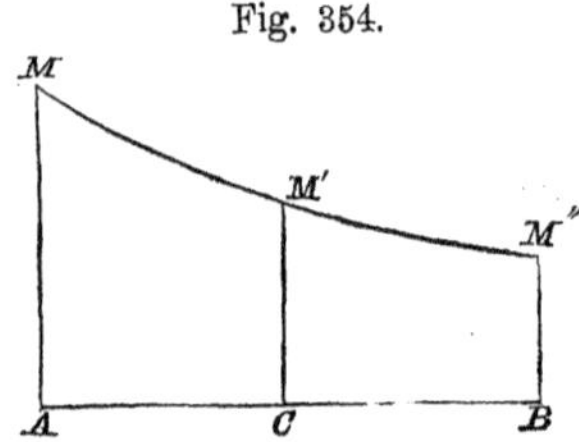

the computation of the total work becomes easy by the rule given in § 46. For this purpose, take

$$A\,C = C B = \frac{p - p'}{2\,p'},$$

and erect the perpendiculars

determination of the work;

$$A\,M = p,$$

$$C\,M' = \frac{2\,p\,p'}{p + p'},$$

$$B M'' = p';$$

join the points M, M', and M''; the area $A\,B\,M''\,M$ will be the value of E: that is to say, the value of the quantity of work performed by the gas during its expansion. But this area is, by the rule just referred to, measured by

its value;

$$\tfrac{1}{3}\,A\,C\,(A\,M + 4\,C\,M' + B M'');$$

and, substituting the values above, we have

$$\tfrac{1}{3} \cdot \frac{p - p'}{2\,p'}\,(p + 4 \cdot \frac{2\,p\,p'}{p + p'} + p') = E;$$

same in other terms;

which, substituted for E, in Eq. (254), gives

$$V = \frac{1}{K}\sqrt{2\,g\,h + \frac{1}{3\,D} \cdot \frac{p - p'}{p'}(p + \frac{8\,p\,p'}{p + p'} + p')} \quad . \quad . \quad (255).$$

value of velocity in terms of pressures:

When the orifice is small, as compared with the area of the piston, the fraction

$$\frac{p'^2\,a^2}{p^2\,A^2}$$

may be neglected, and K will become equal to unity. Moreover, the term $2\,g\,h$, in the case of gases, is scarcely ever appreciable in practice; making these suppositions, Eq. (255) becomes

$$V = \sqrt{\frac{1}{3\,D} \cdot \frac{p - p'}{p'}\,(p + \frac{8\,p\,p'}{p + p'} + p')} \quad . \quad . \quad (256).$$

velocity in case of small orifices:

The pressures p and p' are usually ascertained by means of *gauges*, or *manometers*, as they are sometimes called, and it will be convenient to express the velocity of egress in terms of the indications of these instruments. For this purpose, denote by h the height of a column of mercury resting on a unit of surface, and whose weight is equal to p, and by h' the same for the pressure p'; then, denoting the density of the mercury by $D_{\prime\prime}$, will

use of gauges to determine the pressures;

$$p = g\,h\,D_{\prime\prime}, \quad \text{and} \quad p' = g\,h'\,D_{\prime\prime};$$

which, substituted above, give

$$V = \sqrt{g \cdot \frac{D_{\prime\prime}}{3\,D} \cdot \frac{h - h'}{h'} \cdot (h + \frac{8\,h\,h'}{h + h'} + h')} \quad . \quad . \quad (257);$$

velocity in terms of the indications of the manometer;

in which V will be expressed in feet, g being equal to 32 feet very nearly, and $D_{,,}$ equal to 13.5 nearly.

The expense e, in volume, will be given by the equation

expense;

$$e = a V \quad . \quad . \quad . \quad . \quad . \quad (258);$$

and the quantity Q in volume, discharged in a given time $t_{,,}$ expressed in seconds, will be known from

quantity discharged in volume;

$$Q = a V t_{,} \quad . \quad . \quad . \quad . \quad . \quad (259);$$

in which a must be expressed in square feet. The density D, it will be remembered, is that of the fluid in the vessel near the piston, where the pressure is p; the density D', which the fluid assumes on leaving the orifice, is determined by the pressure p', and is connected with D, according to Mariotte's law, by the relation

density on leaving the orifice;

$$D' = D \frac{p'}{p} = D \frac{h'}{h}.$$

Hence, the expense Q', in weight, will be given by

quantity in weight discharged in unit of time;

$$Q' = D' g a V = D g a V \frac{h'}{h} \quad . \quad . \quad (260);$$

and the quantity Q'' in weight, discharged in the time $t_{,,}$

quantity, in weight, in time $t_{,}$;

$$Q'' = D g a V \frac{h'}{h} t_{,} \quad . \quad . \quad . \quad (261);$$

in which a must be expressed in square feet, as above.

The density D is computed by Eq. (240)'.

example;

Example. The open gauge, connected with a gasometer, containing heavy carbureted hydrogen, shows a difference of level in the mercury of 8

Fig. 355.

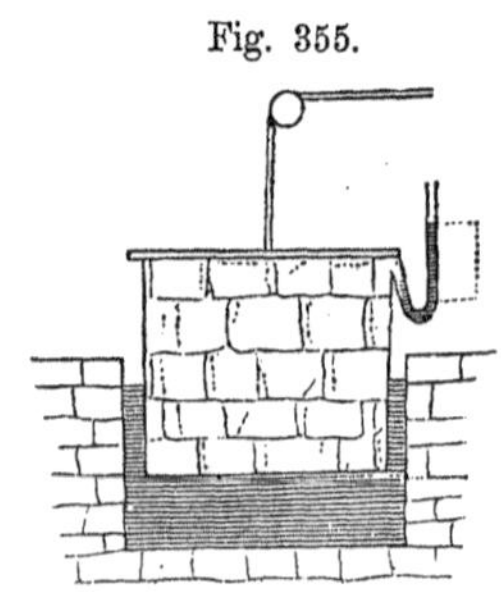

inches; the barometer in the air stands at 28 inches; the thermometer of the gasometer, at 52°: required the velocity with which the gas will flow into the open air, and the volume and weight discharged through an orifice 0.02 of a square foot of area in 20 minutes = 1200 seconds. conditions;

Here,

$$h - h' = 8 \text{ inches} = 0.666 \text{ feet},$$

$$h' = 28 \text{ ''} = 2.333 \text{ ''};$$ data;

whence

$$h = 36 \text{ ''} = 3.000 \text{ ''}$$

$$D_{\prime\prime} = 13.5$$

$$g = 32$$

$$t = 52°;$$

and from Eq. (240)′, after substituting the value of h and t, above, and that of D, in the table, page 533, for heavy carbureted hydrogen, we find

$$D = \frac{0.00127}{30} \times \frac{36}{1 + (52 - 32)\,0.00208} = 0.001465;$$ density;

and these values, in Eq. (257), give

$$V = \sqrt{32 \cdot \frac{13.5}{3 \times 0.00146} \times \frac{0.666}{2.333} \times \left(3 + \frac{8 \times 3 \times 2.333}{3 + 2.333} + 2.333\right)} = 668.02 \text{ ft.}$$ velocity;

Substituting this and the numerical values of a and $t_{\prime}$ in Eq. (259), we find

$$Q = 0.02 \times 668.02 \times 1200 = 16032.00 \text{ cubic feet.}$$ quantity in volume;

The quantity $D g$, in Eq. (261), is the weight of a cubic

foot of the gas, whose density in this case is 0.001465; and as a cubic foot of water weighs 62.5 pounds, the value of Dg becomes $62.5 \times 0.001465 = 0.0916$, nearly; whence

quantity in weight.

$$Q'' = \overset{lbs.}{0.0916} \times 0.02 \times 668.02 \times \frac{28}{36} \times \overset{s}{1200} = \overset{lbs.}{1142.4}.$$

Veinal contraction; theoretical suppositions; results of experience; causes which tend to contract the vein;

§ 295.—A stream flowing through an orifice is called a *vein*. In estimating the quantity of fluid discharged through an orifice, it is supposed, 1st, that the orifice is very small, as compared with a section of the vessel at the upper surface of the fluid; 2d, that there are neither within nor without the vessel any causes to obstruct the free and continuous flow; 3d, that the fluid has no viscosity, and does not adhere to the sides of the vessel and orifice; 4th, that the particles of the fluid reach the upper surface with a common velocity, and also leave the orifice with equal and parallel velocities. None of these conditions are fulfilled in practice, and the theoretical discharge must, therefore, differ from the actual. Experience teaches that the former always exceeds the latter. If we take water, for example, which is far the most important of the liquids in a practical point of view, we shall find it to a certain degree viscous, and always exhibiting a tendency to adhere to ununctuous surfaces with which it may be brought in contact. When water flows through an opening, the adhesion of its particles to the surface will check their motion, and the viscosity of the fluid will transmit this effect towards the interior of the vein; the velocity will, therefore, be greatest at the axis of the latter, and least on and near its surface; the inner particles thus flowing away from those without, the vein will increase in length and diminish in thickness, till, at a certain distance from the orifice, the velocity becomes the same throughout the same cross-section, which usually takes place at a short distance from the aperture. This effect will be increased by the crowding of the particles, arising from the convergence of the paths along which they approach the aperture, every particle, which enters near the edge, tending to pass obliquely across to the opposite side. This diminution of the fluid vein is called *the veinal contraction.* The quantity of fluid discharged must depend upon the degree of veinal contraction, and the velocity of the particles at the section of greatest diminution; and any cause that will diminish the viscosity and adhesion, and draw the particles in the direction of the axis of the vein as they enter the aperture, will increase the discharge.

Veinal contraction;

Experience shows that the greatest contraction takes place at a distance from the vessel varying from a half to once the greatest dimension of the aperture, and that the amount of contraction depends somewhat upon the shape of the vessel about the orifice and the head of fluid. It is further found by experiment, that if a tube of the same shape and size as the vein, from the side of the vessel to the place of greatest contraction, be inserted into the aperture, the actual discharge of fluid may be accurately computed by Eq. (261), provided the smaller base of the tube be substituted for the area of the aperture; and that, generally, without the use of the tube, the actual may be deduced from the theoretical discharge, as given by that equation, by simply multiplying the theoretical discharge into a coefficient whose numerical value depends upon the size of the aperture and head of the fluid. Moreover, all other circumstances being the same, it is ascertained that this coefficient remains constant, whether the aperture be circular, square, or oblong, which embrace all cases of practice, provided that in comparing rectangular with circular orifices, we compare the smallest dimension of the former with the diameter of the latter. The value of this coefficient depends, therefore, when other circumstances are the same, upon the smallest dimension of the rectangular orifice, and upon the diameter of the circle, in the case of circular orifices. But should other circumstances, such as the head of fluid, and the place of

place of greatest contraction;

its amount depends upon;

the actual discharge obtained from the theoretical;

coefficient of discharge;

depends upon;

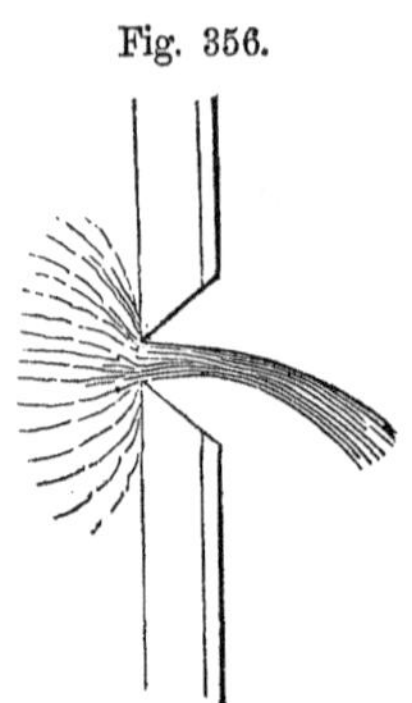
Fig. 356.

discharge through orifices in thin plates;

coefficient deduced from experiments;

the orifice, in respect to the sides and bottom of the vessel, vary, then will the coefficient also vary. When the flow takes place through thin plates, or through orifices whose lips are bevelled externally, the coefficient corresponding to given heads and orifices, may be found in the following table, provided the orifices be remote from the lateral faces of the vessel. This table is deduced from the experiments of Captain Lesbros, of the French engineers, and agrees with the previous experiments of Bossut, Michelotti, and others.

TABLE.

Coefficient values, for the discharge of fluids through thin plates, the orifices being remote from the lateral faces of the vessel.

table of coefficients;

Head of fluid above the centre of the orifice, in feet.	Values of the coefficients for orifices whose smallest dimensions or diameters are—					
	ft. 0.66	*ft.* 0.33	*ft.* 0.16	*ft.* 0.98	*ft.* 0.07	*ft.* 0.03
0.05						0.700
0.07				0.627	0.660	0.696
0.13			0.618	0.632	0.657	0.685
0.20		0.592	0.620	0.640	0.656	0.677
0.26		0.602	0.625	0.638	0.655	0.672
0.33	0.593	0.608	0.630	0.637	0.655	0.667
0.66	0.596	0.613	0.631	0.634	0.654	0.655
1.00	0.601	0.617	0.630	0.632	0.644	0.650
1.64	0.602	0.617	0.628	0.630	0.640	0.644
3.28	0.605	0.615	0.626	0.628	0.633	0.632
5.00	0.603	0.612	0.620	0.620	0.621	0.618
6.65	0.602	0.610	0.615	0.615	0.610	0.610
32.75	0.600	0.600	0.600	0.600	0.600	0.600

coefficients for gas; and for orifices not in the table;

In the instance of gas, the generating head is always greater than 6.65 ft., and the coefficient 0.6, or 0.61, is taken in all cases.

For orifices larger than 0.66 ft., the coefficients are taken as for this dimenson; for orifices smaller than 0.03 ft., the coefficients are the same as for this latter; finally, for orifices between those of the table, we take coefficients whose values are a mean between the latter, corresponding to the given head.

As the orifice approaches one of the lateral faces of the reservoir, the contraction on that side becomes less and less, and will ultimately become nothing, and the coefficient will be greater than those of the table. If the orifice be near two of these faces, the contraction becomes nothing on two sides, and the coefficient will be still greater.

Fig. 357.

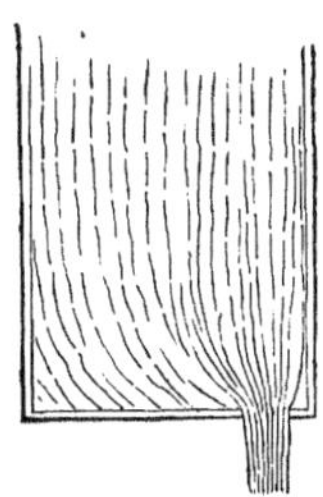

when orifice is near one lateral face;

Fig. 358.

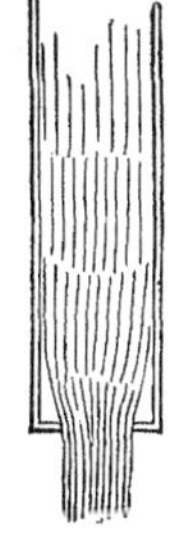

near two lateral faces;

Under these circumstances, we have the following rules: Denote by C the tabular, and by C' the true coefficient corresponding to a given aperture and head, then, if the contraction be nothing on one side, will

coefficient in the first case;

$$C' = 1.03\, C;$$

if nothing on two sides,

coefficient in the second;

$$C' = 1.06\, C;$$

if nothing on three sides,

coefficient for no contraction on three sides.

$$C' = 1.12\, C;$$

and it must be borne in mind, that these results and those of the table are applicable only when the fluid issues through holes in thin plates, or through apertures so bevelled externally that the particles may not be drawn aside by molecular action along their tubular contour.

Discharge through sluice-ways;

§ 296.—When the orifice is rectangular, and has no upper limit, or is open at the top, it is called a *sluice-way*. It is usually a cut made in the edge of a reservoir, through

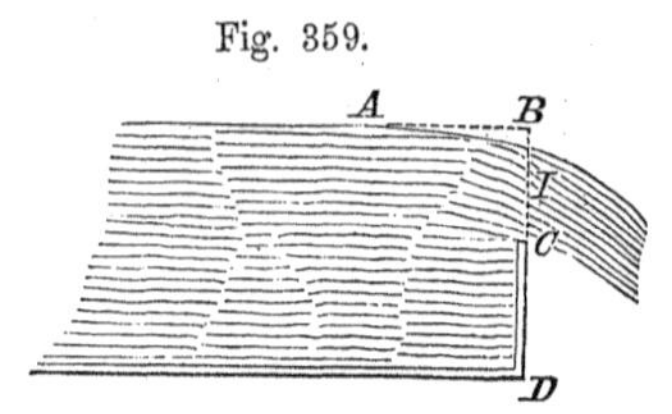

Fig. 359.

which the fluid may flow when it rises above a certain level. The expense is estimated in this wise. Denote by l the length of the horizontal side of the sluice-way; by h the head or distance BI, of the centre of gravity of a transverse section of the flowing fluid below the upper surface of the latter in the reservoir; by H the height of the fluid above the sill C, of the sluice-way; and by V the mean velocity: then, supposing the sluice-way filled to the upper level of the fluid in the reservoir, will

estimate of the expense through a sluice-way;

notation;

$$h = \tfrac{1}{2} H,$$

$$V^2 = 2gh = 2g \times \tfrac{1}{2} H = \tfrac{1}{2}(2gH);$$

whence

value of mean velocity;

$$V = 0.707 \sqrt{2gH};$$

and the theoretical expense will be

theoretical expense;

$$V \times l \times H = 0.707 \cdot \sqrt{2gH} \times l \times H.$$

But this is too great, and experience shows that it should be multiplied by the coefficient 0.57 for all ordinary cases of practice; that is to say, the true expense, denoted by E, will be given by the equation,

practical expense;

$$E = 0.57 \times 0.707 \times l \times H \times \sqrt{2gH} = 0.403\, l \,.\, H \,.\, \sqrt{2gH} \ldots (262).$$

The experiments of Dubuat, Bidone, Eytelwein, and Lesbros, show that the coefficient 0.403 should be reduced about 0.39 when H becomes equal to or greater than 0.66 of a foot, and increased to 0.415 when H becomes less than 0.07 of a foot; but that it remains sensibly the same, whatever be the total contraction or position of the sluice-way in regard to the vertical sides of the reservoir, provided H be measured from the level of the upper surface of the sill to that of a point, as A, in the surface of the fluid in the reservoir which has no sensible velocity. When the sill is on a level with the bottom of the reservoir, the velocity of the upper surface is everywhere sensible, and the coefficient increases to about 0.45. On the contrary, 0.403 is already too large when the sluice-way is prolonged into a trough-like duct, of slight inclination, wherein the fluid may have impressed upon it a whirling or irregular motion by the roughness of the surface.

variation in the value of the coefficient;

The foregoing conclusions suppose that the fluid is discharged through orifices in thin plates, and that, during the flow, the fluid particles are not drawn aside from the converging paths, along which they tend to approach the orifice, by the action of any extraneous cause. When the discharge is through *thick plates without bevel*, or through cylindrical tubes whose lengths are from two to three times the smaller dimension of the orifice, the expense is increased, the mean coefficient, in such cases, augmenting, according to experiment, to about 0.815 for orifices of which the smaller dimension varies from 0.33 to 0.66 of a foot, under heads which give a coefficient 0.619 in the case of thin plates. The cause of this increase is obvious. It is within the observation of every one, that water will wet most surfaces not highly polished or covered with an unctuous coating—in other words, that there exists between the particles of the fluid and those of solids an affinity which will cause the former to spread themselves over the latter and adhere with considerable pertinacity. This affinity becoming effective between the inner surface of the tube and those particles

discharge through thick plates;

values of the coefficients;

explanation;

effects of molecular action.

of the fluid which enter the orifice near its edge, the latter will not only be drawn aside from their converging directions, but will take with them, by the force of viscosity, the other particles, with which they are in sensible contact. The fluid filaments leading through the tube will, therefore, be more nearly parallel than in the case of orifices through thin plates, the contraction of the vein will be less, and the discharge consequently greater.

XVII.

DISCHARGE OF FLUIDS THROUGH PIPES.

Discharge of fluids through pipes;

We have considered the discharge of fluids through thin and thick plates. It remains to discuss the discharge through pipes. When the flow is through pipes whose length does not exceed two or three times their diameter, the quantity discharged in a given time is, as we have seen, greater than through bevelled orifices of the same size; but when the length is increased much beyond this limit, the reverse is the case and, all other things being equal, the discharge will be less as the pipe is longer. The same pipe may be of variable bore, that is to say, it may have a greater cross-section at one point than at another; in which case, the living force of any given portion of the moving fluid cannot be constant throughout. When of considerable length, pipes are rarely perfectly smooth, the fluid particles cannot, therefore, flow through them in parallel filaments, but must be incessantly deflected from their onward course into partial eddies formed by the small irregularities of surface. Moreover, as the pipes increase in length, will the surface exposed to fluid pressure increase, and as the extent of surface, all other things being equal,

less than through orifices;

causes which obstruct the motion;

determines the amount of pressure, the friction, which depends upon the pressure, augments so as greatly to impede the motion. We shall proceed to estimate the value of these influences. friction.

§ 297.—But first of all let us compute the amount of living force resulting from the shock of fluids, flowing with different velocities. For this purpose, let the fluid in the pipe LK flow with the velocity V, and denote by M the mass which flows into the vessel BC in a unit of time; also let the velocity of the fluid in the vessel BC be V', and its mass M'; then will the corresponding living force be

Loss of living force arising from the impact fluids;

Fig. 360.

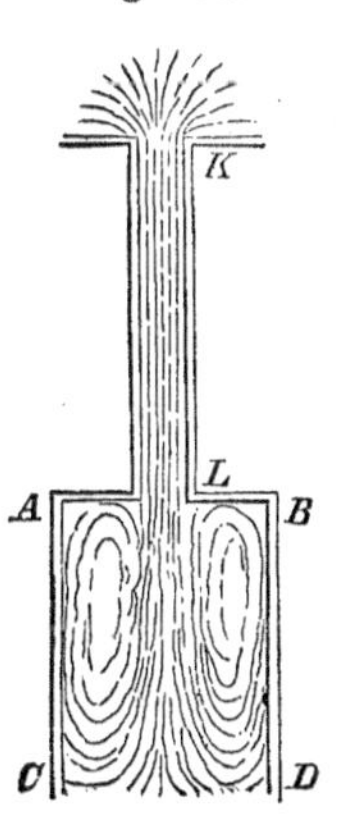

$$M V^2 + M' V'^2;$$

living force before the impact;

and supposing the fluid to be water, which we have regarded as unelastic, the common velocity after impact will be obtained from either of the Eqs. (194) or (195), by making $e = 0$; hence the common velocity denoted by v, will be given by

$$v = \frac{M \,.\, V + M' \,.\, V'}{M + M'};$$

common velocity after the impact;

and the corresponding living force,

$$(M + M')\, v^2 = \left(\frac{M.\, V + M'\, V'}{M + M'}\right)^2 \times (M + M') = \frac{(M\, V + M'\, V')^2}{M + M'};$$

corresponding living force;

and the loss of living force in a unit of time, denoted by L, loss of living force

$$L = MV^2 + M'V'^2 - \frac{(MV + M'V')^2}{M + M'} = \frac{MM'(V - V')^2}{M + M'};$$

and, dividing by M',

same;

$$L = \frac{M(V - V')^2}{1 + \frac{M}{M'}} \quad . \quad . \quad . \quad (263);$$

or when the mass M' is very great as compared to M,

same when a small mass flows into a large mass.

$$L = M(V - V')^2 \quad . \quad . \quad . \quad (264).$$

Loss of living force from contraction of cross-section of a pipe;

§ 298.—It will be an easy matter now to estimate the loss of living force, arising from a contraction of the vessel or pipe through which the fluid may be flowing. Let $A\,B\,C\,D$ be a vessel containing a heavy fluid, of which $A\,B$ is the upper level, and issuing through an opening $a\,b$ in the bottom $C\,D$; and suppose $A'\,B'$ to be a diaphragm, pierced by an opening $a'\,b'$. Denote by A'' the area of the section at $A''\,B''$, by a the area of the contraction at $a\,b$, and by a' that of the contraction at $a'\,b'$. The fluid, in passing through the contraction $a'\,b'$, impinges against that below the diaphragm $A'\,B'$, and if the opening $a\,b$ is beyond the reach of the eddies formed by this conflict, the velocity at either contraction may be computed from that at the other.

Fig. 361.

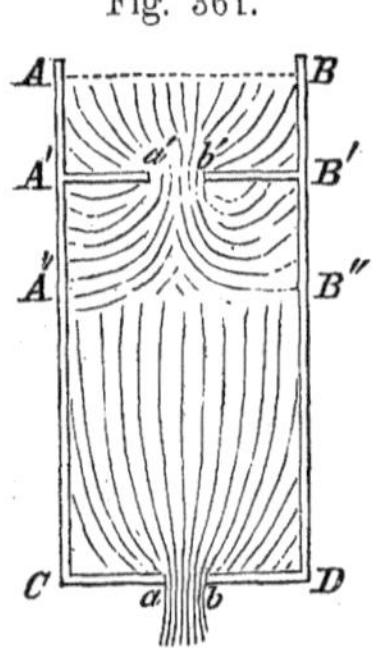

hypothesis;

notation;

Denote by V the velocity of the fluid as it passes the contraction at $a\,b$, by V' that at the contraction $a'\,b'$, and by V'' that at the section $A''\,B''$, supposed beyond the region of eddies; and let m represent the coefficient of the expense at $a\,b$, and m' that at $a'\,b'$: these coefficients may be found from the table. The expense at $a\,b$ will be $m\,a\,V$, that through the section $A''\,B''$ will be $A''\,V''$, and that through the contraction at $a'\,b'$ will be $m'\,a'\,V'$; but as the same quantity of fluid must pass through the sections of $a\,b$, $A''\,B''$, and $a'\,b'$, in the same time, we have

expense through the different sections;

$$m\,a\,V = A''\,V'',$$

$$m\,a\,V = m'\,a'\,V';$$

whence

$$V'' = \frac{m\,a\,V}{A''},$$

velocities;

$$V' = \frac{m\,a\,V}{m'\,a'};$$

and the velocity with which the fluid through $a'\,b'$ impinges against that below the diaphragm, will be

$$V' - V'' = m\,a\left(\frac{1}{m'\,a'} - \frac{1}{A''}\right)V.$$

relative velocity of the impact;

Denoting by w the weight of fluid that passes $a'\,b'$ in any small portion of time, its loss of living force will be

$$\frac{w}{g}(V' - V'')^2 = \frac{w}{g} \cdot m^2\,a^2\left(\frac{1}{m'\,a'} - \frac{1}{A''}\right)^2 \cdot V^2;$$

and denoting the factor $m\,a\left(\frac{1}{m'\,a'} - \frac{1}{A''}\right)$ by K, the quantity of work lost will be

$$\frac{w}{2\,g}\,K^2\,V^2.$$

work lost;

The work of the weight, during the same time, will be $w\,h$, and the quantity of work remaining will be

40

the work remaining;

$$w h - \frac{w}{2 g} K^2 V^2;$$

but this must be equal to half of the living force, hence

which is equal to half the living force;

$$\tfrac{1}{2} \frac{w}{g} V^2 = w h - \frac{w}{2 g} K^2 V^2;$$

whence we find

velocity of egress through $a b$.

$$V = \sqrt{\frac{2 g h}{1 + K^2}} \quad . \quad . \quad . \quad (265);$$

and from which we see that the velocity will be less than that due to the height $A C$, equal to h.

Loss of living force in short pipes;

§ 299.—Let us apply this to the discharge of a fluid through a short pipe, inserted into the orifice in the side of a vessel. The fluid having contracted to its minimum dimensions at n, again dilates, and fills the tube at $a' b'$. Let V be the mean velocity at $a' b'$, where the area of the cross-section of the pipe is a. The fluid particles moving in parallel paths at $a' b'$, the expense will be $a \times V$; while that through a section at $a b$, where the velocity is V', and cross-section a', will be $m a' V'$, in which m is the coefficient corresponding to the area a'; and, as these must be equal, we have

hypothesis and notation;

Fig. 362.

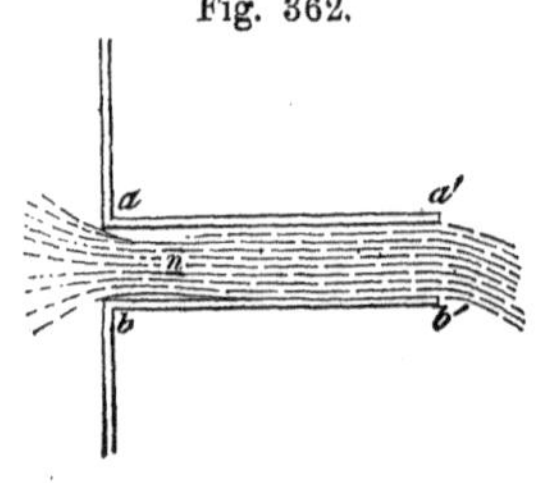

$$a V = m a' V';$$

whence

velocity at the entrance of pipe;

$$V' = \frac{a}{m a'} V;$$

and the loss of living force,

$$\frac{w}{g} \times (V' - V)^2 = \frac{w}{g} \times V^2 \left(\frac{a}{m\,a'} - 1\right)^2.$$ loss of living force;

The quantity of work of the weight, in the same time, is $w \times h$, and this, diminished by half the loss above, must be equal to half the actual living force; and, therefore,

$$\frac{w}{2\,g} V^2 = w\,h - \frac{w}{2\,g} \cdot V^2 \cdot \left(\frac{a}{m\,a'} - 1\right)^2;$$

or making $\frac{a}{m\,a'} - 1 = K$, we find

$$V = \sqrt{\frac{2\,g\,h}{1 + K^2}}.$$ velocity of egress from the pipe;

When the tube is cylindrical $a = a'$, and

$$K = \frac{1}{m} - 1;$$

when the contraction is complete in n, and the head varies from 3 to 7 feet, it is found that m is equal to 0.62 very nearly; whence value of m;

$$K = \frac{1}{0.62} - 1 = 0.613 \text{ very nearly},$$

and

$$\frac{1}{\sqrt{1 + K^2}} = 0.85;$$ value of the constant;

whence

$$V = 0.85\,\sqrt{2\,g\,h}.$$ final value for velocity of egress;

Experiments give the coefficient 0.82, but, in com-

coefficient given by experiment a little less.

puting the foregoing value, no account was taken of friction, which is an additional cause to diminish the work of the weight $w h$.

Flow of fluids through pipes of any length;

§ 300.—When the velocity of a fluid is considerable, and the length of the pipe through which it flows is great, friction, which has thus far been neglected, becomes an effective cause of obstruction, and can never be neglected in estimating the circumstances which determine the quantity discharged. The amount of friction depends, as we have seen in the case of fluids, upon the pressure, and this latter is determined by the extent of surface, and the head which impresses the velocity, so that the length of pipe and the velocity of flow, are the elements from which friction is to be estimated.

case stated;

notation;

Let $a\,b\,b'\,a'$ be a pipe of uniform bore throughout, connecting two reservoirs $A\,C\,D\,B$ and $A'\,C'\,D'\,B'$, partly filled with fluid, the former to the level $A\,B$, and the latter to the level $A'B'$. Denote by H the difference of level between $A\,B$ and $A'\,B'$; by a the area of a cross-section of the bore of the

Fig. 363.

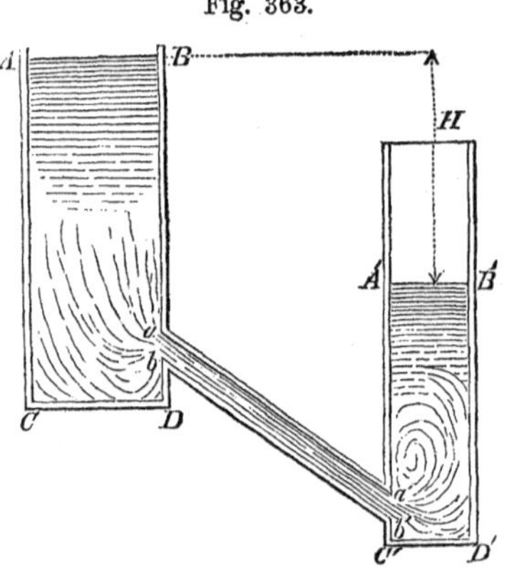

pipe; by C the contour of this section; by L the length of the pipe; and by V the constant velocity of the fluid flowing through it. Experience shows, and the computations of COULOMB, DE MEST, PRONY, EYTELWEIN, and NAVIER,

loss of work from friction in pipes;

teach us, that the loss of work occasioned by friction of pipes, in the time during which a weight of the fluid denoted by w is discharged, is proportional to the value of the expression

$$\frac{w}{g} \cdot \frac{L \times C \times V^2}{a};$$

proportional to this function;

and that this loss is a certain fraction n of this function, or is equal to

$$n \cdot \frac{w}{g} \cdot \frac{L \times C \times V^2}{a}.$$

the loss of work from friction;

If, therefore, there be neither contractions in the pipe, nor sudden turns giving rise to shocks, the only loss of work will be that measured by the above expression, and by that due to a diminution at the orifice $a\,b$, measured by the expression

$$\frac{w}{2\,g} \cdot V^2 \cdot \left(\frac{1}{m} - 1\right)^2 = \frac{w}{2\,g}\, V^2\, K^2,$$

work lost from diminution at the entrance of the pipe;

in which

$$\frac{1}{m} - 1 = K^2;$$

and, because of the principle of fluid level, H is the only distance through which w can act to produce work, we have

$$\frac{w}{2\,g} V^2 = w\,H - \frac{w}{2g} \cdot V^2 \cdot K^2 - n \cdot \frac{w}{g} \cdot \frac{L \,.\, C \,.\, V^2}{a} \,.\,. \;(266);$$

whence

$$V = \sqrt{\frac{2\,g\,H}{1 + K^2 + 2\,n \cdot L \cdot \dfrac{C}{a}}} \;.\;.\quad (267),$$

velocity of egress;

from which the velocity may be found.

The expense, denoted by Q, will be given by

$$Q = a\,V \;.\;.\;.\;.\;.\;.\quad (268).$$

expense;

Taking the value of m equal to 0.60, (see table,) we find

value of the constant;

$$1 + K^2 = 1.4444.$$

Experiment shows that, for water,

value of the coefficient n, for water;

$$n = 0.0035;$$

and for air or gas,

and for gas;

$$n = 0.00324;$$

modification in the formula for gas;

and it is important to remark that, when the question relates to the discharge of gas, we must make

$$H = \frac{1}{2}\frac{D_{\prime\prime}}{3D} \cdot \frac{h - h'}{h'} \cdot \left(h + \frac{8\,h\,h'}{h + h'} + h'\right),$$

as indicated by Eqs. (254), (257), in the latter of which h and h' denote the mercurial altitudes corresponding to the interior and exterior pressures.

Denote by D the internal diameter of the pipe, then will $C = \pi D$, and $a = \frac{\pi D^2}{4}$, so that

$$\frac{C}{a} = \frac{4\,\pi\,D}{\pi\,D^2} = \frac{4}{D}.$$

Substituting these different values and that of gravity, Eq. (22), in the expression for the velocity, we have, after dividing both terms of the fraction by $8n$,

velocity in case of water;

for water, . . . $V = 47.94\sqrt{\frac{D \,.\, H}{L + 51.57 \,.\, D}}$ . . (269),

in case of air;

for air, ... $V = 49.83\sqrt{\frac{D \,.\, H}{L + 55.72 \,.\, D}}$ . . (270);

in which all linear dimensions are expressed in English feet. The first formula may be employed even for gas, because of the small difference between the values of n for the two fluids, provided we employ the proper value for H.

either formula may be employed for gas

Finally, if the aperture $a'b'$ of final egress be smaller than ab, or of less section than a, V being the velocity within the pipe, the expense may still be deduced from a slight modification of the value of the velocity, as given by Eq. (267). For let V' denote the velocity of egress, a' the area of the section at $a'b'$, and m' its coefficient of contraction, then will

Fig. 364.

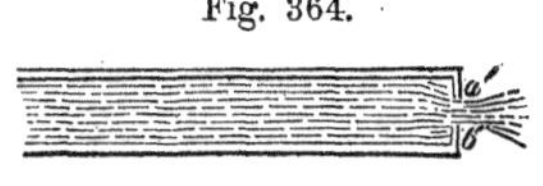

when the aperture of final egress is smaller than section of pipe;

$$a\,V = m'\,a'\,V';$$

condition of permanent flow;

whence

$$V' = \frac{a\,V}{m'\,a'};$$

and the living force of the fluid as it issues through $a'b'$, will be

$$\frac{w}{g}\,V'^2 = \frac{w}{g} \times \frac{a^2}{m'^2\,a'^2} \times V^2;$$

living force of the discharging fluid;

which, being placed equal to the second member of Eq. (266), will give

$$V = \sqrt{\frac{2\,g\,H}{\dfrac{a^2}{m'^2\,a'^2} + K^2 + 2\,n\,L \cdot \dfrac{C}{a}}} \quad . . (271).$$

its velocity;

When a' is very small as compared with a, the value of m' is about 0.60. If the values of a and a' differ but

values of the coefficient m';

slightly, or if the pipe terminates at $a' b'$ in a conical tube, then will the value of m' vary from 0.82 to 0.96.

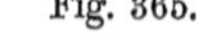

example;

Example. Let the height of the reservoir above the point of delivery be 70 feet, the diameter of the pipe 0.5 of a foot, and its length 1200 feet: required the quantity of water discharged in 24 hours. In this case,

data;

$$D = \overset{ft.}{0.5}; \quad H = \overset{ft.}{70}; \quad L = \overset{ft.}{1200};$$

which, in Eq. (269), give

velocity;

$$V = 47.94 \sqrt{\frac{0.5 \times 70}{1200 + 51.57 \times 0.5}} = 8.102.$$

The value of a, in Eq. (268), will be given by

area of the section of pipe;

$$a = \pi \frac{D^2}{4} = 3.1416 \times \frac{0.25}{4} = 0.196;$$

which, in Eq. (268), gives

expense.

$$Q = a\,V = 0.196 \times 8.102 = 1.6 \text{ nearly};$$

and this multiplied by the number of seconds in 24 hours, equal to 86400, gives 138240 for the number of cubic feet discharged in the given time.

END OF MECHANICS.

CHAMBERS' EDUCATIONAL COURSE.

NATURAL SCIENCES.

The Messrs. Chambers have employed the first professors in Scotland in the preparation of these works. They are now offered to the schools of the United States, under the American revision of D. M. REESE, M. D., LL.D., *late Superintendent of Public Schools in the city and county of New York.*

I. CHAMBERS' TREASURY OF KNOWLEDGE.
II. CLARK'S ELEMENTS OF DRAWING & PERSPECTIVE.
III. CHAMBERS' ELEMENTS OF NATURAL PHILOSOPHY.
IV. REID & BAINS' CHEMISTRY AND ELECTRICITY.
V. HAMILTON'S VEGETABLE AND ANIMAL PHYSIOLOGY.
VI. CHAMBERS' ELEMENTS OF ZOOLOGY.
VII. PAGE'S ELEMENTS OF GEOLOGY.

"It is well known that the original publishers of these works (the Messrs. Chambers of Edinburgh) are able to command the best talent in the preparation of their books, and that it is their practice to deal faithfully with the public. This series will not disappoint the reasonable expectations thus excited. They are elementary works prepared by authors in every way capable of doing justice to their respective undertakings, and who have evidently bestowed upon them the necessary time and labor to adapt them to their purpose. We recommend them to teachers and parents with confidence. If not introduced as class-books in the school, they may be used to excellent advantage in general exercises, and occasional class exercise, for which every teacher ought to provide himself with an ample store of materials. The volumes may be had separately; and the one first named, in the hands of a teacher of the younger classes, might furnish an inexhaustible fund of amusement and instruction. Together, they would constitute a rich treasure to a family of intelligent children, and impart a thirst for knowledge."—*Vermont Chron.*

"Of all the numerous works of this class that have been published, there are none that have acquired a more thoroughly deserved and high reputation than this series. The Chambers of Edinburgh, well known as the careful and intelligent publishers of a vast number of works of much importance in the educational world, are the fathers of this series of books, and the American editor has exercised an unusual degree of judgment in their preparation for the use of schools as well as private families in this country."—*Philadelphia Bulletin.*

"The titles furnish a key to the contents, and it is only necessary for us to say, that the material of each volume is admirably worked up, presenting with sufficient fulness and with much clearness of method the several subjects which are treated."
—*Cincinnati Gazette.*

www.ingramcontent.com/pod-product-compliance
Lightning Source LLC
LaVergne TN
LVHW021059110826
845150LV00001B/121

* 9 7 8 1 4 2 5 5 6 6 4 1 8 *